A TEXTBOOK OF

ENGINEERING MATHEMATICS

A TEXTBOOK OF

ENGINEERING MATHEMATICS

For

B.TECH. 2nd YEAR, *SEMESTER IV*

Strictly as per the New AICTE Model Syllabus (KAS-401)

FOR Dr. A.P.J. ABDUL KALAM TECHNICAL UNIVERSITY, LUCKNOW

By

N.P. BALI
Former Principal
S.B. College, Gurugram
Haryana

Dr. MANISH GOYAL
M.Sc. (Mathematics), Ph.D., CSIR-NET
Associate Professor
Department of Mathematics
Institute of Applied Sciences & Humanities
G.L.A. University, Mathura
U.P.

UNIVERSITY SCIENCE PRESS
(An Imprint of Laxmi Publications Pvt. Ltd.)
An ISO 9001:2015 Company

BENGALURU • CHENNAI • GUWAHATI • HYDERABAD • JALANDHAR
KOCHI • KOLKATA • LUCKNOW • MUMBAI • RANCHI
NEW DELHI

A TEXTBOOK OF ENGINEERING MATHEMATICS

Printed and bound in India
Typeset at Excellent Graphics, Delhi
First Edition : 2003; Second Edition : 2005; Third Edition : 2006; Fourth Edition : 2010; Revised Fourth Edition : 2010
Fifth Edition : 2011; Sixth Edition : 2012; Reprint : 2013, 2014; Seventh Edition : 2015;
Eighth Edition : 2016; Reprint : 2017; Ninth Edition : 2019, Tenth Edition : 2020
ISBN 978-93-80386-53-9

Branches

✆	Bengaluru	080-26 75 69 30	
✆	Chennai	044-24 34 47 26,	24 35 95 07
✆	Guwahati	0361-254 36 69,	251 38 81
✆	Hyderabad	040-27 55 53 83,	27 55 53 93
✆	Jalandhar	0181-222 12 72	
✆	Kochi	0484-237 70 04,	405 13 03
✆	Kolkata	033-22 27 43 84	
✆	Lucknow	0522-220 99 16	
✆	Mumbai	022-24 93 12 61	
✆	Ranchi	0651-220 44 64	

PUBLISHED IN INDIA BY

UNIVERSITY SCIENCE PRESS
(An Imprint of Laxmi Publications Pvt.Ltd.)

An ISO 9001:2015 Company
113, GOLDEN HOUSE, DARYAGANJ,
NEW DELHI - 110002, INDIA
Telephone : 91-11-4353 2500, 4353 2501
Fax : 91-11-2325 2572, 4353 2528
www.laxmipublications.com   info@laxmipublications.com

C—
Printed at:

Dedicated

To

My Mother

Smt. Hiramani Agrawal

—Dr. Manish Goyal

CONTENTS

PREFACE TO THE TENTH EDITION

मय्यावेश्य मनो ये मां नित्ययुक्ता उपासते।
श्रद्धया परयोपेतास्ते मे युक्ततमा मता:।।

–श्रीमद्भागवत्गीता

The Tenth edition of this textbook is an outcome of new syllabus of **Mathematics–IV (KAS 401)** for the students of B. Tech. II year, proposed and implemented by Dr. A.P.J. Abdul Kalam Technical University (A.K.T.U), Lucknow recently. The book has been renovated in the light of the latest syllabus. It will work as the latest ready reckoner for the readers.

The subject matter has been made more lucid and easier to understand. A large number of new solved examples and questions have been added. All the answers have been checked and verified. All the questions of latest university papers have been added in the body of the text. The suggestions from our colleagues and readers have been incorporated at the proper places. An appreciably heavy demand of the book ensures its utility to the users.

Separate exercise (TEST YOUR KNOWLEDGE) have been given at the end of each module.

This book is written with a unique style only meant for the welfare of dear students. We hope that this book will be a strength for them and it will serve their very purpose of attaining excellent results.

We are highly obliged to **Dr. Hari Kishan,** former Head, Department of Mathematics, K.R. (P.G.) College, Mathura, also an eminent author, for his valuable help round the clock in making this book **PERFECT** in all senses.

We are indebted to GOD for shower of blessing. The suggestions with a view to enhance the utility of the book are always welcome.

—AUTHORS

SYLLABUS

DR. A.P.J.A.K.T.U., LUCKNOW

MATHEMATICS-IV

Mathematics-IV (PDE, Probability and Statistics)

<table>
<tr><td>Subject Code</td><td colspan="6">KAS401</td></tr>
<tr><td>Category</td><td colspan="6">Basic Science Course</td></tr>
<tr><td>Subject Name</td><td colspan="6">Mathematics-IV</td></tr>
<tr><td rowspan="3">Scheme and Credits</td><td rowspan="2">L-T-P</td><td rowspan="2">Theory Marks</td><td colspan="2">Sessional</td><td rowspan="2">Total</td><td rowspan="2">Credit</td></tr>
<tr><td>Test</td><td>Assig/Att.</td></tr>
<tr><td>3—1—0</td><td>100</td><td>30</td><td>20</td><td>150</td><td>4</td></tr>
<tr><td>Pre-requisites (if any)</td><td colspan="6">Knowledge of Mathematics I and II of B. Tech or equivalent.</td></tr>
</table>

Course Outcomes

The objective of this course is to familiarize the students with partial differential equation, their application and statistical techniques. It aims to present the students with standard concepts and tools at an intermediate to superior level that will provide them well towards undertaking a variety of problems in the discipline.

The students will learn:

- The idea of partial differential and types of partial differential equations.
- The idea of classification of second partial differential equations, wave, heat equation and transmission lines
- The basic ideas of statistics including measures of central tendency, correlation, regression and their properties.
- The ideas of probability and random variables and various discrete and continuous probability distributions and their properties.
- The statistical methods of studying data samples, hypothesis testing and statistical quality control, control charts and their properties.

Module I: Partial Differential Equations

Origin of Partial Differential Equations, Linear and Non Linear Partial Equations of first order, Lagrange's Equations, Charpit's method, Cauchy's method of Characteristics, Solution of Linear Partial Differential Equation of Higher order with constant coefficients, Equations reducible to linear partial differential equations with constant coefficients.

Module II: Applications of Partial Differential Equations

Classification of linear partial differential equation of second order, Method of separation of variables, Solution of wave and heat conduction equation up to two dimension, Laplace equation in two dimensions, Equations of Transmission lines.

Module III: Statistical Techniques–I

Introduction: Measures of central tendency, Moments, Moment generating function (MGF), Skewness, Kurtosis, Curve Fitting, Method of least squares, Fitting of straight lines, Fitting of second degree parabola, Exponential curves, Correlation and Rank correlation, Regression Analysis: Regression lines of y on x and x on y, regression coefficients, properties of regressions coefficients and non linear regression.

Module IV: Statistical Techniques–II

Probability and Distribution: Introduction, Addition and multiplication law of probability, Conditional probability, Baye's theorem, Random variables (Discrete and Continuous Random variable) Probability mass function and Probability density function, Expectation and variance, Discrete and Continuous Probability distribution: Binomial, Poission and Normal distributions.

Module V: Statistical Techniques–III

Sampling, Testing of Hypothesis and Statistical Quality Control: Introduction, Sampling Theory (Small and Large), Hypothesis, Null hypothesis, Alternative hypothesis, Testing a Hypothesis, Level of significance, Confidence limits, Test of significance of difference of means, T-test, F-test and Chi-square test, One way Analysis of Variance (ANOVA). Statistical Quality Control) SQC, Control Charts, Control Charts for variables ($\overline{X}$ and R Charts), Control Charts for Variables (p, np and C charts).

STANDARD RESULTS

1. $\frac{d}{dx}(x^n) = nx^{n-1}$

2. $\frac{d}{dx}(a^x) = a^x \log_e a$

3. $\frac{d}{dx}(e^x) = e^x$

4. $\frac{d}{dx}(\log_e x) = \frac{1}{x}$

5. $\frac{d}{dx}(\log_{10} x) = \frac{1}{x}\log_{10} e$

6. $\frac{d}{dx}(\sin x) = \cos x$

7. $\frac{d}{dx}(\cos x) = -\sin x$

8. $\frac{d}{dx}(\tan x) = \sec^2 x$

9. $\frac{d}{dx}(\operatorname{cosec} x) = -\operatorname{cosec} x \cot x$

10. $\frac{d}{dx}(\cot x) = -\operatorname{cosec}^2 x$

11. $\frac{d}{dx}(\sec x) = \sec x \tan x$

12. $\frac{d}{dx}(\sin^{-1} x) = \frac{1}{\sqrt{1-x^2}}$

13. $\frac{d}{dx}(\cos^{-1} x) = \frac{-1}{\sqrt{1-x^2}}$

14. $\frac{d}{dx}(\tan^{-1} x) = \frac{1}{1+x^2}$

15. $\frac{d}{dx}(\sec^{-1} x) = \frac{1}{x\sqrt{x^2-1}}$

16. $\frac{d}{dx}(\cot^{-1} x) = \frac{-1}{1+x^2}$

17. $\frac{d}{dx}(\operatorname{cosec}^{-1} x) = -\frac{1}{x\sqrt{x^2-1}}$

18. $\sinh x = \frac{e^x - e^{-x}}{2}$

19. $\cosh x = \frac{e^x + e^{-x}}{2}$

20. $\tanh x = \frac{e^x - e^{-x}}{e^x + e^{-x}}$

21. $\cosh^2 x - \sinh^2 x = 1$, $\operatorname{sech}^2 x + \tanh^2 x = 1$, $\coth^2 x = 1 + \operatorname{cosech}^2 x$

22. $\cosh^2 x + \sinh^2 x = \cosh 2x$

23. $\sinh^{-1} x = \log(x + \sqrt{x^2+1})$, $\cosh^{-1} x = \log(x + \sqrt{x^2-1})$

24. $\frac{d}{dx}(\sinh x) = \cosh x$

25. $\frac{d}{dx}(\cosh x) = \sinh x$

26. $\frac{d}{dx}(\tanh x) = \operatorname{sech}^2 x$

27. $\frac{d}{dx}(\coth x) = -\operatorname{cosech}^2 x$

28. $\frac{d}{dx}(\operatorname{sech} x) = -\operatorname{sech} x \tanh x$

29. $\frac{d}{dx}(\operatorname{cosech} x) = -\operatorname{cosech} x \coth x$

30. Product rule: $\frac{d}{dx}(uv) = u\frac{dv}{dx} + v\frac{du}{dx}$

31. Quotient rule: $\frac{d}{dx}\left(\frac{u}{v}\right) = \frac{v\frac{du}{dx} - u\frac{dv}{dx}}{v^2}$

32. $\frac{dy}{dx} = \frac{dy}{dt}\cdot\frac{dt}{dx}$ if $y = f_1(t)$ and $x = f_2(t)$

33. $\sin^{-1} x + \cos^{-1} x = \frac{\pi}{2}$, $\tan^{-1} x + \cot^{-1} x = \frac{\pi}{2}$, $\sec^{-1} x + \text{cosec}^{-1} x = \frac{\pi}{2}$

34. $\tan^{-1}\left(\frac{a-b}{1+ab}\right) = \tan^{-1} a - \tan^{-1} b$, $\tan^{-1}\left(\frac{a+b}{1-ab}\right) = \tan^{-1} a + \tan^{-1} b$

35. $\tan^{-1}\left(\frac{2x}{1-x^2}\right) = \sin^{-1}\left(\frac{2x}{1+x^2}\right) = 2\tan^{-1} x$

36. $\sin 3x = 3\sin x - 4\sin^3 x$, $\cos 3x = 4\cos^3 x - 3\cos x$, $\tan 3x = \frac{3\tan x - \tan^3 x}{1 - 3\tan^2 x}$

$\sin 2x = 2\sin x \cos x$, $\tan 2x = \frac{2\tan x}{1-\tan^2 x}$,

$\cos 2x = 2\cos^2 x - 1 = 1 - 2\sin^2 x = \cos^2 x - \sin^2 x = \frac{1-\tan^2 x}{1+\tan^2 x}$

37. $\sin x = x - \frac{x^3}{3!} + \frac{x^5}{5!} - \frac{x^7}{7!} + \ldots$, $\cos x = 1 - \frac{x^2}{2!} + \frac{x^4}{4!} - \frac{x^6}{6!} + \ldots$

$e^x = 1 + x + \frac{x^2}{2!} + \frac{x^3}{3!} + \ldots$

$(1-x)^{-1} = 1 + x + x^2 + x^3 + \ldots;\ |x| < 1$ $(1+x)^{-1} = 1 - x + x^2 - x^3 + \ldots$

$(1-x)^{-2} = 1 + 2x + 3x^2 + 4x^3 + \ldots$ $(1+x)^{-2} = 1 - 2x + 3x^2 - 4x^3 + \ldots$

38. $\sin C + \sin D = 2\sin\frac{C+D}{2}\cos\frac{C-D}{2}$, $\sin C - \sin D = 2\cos\frac{C+D}{2}\sin\frac{C-D}{2}$

$\cos C + \cos D = 2\cos\frac{C+D}{2}\cos\frac{C-D}{2}$, $\cos C - \cos D = 2\sin\frac{C+D}{2}\sin\frac{D-C}{2}$

39. $2\cos A\cos B = \cos(A+B) + \cos(A-B)$, $2\sin A\sin B = \cos(A-B) - \cos(A+B)$

$2\sin A\cos B = \sin(A+B) + \sin(A-B)$, $2\cos A\sin B = \sin(A+B) - \sin(A-B)$

40. $\sin(A+B) = \sin A\cos B + \cos A\sin B$, $\sin(A-B) = \sin A\cos B - \cos A\sin B$

$\cos(A+B) = \cos A\cos B - \sin A\sin B$, $\cos(A-B) = \cos A\cos B + \sin A\sin B$

41. $\frac{d}{dx}(\sinh^{-1} x) = \frac{1}{\sqrt{1+x^2}}$, $\frac{d}{dx}(\cosh^{-1} x) = \frac{1}{\sqrt{x^2-1}}$

$\frac{d}{dx}(\tanh^{-1} x) = \frac{1}{1-x^2}$, where $|x| < 1$, $\frac{d}{dx}(\coth^{-1} x) = \frac{1}{x^2-1}$, where $|x| > 1$

$\frac{d}{dx}(\text{sech}^{-1} x) = -\frac{1}{x\sqrt{1-x^2}}$, $\frac{d}{dx}(\text{cosech}^{-1} x) = -\frac{1}{x\sqrt{x^2+1}}$.

42. $(\cos\theta + i\sin\theta)^n = \cos n\theta + i\sin n\theta$, $(\cos\theta + i\sin\theta)^{-n} = \cos n\theta - i\sin n\theta$

43. $\sin^2\theta + \cos^2\theta = 1$, $\sec^2\theta - \tan^2\theta = 1$, $1 + \cot^2\theta = \text{cosec}^2\theta$

44.

θ	0°	30°	45°	60°	90°	180°	270°	360°
sin θ	0	1/2	$1/\sqrt{2}$	$\sqrt{3}/2$	1	0	– 1	0
cos θ	1	$\sqrt{3}/2$	$1/\sqrt{2}$	1/2	0	– 1	0	1
tan θ	0	$1/\sqrt{3}$	1	$\sqrt{3}$	∞	0	∞	0

45.

θ	90° – θ	90° + θ	π – θ	π + θ
sin θ	cos θ	cos θ	sin θ	– sin θ
cos θ	sin θ	– sin θ	– cos θ	– cos θ
tan θ	cot θ	– cot θ	– tan θ	tan θ

46. sine formula: $\dfrac{a}{\sin A} = \dfrac{b}{\sin B} = \dfrac{c}{\sin C}$; cosine formula: $\cos A = \dfrac{b^2 + c^2 - a^2}{2bc}$

47. Area of triangle $\Delta = \sqrt{s(s-a)(s-b)(s-c)}$, where $s = \dfrac{a+b+c}{2}$

48. ${}^nC_r = \dfrac{n!}{r!\,n-r!}$

49. $\int x^n\,dx = \dfrac{x^{n+1}}{n+1} + c;\ n \neq -1$

$\int \dfrac{1}{x}\,dx = \log_e x + c;\ \int e^x\,dx = e^x + c;\ \int a^x\,dx = \dfrac{a^x}{\log_e a} + c$

$\int \sin x\,dx = -\cos x + c;\ \int \cos x\,dx = \sin x + c$

$\int \tan x\,dx = \log \sec x + c;\ \int \cot x\,dx = \log \sin x + c$

$\int \sec x\,dx = \log(\sec x + \tan x) + c = \log \tan\left(\dfrac{\pi}{4} + \dfrac{x}{2}\right) + c$

$\int \text{cosec}\, x\,dx = \log(\text{cosec}\, x - \cot x) + c = \log \tan \dfrac{x}{2} + c$

$\int \sec x \tan x\,dx = \sec x + c;\ \int \text{cosec}\, x \cot x\,dx = -\text{cosec}\, x + c$

50. $\int \dfrac{dx}{\sqrt{a^2 - x^2}} = \sin^{-1}\left(\dfrac{x}{a}\right) + c;\ \int \dfrac{-dx}{\sqrt{a^2 - x^2}} = \cos^{-1}\left(\dfrac{x}{a}\right) + c$

$\int \dfrac{dx}{a^2 + x^2} = \dfrac{1}{a}\tan^{-1}\left(\dfrac{x}{a}\right) + c;\ \int \dfrac{-dx}{a^2 + x^2} = \dfrac{1}{a}\cot^{-1}\left(\dfrac{x}{a}\right) + c$

$\int \dfrac{dx}{a^2 - x^2} = \dfrac{1}{2a}\log\left(\dfrac{a+x}{a-x}\right) + c;\ \int \dfrac{dx}{x^2 - a^2} = \dfrac{1}{2a}\log\left(\dfrac{x-a}{x+a}\right) + c$

$\int \dfrac{dx}{x\sqrt{x^2 - a^2}} = \dfrac{1}{a}\sec^{-1}\left(\dfrac{x}{a}\right) + c;\ \int \dfrac{-dx}{x\sqrt{x^2 - a^2}} = \dfrac{1}{a}\text{cosec}^{-1}\left(\dfrac{x}{a}\right) + c$

51. $\int \operatorname{sech}^2 x\, dx = \tanh x + c, \int \operatorname{cosech}^2 x\, dx = -\coth x + c$

$\int \sinh x\, dx = \cosh x + c, \int \cosh x\, dx = \sinh x + c$

$\int \operatorname{sech} x \tanh x\, dx = -\operatorname{sech} x + c, \int \operatorname{cosech} x \coth x\, dx = -\operatorname{cosech} x + c$

52. $\int \sqrt{a^2 - x^2}\, dx = \frac{1}{2} x \sqrt{a^2 - x^2} + \frac{1}{2} a^2 \sin^{-1} \frac{x}{a} + c$

$\int \sqrt{a^2 + x^2}\, dx = \frac{1}{2} x \sqrt{a^2 + x^2} + \frac{1}{2} a^2 \log (x + \sqrt{a^2 + x^2}) + c$

$\int \sqrt{x^2 - a^2}\, dx = \frac{1}{2} x \sqrt{x^2 - a^2} - \frac{1}{2} a^2 \log (x + \sqrt{x^2 - a^2}) + c$

$\int \frac{dx}{\sqrt{a^2 + x^2}} = \sinh^{-1} \left(\frac{x}{a}\right) + c; \int \frac{dx}{\sqrt{x^2 - a^2}} = \cosh^{-1} \left(\frac{x}{a}\right) + c$

53. $\int_a^b f(x)\, dx = \int_a^b f(y)\, dy; \int_a^b f(x)\, dx = -\int_b^a f(x)\, dx; \int_0^a f(x)\, dx = \int_0^a f(a - x)\, dx$

$$\int_{-a}^{a} f(x)\, dx = \begin{cases} 2\int_0^a f(x)\, dx, & \text{if } f(x) \text{ is even function} \\ 0, & \text{if } f(x) \text{ is odd function} \end{cases}$$

$$\int_0^{2a} f(x)\, dx = \begin{cases} 2\int_0^a f(x)\, dx, & \text{if } f(2a - x) = f(x) \\ 0, & \text{if } f(2a - x) = -f(x) \end{cases}$$

54. Leibnitz rule for differentiation under the integral sign

$$\frac{d}{dx} \int_{\phi(\alpha)}^{\psi(\alpha)} f(x, \alpha)\, dx = \int_{\phi(\alpha)}^{\psi(\alpha)} \frac{\partial}{\partial \alpha}\{f(x, \alpha)\}\, dx + f\{\psi(\alpha), \alpha\} \frac{d\psi(\alpha)}{d\alpha} - f\{\phi(\alpha), \alpha\} \frac{d\phi(\alpha)}{d\alpha}$$

55. If $\vec{r} = x\hat{i} + y\hat{j} + z\hat{k}$ then $|\vec{r}| = \sqrt{x^2 + y^2 + z^2}$ and $\hat{r} = \frac{\vec{r}}{|\vec{r}|} = \frac{x\hat{i} + y\hat{j} + z\hat{k}}{\sqrt{x^2 + y^2 + z^2}}$

56. $\overrightarrow{AB}$ = position vector of B-position vector of A = $\overrightarrow{OB} - \overrightarrow{OA}$

57. $\vec{a} \cdot \vec{b} = |\vec{a}|\, |\vec{b}| \cos \theta$; work done = $\int_c \vec{F} \cdot d\vec{r}$

58. $\vec{a} \times \vec{b} = |\vec{a}|\, |\vec{b}| \sin \theta\, \hat{n}$

59. Area of parallelogram = $\vec{a} \times \vec{b}$, Moment of force = $\vec{r} \times \vec{F}$

60. $\vec{a}\cdot(\vec{b}\times\vec{c}) = [\vec{a}\,\vec{b}\,\vec{c}\,] = \begin{vmatrix} a_1 & a_2 & a_3 \\ b_1 & b_2 & b_3 \\ c_1 & c_2 & c_3 \end{vmatrix} = (\vec{a}\times\vec{b})\,.\,\vec{c}$

where $\vec{a} = \Sigma\, a_1\hat{i}$, $\vec{b} = \Sigma\, b_1\hat{i}$ and $\vec{c} = \Sigma c_1\hat{i}$

If $\vec{a}\cdot(\vec{b}\times\vec{c}) = 0$, then $\vec{a}, \vec{b}, \vec{c}$ are coplanar.

61. $\vec{a}\times(\vec{b}\times\vec{c}) = (\vec{a}\cdot\vec{c})\,\vec{b} - (\vec{a}\cdot\vec{b})\,\vec{c}$

62. $(\vec{a}\times\vec{b})\cdot(\vec{c}\times\vec{d}) = \begin{vmatrix} \vec{a}\cdot\vec{c} & \vec{a}\cdot\vec{d} \\ \vec{b}\cdot\vec{c} & \vec{b}\cdot\vec{d} \end{vmatrix}$

63. $(\vec{a}\times\vec{b})\times(\vec{c}\times\vec{d}) = [\vec{a}\,\vec{b}\,\vec{d}\,]\,\vec{c} - [\vec{a}\,\vec{b}\,\vec{c}\,]\,\vec{d}$

64. $A\,(\text{Adj. }A) = |\,A\,|\,I$

65. $AA^{-1} = I = A^{-1}A$

66. $AI = A = IA$

67. $(ABC)' = C'B'A'$

68. $(AB)C = A(BC)$; $A(B + C) = AB + AC$

69. $A + B = B + A$; $A + (B + C) = (A + B) + C$

70. $(AB)^{-1} = B^{-1}A^{-1}$

71. **Walli's formula**

$$\int_0^{\pi/2} \sin^n\theta\, d\theta = \int_0^{\pi/2} \cos^n\theta\, d\theta = \begin{cases} \dfrac{n-1}{n}\cdot\dfrac{n-3}{n-2}\cdot\dfrac{n-5}{n-4}\cdots\cdots\dfrac{3}{4}\cdot\dfrac{1}{2}\cdot\dfrac{\pi}{2} & \text{if } n \text{ is even} \\[2ex] \dfrac{n-1}{n}\cdot\dfrac{n-3}{n-2}\cdot\dfrac{n-5}{n-4}\cdots\cdots\dfrac{4}{5}\cdot\dfrac{2}{3} & \text{if } n \text{ is odd} \end{cases}$$

72. $$\int e^{ax}\sin bx\, dx = \frac{e^{ax}}{a^2+b^2}(a\sin bx - b\cos bx) + c$$

$$\int e^{ax}\cos bx\, dx = \frac{e^{ax}}{a^2+b^2}(a\cos bx + b\sin bx) + c$$

73. $\Gamma(1/2) = \sqrt{\pi}$, $\Gamma(-1/2) = -2\sqrt{\pi}$

74. $$\log(1+x) = x - \frac{x^2}{2} + \frac{x^3}{3} - \frac{x^4}{4} + \frac{x^5}{5} - \frac{x^6}{6} + \cdots\cdots$$

$$\log(1-x) = -x - \frac{x^2}{2} - \frac{x^3}{3} - \frac{x^4}{4} - \frac{x^5}{5} - \frac{x^6}{6} - \cdots\cdots$$

75. $\sin n\pi = 0$; $\cos n\pi = (-1)^n$, $\sin\left(n+\dfrac{1}{2}\right)\pi = (-1)^n$; $\cos\left(n+\dfrac{1}{2}\right)\pi = 0$, where $n \in I$

76. $x^3 + y^3 + z^3 - 3xyz = (x+y+z)(x^2+y^2+z^2-xy-yz-zx)$.

MODULE 1

Partial Differential Equations

1.1 INTRODUCTION

A differential equation containing partial derivatives of a function of two or more independent variables is called a partial differential equation. *e.g.*,

$$y\frac{\partial z}{\partial x} + x\frac{\partial z}{\partial y} = 0$$

$$\frac{\partial^2 z}{\partial x^2} + \frac{\partial^2 z}{\partial y^2} = 0$$

are the partial differential equations.

When we have a function z of two independent variables x and y, we use the alphabets p, q, r, s, t to denote the partial derivatives as follows:

$$p = \frac{\partial z}{\partial x}, \quad q = \frac{\partial z}{\partial y}$$

$$r = \frac{\partial^2 z}{\partial x^2} = \frac{\partial}{\partial x}\left(\frac{\partial z}{\partial x}\right) = \frac{\partial p}{\partial x}$$

$$s = \frac{\partial^2 z}{\partial x \partial y} = \frac{\partial^2 z}{\partial y \partial x} = \frac{\partial q}{\partial x} = \frac{\partial p}{\partial y}$$

and

$$t = \frac{\partial^2 z}{\partial y} = \frac{\partial}{\partial y}\left(\frac{\partial z}{\partial y}\right) = \frac{\partial q}{\partial y}$$

Partial differential equations generally occur in the problems of Physics and Engineering. Some of the important partial differential equations are

$$\frac{\partial^2 u}{\partial x^2} + \frac{\partial^2 u}{\partial y^2} + \frac{\partial^2 u}{\partial z^2} = 0 \qquad ...(1)$$

$$\frac{\partial^2 u}{\partial t^2} = c^2\left(\frac{\partial^2 u}{\partial x^2} + \frac{\partial^2 u}{\partial y^2} + \frac{\partial^2 u}{\partial z^2}\right) \qquad ...(2)$$

$$\frac{\partial u}{\partial t} = c^2\left(\frac{\partial^2 u}{\partial x^2} + \frac{\partial^2 u}{\partial y^2} + \frac{\partial^2 u}{\partial z^2}\right) \qquad ...(3)$$

Equations (1), (2) and (3) are respectively known as Laplace's equation, wave equation and heat conduction equation.

Sometimes for brevity, the partial differentiation with regard to a variable is denoted by a suffix. *e.g.*, Laplace's equation may be rewritten as $u_{xx} + u_{yy} + u_{zz} = 0$.

If u does not depend on z, then we get two-dimensional Laplace's equation as $u_{xx} + u_{yy} = 0$.

1.2 ORDER AND DEGREE OF PARTIAL DIFFERENTIAL EQUATION

Order of a partial differential equation is the order of the highest ordered derivative present in the equation.

Degree of a partial differential equation is the greatest exponent (power) of the highest ordered derivative present in the equation when it has been made free from radical signs and fractional powers.

1.3 SOLUTION OF PARTIAL DIFFERENTIAL EQUATION

Solution is one which satisfies. The solution of a partial differential equation in a region D is a function having partial derivatives which satisfy the differential equation at every point in D.

The general solution of a p.d.e. contains arbitrary constants or arbitrary functions or both. Consequently, we can say that by the elimination of arbitrary constants or arbitrary functions, partial differential equations can be formed.

1.4 FORMATION OF PARTIAL DIFFERENTIAL EQUATION

(1) By the elimination of arbitrary constants

Let $$f(x, y, z, a, b) = 0 \qquad ...(1)$$

be the given function, where a, b are arbitrary constants. x and y are independent variables and z is a dependent variable. Differentiating eqn. (1) partially w.r.t. x, we get

$$\frac{\partial f}{\partial x} + \frac{\partial f}{\partial z} \cdot \frac{\partial z}{\partial x} = 0 \Rightarrow \frac{\partial f}{\partial x} + p\frac{\partial f}{\partial z} = 0 \qquad ...(2)$$

Again differentiating eqn. (1) partially w.r.t. y, we get

$$\frac{\partial f}{\partial y} + \frac{\partial f}{\partial z} \cdot \frac{\partial z}{\partial y} = 0 \Rightarrow \frac{\partial f}{\partial y} + q\frac{\partial f}{\partial z} = 0 \qquad ...(3)$$

Eliminating a, b from equations (1), (2) and (3), we get

$$F(x, y, z, p, q) = 0$$

which is a partial differential equation of first order.

Note. If the number of arbitrary constants is more than the number of independent variables, then the order of the partial differential equation obtained will be greater than 1.

(2) By the elimination of arbitrary functions

Let u, v be two known functions of x, y, z connected by the relation

$$\phi(u, v) = 0 \qquad ...(1)$$

where ϕ is an arbitrary function.

Differentiating eqn. (1) partially w.r.t. x, we get

$$\frac{\partial \phi}{\partial u}\left(\frac{\partial u}{\partial x} + \frac{\partial u}{\partial z} \cdot \frac{\partial z}{\partial x}\right) + \frac{\partial \phi}{\partial v}\left(\frac{\partial v}{\partial x} + \frac{\partial v}{\partial z} \cdot \frac{\partial z}{\partial x}\right) = 0$$

$$\Rightarrow \qquad \frac{\partial \phi}{\partial u}\left(\frac{\partial u}{\partial x} + p\frac{\partial u}{\partial z}\right) + \frac{\partial \phi}{\partial v}\left(\frac{\partial v}{\partial x} + p\frac{\partial v}{\partial z}\right) = 0$$

$$\Rightarrow \qquad \frac{\partial \phi/\partial u}{\partial \phi/\partial v} = -\left(\frac{\dfrac{\partial v}{\partial x} + p\dfrac{\partial v}{\partial z}}{\dfrac{\partial u}{\partial x} + p\dfrac{\partial u}{\partial z}}\right) \qquad ...(2)$$

Differentiating eq. (1) partially w.r.t. y, we get

$$\frac{\partial \phi}{\partial u}\left(\frac{\partial u}{\partial y} + \frac{\partial u}{\partial z}\frac{\partial z}{\partial y}\right) + \frac{\partial \phi}{\partial v}\left(\frac{\partial v}{\partial y} + \frac{\partial v}{\partial z}\frac{\partial z}{\partial y}\right) = 0$$

$$\Rightarrow \qquad \frac{\partial \phi}{\partial u}\left(\frac{\partial u}{\partial y} + q\frac{\partial u}{\partial z}\right) + \frac{\partial \phi}{\partial v}\left(\frac{\partial v}{\partial y} + q\frac{\partial v}{\partial z}\right) = 0$$

$$\Rightarrow \qquad \frac{\partial \phi/\partial u}{\partial \phi/\partial v} = -\left(\frac{\dfrac{\partial v}{\partial y} + q\dfrac{\partial v}{\partial z}}{\dfrac{\partial u}{\partial y} + q\dfrac{\partial u}{\partial z}}\right) \qquad ...(3)$$

From eqns. (2) and (3), we get

$$\frac{\dfrac{\partial v}{\partial x} + p\dfrac{\partial v}{\partial z}}{\dfrac{\partial u}{\partial x} + p\dfrac{\partial u}{\partial z}} = \frac{\dfrac{\partial v}{\partial y} + q\dfrac{\partial v}{\partial z}}{\dfrac{\partial u}{\partial y} + q\dfrac{\partial u}{\partial z}}$$

$$\Rightarrow \qquad \left(\frac{\partial u}{\partial y}\frac{\partial v}{\partial z} - \frac{\partial u}{\partial z}\frac{\partial v}{\partial y}\right)p + \left(\frac{\partial u}{\partial z}\frac{\partial v}{\partial x} - \frac{\partial u}{\partial x}\frac{\partial v}{\partial z}\right)q = \frac{\partial u}{\partial x}\frac{\partial v}{\partial y} - \frac{\partial u}{\partial y}\frac{\partial v}{\partial x}$$

$$\Rightarrow \qquad \boxed{\mathrm{P}p + \mathrm{Q}q = \mathrm{R}} \qquad ...(4)$$

Eqn. (4) is a partial differential equation of first degree (*i.e.* linear) in p and q.

Note. If the given relation between x, y, z contains two arbitrary functions, then the p.d.e. derived therefrom will be, in general of order greater than one.

ILLUSTRATIVE EXAMPLES

Example 1. *Form partial differential equations from the following equations by eliminating the arbitrary constants:*

(*i*) $z = ax + by + ab$ (*ii*) $z = ax + a^2y^2 + b$

(*iii*) $z = (x + a)(y + b)$ (*iv*) $az + b = a^2x + y.$

Sol. (*i*) Differentiating z partially w.r.t. x and y,

$$p = \frac{\partial z}{\partial x} = a, \quad q = \frac{\partial z}{\partial y} = b$$

Substituting for a and b in the given equation, we get

$$z = px + qy + pq$$

which is a partial differential equation.

(*ii*) Differentiating z partially w.r.t. x and y

$$p = \frac{\partial z}{\partial x} = a, \quad q = \frac{\partial z}{\partial y} = 2a^2 y$$

Eliminating a between these results, we get $q = 2p^2 y$
which is a partial differential equation.

(*iii*) Differentiating z partially w.r.t.x, we get

$$\frac{\partial z}{\partial x} = p = y + b \qquad ...(1)$$

Differentiating z partially w.r.t. y, we get

$$\frac{\partial z}{\partial y} = q = x + a \qquad ...(2)$$

Multiplying eqns. (1) and (2), we get

$$pq = (y + b)(x + a)$$

$$\Rightarrow \qquad pq = z$$

which is a partial differential equation.

(*iv*) Differentiating the given relation w.r.t. x partially, we get

$$a \frac{\partial z}{\partial x} = a^2$$

$$\Rightarrow \qquad \frac{\partial z}{\partial x} = p = a \qquad ...(1)$$

Again differentiating the given relation w.r.t. y partially, we get

$$a \frac{\partial z}{\partial y} = 1$$

$$\Rightarrow \qquad \frac{\partial z}{\partial y} = q = \frac{1}{a} \qquad ...(2)$$

Multiplying eqns.(1) and (2), we get $pq = 1$
which is a partial differential equation.

Example 2. *Form the partial differential equation by eliminating the arbitrary function(s) from the following :*

(*i*) $z = f(x^2 - y^2)$ (U.P.T.U. 2015) (*ii*) $z = \phi(x) \,.\, \psi(y)$

(*iii*) $z = x + y + f(xy)$ (*iv*) $z = f(x + it) + g(x - it)$. (U.K.T.U. 2011)

Sol. (*i*) Differentiating z partially w.r.t. x, we get

$$\frac{\partial z}{\partial x} = p = f'(x^2 - y^2) \,.\, 2x \qquad ...(1)$$

Differentiating z partially w.r.t. y, we get

$$\frac{\partial z}{\partial y} = q = f'(x^2 - y^2) \,.\, (-2y) \qquad ...(2)$$

Dividing eqn. (1) by eqn. (2), we get

$$\frac{p}{q} = \frac{x}{(-y)} \quad \Rightarrow \quad py + qx = 0$$

which is a partial differential equation.

(*ii*) Differentiating z w.r.t. x, partially, we get

$$\frac{\partial z}{\partial x} = p = \phi'(x)\,\psi(y) \quad ...(1)$$

Differentiating z w.r.t. y partially, we get

$$\frac{\partial z}{\partial y} = q = \phi(x)\,\psi'(y) \quad ...(2)$$

Differentiating eqn. (1) partially w.r.t. y, we get

$$\frac{\partial^2 z}{\partial y \partial x} = s = \phi'(x)\,\psi'(y) \quad ...(3)$$

Multiplying eqns. (1) and (2), we get

$$pq = \phi(x)\,\psi(y)\,\phi'(x)\,\psi'(y) = zs \quad | \text{ Using (3)}$$

$$\Rightarrow \quad pq - zs = 0$$

which is a partial differential equation.

(*iii*) Differentiating z w.r.t. x, partially, we get

$$\frac{\partial z}{\partial x} = p = 1 + f'(xy)\,.\,y \quad \Rightarrow \quad p - 1 = yf'(xy) \quad ...(1)$$

Differentiating z w.r.t. y, partially, we get

$$\frac{\partial z}{\partial y} = q = 1 + f'(xy)\,.\,x \quad \Rightarrow \quad q - 1 = x\,f'(xy) \quad ...(2)$$

Dividing eqn. (1) by eqn. (2), we get

$$\frac{p-1}{q-1} = \frac{y}{x} \quad \Rightarrow \quad px - qy = x - y$$

which is a partial differential equation.

(*iv*) Given $z = f(x + it) + g(x - it)$

Differentiating z twice partially w.r.t. x and t, we have

$$\frac{\partial z}{\partial x} = f'(x + it) + g'(x - it)$$

$$\frac{\partial^2 z}{\partial x^2} = f''(x + it) + g''(x - it) \quad ...(1)$$

$$\frac{\partial z}{\partial t} = i\,f'(x + it) - ig'(x - it)$$

$$\frac{\partial^2 z}{\partial t^2} = i^2 f''(x + it) + i^2 g''(x - it)$$

or

$$\frac{\partial^2 z}{\partial t^2} = -f''(x + it) - g''(x - it) \quad ...(2)$$

Adding (1) and (2), we obtain $\dfrac{\partial^2 z}{\partial x^2} + \dfrac{\partial^2 z}{\partial t^2} = 0$

which is a partial differential equation of second order.

1.5 DEFINITIONS

The solution $f(x, y, z, a, b) = 0$ of a first order partial differential equation, which contains two arbitrary constants is called a *complete solution* or *complete integral*.

If in this solution, we put $b = \phi(a)$ and find the envelope of the family of surfaces $f(x, y, z, a, \phi(a)) = 0$, we get a solution involving an arbitrary function ϕ. This is called the *general solution* or *general integral*.

A solution obtained from the complete integral by giving particular values to the arbitrary constant is called a *particular solution* or *particular integral*.

A partial differential equation may have a large number of entirely different solutions. For example, $u = x^2 - y^2$, $u = \log(x^2 + y^2)$, $u = \sin kx \cosh ky$ are solutions of the Laplace equation $\frac{\partial^2 u}{\partial x^2} + \frac{\partial^2 u}{\partial y^2} = 0$. The unique solution of a partial differential equation corresponding to a physical problem must satisfy certain other conditions at the boundary of the region R. These are known as the *boundary conditions*.

If these conditions are given for the time $t = 0$, they are known as the *initial conditions*.

1.6 LINEAR HOMOGENEOUS PARTIAL DIFFERENTIAL EQUATION

A partial differential equation is said to be linear if it is of first degree in the dependent variable and its partial derivatives and also they are not multiplied together.

If, in addition, every term of the equation contains the dependent variable or its derivative, it is called a homogeneous equation.

Note. An equation which is not linear is called a non-linear partial differential equation.

1.7 QUASI-LINEAR PARTIAL DIFFERENTIAL EQUATIONS

A partial differential equation is said to be quasi-linear if degree of highest ordered derivative is one and no products of partial derivatives of the highest order are present.

e.g., $$z\frac{\partial^2 z}{\partial x^2} + \left(\frac{\partial z}{\partial y}\right)^2 = 0 \text{ is a quasi-linear p.d.e. of second order.}$$

The general form of a quasi-linear p.d.e. of the first order is

$$P(x, y, z)\frac{\partial z}{\partial x} + Q(x, y, z)\frac{\partial z}{\partial y} = R(x, y, z) \qquad ...(1)$$

If P and Q are independent of z and R is linear in z then eqn. (1) is linear partial differential equation.

1.8 EQUATIONS SOLVABLE BY DIRECT INTEGRATION

The equations which contain **only one** partial derivative can be solved by direct integration. However, in place of the constants of integration, we must use arbitrary functions of the variable kept constant.

Example 3. *Solve: $t = \sin(xy)$.*

Sol. The given equation is

$$\frac{\partial^2 z}{\partial y^2} = \sin(xy)$$

Integrating w.r.t. y, we get

$$\frac{\partial z}{\partial y} = -\frac{1}{x}\cos(xy) + f(x)$$

The constant is being taken as a function of x.

Again integrating w.r.t. y, we get

$$z = -\frac{1}{x^2}\sin(xy) + yf(x) + \phi(x).$$

Example 4. *Solve: $\frac{\partial^2 z}{\partial x^2} + z = 0$, given that when $x = 0$, $z = e^y$ and $\frac{\partial z}{\partial x} = 1$.*

Sol. If z were a function of **x alone,** the solution would have been $z = c_1 \cos x + c_2 \sin x$, where c_1 and c_2 are arbitrary constants.

But here z is a function of x and y, therefore, c_1 and c_2 are arbitrary functions of y, the independent variable kept constant.

$\therefore$ The solution of the given equation is

$$z = f(y)\cos x + \phi(y)\sin x \text{ and } \frac{\partial z}{\partial x} = -f(y)\sin x + \phi(y)\cos x$$

When $x = 0$, $z = e^y$ $\quad\therefore\quad e^y = f(y)$

When $x = 0$, $\frac{\partial z}{\partial x} = 1$ $\quad\therefore\quad 1 = \phi(y)$

$\therefore$ The required particular solution is $z = e^y \cos x + \sin x$.

1.9 LINEAR PARTIAL DIFFERENTIAL EQUATIONS OF THE FIRST ORDER

A differential equation involving first order partial derivatives p and q only is called a *partial differential equation of the first order*. If p and q both occur in the first degree only and are not multiplied together, then it is called *linear partial differential equation of the first order*.

1.10 LAGRANGE'S EQUATION

The partial differential equation of the form $Pp + Qq = R$, where P, Q and R are functions of x, y, z is the standard form of a quasi-linear partial differential equation of the first order and is called *Lagrange's Equation*.

Now Lagrange's equation $\quad Pp + Qq = R \quad$ ...(1)

is obtained by eliminating an arbitrary function ϕ from $\quad \phi(u, v) = 0 \quad$ ...(2)

where u, v are some definite functions of x, y, z.

Differentiating (2) partially w.r.t. x and y

$$\frac{\partial \phi}{\partial u}\left(\frac{\partial u}{\partial x} + \frac{\partial u}{\partial z}p\right) + \frac{\partial \phi}{\partial v}\left(\frac{\partial v}{\partial x} + \frac{\partial v}{\partial z}p\right) = 0$$

and $$\frac{\partial \phi}{\partial u}\left(\frac{\partial u}{\partial y}+\frac{\partial u}{\partial z} q\right)+\frac{\partial \phi}{\partial v}\left(\frac{\partial v}{\partial y}+\frac{\partial v}{\partial z} q\right)=0$$

Eliminating $\frac{\partial \phi}{\partial u}$ and $\frac{\partial \phi}{\partial v}$, we get

$$\left(\frac{\partial u}{\partial x}+\frac{\partial u}{\partial z} p\right)\left(\frac{\partial v}{\partial y}+\frac{\partial v}{\partial z} q\right)-\left(\frac{\partial v}{\partial x}+\frac{\partial v}{\partial z} p\right)\left(\frac{\partial u}{\partial y}+\frac{\partial u}{\partial z} q\right)=0$$

or $$\left(\frac{\partial u}{\partial y} \cdot \frac{\partial v}{\partial z}-\frac{\partial u}{\partial z} \cdot \frac{\partial v}{\partial y}\right) p+\left(\frac{\partial u}{\partial z} \cdot \frac{\partial v}{\partial x}-\frac{\partial u}{\partial x} \cdot \frac{\partial v}{\partial z}\right) q=\frac{\partial u}{\partial x} \cdot \frac{\partial v}{\partial y}-\frac{\partial u}{\partial y} \cdot \frac{\partial v}{\partial x}$$

which is the same as (1) with

$$\mathrm{P}=\frac{\partial u}{\partial y} \cdot \frac{\partial v}{\partial z}-\frac{\partial u}{\partial z} \cdot \frac{\partial v}{\partial y}, \quad \mathrm{Q}=\frac{\partial u}{\partial z} \cdot \frac{\partial v}{\partial x}-\frac{\partial u}{\partial x} \cdot \frac{\partial v}{\partial z}, \quad \mathrm{R}=\frac{\partial u}{\partial x} \cdot \frac{\partial v}{\partial y}-\frac{\partial u}{\partial y} \cdot \frac{\partial v}{\partial x}.$$

To determine u, v from P, Q, R, suppose $u = a$ and $v = b$, where a, b are constants, so that

$$\frac{\partial u}{\partial x} dx+\frac{\partial u}{\partial y} dy+\frac{\partial u}{\partial z} dz=du=0$$

$$\frac{\partial v}{\partial x} dx+\frac{\partial v}{\partial y} dy+\frac{\partial v}{\partial z} dz=dv=0$$

By cross-multiplication, we have

$$\frac{dx}{\frac{\partial u}{\partial y} \cdot \frac{\partial v}{\partial z}-\frac{\partial u}{\partial z} \cdot \frac{\partial v}{\partial y}}=\frac{dy}{\frac{\partial u}{\partial z} \cdot \frac{\partial v}{\partial x}-\frac{\partial u}{\partial x} \cdot \frac{\partial v}{\partial z}}=\frac{dz}{\frac{\partial u}{\partial x} \cdot \frac{\partial v}{\partial y}-\frac{\partial u}{\partial y} \cdot \frac{\partial v}{\partial x}}$$

or $$\boxed{\frac{dx}{\mathrm{P}}=\frac{dy}{\mathrm{Q}}=\frac{dz}{\mathrm{R}}} \qquad ...(3)$$

The solutions of these equations are $u = a$ and $v = b$. Thus determining u, v from the simultaneous equations (3), we have the solution of the partial differential equation

$\mathrm{P}p + \mathrm{Q}q = \mathrm{R}$ as $\phi(u, v) = 0$ or $v = f(u)$.

Note. Equations (3) are called **Lagrange's auxiliary equations** or **subsidiary equations**.

Remark. In the differential equation $\mathrm{P}p + \mathrm{Q}q = \mathrm{R}$, there are **two independent variables** x and y. To solve it, we have to find **two independent solutions** satisfying the equation $\frac{dx}{\mathrm{P}}=\frac{dy}{\mathrm{Q}}=\frac{dz}{\mathrm{R}}$.

This procedure can be extended to linear partial differential equations of the first order involving more than two independent variables.

If $u_1 = c_1, u_2 = c_2, ..., u_n = c_n$ are n independent solutions of the equations

$$\frac{dx_1}{\mathrm{P}_1}=\frac{dx_2}{\mathrm{P}_2}=\ldots\ldots=\frac{dx_n}{\mathrm{P}_n}=\frac{dz}{\mathrm{R}}$$

then the general solution of the partial differential equation

$$\mathrm{P}_1 \frac{\partial z}{\partial x_1}+\mathrm{P}_2 \frac{\partial z}{\partial x_2}+\ldots\ldots+\mathrm{P}_n \frac{\partial z}{\partial x_n}=\mathrm{R}$$

is $\phi(u_1, u_2, ..., u_n) = 0$, where ϕ is an arbitrary function.

1.11 WORKING RULE

To solve the equation $\quad Pp + Qq = R$

(*i*) Form the auxiliary equations $\dfrac{dx}{P} = \dfrac{dy}{Q} = \dfrac{dz}{R}$

(*ii*) Solve the auxiliary equations by the method of grouping or the method of multipliers or both to get two independent solutions $u = a$ and $v = b$, where a, b are arbitrary constants.

(*iii*) Then $\phi(u, v) = 0$ or $v = f(u)$ is the general solution of the equation $Pp + Qq = R$.

Example 5. *Solve the following partial differential equations:*

(*i*) $yzp - xzq = xy$ (*ii*) $y^2p - xyq = x(z - 2y)$ (U.P.T.U. 2014) (*iii*) $x^2p + y^2q = (x + y)z$ (U.P.T.U. 2015)

Sol. (*i*) The given differential equation is $yzp - xzq = xy$ ...(1)

Comparing equation (1) with $Pp + Qq = R$, we get

$$P = yz, Q = -xz \quad \text{and} \quad R = xy$$

Lagrange's subsidiary equations are

$$\frac{dx}{P} = \frac{dy}{Q} = \frac{dz}{R}$$

$$\Rightarrow \quad \frac{dx}{yz} = \frac{dy}{-xz} = \frac{dz}{xy}$$

From first and second fractions, we get

$$\frac{dx}{yz} = \frac{dy}{-xz} \quad \Rightarrow \quad \frac{dx}{y} = \frac{dy}{-x}$$

$$\Rightarrow \quad xdx + ydy = 0$$

Integrating, we get

$$\frac{x^2}{2} + \frac{y^2}{2} = \frac{c_1}{2} \quad \Rightarrow \quad x^2 + y^2 = c_1 \quad ...(2)$$

From first and third fractions, we get

$$\frac{dx}{yz} = \frac{dz}{xy} \quad \Rightarrow \quad \frac{dx}{z} = \frac{dz}{x} \quad \Rightarrow x\,dx - z\,dz = 0$$

Integrating, we get

$$\frac{x^2}{2} - \frac{z^2}{2} = \frac{c_2}{2} \quad \Rightarrow \quad x^2 - z^2 = c_2$$

Hence the general solution is $\phi(x^2 + y^2, x^2 - z^2) = 0$

(*ii*) Lagrange's subsidiary equations are

$$\frac{dx}{P} = \frac{dy}{Q} = \frac{dz}{R}$$

$$\Rightarrow \quad \frac{dx}{y^2} = \frac{dy}{-xy} = \frac{dz}{x(z - 2y)}$$

From first and second fractions, we have

$$\frac{dx}{y^2} = \frac{dy}{-xy}$$

$$\Rightarrow \qquad x\,dx + y\,dy = 0$$

Integrating, we get

$$\frac{x^2}{2} + \frac{y^2}{2} = \frac{c_1}{2} \quad \Rightarrow \quad x^2 + y^2 = c_1 \qquad \text{...(1)}$$

From second and third fractions, we have

$$\frac{dy}{-xy} = \frac{dz}{x(z-2y)}$$

$$\Rightarrow \qquad -zdy + 2ydy = ydz$$

$$\Rightarrow \qquad 2ydy = ydz + zdy$$

Integrating, we get

$$y^2 = yz + c_2 \qquad \text{...(2)}$$

Hence the general solution is $\phi(x^2 + y^2, y^2 - yz) = 0$

(*iii*) Lagrange's subsidiary equations are

$$\frac{dx}{P} = \frac{dy}{Q} = \frac{dz}{R}$$

$$\Rightarrow \qquad \frac{dx}{x^2} = \frac{dy}{y^2} = \frac{dz}{(x+y)z} \qquad \text{...(1)}$$

From first and second fractions, we have

$$\frac{dx}{x^2} = \frac{dy}{y^2}$$

Integrating, we get

$$-\frac{1}{x} = -\frac{1}{y} - c_1$$

$$\Rightarrow \qquad \frac{1}{x} - \frac{1}{y} = c_1 \qquad \text{...(2)}$$

Again from (1),

$$\frac{dx - dy}{x^2 - y^2} = \frac{dz}{(x+y)z} \quad \Rightarrow \quad \frac{dx - dy}{x - y} = \frac{dz}{z}$$

Integrating, we get

$$\log(x - y) = \log z + \log c_2$$

$$\Rightarrow \qquad \frac{x-y}{z} = c_2 \qquad \text{...(3)}$$

Hence the general solution is

$$\phi\left(\frac{1}{x} - \frac{1}{y}, \frac{x-y}{z}\right) = 0$$

Example 6. *Solve the following differential equations:*

(*i*) $\dfrac{y^2z}{x}p + xzq = y^2$ (*ii*) $pz - qz = z^2 + (x + y)^2$

(*iii*) $(x^2 - yz)p + (y^2 - zx)q = z^2 - xy.$

Sol. (*i*) The given equation can be written as $y^2zp + x^2zq = xy^2$

Comparing with $Pp + Qq = R$, we have $P = y^2z$, $Q = x^2z$, $R = xy^2$

∴ Lagrange's subsidiary equations are

$$\frac{dx}{y^2z} = \frac{dy}{x^2z} = \frac{dz}{xy^2}$$

Taking the first two members, we have $x^2dx = y^2dy$

which on integration gives $x^3 - y^3 = c_1$ ...(1)

Again taking the first and third members, we have $xdx = zdz$

which on integration gives $x^2 - z^2 = c_2$ ...(2)

From (1) and (2), the general solution is

$$\phi(x^3 - y^3, x^2 - z^2) = 0.$$

(*ii*) Here Lagrange's subsidiary equations are

$$\frac{dx}{z} = \frac{dy}{-z} = \frac{dz}{z^2 + (x+y)^2}$$

Taking the first two members, we have $dx + dy = 0$

which on integration gives $x + y = c_1$ ...(1)

Again taking the first and third members, we have

$$dx = \frac{zdz}{z^2 + c_1^2}, \text{ since } x + y = c_1$$

or $$\frac{2z\,dz}{z^2 + c_1^2} = 2dx$$

which on integration gives

$$\log (z^2 + c_1^2) = 2x + c_2$$

or $$\log [z^2 + (x + y)^2] - 2x = c_2$$ ...(2)

From (1) and (2), the general solution is

$$\phi[x + y, \log (x^2 + y^2 + z^2 + 2xy) - 2x] = 0.$$

(*iii*) Here Lagrange's subsidiary equations are

$$\frac{dx}{x^2 - yz} = \frac{dy}{y^2 - zx} = \frac{dz}{z^2 - xy}$$

$$\therefore \quad \frac{dx - dy}{(x-y)(x+y+z)} = \frac{dy - dz}{(y-z)(x+y+z)}$$

$$= \frac{dz - dx}{(z-x)(x+y+z)}$$

Taking the first two members, we have $\dfrac{dx - dy}{x - y} = \dfrac{dy - dz}{y - z}$

which on integration gives

$$\log (x - y) = \log (y - z) + \log c_1$$

or $$\log\left(\frac{x-y}{y-z}\right) = \log c_1$$

or $$\frac{x-y}{y-z} = c_1 \qquad ...(1)$$

Now, using x, y, z as multipliers, we get

$$\text{each fraction} = \frac{x\,dx + y\,dy + z\,dz}{x^3 + y^3 + z^3 - 3xyz}$$

$$= \frac{x\,dx + y\,dy + z\,dz}{(x+y+z)(x^2+y^2+z^2-xy-yz-zx)} \qquad ...(2)$$

Again, using 1, 1, 1 as multipliers, we get

$$\text{each fraction} = \frac{dx + dy + dz}{x^2+y^2+z^2-yz-zx-xy} \qquad ...(3)$$

Hence, from (2) and (3),

$$\frac{x\,dx + y\,dy + z\,dz}{(x+y+z)(x^2+y^2+z^2-yz-zx-xy)} = \frac{dx+dy+dz}{x^2+y^2+z^2-yz-zx-xy}$$

$$\Rightarrow \quad \frac{x\,dx + y\,dy + z\,dz}{x+y+z} = dx + dy + dz$$

$$\Rightarrow \quad x\,dx + y\,dy + z\,dz = (x+y+z)(dx+dy+dz)$$

Intergration gives,

$$\frac{x^2}{2} + \frac{y^2}{2} + \frac{z^2}{2} = \frac{(x+y+z)^2}{2} - c_2$$

$$= \frac{x^2+y^2+z^2}{2} + yz + zx + xy - c_2$$

$$\Rightarrow \quad xy + yz + yx = c_2 \qquad ...(4)$$

Hence the general solution is

$$\phi\left(\frac{x-y}{y-z},\ xy+yz+zx\right) = 0.$$

Example 7. *Solve the following differential equations:*

(*i*) $(mz - ny)p + (nx - lz)q = ly - mx$

(*ii*) $x^2(y-z)p + y^2(z-x)q = z^2(x-y)$. (U.P.T.U. 2013)

Sol. (*i*) Here Lagrange's subsidiary equations are

$$\frac{dx}{mz-ny} = \frac{dy}{nx-lz} = \frac{dz}{ly-mx}$$

Using x, y, z as multipliers, we get

$$\text{each fraction} = \frac{xdx + ydy + zdz}{0}$$

$$\therefore \quad xdx + ydy + zdz = 0$$

which on integration gives $x^2 + y^2 + z^2 = a$ ...(1)

Again using l, m, n as multipliers, we get

$$\text{each fraction} = \frac{ldx + mdy + ndz}{0}$$

$$\therefore \qquad ldx + mdy + ndz = 0$$

which on integration gives $lx + my + nz = b$ ...(2)

From (1) and (2), the general solution is

$$x^2 + y^2 + z^2 = f(lx + my + nz).$$

(*ii*) Here the auxiliary equations are

$$\frac{dx}{x^2(y-z)} = \frac{dy}{y^2(z-x)} = \frac{dz}{z^2(x-y)}$$

Using $\frac{1}{x^2}, \frac{1}{y^2}, \frac{1}{z^2}$ as multipliers, we get

$$\text{each fraction} = \frac{\frac{1}{x^2}dx + \frac{1}{y^2}dy + \frac{1}{z^2}dz}{0}$$

$$\therefore \qquad \frac{dx}{x^2} + \frac{dy}{y^2} + \frac{dz}{z^2} = 0$$

which on integration gives $\frac{1}{x} + \frac{1}{y} + \frac{1}{z} = c_1$ ...(1)

Again using $\frac{1}{x}, \frac{1}{y}, \frac{1}{z}$ as multipliers, we get

$$\text{each fraction} = \frac{\frac{1}{x}dx + \frac{1}{y}dy + \frac{1}{z}dz}{0}$$

$$\therefore \qquad \frac{dx}{x} + \frac{dy}{y} + \frac{dz}{z} = 0$$

which on integration gives

$$\log x + \log y + \log z = \log c_2$$

or $$\log(xyz) = \log c_2$$

or $$xyz = c_2 \qquad \text{...(2)}$$

From (1) and (2), the general solution is

$$\phi\left(\frac{1}{x} + \frac{1}{y} + \frac{1}{z}, xyz\right) = 0.$$

Example 8. *Solve the partial differential equation*

$$x(y^2 + z)\,p - y(x^2 + z)\,q = z(x^2 - y^2) \text{ where } p = \frac{\partial z}{\partial x} \text{ and } q = \frac{\partial z}{\partial y}.$$

Sol. Lagrange's subsidiary equations are

$$\frac{dx}{x(y^2 + z)} = \frac{dy}{-y(x^2 + z)} = \frac{dz}{z(x^2 - y^2)} \qquad \text{...(1)}$$

Using $x, y, -1$ as multipliers, we get

$$\text{each fraction} = \frac{x\,dx + y\,dy - dz}{x^2(y^2+z) - y^2(x^2+z) - z(x^2-y^2)} = \frac{x\,dx + y\,dy - dz}{0}$$

$\therefore \quad x\,dx + y\,dy - dz = 0$

Integrating, we get

$$\frac{x^2}{2} + \frac{y^2}{2} - z = \frac{c_1}{2}$$

$$\Rightarrow \quad x^2 + y^2 - 2z = c_1 \qquad \text{...(2)}$$

Again, using $\frac{1}{x}, \frac{1}{y}$ and $\frac{1}{z}$ as multipliers, we get

$$\text{each fraction} = \frac{\frac{1}{x}dx + \frac{1}{y}dy + \frac{1}{z}dz}{y^2 + z - x^2 - z + x^2 - y^2} = \frac{\frac{1}{x}dx + \frac{1}{y}dy + \frac{1}{z}dz}{0}$$

$$\therefore \quad \frac{1}{x}dx + \frac{1}{y}dy + \frac{1}{z}dz = 0$$

Integrating, we get

$$\log x + \log y + \log z = \log c_2$$

$$\Rightarrow \quad xyz = c_2 \qquad \text{...(3)}$$

Hence the general solution is

$$\phi\,(x^2 + y^2 - 2z, xyz) = 0.$$

Example 9. *Solve the following partial differential equation*

$x(y^2 - z^2)\,p + y(z^2 - x^2)\,q = z(x^2 - y^2)$ *where* $p = \frac{\partial z}{\partial x}$ *and* $q = \frac{\partial z}{\partial y}$.

Sol. Lagrange's subsidiary equations are

$$\frac{dx}{x(y^2 - z^2)} = \frac{dy}{y(z^2 - x^2)} = \frac{dz}{z(x^2 - y^2)} \qquad \text{...(1)}$$

Using multipliers x, y, z, we get

$$\text{each fraction} = \frac{x\,dx + y\,dy + z\,dz}{x^2(y^2 - z^2) + y^2(z^2 - x^2) + z^2(x^2 - y^2)} = \frac{x\,dx + y\,dy + z\,dz}{0}$$

$\therefore \quad x\,dx + y\,dy + z\,dz = 0$

Integrating, we get

$$\frac{x^2}{2} + \frac{y^2}{2} + \frac{z^2}{2} = \frac{c_1}{2}$$

$$\Rightarrow \quad x^2 + y^2 + z^2 = c_1 \qquad \text{...(2)}$$

Again using $\frac{1}{x}, \frac{1}{y}, \frac{1}{z}$ as multipliers, we get

$$\text{each fraction} = \frac{\frac{1}{x}dx + \frac{1}{y}dy + \frac{1}{z}dz}{y^2 - z^2 + z^2 - x^2 + x^2 - y^2} = \frac{\frac{1}{x}dx + \frac{1}{y}dy + \frac{1}{z}dz}{0}$$

$$\therefore \quad \frac{1}{x}dx + \frac{1}{y}dy + \frac{1}{z}dz = 0$$

Integrating, we get

$$\log x + \log y + \log z = \log c_2$$

$$\Rightarrow \quad xyz = c_2 \quad \text{...(3)}$$

Hence the general solution is

$$\phi\,(x^2 + y^2 + z^2, xyz) = 0$$

Example 10. *Solve:*

(*i*) $p + 3q = 5z + \tan (y - 3x)$ (A.K.T.U. 2017)

(*ii*) $(y^2 + z^2 - x^2)\,p - 2xyq + 2xz = 0$

(*iii*) $(y + zx)\,p - (x + yz)q = x^2 - y^2$ (U.P.T.U. 2014)

Sol. (*i*) Lagrange's subsidiary equations are

$$\frac{dx}{1} = \frac{dy}{3} = \frac{dz}{5z + \tan(y - 3x)} \quad \text{...(1)}$$

From first and second fractions, we have

$$\frac{dx}{1} = \frac{dy}{3}$$

Integrating, we get

$$y - 3x = c_1 \quad \text{...(2)}$$

Now, from first and last fractions, we have

$$\frac{dx}{1} = \frac{dz}{5z + \tan c_1}$$

Integrating, we get

$$x = \frac{1}{5}\log(5z + \tan c_1) - \frac{c_2}{5}$$

$$\Rightarrow \quad \log\{5z + \tan(y - 3x)\} - 5x = c_2 \quad \text{...(3)}$$

Hence the general solution is

$$\phi\left[y - 3x, \log\{5z + \tan(y - 3x)\}\right] = 0$$

(*ii*) Lagrange's subsidiary equations are

$$\frac{dx}{y^2 + z^2 - x^2} = \frac{dy}{-2xy} = \frac{dz}{-2xz} \quad \text{...(1)}$$

Taking the last two fractions, we have

$$\frac{dy}{-2xy} = \frac{dz}{-2xz} \quad \Rightarrow \quad \frac{dy}{y} = \frac{dz}{z}$$

Integrating, we get

$$\log y = \log z + \log c_1$$

$\Rightarrow$ $$\frac{y}{z} = c_1 \quad ...(2)$$

Using x, y, z as multipliers, we have

$$\text{each fraction} = \frac{xdx + ydy + zdz}{xy^2 + xz^2 - x^3 - 2xy^2 - 2xz^2} = \frac{xdx + ydy + zdz}{-x(x^2 + y^2 + z^2)}$$

Let $$\frac{xdx + ydy + zdz}{-x(x^2 + y^2 + z^2)} = \frac{dz}{-2xz}$$

$\Rightarrow$ $$\frac{2xdx + 2ydy + 2zdz}{x^2 + y^2 + z^2} = \frac{dz}{z}$$

Integrating, we get

$$\log\left(x^2 + y^2 + z^2\right) = \log z + \log c_2$$

$\Rightarrow$ $$\frac{x^2 + y^2 + z^2}{z} = c_2$$

Hence the general solution is

$$\phi\left(\frac{y}{z}, \frac{x^2 + y^2 + z^2}{z}\right) = 0.$$

(*iii*) Lagrange's subsidiary equations are

$$\frac{dx}{y + zx} = \frac{dy}{-(x + yz)} = \frac{dz}{x^2 - y^2}$$

Using multipliers x, y and $-z$, we get

$$\text{each fraction} = \frac{xdx + ydy - zdz}{xy + x^2z - yx - y^2z - zx^2 + zy^2} = \frac{xdx + ydy - zdz}{0}$$

$\therefore$ $$xdx + ydy - zdz = 0$$

Integrating, we get

$$\frac{x^2}{2} + \frac{y^2}{2} - \frac{z^2}{2} = \frac{c_1}{2}$$

$\Rightarrow$ $$x^2 + y^2 - z^2 = c_1 \quad ...(1)$$

Again, using multipliers $y, x, 1$; we get

$$\text{each fraction} = \frac{ydx + xdy + dz}{y^2 + yzx - x^2 - xyz + x^2 - y^2} = \frac{ydx + xdy + dz}{0}$$

$\therefore$ $$ydx + xdy + dz = 0$$

Integrating, we get

$$xy + z = c_2 \qquad ...(2)$$

Hence the general solution is

$$\phi(x^2 + y^2 - z^2, xy + z) = 0.$$

Example 11. *Solve:*

(i) $(x^2 - y^2 - yz)\, p + (x^2 - y^2 - zx)\, q = z\,(x - y)$

(ii) $(x^2 + xy)\, p - (xy + y^2)\, q = -\,(x - y)\,(2x + 2y + z)$

Sol. (*i*) Lagrange's subsidiary equations are

$$\frac{dx}{x^2 - y^2 - yz} = \frac{dy}{x^2 - y^2 - zx} = \frac{dz}{z\,(x - y)}$$

Using 1, –1, –1 as multipliers, we have

$$\text{each fraction} = \frac{dx - dy - dz}{x^2 - y^2 - yz - x^2 + y^2 + zx - zx + zy} = \frac{dx - dy - dz}{0}$$

$\therefore \qquad dx - dy - dz = 0$

Integrating, we get

$$x - y - z = c_1 \qquad ...(1)$$

Again, from first and second, we have

$$\frac{xdx - ydy}{x^3 - xy^2 - xyz - yx^2 + y^3 + yzx} = \frac{dz}{z\,(x - y)}$$

$$\Rightarrow \qquad \frac{xdx - ydy}{x^3 - xy^2 - yx^2 + y^3} = \frac{dz}{z\,(x - y)}$$

$$\Rightarrow \qquad \frac{xdx - ydy}{(x - y)(x^2 - y^2)} = \frac{dz}{z\,(x - y)}$$

$$\Rightarrow \qquad \frac{xdx - ydy}{x^2 - y^2} = \frac{dz}{z}$$

Integrating, we get

$$\frac{1}{2}\log(x^2 - y^2) = \log z + \frac{1}{2}\log c_2$$

$$\Rightarrow \qquad \frac{x^2 - y^2}{z^2} = c_2 \qquad ...(2)$$

Hence the general solution is

$$\phi\left(x - y - z, \frac{x^2 - y^2}{z^2}\right) = 0$$

(*ii*) Lagrange's subsidiary equations are

$$\frac{dx}{x\,(x + y)} = \frac{dy}{-\,y\,(x + y)} = \frac{dz}{-\,(x - y)\,(2x + 2y + z)}$$

Taking first and second fractions, we have

$$\frac{dx}{x} = \frac{dy}{-\,y}$$

Integrating, we get

$$\log x = -\log y + \log c_1$$

$$\Rightarrow \qquad xy = c_1 \qquad \text{...(1)}$$

Again from first and second,

$$\frac{dx + dy}{(x-y)(x+y)} = \frac{dz}{-(x-y)(2x+2y+z)}$$

$$\Rightarrow \qquad (2x + 2y + z)(dx + dy) + (x + y)\,dz = 0$$

$$\Rightarrow \qquad (x + y + z)\,d(x + y) + (x + y)(dx + dy) + (x + y)\,dz = 0$$

$$\Rightarrow \qquad (x + y + z)\,d(x + y) + (x + y)\,d(x + y + z) = 0$$

$$\Rightarrow \qquad d\{(x + y)(x + y + z)\} = 0$$

Integrating, we get

$$(x + y)(x + y + z) = c_2$$

Hence the general solution is

$$\phi\,\{xy, (x + y)(x + y + z)\} = 0$$

Example 12. *Solve the partial differential equations:*

(*i*) $\left(\dfrac{y-z}{yz}\right) p + \left(\dfrac{z-x}{zx}\right) q = \left(\dfrac{x-y}{xy}\right)$

(*ii*) $(y^2 + z^2)\,p - xyq = -zx$ (A.K.T.U. 2016, 20017)

Sol. (*i*) The given equation can also be written as

$$x(y - z)\,p + y(z - x)\,q = z(x - y) \qquad \text{...(1)}$$

Lagrange's subsidiary equations are

$$\frac{dx}{x(y-z)} = \frac{dy}{y(z-x)} = \frac{dz}{z(x-y)}$$

Using multipliers as $\dfrac{1}{x}, \dfrac{1}{y}, \dfrac{1}{z}$, we have

$$\text{each fraction} = \frac{\frac{1}{x}dx + \frac{1}{y}dy + \frac{1}{z}dz}{y - z + z - x + x - y} = \frac{\frac{1}{x}dx + \frac{1}{y}dy + \frac{1}{z}dz}{0}$$

$$\therefore \qquad \frac{1}{x}dx + \frac{1}{y}dy + \frac{1}{z}dz = 0$$

Integrating, we get

$$\log x + \log y + \log z = \log c_1 \qquad \Rightarrow \quad xyz = c_1 \qquad \text{...(2)}$$

Again, using 1, 1, 1 as multipliers, we get

$$\text{each fraction} = \frac{dx + dy + dz}{x(y-z) + y(z-x) + z(x-y)} = \frac{dx + dy + dz}{0}$$

$$\therefore \qquad dx + dy + dz = 0$$

Integrating, we get

$$x + y + z = c_2$$

Hence the general solution is

$$\phi\,(xyz, x + y + z) = 0.$$

(*ii*) Lagrange's subsidiary eqns. are

$$\frac{dx}{y^2 + z^2} = \frac{dy}{-xy} = \frac{dz}{-zx} \qquad ...(1)$$

From second and third fractions,

$$\frac{dy}{y} = \frac{dz}{z}$$

Integrating, we get $\qquad \frac{y}{z} = c_1$

Now, using multipliers as x, y, z, we get

$$\text{each fraction} = \frac{x\,dx + y\,dy + z\,dz}{0}$$

$$\therefore \qquad x\,dx + y\,dy + z\,dz = 0$$

Integrating, we get

$$\frac{x^2}{2} + \frac{y^2}{2} + \frac{z^2}{2} = \frac{c_2}{2} \quad \Rightarrow \quad x^2 + y^2 + z^2 = c_2$$

Hence the general solution is

$$\phi\left(\frac{y}{z}, x^2 + y^2 + z^2\right) = 0.$$

TEST YOUR KNOWLEDGE

Solve the following partial differential equations:

1. $2p + 3q = 1$
2. $xp + yq = 3z$
3. $(y - z)p + (x - y)q = z - x$
4. $p \tan x + q \tan y = \tan z$
5. $p\sqrt{x} + q\sqrt{y} = \sqrt{z}$
6. $(z - y)p + (x - z)q = y - x$
7. $yzp + zxq = xy$
8. $(y + z)p + (z + x)q = x + y$
9. $2yzp + zxq = 3xy$
10. $z(xp - yq) = y^2 - x^2$
11. $2xzp + 2yzq = z^2 - x^2 - y^2$
12. $(3 - 2yz)\, p + x(2z - 1)\, q = 2x\,(y - 3)$
13. $x\,(z - 3y^3)\, p + y\,(3x^3 - z)\, q = 3\,(y^3 - x^3)\, z$ [**Hint:** Multipliers $\to (x^2, y^2, 1/3)$ and $(1/x, 1/y, 1/z)$]
14. $(z^2 - 2yz - y^2)\, p + x(y + z)\, q = x\,(y - z)$
15. $y^2\,(x + y)\, p + x^2\,(x + y)\, q = (x^2 + y^2)\, z$ [**Hint:** Multipliers $\to (1, 1, 0)$ and $(x^2, -y^2, 0)$]
16. $(x + 2y^2 + z)\, p + y\,(2x - 1)\, q + 2\,(x^2 + y^2 + xz) = 0$ [**Hint:** Multipliers $\to (x, -y, 1/2)$ and $(y, x + z, y)$]
17. $yp + xq = xyz^2\,(x^2 - y^2)$ [*U.P.T.U. 2015*]
18. $(x - y)\, y^2p + (y - x)\, x^2\, q = (x^2 + y^2)z$
19. $x\frac{\partial u}{\partial x} + y\frac{\partial u}{\partial y} + z\frac{\partial u}{\partial z} = xyz$
20. $x_1p_1 + 2x_2p_2 + 3x_3p_3 + 4x_4p_4 = 0$

Answers

1. $\phi(3x - 2y, y - 3z) = 0$
2. $\phi\left(\frac{x}{y}, \frac{x^3}{z}\right) = 0$
3. $\phi\left(x + y + z, \frac{x^2}{2} + yz\right) = 0$
4. $\phi\left(\frac{\sin x}{\sin y}, \frac{\sin y}{\sin z}\right) = 0$
5. $\sqrt{x} - \sqrt{y} = f(\sqrt{x} - \sqrt{z})$
6. $x^2 + y^2 + z^2 = f(x + y + z)$
7. $\phi(x^2 - y^2, x^2 - z^2) = 0$
8. $\phi\left[\frac{x - y}{y - z}, (x - y)^2 (x + y + z)\right] = 0$
9. $\phi(x^2 - 2y^2, 3y^2 - z^2) = 0$
10. $\phi(xy, x^2 + y^2 + z^2) = 0$
11. $\phi\left(\frac{y}{x}, \frac{x^2 + y^2 + z^2}{x}\right) = 0$
12. $\phi(y^2 - 6y - z^2 + z, x^2 + z^2 + 6y) = 0$
13. $\phi(x^3 + y^3 + z, xyz) = 0$
14. $\phi(x^2 + y^2 + z^2, y^2 - 2yz - z^2) = 0$
15. $\phi\left(x^3 - y^3, \frac{x + y}{z}\right) = 0$
16. $\phi(x^2 - y^2 + z, xy + yz) = 0$
17. $\phi\left\{x^2 - y^2, \frac{x^4}{4} - \frac{y^4}{4} + \frac{1}{z}\right\} = 0$
18. $\phi\left\{x^3 + y^3, \frac{x - y}{z}\right\} = 0$
19. $\phi\left(\frac{x}{y}, \frac{y}{z}, xyz - 3u\right) = 0$
20. $z = \phi\left(\frac{x_1^2}{x_2}, \frac{x_1^3}{x_3}, \frac{x_1^4}{x_4}\right)$

1.12 NON-LINEAR PDE OF FIRST ORDER

A partial differential equation which involves first order partial derivatives p and q with degree higher than one and the products of p and q is called a non-linear partial differential equation of the first order. The complete solution of such an equation involves only two arbitrary constants (*i.e.*, equal to the number of independent variables). There is a general method for solving such equations. Before giving this method, we shall deal with some special types of such equations which can be solved easily by methods other than the general method.

1.12.1. Equations of the Form *f*(*p*, *q*) = 0, *i.e.*,

equations involving only ***p*** *and* ***q*** *and no* ***x*, *y*, *z***

The complete solution is $z = ax + by + c$ ...(1)

where a and b are connected by the relation

$$f(a, b) = 0 \qquad \text{...(2)} \qquad \left[\text{since } p = \frac{\partial z}{\partial x} = a \text{ and } q = \frac{\partial z}{\partial y} = b\right]$$

From (2), we can find b in terms of a. Let $b = \phi(a)$.

Putting this value of b in (1), the complete solution is

$$z = ax + \phi(a)y + c, \quad \text{where } a \text{ and } c \text{ are arbitrary constants.}$$

ILLUSTRATIVE EXAMPLES

Example 1. *Solve:*

(*i*) $\sqrt{p} + \sqrt{q} = 1$ (*ii*) $pq = p + q$.

Sol. (*i*) The equation is of the form $f(p, q) = 0$

The complete solution is $z = ax + by + c$ ...(1)

where $\sqrt{a} + \sqrt{b} = 1$ or $b = (1 - \sqrt{a})^2$

$\therefore$ From (1), the complete solution is

$$z = ax + (1 - \sqrt{a})^2 y + c$$

(*ii*) The equation is of the form $f(p, q) = 0$

The complete solution is $z = ax + by + c$ ...(1)

where $ab = a + b$ or $b = \dfrac{a}{a-1}$

$\therefore$ From (1), the complete solution is $z = ax + \dfrac{a}{a-1}y + c$.

Example 2. *Solve* $x^2p^2 + y^2q^2 = z^2$.

Sol. The given equation can be written as

$$\left(\frac{x}{z} \cdot \frac{\partial z}{\partial x}\right)^2 + \left(\frac{y}{z} \cdot \frac{\partial z}{\partial y}\right)^2 = 1 \quad ...(1)$$

Let $\dfrac{\partial x}{x} = \partial X, \dfrac{\partial y}{y} = \partial Y, \dfrac{\partial z}{z} = \partial Z$ then $X = \log x, Y = \log y, Z = \log z$

$$\frac{\partial Z}{\partial X} = \frac{x}{z} \cdot \frac{\partial z}{\partial x} \text{ and } \frac{\partial Z}{\partial Y} = \frac{y}{z} \cdot \frac{\partial z}{\partial y}$$

$\therefore$ Eqn. (1) can be written as $\left(\dfrac{\partial Z}{\partial X}\right)^2 + \left(\dfrac{\partial Z}{\partial Y}\right)^2 = 1$

or $P^2 + Q^2 = 1$, where $P = \dfrac{\partial Z}{\partial X}$ and $Q = \dfrac{\partial Z}{\partial Y}$

It is of the form $f(P, Q) = 0$

Its complete solution is $Z = aX + bY + c$ ...(2)

where $a^2 + b^2 = 1$ or $b = \sqrt{1 - a^2}$

$\therefore$ From (2), the complete solution is $\log z = a \log x + \sqrt{1-a^2} \log y + c$.

Example 3. *Solve* $(y - x)(qy - px) = (p - q)^2$.

Sol. Let $x + y = X$ and $xy = Y$

so that $$p = \frac{\partial z}{\partial x} = \frac{\partial z}{\partial X} \cdot \frac{\partial X}{\partial x} + \frac{\partial z}{\partial Y} \cdot \frac{\partial Y}{\partial x} = \frac{\partial z}{\partial X} + y\frac{\partial z}{\partial Y}$$

and $$q = \frac{\partial z}{\partial y} = \frac{\partial z}{\partial X} \cdot \frac{\partial X}{\partial y} + \frac{\partial z}{\partial Y} \cdot \frac{\partial Y}{\partial y} = \frac{\partial z}{\partial X} + x\frac{\partial z}{\partial Y}$$

$$\therefore \qquad qy - px = (y - x)\frac{\partial z}{\partial X} \text{ and } p - q = (y - x)\frac{\partial z}{\partial Y}$$

$\therefore$ The given equation can be written as $\dfrac{\partial z}{\partial X} = \left(\dfrac{\partial z}{\partial Y}\right)^2$

or
$$P = Q^2, \text{ where } P = \frac{\partial z}{\partial X} \text{ and } Q = \frac{\partial z}{\partial Y}$$

It is of the form $f(P, Q) = 0$

Its complete solution is $z = aX + bY + c$ ...(1)

where $a = b^2$ or $b = \sqrt{a}$

$\therefore$ From (1), the complete solution is $z = a(x + y) + \sqrt{a}\, xy + c$.

1.12.2. Equations of the Form $z = px + qy + f(p, q)$

The complete solution is $z = ax + by + f(a, b)$

obtained by writing a for p and b for q.

Example 4. *Solve:* $z = px + qy + \sqrt{1 + p^2 + q^2}$.

Sol. The given equation is of the form $z = px + qy + f(p, q)$

Its complete solution is $z = ax + by + \sqrt{1 + a^2 + b^2}$.

Example 5. *Find the singular integral of* $z = px + qy + pq$.

Sol. The equation is of the form

$$z = px + qy + f(p, q)$$

Its complete integral is

$$z = ax + by + ab \qquad ...(1)$$

Diff. (1) partially w.r.t. a and b,

$$0 = x + b \quad \text{and} \quad 0 = y + a \qquad ...(2)$$

Eliminating a and b between eqns. (1) and (2), we get

$$z = -xy - xy + xy$$

$$\Rightarrow \qquad z = -xy$$

which is the required singular solution, for it satisfies the given equation.

Example 6. *Solve:* $4xyz = pq + 2px^2y + 2qxy^2$.

Sol. Let $x^2 = X$ and $y^2 = Y$

so that
$$p = \frac{\partial z}{\partial x} = \frac{\partial z}{\partial X}\cdot\frac{\partial X}{\partial x} = 2x\frac{\partial z}{\partial X} \quad \text{and} \quad q = \frac{\partial z}{\partial y} = \frac{\partial z}{\partial Y}\cdot\frac{\partial Y}{\partial y} = 2y\frac{\partial z}{\partial Y}$$

$\therefore$ The given equation becomes

$$4xyz = 4xy\frac{\partial z}{\partial X}\cdot\frac{\partial z}{\partial Y} + 4x^3y\frac{\partial z}{\partial X} + 4xy^3\frac{\partial z}{\partial Y}$$

or
$$z = x^2\frac{\partial z}{\partial X} + y^2\frac{\partial z}{\partial Y} + \frac{\partial z}{\partial X}\cdot\frac{\partial z}{\partial Y} = X\frac{\partial z}{\partial X} + Y\frac{\partial z}{\partial Y} + \frac{\partial z}{\partial X}\cdot\frac{\partial z}{\partial Y}$$

or $$z = PX + QY + PQ, \quad \text{where } P = \frac{\partial z}{\partial X} \quad \text{and} \quad Q = \frac{\partial z}{\partial Y}$$

It is of the form $z = PX + QY + f(P, Q)$

Its complete solution is $z = aX + bY + ab$ or $z = ax^2 + by^2 + ab$.

1.12.3. Equations of the Form $f(z, p, q) = 0$,

i.e., equations not containing x *and* y.

Assume $z = \phi(x + ay) = \phi(u)$, where $u = x + ay$ as a trial solution of the given equation.

$$\therefore \quad p = \frac{\partial z}{\partial x} = \phi'(x + ay) = \phi'(u) = \frac{dz}{du}$$

$$q = \frac{dz}{dy} = \phi'(x + ay)a = a\phi'(u) = a\frac{dz}{du}$$

Substituting these values of p and q in the given equation, we get $f\left(z, \frac{dz}{du}, a\frac{dz}{du}\right) = 0$

which is an ordinary differential equation of the first order. Integrating it, we get the complete solution.

Method. (*i*) Assume $u = x + ay$ so that $p = \frac{dz}{du}$ and $q = a\frac{dz}{du}$.

(*ii*) Substitute these values of p and q in the given equation.

(*iii*) Solve the resulting ordinary differential equation in z and u.

(*iv*) Replace u by $x + ay$.

Example 7. *Solve:* $z^2(p^2 + q^2 + 1) = a^2$.

Sol. The given equation is of the form $f(z, p, q) = 0$

Let $u = x + by$ (note the use of b instead of a, since a is a given constant)

so that $$p = \frac{dz}{du} \quad \text{and} \quad q = b\frac{dz}{du}.$$

Substituting these values of p and q in the given equation, we get

$$z^2\left[\left(\frac{dz}{du}\right)^2 + b^2\left(\frac{dz}{du}\right)^2 + 1\right] = a^2$$

or $$z^2(1 + b^2)\left(\frac{dz}{du}\right)^2 = a^2 - z^2$$

or $$z\sqrt{1+b^2}\,\frac{dz}{du} = \pm\sqrt{a^2 - z^2}$$

or $$\pm\sqrt{1+b^2}\,.\,\frac{z}{\sqrt{a^2 - z^2}}\,dz = du$$

Integrating, we have

$$\pm\sqrt{1+b^2}\,\sqrt{a^2 - z^2} = u + c$$

or $$(1 + b^2)(a^2 - z^2) = (x + by + c)^2$$

Example 8. *Solve:* $z^2 (p^2x^2 + q^2) = 1$.

Sol. The given equation can be written as

$$z^2 \left[\left(x \frac{\partial z}{\partial x} \right)^2 + \left(\frac{\partial z}{\partial y} \right)^2 \right] = 1 \quad \text{...(1)}$$

Let $\quad \text{X} = \log x$

so that $\quad \frac{\partial z}{\partial x} = \frac{\partial z}{\partial \text{X}} \cdot \frac{\partial \text{X}}{\partial x} = \frac{\partial z}{\partial \text{X}} \cdot \frac{1}{x} \quad$ *i.e.,* $\quad x \frac{\partial z}{\partial x} = \frac{\partial z}{\partial \text{X}}$

$\therefore$ Equation (1) reduces to $z^2 \left[\left(\frac{\partial z}{\partial \text{X}} \right)^2 + \left(\frac{\partial z}{\partial y} \right)^2 \right] = 1$

or $$z^2 (\text{P}^2 + q^2) = 1, \text{ where } \text{P} = \frac{\partial z}{\partial \text{X}} \quad \text{...(2)}$$

Let $\quad u = \text{X} + ay \quad$ so that $\quad \text{P} = \frac{dz}{du} \quad$ and $\quad q = a \frac{dz}{du}$

$\therefore$ From (2), we have $z^2 \left[\left(\frac{dz}{du} \right)^2 + a^2 \left(\frac{dz}{du} \right)^2 \right] = 1$

or $$(1 + a^2)z^2 \left(\frac{dz}{du} \right)^2 = 1 \quad \text{or} \quad \sqrt{1 + a^2} \,.\, z\, dz = \pm\, du$$

Integrating, $$\sqrt{1 + a^2} \cdot \frac{z^2}{2} = \pm\, u + b$$

or $$\sqrt{1 + a^2} \cdot z^2 = \pm\, 2(\text{X} + ay) + 2b$$

or $$\sqrt{1 + a^2} \cdot z^2 = \pm\, 2\, (\log x + ay) + c$$

1.12.4. Equations of the Form $f_1 (x, p) = f_2 (y, q)$

i.e., equations in which z *is absent and the terms involving* x *and* p *can be separated from those involving* y *and* q.

As a trial solution, let us put each side equal to an arbitrary constant a. Then

$$f_1 (x, p) = f_2 (y, q) = a$$

Solving these equations for p and q, let $p = \text{F}_1 (x)$ and $q = \text{F}_2 (y)$

Since $\quad dz = \frac{\partial z}{\partial x} dx + \frac{\partial z}{\partial y} dy = p dx + q dy$

$\therefore \quad dz = \text{F}_1(x)\, dx + \text{F}_2(y)\, dy$

Integrating, $\quad z = \int \text{F}_1 (x)\, dx + \int \text{F}_2 (y)\, dy + b$

Example 9. *Solve:* $p^2 - q^2 = x - y$.

Sol. The given equation is $p^2 - x = q^2 - y$ which is of the form $f_1(x, p) = f_2(y, q)$

Let $p^2 - x = q^2 - y = a$

then $p^2 = x + a$ and $q^2 = y + a$ *i.e.,* $p = \sqrt{x+a}$ and $q = \sqrt{y+a}$

Substituting these values of p and q in $dz = pdx + qdy$, we get

$$dz = \sqrt{x+a}\, dx + \sqrt{y+a}\, dy$$

Integrating, $z = \frac{2}{3}(x+a)^{3/2} + \frac{2}{3}(y+a)^{3/2} + b$

Example 10. *Solve:* $yp = 2yx + \log q$.

Sol. The given equation can be written as

$$p = 2x + \frac{1}{y}\log q \quad \text{or} \quad p - 2x = \frac{1}{y}\log q$$

which is of the form $f_1(x, p) = f_2(y, q)$.

Let $p - 2x = \frac{1}{y}\log q = a$ then $p = 2x + a$ and $\log q = ay$, *i.e.,* $q = e^{ay}$

Substituting these values of p and q in $dz = pdx + qdy$, we get

$$dz = (2x + a)\, dx + e^{ay}\, dy$$

Integrating, $z = x^2 + ax + \frac{1}{a}e^{ay} + b$

TEST YOUR KNOWLEDGE

Solve the following equations:

1. $p^2 + q^2 = 2$
2. $p^3 - q^3 = 0$
3. $p^2 + p = q^2$
4. $p = e^q$
5. $(x + y)(p + q)^2 + (x - y)(p - q)^2 = 1$
6. $z = px + qy - 2\sqrt{pq}$
7. $z = px + qy + \sin(p + q)$
8. $(pq - p - q)(z - px - qy) = pq$
9. $p^2 = zq$
10. $p(1 + q^2) = q(z - a)$
11. $p(1 + q) = qz$
12. $z^2(p^2 + q^2 + 1) = 1$
13. $z = p^2 + q^2$
14. $p + q = z$
15. $z^2(p^2z^2 + q^2) = 1$
16. $9(p^2z + q^2) = 4$
17. $p^2 + q^2 = x + y$
18. $p - x^2 = q + y^2$
19. $\sqrt{p} + \sqrt{q} = x + y$
20. $\sqrt{p} + \sqrt{q} = 2x$
21. $pe^y = qe^x$
22. $q = xyp^2$
23. $q(p - \cos x) = \cos y$
24. $yp + xq + pq = 0$
25. $z(p^2 - q^2) = x - y$
26. $p + q = \sin x + \sin y$
27. $z^2(p^2 + q^2) = x^2 + y^2$.

Answers

1. $z = ax + \sqrt{2-a^2}\, y + c$
2. $z = a(x + y) + c$
3. $z = ax + \sqrt{a^2+a}\, y + c$
4. $z = ax + y\log a + c$
5. $z = a\sqrt{x+y} + \sqrt{1-a^2}\sqrt{x-y} + c$
6. $z = ax + by - 2\sqrt{ab}$
7. $z = ax + by + \sin(a + b)$
8. $z = ax + by + \frac{ab}{ab - a - b}$
9. $z = be^{ax + a^2 y}$
10. $4(bz - ab - 1) = (x + by + c)^2$
11. $\log(az - 1) = x + ay + b$

12. $(1 + a^2)(1 - z^2) = (x + ay + b)^2$

13. $4z(1 + a^2) = (x + ay + b)^2$

14. $(1 + a)\log z = x + ay + b$

15. $(z^2 + a^2)^3 = 9(x + ay + b)^2$

16. $(z + a^2)^3 = (x + ay + b)^2$

17. $z = \frac{2}{3}(x + a)^{3/2} + \frac{2}{3}(y - a)^{3/2} + b$

18. $z = \frac{1}{3}(x^3 - y^3) + a(x + y) + b$

19. $z = \frac{1}{3}(x + a)^3 + \frac{1}{3}(y - a)^3 + b$

20. $z = \frac{1}{6}(2x + a)^3 + a^2y + b$

21. $z = ae^x + ae^y + b$

22. $z = 2\sqrt{ax} + \frac{1}{2}ay^2 + b$

23. $z = ax + \sin x + \frac{1}{a}\sin y + b$

24. $2z = ax^2 - \frac{a}{a+1}y^2 + b$

25. $z^{3/2} = (x + a)^{3/2} + (y + a)^{3/2} + b$

26. $z = a(x - y) - (\cos x + \cos y) + c$

27. $z^2 = x\sqrt{x^2 + a} + y\sqrt{y^2 - a} + a\log\frac{x + \sqrt{x^2 + a}}{y + \sqrt{y^2 - a}} + c.$

1.13 CHARPIT'S METHOD

This is a general method for finding the complete solution of non-linear partial differential equations of the first order.

Let the given equation be $f(x, y, z, p, q) = 0$ ...(1)

If we can find another relation $F(x, y, z, p, q) = 0$ ...(2)

involving x, y, z, p and q, then we can solve (1) and (2) for p and q and substitute in

$$dz = pdx + qdy \qquad ...(3)$$

Solution of (3), if it exists, is the complete solution of (1).

To determine F, we differentiate (1) and (2) partially w.r.t. x and y. Thus

$$\frac{\partial f}{\partial x} + \frac{\partial f}{\partial z}\cdot p + \frac{\partial f}{\partial p}\cdot\frac{\partial p}{\partial x} + \frac{\partial f}{\partial q}\cdot\frac{\partial q}{\partial x} = 0 \qquad ...(4)$$

$$\frac{\partial F}{\partial x} + \frac{\partial F}{\partial z}\cdot p + \frac{\partial F}{\partial p}\cdot\frac{\partial p}{\partial x} + \frac{\partial F}{\partial q}\cdot\frac{\partial q}{\partial x} = 0 \qquad ...(5)$$

$$\frac{\partial f}{\partial y} + \frac{\partial f}{\partial z}\cdot q + \frac{\partial f}{\partial p}\cdot\frac{\partial p}{\partial y} + \frac{\partial f}{\partial q}\cdot\frac{\partial q}{\partial y} = 0 \qquad ...(6)$$

$$\frac{\partial F}{\partial y} + \frac{\partial F}{\partial z}\cdot q + \frac{\partial F}{\partial p}\cdot\frac{\partial p}{\partial y} + \frac{\partial F}{\partial q}\cdot\frac{\partial q}{\partial y} = 0 \qquad ...(7)$$

Eliminating $\frac{\partial p}{\partial x}$ between (4) and (5), we get

$$\left(\frac{\partial f}{\partial x}\cdot\frac{\partial F}{\partial p} - \frac{\partial F}{\partial x}\cdot\frac{\partial f}{\partial p}\right) + \left(\frac{\partial f}{\partial z}\cdot\frac{\partial F}{\partial p} - \frac{\partial F}{\partial z}\cdot\frac{\partial f}{\partial p}\right)p + \left(\frac{\partial f}{\partial q}\cdot\frac{\partial F}{\partial p} - \frac{\partial F}{\partial q}\cdot\frac{\partial f}{\partial p}\right)\frac{\partial q}{\partial x} = 0 \qquad ...(8)$$

Eliminating $\frac{\partial q}{\partial y}$ between (6) and (7), we get

$$\left(\frac{\partial f}{\partial y}\cdot\frac{\partial F}{\partial q} - \frac{\partial F}{\partial y}\cdot\frac{\partial f}{\partial q}\right) + \left(\frac{\partial f}{\partial z}\cdot\frac{\partial F}{\partial q} - \frac{\partial F}{\partial z}\cdot\frac{\partial f}{\partial q}\right)q + \left(\frac{\partial f}{\partial p}\cdot\frac{\partial F}{\partial q} - \frac{\partial F}{\partial p}\cdot\frac{\partial f}{\partial q}\right)\frac{\partial p}{\partial y} = 0 \qquad ...(9)$$

Since $\frac{\partial q}{\partial x} = \frac{\partial^2 z}{\partial x \partial y} = \frac{\partial^2 z}{\partial y \partial x} = \frac{\partial p}{\partial y}$, the last terms in (8) and (9) differ in sign only. Adding (8) and (9), we get

$$\left(\frac{\partial f}{\partial x} + p\frac{\partial f}{\partial z}\right)\frac{\partial F}{\partial p} + \left(\frac{\partial f}{\partial y} + q\frac{\partial f}{\partial z}\right)\frac{\partial F}{\partial q} + \left(-p\frac{\partial f}{\partial p} - q\frac{\partial f}{\partial q}\right)\frac{\partial F}{\partial z} + \left(-\frac{\partial f}{\partial p}\right)\frac{\partial F}{\partial x} + \left(-\frac{\partial f}{\partial q}\right)\frac{\partial F}{\partial y} = 0 \quad \text{...(10)}$$

which is a linear partial differential equation of the first order with x, y, z, p, q as independent variables and F as the dependent variable.

∴ The auxiliary equations are

$$\frac{dp}{\frac{\partial f}{\partial x} + p\frac{\partial f}{\partial z}} = \frac{dq}{\frac{\partial f}{\partial y} + q\frac{\partial f}{\partial z}} = \frac{dz}{-p\frac{\partial f}{\partial p} - q\frac{\partial f}{\partial q}} = \frac{dx}{-\frac{\partial f}{\partial p}} = \frac{dy}{-\frac{\partial f}{\partial q}} = \frac{dF}{0} \quad \text{...(11)}$$

Any integral of (11) which involves p or q or both can be taken as the assumed relation (2). In practice, we choose the simplest of the integrals of (11).

ILLUSTRATIVE EXAMPLES

Example 1. *Solve:* $2zx - px^2 - 2pxy + pq = 0.$

Sol. Here $f \equiv 2zx - px^2 - 2qxy + pq = 0$ ...(1)

∴ $\frac{\partial f}{\partial x} = 2z - 2px - 2qy, \frac{\partial f}{\partial y} = -2qx, \frac{\partial f}{\partial z} = 2x, \frac{\partial f}{\partial p} = -x^2 + q, \frac{\partial f}{\partial q} = -2xy + p$

Charpit's auxiliary equations are

$$\frac{dp}{\frac{\partial f}{\partial x} + p\frac{\partial f}{\partial z}} = \frac{dq}{\frac{\partial f}{\partial y} + q\frac{\partial f}{\partial z}} = \frac{dz}{-p\frac{\partial f}{\partial p} - q\frac{\partial f}{\partial q}} = \frac{dx}{-\frac{\partial f}{\partial p}} = \frac{dy}{-\frac{\partial f}{\partial q}} = \frac{dF}{0}$$

or $$\frac{dp}{2z - 2qy} = \frac{dq}{0} = \frac{dz}{px^2 - 2pq + 2qxy} = \frac{dx}{x^2 - q} = \frac{dy}{2xy - p} = \frac{dF}{0}$$

∴ $dq = 0$ or $q = a$

Putting $q = a$ in (1), we get $p = \frac{2x(z - ay)}{x^2 - a}$

∴ $$dz = pdx + qdy = \frac{2x(z - ay)}{x^2 - a}dx + ady$$

or $$\frac{dz - ady}{z - ay} = \frac{2x}{x^2 - a}dx$$

Integrating, $\log(z - ay) = \log(x^2 - a) + \log b$ or $z - ay = b(x^2 - a)$

∴ $z = ay + b(x^2 - a)$ which is the required complete integral.

Example 2. *Solve:* $(p^2 + q^2)y = qz.$ (M.T.U. 2013)

Sol. Here $f \equiv (p^2 + q^2)y - qz = 0$ ...(1)

∴ $\frac{\partial f}{\partial x} = 0, \frac{\partial f}{\partial y} = p^2 + q^2, \frac{\partial f}{\partial z} = -q, \frac{\partial f}{\partial p} = 2py, \frac{\partial f}{\partial q} = 2qy - z$

Charpit's auxiliary equations are

$$\frac{dp}{\frac{\partial f}{\partial x}+p\frac{\partial f}{\partial z}}=\frac{dq}{\frac{\partial f}{\partial y}+q\frac{\partial f}{\partial z}}=\frac{dz}{-p\frac{\partial f}{\partial p}-q\frac{\partial f}{\partial q}}=\frac{dx}{-\frac{\partial f}{\partial p}}=\frac{dy}{-\frac{\partial f}{\partial q}}=\frac{dF}{0}$$

or
$$\frac{dp}{-pq}=\frac{dq}{p^2}=\frac{dz}{-qz}=\frac{dx}{-2py}=\frac{dy}{-2qy+z}=\frac{dF}{0}$$

From the first two members, we have $pdp + qdq = 0$

Integrating, $p^2 + q^2 = a^2$ ...(2)

Putting in (1), $q = \dfrac{a^2 y}{z}$

$\therefore$ From (2), $p = \sqrt{a^2 - q^2} = \sqrt{a^2 - \dfrac{a^4y^2}{z^2}} = \dfrac{a}{z}\sqrt{z^2 - a^2y^2}$

$\therefore$ $dz = pdx + qdy = \dfrac{a}{z}\sqrt{z^2 - a^2y^2}\, dx + \dfrac{a^2y}{z}\, dy$

or $zdz - a^2ydy = a\sqrt{z^2 - a^2y^2}\, dx$ or $\dfrac{\frac{1}{2}d(z^2 - a^2y^2)}{\sqrt{z^2 - a^2y^2}} = a\, dx$

Integrating, we get $\sqrt{z^2 - a^2y^2} = ax + b$

or $z^2 = (ax + b)^2 + a^2y^2$

TEST YOUR KNOWLEDGE

Solve the following equations by Charpit's method:

1. $pz + qy = pq$ **2.** $z^2 = pqxy$ **3.** $pxy + pq + qy = yz$

4. $z = p^2x + q^2y$ **5.** $2(z + xp + yp) = yp^2$ **6.** $q + xp = p^2$.

Answers

1. $az = \dfrac{1}{2}(y + ax)^2 + b$ **2.** $z = ax^b y^{1/b}$

3. $\log(z - ax) = y - a\log(a + y) + b$ **4.** $\sqrt{(1+a)z} = \sqrt{ax} + \sqrt{y} + b$ **5.** $z = \dfrac{ax}{y^2} - \dfrac{a^2}{4y^3} + \dfrac{b}{y}$

6. $z = axe^{-y} - \dfrac{1}{2}a^2e^{-2y} + b$.

1.14 CAUCHY'S METHOD OF CHARACTERISTICS

Consider a first order partial differential equation

$$a\frac{\partial u}{\partial x} + b\frac{\partial u}{\partial y} = f(x, y) + ku \quad ; \quad u(0, y) = h(y) \qquad ...(1)$$

Here a, b and f depend on x, y and u but not on the derivatives of u.

Let $u(x, y)$ be the solution of eq. (1) then by chain rule,

$$du = \frac{\partial u}{\partial x}dx + \frac{\partial u}{\partial y}dy \qquad ...(2)$$

From (1) and (2), we get

$$\frac{dx}{a} = \frac{dy}{b} = \frac{du}{f(x, y) + ku} \quad ...(3)$$

Taking,

$$\frac{dx}{a} = \frac{dy}{b} \quad ...(4)$$

We get a solution,

$$bx - ay = C \quad ...(5)$$

Where c is a constant.
Now, consider

$$\frac{dx}{a} = \frac{du}{f(x, y) + ku} \quad ...(6)$$

$$\Rightarrow \quad \frac{dx}{a} = \frac{du}{f\left(x, \frac{bx - c}{a}\right) + ku} \quad ...(7)$$

Eqn. (7) can be written as

$$\frac{du}{dx} - \frac{k}{a}u = \frac{1}{a}f\left(x, \frac{bx - c}{a}\right) \quad ...(8)$$

which is a first order linear ordinary differential equation.

$$\text{I.F.} = e^{-\frac{k}{a}x}$$

Solution of eqn. (8) is of the form,

$$u = G(x, c) + c_1 \quad ...(9)$$

Where $c_1 = g(c)$; g is an arbitrary function.
Hence eqn. (9) is

$$u = G(x, c) + g(c)$$

$g(c)$ can be determined by using the given conditions.

ILLUSTRATIVE EXAMPLES

Example 1. *Use Cauchy's method of characteristics to solve the partial differential equaiton*

$$\frac{\partial u}{\partial x} + \frac{du}{\partial y} = x + y \quad ; \quad u(x, 0) = 0$$

Sol. The system of equations is

$$\frac{dx}{1} = \frac{dy}{1} = \frac{du}{x + y} \quad ...(1)$$

Taking the first two, we get

$$x - y = c \quad ...(2)$$

Where c is a constant.
Now, taking,

$$\frac{dy}{1} = \frac{du}{x + y}$$

$\Rightarrow$ $$\frac{dy}{1} = \frac{du}{2y+c}$$

We get, $$u(x, y) = y^2 + cy + c_1 \qquad ...(3)$$

Let $c_1 = g(c)$ then,

From (3),

$$u(x, y) = y^2 + cy + g(c)$$

$$= y^2 + y(x - y) + g(x - y) \qquad ...(4)$$

where $g(x - y)$ is an arbitrary function.

Applying the condition $u(x, 0) = 0$ on eq. (4), we get

$$0 = g(x)$$

$\therefore$ $$g(x - y) = 0$$

Consequently, the solution is,

$$u(x, y) = xy$$

Example 2. *Use the method of characteristics to solve:*

$$u_x - u_y = 0 \quad ; \qquad u(x, 0) = x$$

Sol. We have,

$$\frac{dx}{1} = \frac{dy}{-1} = \frac{du}{0} \qquad ...(1)$$

From the first two, we get

$$x + y = c \qquad ...(2)$$

where c is a constant.

Also, from eq. (1)

$$du = 0$$

$\therefore$ $$u(x, y) = c_1$$

Hence,

$$u(x, y) = c_1 = g(c) = g(x + y) \qquad ...(3)$$

where $g(x + y)$ is an arbitrary function.

From the given condition,

$$u(x, 0) = x = g(x)$$

$\therefore$ $$g(x + y) = x + y$$

Therefore, from eq. (3)

$$u(x, y) = x + y$$

Example 3. *Use the method of characteristics to solve the first order PDE.*

$$u_x - yu = 0 \qquad ; \qquad u(0, y) = 1$$

Sol. We have,

$$\frac{dx}{1} = \frac{du}{yu} = \frac{dy}{0} \qquad ...(1)$$

One solution is,

$$y = c \qquad ...(2)$$

Taking the first two,

$$\frac{dx}{1} = \frac{du}{uc}$$

Integration yields,

$$\log u = cx + g(c)$$

or $$\log u = cx + g(y) \qquad ...(3)$$

Applying the condition, we get

$$0 = g(y)$$

Hence the solution is,

$$u = e^{xy}$$

TEST YOUR KNOWLEDGE

Use Cauchy's method of characteristics to solve the following first order partial differential equations:

1. $u_x + u_y = u,$ $u(x, 0) = 1 + e^x$
2. $x u_x + u_y = x \sinh y + u,$ $u(0, y) = 0$
3. $u_x + yu = 0,$ $u(0, y) = 1$
4. $u_x + u_y = 2x + 2y,$ $u(x, 0) = x^2$
5. $u_x + u_y = 2u,$ $u(x, 0) = e^x$
6. $u_x + yu_y = 2u,$ $u(0, y) = y$
7. $x u_x + u_y = 2u,$ $u(x, 0) = x$
8. $u_x + u_y = \sinh x + \sinh y,$ $u(0, y) = 1 + \cosh y$
9. $u_x + u_y = 1 + \cos y,$ $u(0, y) = \sin y$
10. $u_x - u_y = 2,$ $u(0, y) = -y.$

Answers

1. $u = e^x + e^y$ **2.** $u = x \cosh y$ **3.** $u = e^{-xy}$
4. $u = x^2 + y^2$ **5.** $u = e^{x+y}$ **6.** $u = y\, e^x$
7. $u = x\, e^y$ **8.** $u = \cosh x + \cosh y$ **9.** $u = x + \sin y$
10. $u = x - y$

1.15 PARTIAL DIFFERENTIAL EQUATIONS OF THE SECOND ORDER

An equation is said to be a partial differential equation of the second order when it includes at least one of the partial derivatives r, s, t but none of the higher order.

1.16 LINEAR PARTIAL DIFFERENTIAL EQUATION WITH CONSTANT COEFFICIENTS

A partial differential equation in which the dependent variable and its derivatives appear only in the first degree and are not multiplied together is called a linear partial differential equation.

General form of such an equation is

$$A_0 \frac{\partial^n z}{\partial x^n} + A_1 \frac{\partial^n z}{\partial x^{n-1}\partial y} + \cdots\cdots + A_n \frac{\partial^n z}{\partial y^n} + B_0 \frac{\partial^{n-1} z}{\partial x^{n-1}}$$

$$+ B_1 \frac{\partial^{n-1} z}{\partial x^{n-2}\partial y} + \cdots\cdots + B_{n-1} \frac{\partial^{n-1} z}{\partial y^{n-1}} + C_0 \frac{\partial z}{\partial x} + C_1 \frac{\partial z}{\partial y} + P_0 z = F(x, y) \qquad ...(1)$$

where the coefficients $A_0, A_1, ..., A_n; B_0, B_1, ..., B_{n-1}; C_0, C_1, P_0$ are constants or functions of x and y.

If these coefficients are constants, then eqn. (1) is called as linear partial differential equation with constant coefficients.

If all the derivatives appearing in eqn. (1) are of the same order then the differential equation is called a linear homogeneous p.d.e. with constant coefficients while if all the derivatives are not of same order, then it is called a non-homogeneous linear p.d.e. with constant coefficients.

1.17 HOMOGENEOUS LINEAR PARTIAL DIFFERENTIAL EQUATIONS WITH CONSTANT COEFFICIENTS

An equation of the form $a_0 \frac{\partial^n z}{\partial x^n} + a_1 \frac{\partial^n z}{\partial x^{n-1} \partial y} + a_2 \frac{\partial^n z}{\partial x^{n-2} \partial y^2} + ... + a_n \frac{\partial^n z}{\partial y^n} = F(x, y)$ ...(1)

where $a_0, a_1, a_2, ..., a_n$ are constants, is called a *homogeneous linear partial differential equation of the nth order with constant coefficients.*

Here, all the partial derivatives are of the nth order

Writing D for $\frac{\partial}{\partial x}$ and D′ for $\frac{\partial}{\partial y}$, eqn. (1) can be written as

$$(a_0 D^n + a_1 D^{n-1}D' + a_2 D^{n-2}D'^2 + \cdots + a_n D'^n)z = F(x, y)$$

or $$\boxed{\phi(D, D')z = F(x, y)}$$ ...(2)

As in the case of ordinary linear differential equations with constant coefficients, the complete solution of (2) consists of two parts:

***(i)* the complementary function** (C.F.) which is the complete solution of the equation $\phi(D, D')z = 0$. It must contain n arbitrary functions, where n is the order of the differential equation.

***(ii)* the particular integral** (P.I.) which is a particular solution (free from arbitrary constants) of $\phi(D, D')\, z = F(x, y)$

The complete solution of (2) is $\boxed{z = \text{C.F.} + \text{P.I.}}$

1.18 RULES FOR FINDING C.F.

We explain the procedure for finding the complementary function of a differential equation of the second order. It can be easily extended to differential equations of higher order.

Consider the equation $\frac{\partial^2 z}{\partial x^2} + a_1 \frac{\partial^2 z}{\partial x \partial y} + a_2 \frac{\partial^2 z}{\partial y^2} = 0$

which in symbolic form is $(D^2 + a_1 DD' + a_2 D'^2)z = 0$ ...(1)

Its symbolic operator equated to zero, *i.e.*, $D^2 + a_1 DD' + a_2 D'^2 = 0$ ...(2)

is called the **auxiliary equation** (A.E.)

Considered as a quadratic in D/D′, let its roots be m_1, m_2.

Case I. ***When the A.E. has distinct roots*** *i.e.*, $m_1 \neq m_2$, then (2) can be written as

$$(D - m_1D')(D - m_2D')z = 0 \quad ...(3)$$

Now the solution of $(D - m_2D')z = 0$ will also be a solution of (3).

But $\quad (D - m_2D')z = 0$

$\Rightarrow \quad p - m_2q = 0$

which is of Lagrange's form and the auxiliary equations are $\dfrac{dx}{1} = \dfrac{dy}{-m_2} = \dfrac{dz}{0}$

The first two members give $dy + m_2dx = 0$ or $y + m_2x = a$

Also, $\quad dz = 0$ or $z = b$

$\therefore \quad z = f_2(y + m_2x)$ is a solution of $(D - m_2D')z = 0$

Similarly, (3) will also be satisfied by the solution of

$(D - m_1D')z = 0$ *i.e.*, by $z = f_1(y + m_1x)$.

Hence the complete solution of (2) is

$$\boxed{z = f_1(y + m_1x) + f_2(y + m_2x)}$$

Case II. ***When the A.E. has equal roots,*** each $= m$, then (2) can be written as

$$(D - mD')(D - mD')z = 0 \quad ...(4)$$

Let $(D - mD')z = u$, then (4) becomes $(D - mD')u = 0$

Its solution is $u = f(y + mx)$, as proved in Case I.

$\therefore \quad (D - mD')z = u$ takes the form $(D - mD')z = f(y + mx)$

or $\quad p - mq = f(y + mx)$ [Lagrange's form]

The Auxiliary equations are $\dfrac{dx}{1} = \dfrac{dy}{-m} = \dfrac{dz}{f(y+mx)}$

Now, $\quad \dfrac{dx}{1} = \dfrac{dy}{-m}$ gives $dy + mdx = 0$ or $y + mx = a$

Also, $\quad \dfrac{dx}{1} = \dfrac{dz}{f(a)}$ gives $dz = f(a)\, dx$

or $\quad z = f(a)x + b$ *i.e.*, $z - xf(y + mx) = b$

$\therefore$ The complete solution of eqn. (2) is

$$z - xf(y + mx) = \phi(y + mx)$$

or

$$\boxed{z = \phi(y + mx) + xf(y + mx)}$$

Note 1. Auxiliary eqn. of $\phi(D, D')z = F(x, y)$ is obtained by putting $D = m$ and $D' = 1$ in $\phi(D, D') = 0$.

Hence the A.E. is $\phi(m, 1) = 0$.

Note 2. Generalising the results of Case I and Case II, we have

(*i*) if the roots of A.E. are $m_1, m_2, m_3, ...$ (all distinct roots), then

$$\text{C.F.} = f_1(y + m_1x) + f_2(y + m_2x) + f_3(y + m_3x) + \cdots\cdots$$

(*ii*) if the roots of A.E. are $m_1, m_1, m_2, \ldots$ (two equal roots), then

$$\text{C.F.} = f_1(y + m_1x) + xf_2(y + m_1x) + f_3(y + m_2x) + \ldots\ldots$$

(*iii*) if the roots of A.E. are $m_1, m_1, m_1, \ldots$ (three equal roots), then

$$\text{C.F.} = f_1(y + m_1x) + xf_2(y + m_1x) + x^2f_3(y + m_1x) + \ldots\ldots$$

Note 3. Corresponding to a non-repeated factor D′ on L.H.S., the part of C.F. is taken as $\phi(x)$ and for D'^2 on LHS, the part of C.F. is $y\,\phi(x) + \psi(x)$.

ILLUSTRATIVE EXAMPLES

Example 1. *Solve:*

(*i*) $\dfrac{\partial^2 z}{\partial x^2} - \dfrac{\partial^2 z}{\partial x \partial y} - 6\dfrac{\partial^2 z}{\partial y^2} = 0$ (*ii*) $(D + 2D')(D - 3D')^2 z = 0$ (*iii*) $4r - 12s + 9t = 0.$

Sol. (*i*) The given equation is

$$(D^2 - DD' - 6D'^2)z = 0, \quad \text{where } D \equiv \frac{\partial}{\partial x} \text{ and } D' \equiv \frac{\partial}{\partial y}$$

Auxiliary equation is

$$m^2 - m - 6 = 0$$

$$\Rightarrow \quad (m - 3)(m + 2) = 0 \Rightarrow m = 3, -2$$

$$\therefore \quad \text{C.F.} = f_1(y + 3x) + f_2(y - 2x)$$

$$\text{P.I.} = 0$$

Hence the complete solution is

$$z = \text{C.F.} + \text{P.I.} = f_1(y + 3x) + f_2(y - 2x)$$

where f_1 and f_2 are arbitrary functions.

(*ii*) Auxiliary equation is

$$(m + 2)(m - 3)^2 = 0 \Rightarrow m = -2, 3, 3$$

$$\therefore \quad \text{C.F.} = f_1(y - 2x) + f_2(y + 3x) + xf_3(y + 3x)$$

$$\text{P.I.} = 0$$

Hence the complete solution is

$$z = \text{C.F.} + \text{P.I.} = f_1(y - 2x) + f_2(y + 3x) + xf_3(y + 3x)$$

where f_1, f_2 and f_3 are arbitrary functions.

(*iii*) The given equation is

$$4\frac{\partial^2 z}{\partial x^2} - 12\frac{\partial^2 z}{\partial x \partial y} + 9\frac{\partial^2 z}{\partial y^2} = 0$$

$$\Rightarrow \quad (4D^2 - 12DD' + 9D'^2)z = 0$$

Auxiliary equation is $\quad 4m^2 - 12m + 9 = 0$

$$\Rightarrow \quad (2m - 3)^2 = 0 \Rightarrow m = \frac{3}{2}, \frac{3}{2}$$

$$\therefore \quad \text{C.F.} = f_1\left(y + \frac{3}{2}x\right) + xf_2\left(y + \frac{3}{2}x\right)$$

$$\text{P.I.} = 0$$

Hence the complete solution is

$$z = \text{C.F.} + \text{P.I.} = f_1\left(y + \frac{3}{2}x\right) + x f_2\left(y + \frac{3}{2}x\right)$$

where f_1 and f_2 are arbitrary functions.

Example 2. *Solve:*

(*i*) $\dfrac{\partial^3 z}{\partial x^3} - 3\dfrac{\partial^3 z}{\partial x^2 \partial y} + 2\dfrac{\partial^3 z}{\partial x \partial y^2} = 0$ (*ii*) $r = a^2 t$

(*iii*) $(D^3D'^2 + D^2D'^3)\, z = 0$ (*iv*) $\dfrac{\partial^4 z}{\partial x^4} - \dfrac{\partial^4 z}{\partial y^4} = 0.$

Sol. (*i*) The given equation is

$$(D^3 - 3D^2D' + 2DD'^2)z = 0$$

The auxiliary equation is $m^3 - 3m^2 + 2m = 0$

$\Rightarrow$ $m(m - 1)(m - 2) = 0$

$\Rightarrow$ $m = 0, 1, 2$

$\therefore$ $\text{C.F.} = f_1(y) + f_2(y + x) + f_3(y + 2x)$

$\text{P.I.} = 0$

Hence the complete solution is $z = \text{C.F.} + \text{P.I.} = f_1(y) + f_2(y + x) + f_3(y + 2x)$.

where f_1, f_2 and f_3 are arbitrary functions.

(*ii*) The given equation is $(D^2 - a^2D'^2)z = 0$

The auxiliary equation is $m^2 - a^2 = 0 \Rightarrow m = \pm a$

$\therefore$ $\text{C.F.} = f_1(y + ax) + f_2(y - ax)$

$\text{P.I.} = 0$

Hence the complete solution is

$$z = \text{C.F.} + \text{P.I.} = f_1(y + ax) + f_2(y - ax).$$

where f_1 and f_2 are arbitrary functions.

(*iii*) The given equation is $D^2D'^2(D + D')z = 0$

$\therefore$ $\text{C.F.} = f_1(y) + xf_2(y) + f_3(x) + yf_4(x) + f_5(y - x)$

$\text{P.I.} = 0$

Hence the complete solution is

$$z = \text{C.F.} + \text{P.I.} = f_1(y) + xf_2(y) + f_3(x) + yf_4(x) + f_5(y - x).$$

where f_1, f_2, f_3, f_4 and f_5 are arbitrary functions.

(*iv*) The given equation is $(D^4 - D'^4)z = 0$

The auxiliary equation is $m^4 - 1 = 0$

$\Rightarrow$ $(m^2 - 1)(m^2 + 1) = 0$

$\Rightarrow$ $m = \pm 1, \pm i$

$\therefore$ $\text{C.F.} = f_1(y + x) + f_2(y - x) + f_3(y + ix) + f_4(y - ix)$

$\text{P.I.} = 0$

$\therefore$ The complete solution is

$$z = \text{C.F.} + \text{P.I.} = f_1(y + x) + f_2(y - x) + f_3(y + ix) + f_4(y - ix).$$

where f_1, f_2, f_3 and f_4 are arbitrary functions.

Example 3. *Solve:*

(*i*) $(D^3 - 6D^2D' + 11DD'^2 - 6D'^3)z = 0$ (*ii*) $(D^3 - 6D^2D' + 12DD'^2 - 8D'^3)z = 0$

(*iii*) $r - 4s + 4t = 0$.

Sol. (*i*) Auxiliary equation is

$$m^3 - 6m^2 + 11m - 6 = 0$$

$$\Rightarrow \quad (m-1)(m-2)(m-3) = 0$$

$$\Rightarrow \quad m = 1, 2, 3$$

$$\therefore \quad \text{C.F.} = f_1(y + x) + f_2(y + 2x) + f_3(y + 3x)$$

$$\text{P.I.} = 0$$

Hence the complete solution is

$$z = \text{C.F.} + \text{P.I.} = f_1(y + x) + f_2(y + 2x) + f_3(y + 3x)$$

where f_1, f_2 and f_3 are arbitrary functions.

(*ii*) Auxiliary equation is

$$m^3 - 6m^2 + 12m - 8 = 0$$

$$\Rightarrow \quad (m-2)^3 = 0 \quad \Rightarrow \quad m = 2, 2, 2$$

$$\therefore \quad \text{C.F.} = f_1(y + 2x) + x\, f_2(y + 2x) + x^2 f_3(y + 2x)$$

$$\text{P.I.} = 0$$

Hence the complete solution is

$$z = \text{C.F.} + \text{P.I.} = f_1(y + 2x) + x\, f_2(y + 2x) + x^2 f_3(y + 2x)$$

where f_1, f_2 and f_3 are arbitrary functions.

(*iii*) Auxiliary equation is

$$m^2 - 4m + 4 = 0 \quad \Rightarrow \quad m = 2, 2$$

$$\therefore \quad \text{C.F.} = f_1(y + 2x) + x\, f_2(y + 2x)$$

$$\text{P.I.} = 0$$

Hence the complete solution is

$$z = \text{C.F.} + \text{P.I.} = f_1(y + 2x) + x\, f_2(y + 2x)$$

where f_1 and f_2 are arbitrary functions.

Example 4. *Solve the linear partial differential equation* $\dfrac{\partial^4 z}{\partial x^4} + \dfrac{\partial^4 z}{\partial y^4} = 0$.

Sol. The given equation is $\quad (D^4 + D'^4)z = 0$

Auxiliary equation is $\quad m^4 + 1 = 0$

$$\Rightarrow \quad m^4 + 1 + 2m^2 = 2m^2$$

$$\Rightarrow \quad (m^2 + 1)^2 - (m\sqrt{2})^2 = 0$$

$$\Rightarrow \quad (m^2 + \sqrt{2}m + 1)(m^2 - \sqrt{2}m + 1) = 0$$

so that $\quad m^2 + \sqrt{2}m + 1 = 0 \quad \text{or} \quad m^2 - \sqrt{2}m + 1 = 0$

$\Rightarrow \qquad m = \dfrac{-1 \pm i}{\sqrt{2}}, \dfrac{1 \pm i}{\sqrt{2}}$

Let $\qquad z_1 = \dfrac{-1+i}{\sqrt{2}}$ and $z_2 = \dfrac{1+i}{\sqrt{2}}$

then, $\qquad m = z_1, \overline{z_1}, z_2, \overline{z_2}$

Here $\overline{z_1}$ and $\overline{z_2}$ denote complex conjugate of z_1 and z_2 respectively.

$\therefore \qquad \text{C.F.} = f_1(y + z_1 x) + f_2(y + \overline{z_1}\, x) + f_3(y + z_2 x) + f_4(y + \overline{z_2}\, x)$

$\text{P.I.} = 0$

Hence the complete solution is

$$z = \text{C.F.} + \text{P.I.} = f_1(y + z_1 x) + f_2(y + \overline{z_1} x) + f_3(y + z_2 x) + f_4(y + \overline{z_2} x)$$

where f_1, f_2, f_3 and f_4 are arbitrary functions.

TEST YOUR KNOWLEDGE

Solve the following partial differential equations:

1. $2\dfrac{\partial^2 z}{\partial x^2} + 5\dfrac{\partial^2 z}{\partial x\, \partial y} + 2\dfrac{\partial^2 z}{\partial y^2} = 0$
2. $\dfrac{\partial^3 z}{\partial x^3} - 4\dfrac{\partial^3 z}{\partial x^2\, \partial y} + 4\dfrac{\partial^3 z}{\partial x \partial y^2} = 0$
3. $\dfrac{\partial^2 z}{\partial x^2} + \dfrac{\partial^2 z}{\partial x\, \partial y} = 0$
4. $\dfrac{\partial^3 z}{\partial x^3} - 4\dfrac{\partial^3 z}{\partial x^2\, \partial y} + 3\dfrac{\partial^3 z}{\partial x \partial y^2} = 0$
5. $\dfrac{\partial^4 z}{\partial x^4} + \dfrac{\partial^4 z}{\partial y^4} = 2\dfrac{\partial^4 z}{\partial x^2\, \partial y^2}$
6. $\dfrac{\partial^3 z}{\partial x^3} + 3\dfrac{\partial^3 z}{\partial x^2\, \partial y} + 2\dfrac{\partial^3 z}{\partial x \partial y^2} = 0$
7. $\dfrac{\partial^2 z}{\partial x^2} + a^2\dfrac{\partial^2 z}{\partial y^2} = 0$
8. $\dfrac{\partial^2 z}{\partial x^2} - 3a\dfrac{\partial^2 z}{\partial x \partial y} + 2a^2\dfrac{\partial^2 z}{\partial y^2} = 0$
9. $25r - 40s + 16t = 0$
10. $r + 2s + t = 0.$

Answers

1. $z = f_1(y - 2x) + f_2\left(y - \dfrac{x}{2}\right)$
2. $z = f_1(y) + f_2(y + 2x) + xf_3(y + 2x)$
3. $z = f_1(y) + f_2(y - x)$
4. $z = f_1(y) + f_2(y + x) + f_3(y + 3x)$
5. $z = f_1(y - x) + x\, f_2(y - x) + f_3(y + x) + xf_4(y + x)$
6. $z = f_1(y) + f_2(y - x) + f_3(y - 2x)$
7. $z = f_1(y + aix) + f_2(y - aix)$
8. $z = f_1(y + ax) + f_2(y + 2ax)$
9. $z = f_1\left(y + \dfrac{4}{5}x\right) + x\, f_2\left(y + \dfrac{4}{5}x\right)$
10. $z = f_1(y - x) + x\, f_2(y - x).$

1.19 RULES FOR FINDING P.I.

P.I. of the equation $F(D, D')z = \phi(x, y)$ is given by $\dfrac{1}{F(D, D')}\, \phi(x, y)$.

Short methods:

Case: When $\phi(x, y)$ is a function of $ax + by$

To find out P.I. of the equation

$$F(D, D')z = \phi(ax + by);\ F(a, b) \neq 0$$

where F(D, D′) is a homogeneous function of D and D′ of degree n.

Steps involved

Step 1. Replace D by a, D′ by b in F(D, D′) to get F(a, b).

Step 2. Put $ax + by = u$ and integrate $\phi(u)$, n times w.r.t. u.

Then,
$$\text{P.I.} = \frac{1}{F(a, b)} \underbrace{\iiint \ldots\ldots \int}_{(n \text{ times})} \phi(u)\, \underbrace{du\, du\, du \ldots\ldots du}_{(n \text{ times})}.$$

Step 3. Replace u by $ax + by$ at last.

Remark. It should be noted that here F(a, b) ≠ 0. If F(a, b) = 0, then method fails.

Cases of failure

(i) To find out P.I. of the equation

$$F(D, D')z = \phi(ax + by)\ ;\ F(a, b) = 0$$

where F(D, D′) is a homogeneous function of D and D′ of degree n.

Steps involved

Step 1. Differentiate F(D, D′) partially w.r.t. D and simultaneously multiply the expression by x.

Step 2. Check whether F′(a, b) ≠ 0 ?

Then,
$$\text{P.I.} = \frac{1}{F(D, D')} \phi(ax + by)\ ;\ F(a, b) = 0$$
$$= x \cdot \frac{1}{\frac{\partial}{\partial D}\{F(D, D')\}} \phi(ax + by)$$

This method fails if F′(a, b) = 0.

(ii) To find out P.I. of the equation

$$F(D, D')z = \phi(ax + by)\ ;\ F'(a, b) = 0$$

where F(D, D′) is a homogeneous function of D and D′ of degree n.

Steps Involved:

Step 1. Repeat the procedure as again differentiate F′(D, D′) partially w.r.t. D and simultaneously multiply the expression by x.

Step 2. Check whether F″(a, b) ≠ 0 ?

Then
$$\text{P.I.} = x \cdot \frac{1}{F'(D, D')} \phi(ax + by)$$
$$= x^2 \cdot \frac{1}{\frac{\partial}{\partial D}\{F'(D, D')\}} \phi(ax + by)$$

This method also fails if F″(a, b) = 0.

We proceed this method as long as the derivative of F(D, D′) vanishes when D is replaced by a and D′ by b.

If $F^r(a, b) \neq 0$, then $\dfrac{1}{F^r(D, D')}$ can be evaluated for particular integral.

Special Case: As a Particular Case when in F(D, D′), Power of D′ in Highest Degree Terms is Greater than that of D and Case of Failure Occurs:

Step 1. Differentiate F(D, D′) partially w.r.t. D′ and simultaneously multiply the expression by y.

Step 2. Check whether $F'(a, b) \neq 0$?

Then,
$$\text{P.I.} = \frac{1}{F(D, D')} \cdot \phi(ax + by);\ F(a, b) \neq 0$$
$$= y \cdot \frac{1}{\frac{\partial}{\partial D'}\{F(D, D')\}}\, \phi(ax + by).$$

Step 3. This method fails if $F'(a, b) = 0$.

In such case, we follow the same procedure as given below:

$$\text{P.I.} = y \cdot \frac{1}{F'(D, D')}\, \phi(ax + by);\ F'(a, b) = 0$$
$$= y^2 \cdot \frac{1}{\frac{\partial}{\partial D'}\{F'(D, D')\}}\, \phi(ax + by).$$

Step 4. This method also fails if $F''(a, b) = 0$. We repeat the method as long as derivative of F(D, D′) vanishes when D is replaced by a and D′ by b.

ILLUSTRATIVE EXAMPLES

Example 1. *Solve the linar partial differential equation*

$$\frac{\partial^3 u}{\partial x^3} - 3\frac{\partial^3 u}{\partial x^2 \partial y} + 4\frac{\partial^3 u}{\partial y^3} = e^{x+2y}.$$

Sol. The given equation is

$$(D^3 - 3D^2D' + 4D'^3)\, u = e^{x+2y} \text{ where } D \equiv \frac{\partial}{\partial x} \text{ and } D' \equiv \frac{\partial}{\partial y}$$

Auxiliary equation is

$$m^3 - 3m^2 + 4 = 0$$
$$m^2(m+1) - 4m(m+1) + 4(m+1) = 0$$
$$(m-2)^2(m+1) = 0$$
$$\Rightarrow \quad m = 2, 2, -1$$
$$\therefore \quad \text{C.F.} = f_1(y - x) + f_2(y + 2x) + x f_3(y + 2x)$$

$$\text{P.I.} = \frac{1}{D^3 - 3D^2D' + 4D'^3} e^{x+2y}$$

$$= \frac{1}{(1)^3 - 3(1)^2(2) + 4(2)^3} \iiint e^u \, du \, du \, du \qquad \text{where } x + 2y = u$$

$$= \frac{1}{27} e^{x+2y}$$

Hence the complete solution is

$$u = \text{C.F.} + \text{P.I.} = f_1(y - x) + f_2(y + 2x) + x f_3(y + 2x) + \frac{1}{27} e^{x+2y}$$

where f_1, f_2 and f_3 are arbitrary functions.

Example 2. *Solve the linear partial differential equation*

$$\frac{\partial^2 z}{\partial x^2} + 3\frac{\partial^2 z}{\partial x \partial y} + 2\frac{\partial^2 z}{\partial y^2} = x + y.$$

Sol. The given equation is

$$(D^2 + 3DD' + 2D'^2) z = x + y.$$

The auxiliary equation is $\quad m^2 + 3m + 2 = 0$

$\Rightarrow \quad (m + 1)(m + 2) = 0$

$\Rightarrow \quad m = -1, -2$

$\therefore \quad \text{C.F.} = f_1(y - x) + f_2(y - 2x)$

$$\text{P.I.} = \frac{1}{D^2 + 3DD' + 2D'^2}(x + y)$$

$$= \frac{1}{(1)^2 + 3(1)(1) + 2(1)^2} \iint u \, du \, du, \qquad \text{where } x + y = u$$

$$= \frac{1}{6} \cdot \frac{u^3}{6} = \frac{u^3}{36} = \frac{(x + y)^3}{36}$$

Hence the complete solution is

$$z = \text{C.F.} + \text{P.I.} = f_1(y - x) + f_2(y - 2x) + \frac{(x + y)^3}{36}$$

where f_1 and f_2 are arbitrary functions.

Example 3. *Solve the linear partial differential equation*

$$\frac{\partial^2 z}{\partial x^2} - 2\frac{\partial^2 z}{\partial x \partial y} + \frac{\partial^2 z}{\partial y^2} = \sin(2x + 3y).$$

Sol. The given equation is

$$(D^2 - 2DD' + D'^2) z = \sin(2x + 3y)$$

The auxiliary equation is

$$m^2 - 2m + 1 = 0$$

$\Rightarrow \quad (m - 1)^2 = 0 \quad \Rightarrow \quad m = 1, 1.$

$\therefore$ C.F. $= f_1(y + x) + xf_2(y + x)$

$$\text{P.I.} = \frac{1}{D^2 - 2DD' + D'^2} \sin(2x + 3y) = \frac{1}{(D - D')^2} \sin(2x + 3y)$$

$$= \frac{1}{(2-3)^2} \iint \sin u \, du \, du \qquad \text{where } 2x + 3y = u$$

$$= -\sin u = -\sin(2x + 3y).$$

Hence the complete solution is

$$z = \text{C.F.} + \text{P.I.} = f_1(y + x) + xf_2(y + x) - \sin(2x + 3y)$$

where f_1 and f_2 are arbitrary functions.

Example 4. *Solve:* $(2D^2 - 5DD' + 2D'^2)z = 24(y - x)$.

Sol. The auxiliary equation is

$$2m^2 - 5m + 2 = 0$$

$$\Rightarrow \qquad (2m - 1)(m - 2) = 0 \quad \Rightarrow \quad m = \tfrac{1}{2}, 2$$

Hence, $\quad$ C.F. $= f_1\left(y + \dfrac{1}{2}x\right) + f_2(y + 2x) = \phi_1(2y + x) + f_2(y + 2x)$

$$\text{P.I.} = \frac{1}{2D^2 - 5DD' + 2D'^2} 24(y - x) \qquad | \text{ Here } a = -1, b = 1$$

$$= \frac{24}{2(-1)^2 - 5(-1)(1) + 2(1)^2} \iint u \, du \, du \qquad \text{where } y - x = u$$

$$= \frac{4}{9} u^3 = \frac{4}{9}(y - x)^3$$

$\therefore$ Complete solution is

$$z = \text{C.F.} + \text{P.I.} = \phi_1(2y + x) + f_2(y + 2x) + \frac{4}{9}(y - x)^3.$$

where ϕ_1 and f_2 are arbitrary functions.

Example 5. *Solve:* $r + s - 2t = \sqrt{2x + y}$. (U.P.T.U. 2015)

Sol. The given equation is

$$\frac{\partial^2 z}{\partial x^2} + \frac{\partial^2 z}{\partial x \partial y} - 2\frac{\partial^2 z}{\partial y^2} = (2x + y)^{1/2}$$

or $\quad (D^2 + DD' - 2D'^2)z = (2x + y)^{1/2}$

Auxiliary equation is

$$m^2 + m - 2 = 0$$

$$\Rightarrow \qquad (m - 1)(m + 2) = 0 \quad \Rightarrow \quad m = 1, -2$$

$\therefore$ C.F. $= f_1(y + x) + f_2(y - 2x)$

$$\text{P.I.} = \frac{1}{D^2 + DD' - 2D'^2}(2x + y)^{1/2} \qquad | \text{ Here } a = 2, b = 1$$

$$= \frac{1}{(2)^2 + (2)(1) - 2(1)^2} \iint \sqrt{u}\, du\, du \qquad \text{where } u = 2x + y$$

$$= \frac{1}{4} \cdot \frac{4}{15} u^{5/2} = \frac{1}{15} (2x + y)^{5/2}$$

Hence the required complete solution is

$$z = \text{C.F.} + \text{P.I.} = f_1(y + x) + f_2(y - 2x) + \frac{1}{15} (2x + y)^{5/2}.$$

where f_1 and f_2 are arbitrary functions.

Example 6. *Solve the linear partial differential equation*

$$\frac{\partial^2 z}{\partial x^2} + \frac{\partial^2 z}{\partial y^2} = \cos mx \cos ny + 30\, (2x + y).$$ [G.B.T.U. (AG) 2011, 2012]

Sol. The given equation is

$$(D^2 + D'^2)z = \cos mx \cos ny + 30\, (2x + y).$$

Auxiliary equation is

$$m^2 + 1 = 0 \Rightarrow m = \pm i$$

$$\therefore \qquad \text{C.F.} = f_1 (y + ix) + f_2(y - ix)$$

$$\text{P.I.} = \frac{1}{D^2 + D'^2} (\cos mx \cos ny) + \frac{1}{D^2 + D'^2}\, 30\, (2x + y) = P_1 + P_2$$

where
$$P_1 = \frac{1}{D^2 + D'^2} \cos mx \cos ny$$

$$= \frac{1}{2} \frac{1}{D^2 + D'^2} [\cos (mx + ny) + \cos (mx - ny)]$$

$$= \frac{1}{2}\left[\frac{1}{D^2 + D'^2} \cos (mx + ny) + \frac{1}{D^2 + D'^2} \cos (mx - ny)\right]$$

$$= \frac{1}{2}\left[\frac{1}{m^2 + n^2} \iint \cos u\, du\, du + \frac{1}{m^2 + n^2} \iint \cos v\, dv\, dv\right]$$

$$= \frac{1}{2}\left[\frac{-1}{m^2 + n^2} \cos u - \frac{1}{m^2 + n^2} \cos v\right] \text{ where } \quad mx + ny = u \text{ and } mx - ny = v$$

$$= \frac{-1}{2(m^2 + n^2)} [\cos (mx + ny) + \cos (mx - ny)]$$

and
$$P_2 = \frac{1}{D^2 + D'^2}\, 30\, (2x + y) = \frac{30}{(2)^2 + (1)^2} \iint \omega\, d\omega\, d\omega, \qquad \text{where } 2x + y = \omega$$

$$= 6 \frac{\omega^3}{6} = \omega^3 = (2x + y)^3.$$

$$\therefore \quad \text{P.I.} = -\frac{1}{2(m^2 + n^2)} [\cos (mx + ny) + \cos (mx - ny)] + (2x + y)^3$$

$$= -\frac{1}{m^2 + n^2} \cos mx \cos ny + (2x + y)^3.$$

Hence the complete solution is

$$z = \text{C.F.} + \text{P.I.}$$

$$\Rightarrow \quad z = f_1(y + ix) + f_2(y - ix) - \frac{1}{m^2 + n^2} \cos mx \cos ny + (2x + y)^3$$

where f_1 and f_2 are arbitrary functions.

Example 7. *Solve :* $\dfrac{\partial^2 z}{\partial x^2} - 3\dfrac{\partial^2 z}{\partial x \partial y} + 2\dfrac{\partial^2 z}{\partial y^2} = e^{2x + 3y} + \sin(x - 2y).$

Sol. The given equation is

$$(D^2 - 3DD' + 2D'^2)z = e^{2x + 3y} + \sin(x - 2y)$$

Auxiliary equation is

$$m^2 - 3m + 2 = 0$$

$$\Rightarrow \quad (m - 1)(m - 2) = 0 \quad \Rightarrow \quad m = 1, 2$$

$$\therefore \quad \text{C.F.} = f_1(y + x) + f_2(y + 2x)$$

$$\text{P.I.} = \frac{1}{(D^2 - 3DD' + 2D'^2)} e^{2x + 3y} + \frac{1}{(D^2 - 3DD' + 2D'^2)} \sin(x - 2y)$$

$$= P_1 + P_2$$

where

$$P_1 = \frac{1}{D^2 - 3DD' + 2D'^2} e^{2x + 3y}$$

$$= \frac{1}{(2)^2 - 3(2)(3) + 2(3)^2} \iint e^u \, du \, du, \qquad \text{where } 2x + 3y = u$$

$$= \frac{1}{4} e^u = \frac{1}{4} e^{2x + 3y}$$

$$P_2 = \frac{1}{D^2 - 3DD' + 2D'^2} \sin(x - 2y) = \frac{1}{(1)^2 - 3(1)(-2) + 2(-2)^2} \iint \sin v \, dv \, dv$$

where $x - 2y = v$

$$= \frac{1}{15}(-\sin v) = -\frac{1}{15} \sin(x - 2y)$$

$$\therefore \quad \text{P.I.} = P_1 + P_2 = \frac{1}{4} e^{2x + 3y} - \frac{1}{15} \sin(x - 2y)$$

Hence the complete solution is

$$z = \text{C.F.} + \text{P.I.} = f_1(y + x) + f_2(y + 2x) + \frac{1}{4} e^{2x + 3y} - \frac{1}{15} \sin(x - 2y)$$

where f_1 and f_2 are arbitrary functions.

Example 8. *Solve :* $\dfrac{\partial^2 z}{\partial x^2} - 2\dfrac{\partial^2 z}{\partial x \, \partial y} + \dfrac{\partial^2 z}{\partial y^2} = \sin x.$ (M.T.U. 2013)

Sol. The given equation is

$$(D^2 - 2DD' + D'^2)z = \sin x$$

Auxiliary equation is

$$m^2 - 2m + 1 = 0 \quad \Rightarrow \quad m = 1, 1$$

$\therefore \quad \text{C.F.} = f_1(y + x) + x f_2(y + x)$

$$\text{P.I.} = \frac{1}{(D - D')^2} \sin(x + 0.y)$$

$$= \frac{1}{(1-0)^2} \iint \sin u \, du \, du, \quad \text{where } x = u$$

$$= -\sin u = -\sin x$$

Hence the complete solution is

$$z = \text{C.F.} + \text{P.I.} = f_1(y + x) + x f_2(y + x) - \sin x$$

where f_1 and f_2 are arbitrary functions.

Example 9. *Solve the linear partial differential equation:*

$$\frac{\partial^3 z}{\partial x^3} - 4\frac{\partial^3 z}{\partial x^2 \partial y} + 4\frac{\partial^3 z}{\partial x \partial y^2} = 4 \sin(2x + y).$$

Sol. The given equation is

$$(D^3 - 4D^2 D' + 4DD'^2)z = 4 \sin(2x + y)$$

The auxiliary equation is $\quad m^3 - 4m^2 + 4m = 0$

$\Rightarrow \quad m(m^2 - 4m + 4) = 0$

$\Rightarrow \quad m = 0, 2, 2.$

$\therefore \quad \text{C.F.} = f_1(y) + f_2(y + 2x) + x f_3(y + 2x)$

$$\text{P.I.} = \frac{1}{D^3 - 4D^2D' + 4DD'^2} 4 \sin(2x + y)$$

$$= \frac{4}{D}\left[\frac{1}{D^2 - 4DD' + 4D'^2} \sin(2x + y)\right] = \frac{4}{D}\left[\frac{1}{(D - 2D')^2} \sin(2x + y)\right]$$

$$= x \cdot \frac{4}{D}\left[\frac{1}{2(D - 2D')} \sin(2x + y)\right] = 4x^2 \cdot \frac{1}{D}\left[\frac{1}{2} \sin(2x + y)\right]$$

$$= 2x^2 \frac{1}{D} \sin(2x + y) = -2x^2 \frac{\cos(2x + y)}{2} = -x^2 \cos(2x + y).$$

Hence the complete solution is

$$z = \text{C.F.} + \text{P.I.} = f_1(y) + f_2(y + 2x) + x f_3(y + 2x) - x^2 \cos(2x + y)$$

where f_1, f_2 and f_3 are arbitrary functions.

Example 10. *Solve the linear partial differential equation*

$$2\frac{\partial^2 z}{\partial x^2} - 5\frac{\partial^2 z}{\partial x \partial y} + 2\frac{\partial^2 z}{\partial y^2} = 5 \sin(2x + y).$$

Sol. The given equation is

$$(2D^2 - 5DD' + 2D'^2) z = 5 \sin(2x + y)$$

The auxiliary equation is

$$2m^2 - 5m + 2 = 0 \quad \Rightarrow \quad m = \frac{1}{2}, 2.$$

$$\therefore \quad \text{C.F.} = f_1\left(y + \frac{1}{2}x\right) + f_2(y + 2x)$$

$$\text{P.I.} = \frac{1}{2D^2 - 5DD' + 2D'^2} 5 \sin (2x + y)$$

$$= 5x \cdot \frac{1}{4D - 5D'} \sin (2x + y)$$

$$= 5x \cdot \frac{1}{4(2) - 5(1)} \int \sin u \, du, \quad \text{where } 2x + y = u$$

$$= 5x \cdot \frac{1}{3} (- \cos u) = - \frac{5x}{3} \cos (2x + y).$$

Hence the complete solution is

$$z = \text{C.F.} + \text{P.I.} = f_1 \left(y + \frac{1}{2} x \right) + f_2 (y + 2x) - \frac{5x}{3} \cos (2x + y)$$

where f_1 and f_2 are arbitrary functions.

Example 11. *Solve:* $(D^2 + 5DD' + 6D'^2) z = \dfrac{1}{y - 2x}$.

Sol. Auxiliary equation is $m^2 + 5m + 6 = 0$

$\Rightarrow$ $(m + 2)(m + 3) = 0$

$\Rightarrow$ $m = -2, -3$

$\therefore$ $\text{C.F.} = f_1 (y - 2x) + f_2 (y - 3x)$

$$\text{P.I.} = \frac{1}{D^2 + 5DD' + 6D'^2} \left(\frac{1}{y - 2x} \right)$$

$$= x \cdot \frac{1}{2D + 5D'} \left(\frac{1}{y - 2x} \right)$$

$$= x \cdot \frac{1}{2(-2) + 5(1)} \int \frac{1}{u} du, \quad \text{where } y - 2x = u$$

$$= x \log u = x \log (y - 2x)$$

Hence the complete solution is

$$z = \text{C.F.} + \text{P.I.} = f_1 (y - 2x) + f_2 (y - 3x) + x \log (y - 2x)$$

where f_1 and f_2 are arbitrary functions.

Example 12. *Solve the linear partial differential equations:*

(*i*) $\dfrac{\partial^2 z}{\partial x^2} - \dfrac{\partial^2 z}{\partial x \partial y} = \sin x \cos y$

(*ii*) $\dfrac{\partial^2 z}{\partial x^2} - 2 \dfrac{\partial^2 z}{\partial x \partial y} = \sin x \cos 2y$ (A.K.T.U. 2018)

(*iii*) $(D^2 - DD')z = \cos 2y(\sin x + \cos x)$.

Sol. (*i*) The given equation is

$$(D^2 - DD') z = \sin x \cos y$$

The auxiliary equation is $m^2 - m = 0 \Rightarrow m = 0, 1.$

$\therefore \quad \text{C.F.} = f_1(y) + f_2(y+x)$

$$\text{P.I.} = \frac{1}{D^2 - DD'} \sin x \cos y$$

$$= \frac{1}{2(D^2 - DD')} [\sin(x+y) + \sin(x-y)]$$

$$= \frac{1}{2}\left[\frac{1}{D^2 - DD'} \sin(x+y) + \frac{1}{D^2 - DD'} \sin(x-y)\right] = \frac{1}{2}[P_1 + P_2]$$

where $P_1 = \dfrac{1}{D^2 - DD'} \sin(x+y)$

$$= x \cdot \frac{1}{2D - D'} \sin(x+y) = x \cdot \frac{1}{2(1)-(1)} \int \sin u \, du, \qquad \text{where } x + y = u$$

$$= x \cdot (-\cos u) = -x \cos(x+y)$$

$$P_2 = \frac{1}{D^2 - DD'} \sin(x-y)$$

$$= \frac{1}{(1)^2 - (1)(-1)} \iint \sin \omega \, d\omega \, d\omega, \qquad \text{where } x - y = \omega$$

$$= \frac{1}{2}(-\sin \omega) = -\frac{1}{2} \sin(x-y)$$

$$\therefore \quad \text{P.I.} = \frac{1}{2}\left[-x\cos(x+y) - \frac{1}{2}\sin(x-y)\right].$$

Hence the complete solution is

$$z = \text{C.F.} + \text{P.I.} = f_1(y) + f_2(y+x) - \frac{x}{2}\cos(x+y) - \frac{1}{4}\sin(x-y)$$

where f_1 and f_2 are arbitrary functions.

(*ii*) The given equation is

$$(D^2 - 2DD')z = \sin x \cos 2y$$

Auxiliary equation is

$$m^2 - 2m = 0 \quad \Rightarrow \quad m = 0, 2$$

$\therefore \quad \text{C.F.} = f_1(y) + f_2(y + 2x)$

$$\text{P.I.} = \frac{1}{D^2 - 2DD'} \sin x \cos 2y$$

$$= \frac{1}{2(D^2 - 2DD')}[\sin(x+2y) + \sin(x-2y)]$$

$$= \frac{1}{2}\left[\frac{1}{D^2 - 2DD'}\sin(x+2y) + \frac{1}{D^2 - 2DD'}\sin(x-2y)\right]$$

$$= \frac{1}{2}(P_1 + P_2) \qquad \text{...(1)}$$

where, $P_1 = \dfrac{1}{D^2 - 2DD'} \sin(x+2y)$

$$= \frac{1}{(1)^2 - 2(1)(2)} \iint \sin u \, du \, du \qquad \text{where } x + 2y = u$$

$$= -\frac{1}{3}(-\sin u) = \frac{1}{3}\sin(x + 2y)$$

$$P_2 = \frac{1}{D^2 - 2DD'} \sin(x - 2y)$$

$$= \frac{1}{(1)^2 - 2(1)(-2)} \iint \sin v \, dv \, dv \qquad \text{where } x - 2y = v$$

$$= -\frac{1}{5}\sin v = -\frac{1}{5}\sin(x - 2y)$$

$\therefore$ From (1),

$$\text{P.I.} = \frac{1}{6}\sin(x + 2y) - \frac{1}{10}\sin(x - 2y)$$

Hence the complete solution is

$$z = \text{C.F.} + \text{P.I.} = f_1(y) + f_2(y + x) + \frac{1}{6}\sin(x + 2y) - \frac{1}{10}\sin(x - 2y)$$

where f_1 and f_2 are arbitrary functions.

(*iii*) Auxiliary equation is

$$m^2 - m = 0 \quad \Rightarrow \quad m = 0, 1$$

$$\therefore \qquad \text{C.F.} = f_1(y) + f_2(y + x)$$

$$\text{P.I.} = \frac{1}{D^2 - DD'} \cos 2y (\sin x + \cos x)$$

$$= \frac{1}{2}\left[\frac{1}{D^2 - DD'}\{\sin(x + 2y) + \sin(x - 2y)\} + \frac{1}{D^2 - DD'}\{\cos(x + 2y) + \cos(x - 2y)\}\right]$$

$$= \frac{1}{2}\left[\frac{1}{D^2 - DD'}\{\sin(x + 2y) + \cos(x + 2y)\} + \frac{1}{D^2 - DD'}\{\sin(x - 2y) + \cos(x - 2y)\}\right]$$

$$= \frac{1}{2}\left[\frac{1}{(1)^2 - (1)(2)}\iint (\sin u + \cos u)\, du\, du + \frac{1}{(1)^2 - (1)(-2)}\iint (\sin v + \cos v)\, dv\, dv\right]$$

where $x + 2y = u$ and $x - 2y = v$

$$= \frac{1}{2}\left[\sin(x + 2y) + \cos(x + 2y) - \frac{1}{3}\sin(x - 2y) - \frac{1}{3}\cos(x - 2y)\right]$$

$$= \frac{1}{2}[\sin(x + 2y) + \cos(x + 2y)] - \frac{1}{6}[\sin(x - 2y) + \cos(x - 2y)]$$

Hence the complete solution is

$$z = \text{C.F.} + \text{P.I.} = f_1(y) + f_2(y + x) + \frac{1}{2}[\sin(x + 2y) + \cos(x + 2y)]$$

$$- \frac{1}{6}[\sin(x - 2y) + \cos(x - 2y)]$$

where f_1 and f_2 are arbitrary functions.

Example 13. *Solve:*

(*i*) $4r - 4s + t = 16 \log (x + 2y)$ (*ii*) $r + 2s + t = 2(y - x) + \sin (x - y)$

[M.T.U. (SUM) 2011]

(*iii*) $2r - s - 3t = 5 \dfrac{e^x}{e^y}$ (*iv*) $(D - D')^2 z = x + \phi(x + y)$.

Sol. (*i*) The given equation is

$$(4D^2 - 4DD' + D'^2)z = 16 \log (x + 2y)$$

Auxiliary equation is $\quad 4m^2 - 4m + 1 = 0 \Rightarrow m = \frac{1}{2}, \frac{1}{2}$

$$\therefore \quad \text{C.F.} = f_1\left(y + \frac{1}{2}x\right) + x f_2\left(y + \frac{1}{2}x\right)$$

$$\text{P.I.} = \frac{1}{(2D - D')^2} 16 \log (x + 2y)$$

$$= 16 \cdot x \cdot \frac{1}{2(2D - D').2} \log (x + 2y) = 4x \cdot \frac{1}{2D - D'} \log (x + 2y)$$

$$= 4x^2 \cdot \frac{1}{2} \log (x + 2y) = 2x^2 \log (x + 2y)$$

Hence the complete solution is

$$z = \text{C.F.} + \text{P.I.} = f_1\left(y + \frac{1}{2}x\right) + x f_2\left(y + \frac{1}{2}x\right) + 2x^2 \log (x + 2y)$$

where f_1 and f_2 are arbitrary functions.

(*ii*) The given equation is

$$(D^2 + 2DD' + D'^2) z = 2(y - x) + \sin (x - y)$$

Auxiliary equation is $\quad m^2 + 2m + 1 = 0$

$$\Rightarrow \quad m = -1, -1$$

$$\therefore \quad \text{C.F.} = f_1 (y - x) + x f_2 (y - x)$$

$$\text{P.I.} = \frac{1}{(D + D')^2} 2(y - x) + \frac{1}{(D + D')^2} \sin (x - y) = P_1 + P_2$$

where $$P_1 = \frac{1}{(D + D')^2} \cdot 2(y - x) = 2 \cdot x \cdot \frac{1}{2(D + D')} (y - x)$$

$$= x \cdot \frac{1}{D + D'} (y - x) = x^2 \cdot (y - x)$$

$$P_2 = \frac{1}{(D + D')^2} \sin (x - y) = x \,.\, \frac{1}{2(D + D')} \sin (x - y) = x^2 \,.\, \frac{1}{2} \sin (x - y)$$

$$\therefore \quad \text{P.I.} = x^2 (y - x) + \frac{x^2}{2} \sin (x - y)$$

Hence the complete solution is

$$z = \text{C.F.} + \text{P.I.} = f_1\,(y - x) + x\, f_2\,(y - x) + x^2\,(y - x) + \frac{x^2}{2}\sin\,(x - y)$$

where f_1 and f_2 are arbitrary functions.

(*iii*) The given equation is

$$(2D^2 - DD' - 3D'^2)z = 5e^{x-y}$$

Auxiliary equation is $\quad 2m^2 - m - 3 = 0 \;\Rightarrow\; m = -1, 3/2$

$$\therefore \quad \text{C.F.} = f_1\,(y - x) + f_2\left(y + \frac{3}{2}x\right)$$

$$\text{P.I.} = \frac{1}{2D^2 - DD' - 3D'^2}\,(5e^{x-y}) = 5x \cdot \frac{1}{4D - D'}\,e^{x-y}$$

$$= 5x \,.\, \frac{1}{4(1) - (-1)} \int e^u \, du = xe^{x-y}, \qquad \text{where } x - y = u.$$

Hence the complete solution is

$$z = \text{C.F.} + \text{P.I.} = f_1\,(y - x) + f_2\left(y + \frac{3}{2}x\right) + x\,e^{x-y}$$

where f_1 and f_2 are arbitrary functions.

(*iv*) The given equation is $\quad (D - D')^2\, z = x + \phi(x + y)$

Auxiliary equation is $\quad (m - 1)^2 = 0, \;\Rightarrow\; m = 1, 1$

$$\therefore \quad \text{C.F.} = f_1\,(y + x) + x\, f_2(y + x)$$

$$\text{P.I.} = \frac{1}{(D - D')^2}\,x + \frac{1}{(D - D')^2}\,\phi(x + y) = P_1 + P_2$$

where

$$P_1 = \frac{1}{(D - D')^2}\,(x + 0.y) = \frac{1}{(1 - 0)^2}\iint u\, du\, du, \qquad \text{where } x = u.$$

$$= \frac{u^3}{6} = \frac{x^3}{6}$$

$$P_2 = \frac{1}{(D - D')^2}\,\phi(x + y) = x \,.\, \frac{1}{2(D - D')}\,\phi(x + y) = \frac{x^2}{2}\,\phi(x + y)$$

$$\therefore \quad \text{P.I.} = P_1 + P_2 = \frac{x^3}{6} + \frac{x^2}{2}\,\phi(x + y)$$

Hence the complete solution is

$$z = \text{C.F.} + \text{P.I.} = f_1\,(y + x) + x\, f_2\,(y + x) + \frac{x^3}{6} + \frac{x^2}{2}\,\phi(x + y)$$

where f_1 and f_2 are arbitrary functions.

Example 14. *Solve the differential equation:*

$$\frac{\partial^3 z}{\partial x^2 \partial y} - 2\frac{\partial^3 z}{\partial x \partial y^2} + \frac{\partial^3 z}{\partial y^3} = \frac{1}{x^2}.$$

Sol. The given equation is

$$(D^2D' - 2DD'^2 + D'^3)\, z = \frac{1}{x^2}$$

$$\Rightarrow \qquad D'(D^2 - 2DD' + D'^2)z = \frac{1}{x^2}$$

Part of C.F. corresponding to $D' \left(\equiv \frac{\partial}{\partial y}\right)$ is $f_1(x)$

Auxiliary equation of the remaining factor is

$$m^2 - 2m + 1 = 0 \quad \Rightarrow \quad m = 1, 1$$

$\therefore$ Another part of C.F. $= f_2(y + x) + x f_3(y + x)$

$\therefore$ Complete C.F. $= f_1(x) + f_2(y + x) + x f_3(y + x)$

$$\text{P.I.} = \frac{1}{D'^3 + D^2D' - 2DD'^2}\left(\frac{1}{x^2}\right) \qquad | \; x^{-2} = (x + 0.y)^{-2}$$

$$= y \cdot \frac{1}{3D'^2 + D^2 - 4DD'}\left(\frac{1}{x^2}\right) \qquad | \text{ Using Special Case of art. 1.19}$$

$$= y \cdot \frac{1}{3(0)^2 + (1)^2 - 4(1)(0)} \iint \frac{1}{u^2}\, du\, du, \qquad \text{where } x = u$$

$$= y \int \frac{-1}{u}\, du = y(-\log u) = -y \log x$$

Hence the complete solution is

$$z = \text{C.F.} + \text{P.I.} = f_1(x) + f_2(y + x) + x f_3(y + x) - y \log x$$

where f_1, f_2 and f_3 are arbitrary functions.

Example 15. *Solve:* $\frac{\partial^3 z}{\partial x^3} - 7\frac{\partial^3 z}{\partial x \partial y^2} - 6\frac{\partial^3 z}{\partial y^3} = \sin(x + 2y) + e^{3x+y}.$ (U.P.T.U. 2014)

Sol. The given equation is

$$(D^3 - 7DD'^2 - 6D'^3)z = \sin(x + 2y) + e^{3x+y}$$

Auxiliary equation is

$$m^3 - 7m - 6 = 0 \quad \Rightarrow \quad m = -1, -2, 3$$

$$\therefore \qquad \text{C.F.} = f_1(y - x) + f_2(y - 2x) + f_3(y + 3x)$$

$$\text{P.I.} = \frac{1}{D^3 - 7DD'^2 - 6D'^3}\sin(x + 2y) + \frac{1}{D^3 - 7DD'^2 - 6D'^3}e^{3x+y}.$$

P.I. corresponding to $\sin(x + 2y)$

$$= \frac{1}{(1)^3 - 7(1)(2)^2 - 6(2)^3}\iiint \sin u\, du\, du\, du, \qquad \text{where } x + 2y = u$$

$$= -\frac{1}{75}\cos u = -\frac{1}{75}\cos(x + 2y)$$

P.I. corresponding to e^{3x+y}

$$= \frac{1}{D^3 - 7DD'^2 - 6D'^3} (e^{3x+y})$$

$$= x \cdot \frac{1}{\frac{\partial}{\partial D}(D^3 - 7DD'^2 - 6D'^3)} e^{3x+y} = x \cdot \frac{1}{3D^2 - 7D'^2} e^{3x+y}$$

$$= x \cdot \frac{1}{3(3)^2 - 7(1)^2} e^{3x+y} = x \cdot \frac{1}{20} e^{3x+y}$$

$\therefore$ Required P.I. $= -\frac{1}{75}\cos(2y + x) + \frac{x}{20} e^{3x+y}$

$\therefore$ Complete solution is

$$z = \text{C.F.} + \text{P.I.} = f_1(y - x) + f_2(y - 2x) + f_3(y + 3x) - \frac{1}{75}\cos(x + 2y) + \frac{x}{20} e^{3x+y}$$

where f_1, f_2 and f_3 are arbitrary functions.

TEST YOUR KNOWLEDGE

Solve the following partial differential equations:

1. (*i*) $\frac{\partial^2 z}{\partial x^2} - 7\frac{\partial^2 z}{\partial x \partial y} + 12\frac{\partial^2 z}{\partial y^2} = e^{x-y}$ (*ii*) $(D^2 + 2DD' + D'^2) z = e^{2x+3y}$ (*U.P.T.U. 2015*)

2. (*i*) $\frac{\partial^2 z}{\partial x^2} + \frac{\partial^2 z}{\partial x \partial y} - 6\frac{\partial^2 z}{\partial y^2} = \cos(3x + y)$ (*ii*) $(D^3 - 4D^2D' + 4DD'^2)z = 6\sin(3x + 2y)$ (*U.K.T.U. 2012*)

3. $(D^2 - 4DD' + 4D'^2) z = e^{2x+y}$

4. $\frac{\partial^2 V}{\partial x^2} + \frac{\partial^2 V}{\partial y^2} = 12(x + y)$

5. $(D^2 - 8DD' + 7D'^2)z = \sin(7x + y)$

6. $\frac{\partial^2 z}{\partial x^2} - \frac{\partial^2 z}{\partial x \partial y} = \cos x \cos 2y$

7. $(D^2 - 6DD' + 9D'^2)z = 6x + 2y$

8. $(D^2 - 2DD' + D'^2)z = e^{x+y}$

9. (*i*) $4r + 12s + 9t = e^{3x-2y}$ (*ii*) $(D^2 + 3DD' + 2D'^2) z = e^{2x-3y}$

10. (*i*) $\frac{\partial^3 z}{\partial x^2 \partial y} - 5\frac{\partial^3 z}{\partial x \partial y^2} + 6\frac{\partial^3 z}{\partial y^3} = e^x$ (*ii*) $(D^2 + 7DD' + 12D'^2) z = \sinh x$ (*A.K.T.U. 2017*)

11. (*i*) $r + s - 6t = \cos(2x + y)$ (*ii*) $(D^2 - 5DD' + 4D'^2)z = \sin(4x + y)$ [*G.B.T.U. (AG) 2011*]

12. (*i*) $(D - D')^2 z = \tan(y + x)$ (*ii*) $(D^2 - DD')z = \sin(x + 2y)$ [*G.B.T.U. (AG) 2012*]

13. $(D^2 + 3DD' - 4D'^2)z = x + \sin y$

14. $(D^3 - 4D^2D' + 5DD'^2 - 2D'^3)z = e^{y+2x} + (y + x)^{1/2}$ (*A.K.T.U. 2017*)

15. $(D^2 - 3DD' + 2D'^2) z = e^{2x-y} + e^{x+y} + \cos(x + 2y)$ (*G.B.T.U. 2012*)

Answers

1. (*i*) $z = f_1(y + 3x) + f_2(y + 4x) + \frac{1}{20} e^{x-y}$ (*ii*) $z = f_1(y - x) + x f_2(y - x) + \frac{1}{25} e^{2x+3y}$

2. (*i*) $z = f_1(y - 3x) + f_2(y + 2x) - \frac{1}{6} \cos(3x + y)$ (*ii*) $z = f_1(y) + f_2(y + 2x) + x f_3(y + 2x) + 2\cos(3x + 2y)$

3. $z = f_1(y + 2x) + x f_2(y + 2x) + \frac{x^2}{2} e^{2x+y}$ 4. $V = f_1(y + ix) + f_2(y - ix) + (x + y)^3$.

5. $z = f_1(y + x) + f_2(y + 7x) - \frac{x}{6} \cos(7x + y)$ 6. $z = f_1(y) + f_2(y + x) + \frac{1}{2} \cos(x + 2y) - \frac{1}{6} \cos(x - 2y)$

7. $z = f_1(y + 3x) + x f_2(y + 3x) + x^2(3x + y)$ 8. $z = f_1(y + x) + x f_2(y + x) + \frac{1}{2} x^2 e^{x+y}$

9. (*i*) $z = f_1(2y - 3x) + x f_2(2y - 3x) + \frac{1}{8} x^2 e^{3x-2y}$ (*ii*) $z = f_1(y - x) + f_2(y - 2x) + \frac{1}{4} e^{2x-3y}$

10. (*i*) $z = f_1(x) + f_2(y + 2x) + f_3(y + 3x) + ye^x$ (*ii*) $z = f_1(y - 3x) + f_2(y - 4x) + \sinh x$

11. (*i*) $z = f_1(y - 3x) + f_2(y + 2x) + \frac{x}{5} \sin(2x + y)$ (*ii*) $z = f_1(y + x) + f_2(y + 4x) - \frac{x}{3} \cos(4x + y)$

12. (*i*) $z = f_1(y + x) + x f_2(y + x) + \frac{x^2}{2} \tan(y + x)$ (*ii*) $z = f_1(y) + f_2(y + x) + \sin(x + 2y)$

13. $z = f_1(y + x) + f_2(y - 4x) + \frac{x^3}{6} + \frac{1}{4} \sin y$

14. $z = f_1(y + x) + x f_2(y + x) + f_3(y + 2x) + x e^{y+x} - \frac{x^2}{3}(y + x)^{3/2}$

15. $z = f_1(y + x) + f_2(y + 2x) + \frac{1}{12} e^{2x-y} - x e^{x+y} - \frac{1}{3} \cos(x + 2y)$

1.20 WHEN $\phi(x, y)$ IS OF THE FORM $x^m y^n$ OR A RATIONAL INTEGRAL ALGEBRAIC FUNCTION OF x AND y

In this case, P.I. is obtained by expanding F(D, D′) in an infinite series of ascending powers of D or D′ or $\frac{D'}{D}$. It is to be noted that the difference in the two answers of P.I. is immaterial because it can be incorporated in the arbitrary functions occurring in C.F. of the given equation.

ILLUSTRATIVE EXAMPLES

Example 1. *Solve*: $\frac{\partial^3 z}{\partial x^3} - \frac{\partial^3 z}{\partial y^3} = x^3 y^3$.

Sol. The given equation is $(D^3 - D'^3)z = x^3y^3$

Auxiliary equation is $m^3 - 1 = 0$

$\Rightarrow$ $m = 1, \omega, \omega^2$

where ω is one of the cube roots of unity.

$\therefore$ $\text{C.F.} = f_1(y + x) + f_2(y + \omega x) + f_3(y + \omega^2 x)$

$$\text{P.I.} = \frac{1}{D^3 - D'^3}(x^3y^3) = \frac{1}{D^3\left(1 - \dfrac{D'^3}{D^3}\right)}(x^3y^3)$$

$$= \frac{1}{D^3}\left(1 - \frac{D'^3}{D^3}\right)^{-1}(x^3y^3) = \frac{1}{D^3}\left(1 + \frac{D'^3}{D^3}\right)(x^3y^3)$$

$$= \frac{1}{D^3}\left[x^3y^3 + \frac{1}{D^3}D'^3(x^3y^3)\right] = \frac{1}{D^3}\left[x^3y^3 + \frac{1}{D^3}(6x^3)\right]$$

$$= \frac{1}{D^3}(x^3y^3) + \frac{1}{D^6}(6x^3) = \frac{x^6y^3}{6\cdot5\cdot4} + \frac{6x^9}{4\cdot5\cdot6\cdot7\cdot8\cdot9} = \frac{x^6y^3}{120} + \frac{x^9}{10080}$$

Hence the complete solution is

$$z = \text{C.F.} + \text{P.I.} = f_1(y + x) + f_2(y + \omega x) + f_3(y + \omega^2 x) + \frac{x^6y^3}{120} + \frac{x^9}{10080}.$$

where f_1, f_2 and f_3 are arbitrary functions.

Example 2. *Find a real function V of x and y, reducing to zero when y = 0 and satisfying*

$$\frac{\partial^2 V}{\partial x^2} + \frac{\partial^2 V}{\partial y^2} = -4\pi(x^2 + y^2).$$

Sol. The given equation is $(D^2 + D'^2)\,V = -4\pi(x^2 + y^2)$

The auxiliary equation is $m^2 + 1 = 0$

$\Rightarrow$ $m = \pm i$

$\therefore$ C.F. $= f_1(y + ix) + f_2(y - ix)$, which is imaginary.

Now,

$$\text{P.I.} = \frac{1}{D^2 + D'^2}[-4\pi(x^2 + y^2)] = -\frac{4\pi}{D^2}\left(1 + \frac{D'^2}{D^2}\right)^{-1}(x^2 + y^2)$$

$$= -\frac{4\pi}{D^2}\left(1 - \frac{D'^2}{D^2}\right)(x^2 + y^2) = -\frac{4\pi}{D^2}\left[x^2 + y^2 - \frac{1}{D^2}(2)\right]$$

$$= -\frac{4\pi}{D^2}\{x^2 + y^2 - x^2\} = -\frac{4\pi}{D^2}(y^2)$$

$$= -\frac{4\pi}{D}(xy^2) = -4\pi\left(\frac{x^2}{2}y^2\right)$$

$$\text{P.I.} = -2\pi x^2y^2$$

Since V is a real function, hence $V = -2\pi x^2y^2$.

Example 3. *Solve:* $(D^2 - 6DD' + 9D'^2)\,z = 12x^2 + 36xy$.

Sol. Auxiliary equation is

$$m^2 - 6m + 9 = 0$$

$\Rightarrow$ $m = 3, 3$

$\therefore$ C.F. $= f_1(y + 3x) + x f_2(y + 3x)$

$$\text{P.I.} = \frac{1}{D^2 - 6DD' + 9D'^2}(12x^2) + \frac{1}{D^2 - 6DD' + 9D'^2}(36xy) = P_1 + P_2$$

where

$$P_1 = \frac{1}{D^2 - 6DD' + 9D'^2}(12x^2)$$

$$= \frac{12}{(1)^2 - 6(1)(0) + 9(0)^2} \iint u^2 \, du \, du, \qquad \text{where } x = u$$

$$= 12 \, \frac{u^4}{12} = u^4 = x^4$$

$$P_2 = \frac{1}{D^2 - 6DD' + 9D'^2} (36xy)$$

$$= \frac{36}{(D - 3D')^2} (xy) = \frac{36}{D^2} \left(1 - \frac{3D'}{D}\right)^{-2} (xy)$$

$$= \frac{36}{D^2} \left(1 + \frac{6D'}{D}\right) (xy) \qquad \text{| Leaving higher power terms}$$

$$= \frac{36}{D^2} \left[xy + \frac{6}{D}(x)\right] = \frac{36}{D^2} (xy + 3x^2)$$

$$= 36 \left(\frac{x^3}{6} y + \frac{x^4}{4}\right) = 6x^3y + 9x^4$$

$\therefore$ P.I. $= P_1 + P_2 = 6x^3y + 10x^4$

Hence the complete solution is

$$z = \text{C.F.} + \text{P.I.} = f_1(y + 3x) + xf_2 (y + 3x) + 6x^3y + 10x^4$$

where f_1 and f_2 are arbitrary functions.

TEST YOUR KNOWLEDGE

Solve:

1. $\dfrac{\partial^2 z}{\partial x^2} + 3 \dfrac{\partial^2 z}{\partial x \, \partial y} + 2 \dfrac{\partial^2 z}{\partial y^2} = 12\,xy$ *(U.P.T.U. 2015)*

2. $\dfrac{\partial^2 z}{\partial x^2} + 2 \dfrac{\partial^2 z}{\partial x \, \partial y} + \dfrac{\partial^2 z}{\partial y^2} = x^2 + xy + y^2$

3. $\dfrac{\partial^3 z}{\partial x^3} - 2 \dfrac{\partial^3 z}{\partial x^2 \, \partial y} = 2e^{2x} + 3x^2y$

4. $(D^2 - a^2 D'^2)z = x$

5. $[D^2 + (a + b)\, DD' + ab\, D'^2]z = xy$ *[G.B.T.U. (AG) 2011, G.B.T.U. 2013]*

6. $(D^2 - 2DD' + D'^2)z = e^{x + 2y} + x^3$.

7. $\dfrac{\partial^2 z}{\partial x^2} - 2 \dfrac{\partial^2 z}{\partial x \partial y} - 15 \dfrac{\partial^2 z}{\partial y^2} = 12xy$ *(U.P.T.U. 2013)*

Answers

1. $z = f_1 (y - x) + f_2(y - 2x) + 2x^3y - \dfrac{3}{2} x^4$

2. $z = f_1 (y - x) + x\, f_2 (y - x) + \dfrac{1}{4} (x^4 - 2x^3y + 2x^2y^2)$

3. $z = f_1(y) + x\, f_2(y) + f_3 (y + 2x) + \dfrac{1}{4} e^{2x} + \dfrac{1}{60} (3x^5y + x^6)$

4. $z = f_1 (y + ax) + f_2 (y - ax) + \dfrac{x^3}{6}$

5. $z = f_1 (y - ax) + f_2 (y - bx) + \dfrac{1}{6} x^3y - \left(\dfrac{a + b}{24}\right) x^4$

6. $z = f_1 (y + x) + x\, f_2 (y + x) + e^{x + 2y} + \dfrac{x^5}{20}$

7. $z = f_1(y - 3x) + f_2(y + 5x) + 2x^3y + x^4$.

1.21 GENERAL METHOD TO FIND THE P.I.

$\phi(x, y)$ is not always of the form given above. The general method is applicable to all cases, where $\phi(x, y)$ is not of the form given above.

Now, F(D, D′) can be factorised, in general, into n linear factors.

$$\therefore \quad \text{P.I.} = \frac{1}{F(D, D')} \phi(x, y)$$

$$= \frac{1}{(D - m_1D')(D - m_2D') \ldots (D - m_nD')} \phi(x, y)$$

$$= \frac{1}{D - m_1D'} \cdot \frac{1}{D - m_2D'} \cdots\cdots \frac{1}{D - m_nD'} \phi(x, y)$$

Let us evaluate $\frac{1}{D - mD'} \phi(x, y)$

Consider the equation,

$(D - mD')z = \phi(x, y)$ or $p - mq = \phi(x, y)$ [Lagrange's form]

The subsidiary equations are $\frac{dx}{1} = \frac{dy}{-m} = \frac{dz}{\phi(x, y)}$

From the first two members $dy + mdx = 0$ or $y + mx = c$

From the first and last members, we have

$$dz = \phi(x, y)dx = \phi(x, c - mx)dx$$

$$\therefore \quad z = \int \phi(x, c - mx)dx$$

or

$$\boxed{\frac{1}{D - mD'} \phi(x, y) = \int \phi(x, c - mx)\, dx}$$

where c is replaced by $y + mx$ after integration.

By repeated application of the above rule, the P.I. can be evaluated.

ILLUSTRATIVE EXAMPLES

Example 1. *Solve the linear partial differential equation*

$$\frac{\partial^2 z}{\partial x^2} + \frac{\partial^2 z}{\partial x \partial y} - 6\frac{\partial^2 z}{\partial y^2} = y \cos x.$$

Sol. The given equation is

$$(D^2 + DD' - 6D'^2)z = y \cos x$$

The auxiliary equation is

$$m^2 + m - 6 = 0$$

$$\Rightarrow \quad (m - 2)(m + 3) = 0$$

$$\Rightarrow \quad m = 2, -3.$$

$$\therefore \quad \text{C.F.} = f_1(y + 2x) + f_2(y - 3x)$$

$$\text{P.I.} = \frac{1}{D^2 + DD' - 6D'^2} y \cos x = \frac{1}{(D - 2D')(D + 3D')} y \cos x$$

$$= \frac{1}{D - 2D'} \int (c + 3x) \cos x \, dx, \quad \text{where } y = c + 3x$$

$$= \frac{1}{D - 2D'} \{c \sin x + 3 \int x \cos x \, dx\}$$

$$= \frac{1}{D - 2D'} [c \sin x + 3 \{x \sin x - \int 1 . \sin x \, dx\}]$$

$$= \frac{1}{D - 2D'} [(c + 3x) \sin x + 3 \cos x]$$

$$= \frac{1}{D - 2D'} (y \sin x + 3 \cos x), \quad \text{where } c = y - 3x$$

$$= \int (b - 2x) \sin x \, dx + 3 \sin x, \quad \text{where } y = b - 2x$$

$$= - b \cos x - 2 \{x (- \cos x) - \int 1 . (- \cos x) \, dx\} + 3 \sin x$$

$$= - b \cos x + 2x \cos x - 2 \sin x + 3 \sin x$$

$$= - (b - 2x) \cos x + \sin x$$

$$= - y \cos x + \sin x, \quad \text{where } b = y + 2x$$

Hence the complete solution is

$$z = \text{C.F.} + \text{P.I.} = f_1 (y + 2x) + f_2 (y - 3x) - y \cos x + \sin x$$

where f_1 and f_2 are arbitrary functions.

Example 2. *Solve the linear partial differential equation*

$$\frac{\partial^2 z}{\partial x^2} + \frac{\partial^2 z}{\partial x \partial y} - 2 \frac{\partial^2 z}{\partial y^2} = (y - 1) e^x.$$

Sol. The given equation is

$$(D^2 + DD' - 2D'^2) z = (y - 1) e^x$$

The auxiliary equation is

$$m^2 + m - 2 = 0$$

$$\Rightarrow \qquad (m - 1)(m + 2) = 0$$

$$\Rightarrow \qquad m = 1, - 2$$

$$\text{C.F.} = f_1 (y + x) + f_2 (y - 2x)$$

$$\text{P.I.} = \frac{1}{D^2 + DD' - 2D'^2} (y - 1) e^x$$

$$= \frac{1}{(D - D')(D + 2D')} (y - 1) e^x$$

$$= \frac{1}{D - D'} \int (c + 2x - 1) e^x \, dx, \quad \text{where } y = c + 2x$$

$$= \frac{1}{D - D'} [(c - 1) e^x + 2 (x - 1) e^x]$$

$$= \frac{1}{D - D'} [(c + 2x) e^x - 3e^x]$$

$$= \frac{1}{D - D'} (ye^x - 3e^x), \qquad \text{where } c = y - 2x$$

$$= \int (b - x)\, e^x\, dx - 3e^x, \qquad \text{where } y = b - x$$

$$= be^x - (x - 1)\, e^x - 3e^x$$

$$= (b - x - 2)\, e^x = (y - 2)\, e^x, \qquad \text{where } b = y + x$$

Hence the complete solution is

$$z = \text{C.F.} + \text{P.I.} = f_1 (y + x) + f_2 (y - 2x) + (y - 2)\, e^x$$

where f_1 and f_2 are arbitrary functions.

Example 3. *Solve the partial differential equation:*

$$r - t = \tan^3 x \tan y - \tan x \tan^3 y.$$

Sol. The given equation is

$$\frac{\partial^2 z}{\partial x^2} - \frac{\partial^2 z}{\partial y^2} = \tan x \tan y\, (\sec^2 x - \sec^2 y).$$

Auxiliary equation is $m^2 - 1 = 0 \Rightarrow m = \pm 1.$

$$\text{C.F.} = f_1 (y + x) + f_2 (y - x)$$

$$\text{P.I.} = \frac{1}{D^2 - D'^2} \tan x \tan y\, (\sec^2 x - \sec^2 y)$$

$$= \frac{1}{D + D'} \left[\frac{1}{D - D'} (\tan x \sec^2 x \tan y - \tan x \tan y \sec^2 y) \right]$$

$$= \frac{1}{D + D'} \left[\int \tan x \sec^2 x \tan (c - x)\, dx - \int \tan x \tan (c - x) \sec^2 (c - x)\, dx \right]$$

where $y = c - x$

$$= \frac{1}{D + D'} \left[\tan (c - x) \frac{\tan^2 x}{2} + \int \sec^2 (c - x) \frac{\tan^2 x}{2}\, dx + \tan x \,.\, \frac{\tan^2 (c - x)}{2} - \int \sec^2 x \frac{\tan^2 (c - x)}{2}\, dx \right]$$

$$= \frac{1}{2(D + D')} [\tan (c - x) \tan^2 x + \tan x \tan^2 (c - x] + \int \sec^2 (c - x) (\sec^2 x - 1)\, dx - \int \sec^2 x \{\sec^2 (c - x) - 1\}\, dx]$$

$$= \frac{1}{2(D + D')} [\tan (c - x) \tan^2 x + \tan x \tan^2 (c - x) + \int \{\sec^2 x - \sec^2 (c - x)\}\, dx]$$

$$= \frac{1}{2(D + D')} [\tan (c - x) \tan^2 x + \tan x \tan^2 (c - x) + \tan x + \tan (c - x)]$$

$$= \frac{1}{2(D + D')} [\tan x \sec^2 (c - x) + \tan (c - x) \sec^2 x]$$

$$= \frac{1}{2(D + D')} (\tan x \sec^2 y + \tan y \sec^2 x) \quad \text{where } c = y + x$$

$$= \frac{1}{2} [\int \tan x \sec^2 (b + x)\, dx + \int \tan (b + x) \sec^2 x\, dx], \quad \text{where } y = b + x$$

$$= \frac{1}{2} [\tan x \,.\, \tan (b + x) - \int \sec^2 x \,.\, \tan (b + x)\, dx + \int \tan (b + x) \sec^2 x\, dx]$$

$$= \frac{1}{2} \tan x \tan (b + x) = \frac{1}{2} \tan x \tan y \quad \text{where } b = y - x.$$

Hence the complete solution is

$$z = \text{C.F.} + \text{P.I.} = f_1 (y + x) + f_2 (y - x) + \frac{1}{2} \tan x \tan y$$

where f_1 and f_2 are arbitrary functions.

Example 4. *Solve:* $r - s - 2t = (2x^2 + xy - y^2) \sin xy - \cos xy.$

Sol. The given equation is

$$(D^2 - DD' - 2D'^2)z = (2x^2 + xy - y^2) \sin xy - \cos xy$$

Auxiliary equation is $\quad m^2 - m - 2 = 0$

$\Rightarrow \quad (m + 1)(m - 2) = 0 \quad \Rightarrow \quad m = -1, 2$

$\therefore \quad \text{C.F.} = f_1(y - x) + f_2(y + 2x)$

$$\text{P.I.} = \frac{1}{(D + D')(D - 2D')} [(2x - y)(x + y) \sin xy - \cos xy]$$

$$= \frac{1}{(D + D')} \int [(4x - c)(c - x) \sin (cx - 2x^2) - \cos (cx - 2x^2)]\, dx, \text{ where } y = c - 2x$$

$$= \frac{1}{(D + D')} \int [(x - c)(c - 4x) \sin (cx - 2x^2) - \cos (cx - 2x^2)]\, dx$$

$$= \frac{1}{(D + D')} \left[(x - c)\{-\cos (cx - 2x^2)\} + \int \cos (cx - 2x^2)\, dx - \int \cos (cx - 2x^2)\, dx\right]$$

$$= \frac{1}{(D + D')} (c - x) \cos \{x (c - 2x)\}$$

$$= \frac{1}{D + D'} (y + x) \cos xy, \qquad \text{where } c = y + 2x$$

$$= \int (b + 2x) \cos (bx + x^2)\, dx, \qquad \text{where } y = b + x$$

$$= \sin (bx + x^2) = \sin xy, \qquad \text{where } b = y - x$$

$\therefore$ Complete solution is

$$z = \text{C.F.} + \text{P.I.} = f_1(y - x) + f_2(y + 2x) + \sin xy$$

where f_1 and f_2 are arbitrary functions.

Example 5. *Solve:* $(D^2 + 2DD' + D'^2)z = 2 \cos y - x \sin y.$

Sol. Auxiliary equation is

$$m^2 + 2m + 1 = 0$$

$\Rightarrow \quad m = -1, -1$

$\therefore \quad \text{C.F.} = f_1 (y - x) + x f_2 (y - x)$

$$\text{P.I.} = \frac{1}{(D+D')^2}\, 2\cos y - \frac{1}{(D+D')^2}\,(x \sin y) = P_1 - P_2$$

where $$P_1 = \frac{1}{(D+D')^2}\, 2\cos y = \frac{2}{(0+1)^2} \iint \cos u\, du\, du, \text{ where } y = u$$

$$= 2(-\cos y) = -2\cos y$$

$$P_2 = \frac{1}{(D+D')^2}\, x \sin y = \frac{1}{D+D'} \int x \sin(c+x)\, dx, \text{ where } y = c + x$$

$$= \frac{1}{D+D'}\left[x\{-\cos(c+x)\} - \int 1.\{-\cos(c+x)\}\, dx\right]$$

$$= \frac{1}{D+D'}\,[-x\cos(c+x) + \sin(c+x)]$$

$$= \frac{1}{D+D'}\,(-x\cos y + \sin y), \quad \text{where } c = y - x$$

$$= \int -x\cos(b+x)\, dx + \int \sin(b+x)\, dx, \quad \text{where } y = b + x$$

$$= (-x)\,.\,\sin(b+x) - \int (-1)\,.\,\sin(b+x)\, dx + \int \sin(b+x)\, dx$$

$$= -x\sin(b+x) - 2\cos(b+x)$$

$$= -x\sin y - 2\cos y, \quad \text{where } b = y - x$$

$\therefore$ P.I. $= P_1 - P_2 = -2\cos y + x\sin y + 2\cos y = x \sin y$

Hence the complete solution is

$$z = \text{C.F.} + \text{P.I.} = f_1(y-x) + x f_2(y-x) + x\sin y$$

where f_1 and f_2 are arbitrary functions.

TEST YOUR KNOWLEDGE

Solve:

1. $(D^2 - DD' - 2D'^2)z = (y-1)e^x$ *(U.P.T.U. 2014)*
2. $(D - D')(D + 2D')z = (y+1)e^x$
3. $(D^3 + 2D^2D' - DD'^2 - 2D'^3)z = (y+2)e^x$
4. $r - 4t = \dfrac{4x}{y^2} - \dfrac{y}{x^2}$
5. $(D^3 + D^2D' - DD'^2 - D'^3)z = e^x \cos 2y$
6. $(D^2 + DD' - 6D'^2)z = y \sin x$

[*M.T.U. (SUM) 2011; M.T.U. 2012*]

Answers

1. $z = f_1(y-x) + f_2(y+2x) + ye^x$
2. $z = f_1(y+x) + f_2(y-2x) + ye^x$
3. $z = f_1(y+x) + f_2(y-x) + f_3(y-2x) + ye^x$
4. $z = f_1(y+2x) + f_2(y-2x) + x\log y + y\log x + 3x$
5. $z = f_1(y-x) + x f_2(y-x) + f_3(y+x) + \dfrac{e^x}{25}(\cos 2y + 2\sin 2y)$
6. $z = f_1(y+2x) + f_2(y-3x) - y\sin x - \cos x.$

1.22 NON-HOMOGENEOUS LINEAR PARTIAL DIFFERENTIAL EQUATIONS WITH CONSTANT COEFFICIENTS

In the equation $\phi(D, D')z = F(x, y)$ ...(1)

if the polynomial $\phi(D, D')$ in D, D′ is not homogeneous, then (1) is called a non-homogeneous linear partial differential equation.

Its complete solution is $z = \text{C.F.} + \text{P.I.}$

1.23 METHODS FOR FINDING OUT C.F.

(*a*) We resolve $\phi(D, D')$ into linear factors of the form **D – *m*D′ – *a*.**

Now consider the equation $(D - mD' - a)z = 0$ ...(2)

or $$p - mq = az$$

Lagrange's auxiliary equations are

$$\frac{dx}{1} = \frac{dy}{-m} = \frac{dz}{az}$$

From the first two members $dy + mdx = 0 \quad \therefore \quad y + mx = b$

From the first and last members $\frac{dz}{z} = adx \quad \therefore \quad \log z = ax + \log c$ or $z = ce^{ax}$

$\therefore$ The complete solution of (2) is $z = e^{ax} f(y + mx)$

Hence the C.F. of (1), *i.e.*, the complete solution of

$$(D - m_1D' - a_1)(D - m_2D' - a_2)......(D - m_nD' - a_n)z = 0 \text{ is}$$

$$z = e^{a_1x} f_1(y + m_1x) + e^{a_2x} f_2(y + m_2x) + ... + e^{a_nx} f_n(y + m_nx).$$

1.23.1. In the Case of Repeated Factors, *e.g.*, $(D - mD' - a)^3z = 0$

We have $z = e^{ax} f_1(y + mx) + xe^{ax} f_2(y + mx) + x^2e^{ax} f_3(y + mx).$

1.23.2. If the Equation is of the Form

$$(\alpha D + \beta D' + \gamma)z = 0 \qquad ...(3)$$

$$\Rightarrow \qquad \alpha p + \beta q = -\gamma z$$

It is of Lagrange's form.

Lagrange's subsidiary equations are

$$\frac{dx}{\alpha} = \frac{dy}{\beta} = \frac{dz}{-\gamma z} \qquad ...(4)$$

First two will give $\alpha y - \beta x = c_1$ ...(5)

First and last will give, $\frac{dz}{z} = -\frac{\gamma}{\alpha} dx$

Integration gives, $\log z = -\frac{\gamma}{\alpha} x + \log c_2$

$$\Rightarrow \qquad z = c_2 \, e^{-\frac{\gamma}{\alpha}x} = \phi(c_1) \, e^{-\frac{\gamma}{\alpha}x}$$

$$\Rightarrow \qquad z = e^{-\frac{\gamma}{\alpha}x}\ \phi(\alpha y - \beta x)$$

where ϕ is an arbitrary function.

Note. The above result is not applicable in the absence of the first term *i.e.,* D or αD and also when $\alpha = 0$.

Remark 1. Corresponding to each non-repeated factor $(a\text{D}' + b)$, the part of C.F. is $e^{-(by/a)}\ \phi(ax)$ when $a \neq 0$.

Remark 2. Corresponding to repeated factor $(a\text{D}' + b)^r$, the part of C.F. is

$$e^{-\left(\frac{by}{a}\right)}\ [\phi_1(ax) + y\ \phi_2(ax) + y^2\ \phi_3(ax) + \ldots\ldots + y^{r-1}\ \phi_r(ax)].$$

Remark 3. As a particular case of remark 1 with $b = 0$, $a = 1$, corresponding to non-repeated factor D′, the part of C.F. $= \phi_1(x)$.

Remark 4. As a particular case of remark 2 with $b = 0$, $a = 1$, corresponding to repeated factor D'^r, part of C.F. $= \phi_1(x) + y\ \phi_2(x) + y^2\ \phi_3(x) + \ldots\ldots + y^{r-1}\ \phi_r(x)$

(*b*) **When F(D, D′) cannot be factorized into linear factors:**

In such cases, we use a trial method.

Let the equation be $\qquad (\text{D} - \text{D}'^2)z = 0 \qquad \ldots(1)$

Let the trial solution of (1) be $\qquad z = \text{A}e^{hx+ky}$, where A, h and k are constants. $\qquad \ldots(2)$

From (2),

$$\text{D}z = \frac{\partial z}{\partial x} = \text{A}he^{hx+ky}$$

$$\text{D}'z = \frac{\partial z}{\partial y} = \text{A}\ ke^{hx+ky}$$

$$\text{D}'^2 z = \frac{\partial^2 z}{\partial y^2} = \text{A}\ k^2\ e^{hx+ky}$$

Putting in (1), we get $\text{A}he^{hx+ky} - \text{A}\ k^2\ e^{hx+ky} = 0$

$$\Rightarrow \qquad \text{A}(h - k^2)\ e^{hx+ky} = 0$$

or

$$h = k^2 \qquad \ldots(3)$$

Equation (2) gives,

$$z = \text{A}e^{k^2x + ky} \qquad \ldots(4)$$

Since all values of k will satisfy eqn. (1), a more general solution of (1) is given by

$$z = \Sigma\ \text{A}e^{k^2x + ky}$$

where A and k are arbitrary constants and Σ denotes that any number of terms may be taken.

ILLUSTRATIVE EXAMPLES

Example 1. *Solve the linear partial differential equation*

$$(D + D' - 1)\ (D + 2D' - 2)z = 0.$$

Sol. The given equation is

$$(\text{D} + \text{D}' - 1)\ (\text{D} + 2\text{D}' - 2)z = 0.$$

Its $\quad$ C.F. $= e^x f_1\ (y - x) + e^{2x} f_2\ (y - 2x)$

and $\quad$ P.I. $= 0$

Hence complete solution is

$$z = \text{C.F.} + \text{P.I.} = e^x f_1(y - x) + e^{2x} f_2(y - 2x)$$

where f_1 and f_2 are arbitrary functions.

Example 2. *Solve:* $DD'(D + 2D' + 1)z = 0.$

Sol. The given equation is

$$DD'(D + 2D' + 1)z = 0$$

Corresponding to the factor D, part of C.F. $= f_1(y)$

Corresponding to the factor D′, part of C.F. $= f_2(x)$

Corresponding to the factor (D + 2D′ + 1) part of C.F. $= e^{-x} f_3(y - 2x)$

Hence combined C.F. $= f_1(y) + f_2(x) + e^{-x} f_3(y - 2x)$

P.I. = 0

Hence complete solution is

$$z = \text{C.F.} + \text{P.I.} = f_1(y) + f_2(x) + e^{-x} f_3(y - 2x)$$

where f_1, f_2 and f_3 are arbitrary functions.

Example 3. *Solve:* $r + 2s + t + 2p + 2q + z = 0.$

Sol. The given equation is

$$[D^2 + 2DD' + D'^2 + 2D + 2D' + 1]z = 0$$

$$\Rightarrow \quad \{(D + D')^2 + 2(D + D') + 1\}z = 0$$

$$\Rightarrow \quad (D + D' + 1)^2 z = 0$$

Its C.F. $= e^{-x} f_1(y - x) + x\, e^{-x} f_2(y - x)$

P.I. = 0

Hence complete solution is

$$z = \text{C.F.} + \text{P.I.} = e^{-x} f_1(y - x) + x\, e^{-x} f_2(y - x)$$

where f_1 and f_2 are arbitrary functions.

Example 4. *Solve:*

(*i*) $r - t + p - q = 0$ (U.P.T.U. 2013, 2014)

(*ii*) $(D + 4D' + 5)^2 z = 0$ [A.K.T.U. 2018]

(*iii*) $(D^2 - DD' - 2D)z = 0$

(*iv*) $(D + 1)(D + D' - 1)z = 0.$

Sol. (*i*) The given equation is

$$(D^2 - D'^2 + D - D')z = 0$$

$$\Rightarrow \quad (D - D')(D + D' + 1)z = 0$$

Its C.F. $= f_1(y + x) + e^{-x} f_2(y - x)$

P.I. = 0

Hence complete solution is

$$z = \text{C.F.} + \text{P.I.} = f_1(y + x) + e^{-x} f_2(y - x)$$

where f_1 and f_2 are arbitrary functions.

(*ii*) The given equation is

$$(D + 4D' + 5)^2 z = 0$$

Its C.F. $= e^{-5x} f_1(y - 4x) + x\, e^{-5x} f_2(y - 4x)$

P.I. = 0

Hence complete solution is

$$z = \text{C.F.} + \text{P.I.} = e^{-5x} f_1 (y - 4x) + x\, e^{-5x} f_2 (y - 4x)$$

where f_1 and f_2 are arbitrary functions.

(*iii*) The given equation is

$$(D^2 - DD' - 2D)z = 0$$

$$\Rightarrow \qquad D (D - D' - 2)z = 0$$

Its $\quad \text{C.F.} = f_1(y) + e^{2x} f_2 (y + x)$

$\text{P.I.} = 0$

Hence complete solution is

$$z = \text{C.F.} + \text{P.I.} = f_1(y) + e^{2x} f_2 (y + x)$$

where f_1 and f_2 are arbitrary functions.

(*iv*) The given equation is

$$(D + 1) (D + D' - 1)z = 0$$

Its $\quad \text{C.F.} = e^{-x} f_1(y) + e^{x} f_2 (y - x)$

$\text{P.I.} = 0$

Hence complete solution is

$$z = \text{C.F.} + \text{P.I.} = e^{-x} f_1(y) + e^{x} f_2 (y - x)$$

where f_1 and f_2 are arbitrary functions.

Example 5. *Solve:* (*i*) $(D^2 - D'^2 + D + 3D' - 2)z = 0$

(*ii*) $(D^2 - DD' - 2D'^2 + 2D + 2D')z = 0.$

Sol. (*i*) The given equation is

$$(D^2 - D'^2 + D + 3D' - 2)z = 0$$

$$\Rightarrow \qquad (D^2 - D'^2 + 2D - D + 2D' + D' - 2 + DD' - DD')z = 0$$

$$\Rightarrow \qquad [D^2 - DD' + 2D + DD' - D'^2 + 2D' - D + D' - 2]z = 0$$

$$\Rightarrow \qquad (D - D' + 2) (D + D' - 1)z = 0$$

Its $\quad \text{C.F.} = e^{-2x} f_1 (y + x) + e^{x} f_2 (y - x)$

$\text{P.I.} = 0$

Hence complete solution is $\quad z = \text{C.F.} + \text{P.I.} = e^{-2x} f_1 (y + x) + e^{x} f_2 (y - x)$

where f_1 and f_2 are arbitrary functions.

(*ii*) The given equation is

$$(D^2 - DD' - 2D'^2 + 2D + 2D')z = 0$$

$$\Rightarrow \qquad (D + D') (D - 2D') + 2(D + D')]z = 0$$

$$\Rightarrow \qquad (D + D')(D - 2D' + 2)z = 0$$

Its $\quad \text{C.F.} = f_1 (y - x) + e^{-2x} f_2 (y + 2x)$

$\text{P.I.} = 0$

Hence complete solution is

$$z = \text{C.F.} + \text{P.I.} = f_1 (y - x) + e^{-2x} f_2 (y + 2x)$$

where f_1 and f_2 are arbitrary functions.

Example 6. *Solve:* $(D^2 + D'^2 - p^2)z = 0.$

Sol. Here $D^2 + D'^2 - p^2$ cannot be resolved into linear factors in D and D′.

Let $z = Ae^{hx+ky}$

$\therefore$ $D^2z = Ah^2 e^{hx+ky}$

$D'^2z = Ak^2 e^{hx+ky}$

$\therefore$ $(D^2 + D'^2 - p^2)z = A(h^2 + k^2 - p^2) e^{hx+ky}$

Then, $(D^2 + D'^2 - p^2)z = 0$

$\Rightarrow$ $A(h^2 + k^2 - p^2) e^{hx+ky} = 0$

$\Rightarrow$ $h^2 + k^2 - p^2 = 0$ or $h^2 + k^2 = p^2$

$\therefore$ C.F. $= \Sigma Ae^{hx+ky}$, where $h^2 + k^2 - p^2 = 0$

P.I. = 0

$\therefore$ $z = \Sigma Ae^{hx+ky}$, where $h^2 + k^2 - p^2 = 0$

Now, h may be taken as $p \cos \alpha$ and k may be taken as $p \sin \alpha$. Therefore The complete solution is

$$z = \Sigma Ae^{p(x \cos \alpha + y \sin \alpha)}$$

where A is arbitrary constant and Σ denotes that any number of terms may be taken.

TEST YOUR KNOWLEDGE

Solve the following partial differential equations:

1. $(D + D' - 1)(D + 2D' - 3)z = 0$

2. $r - 3s + 2t - p + 2q = 0$

3. $DD'(D - 2D' - 3)z = 0$

4. $t + s + q = 0$

5. $(D^2 - a^2D'^2 + 2abD + 2a^2bD')z = 0$

6. $(D - 2D' - 1)(D - 2D'^2 - 1)z = 0$

7. $(2D^4 - 3D^2D' + D'^2)z = 0$

8. $(D^3 - 3DD' + D' + 4)z = 0$

9. $2s + t - 3q = 0$

10. $(DD' + aD + bD' + ab)z = 0.$

Answers

1. $z = e^x f_1(y - x) + e^{3x} f_2(y - 2x)$

2. $z = f_1(y + 2x) + e^x f_2(y + x)$

3. $z = f_1(y) + f_2(x) + e^{3x} f_3(y + 2x)$

4. $z = f_1(x) + e^{-x} f_2(y - x)$

5. $z = f_1(y - ax) + e^{-2abx} f_2(y + ax)$

6. $z = e^x f_1(y + 2x) + \Sigma Ae^{(2k^2+1)x+ky}$

7. $z = \Sigma Ae^{hx+h^2y} + \Sigma A'e^{h'x+h'^2y}$

8. $z = \Sigma A e^{hx+ky}$, where $h^3 - 3hk + k + 4 = 0$

9. $z = f_1(y) + e^{(3x/2)} f_2(2y - x).$

10. $z = e^{-ay} f_1(x) + e^{-bx} f_2(y)$

1.24 P.I. OF NON-HOMOGENEOUS LINEAR PARTIAL DIFFERENTIAL EQUATION WITH CONSTANT COEFFICIENTS

Let the given equation be $\phi(D, D')z = F(x, y)$

then $$\text{P.I.} = \frac{1}{\phi(D, D')} F(x, y)$$

The methods of finding out P.I. of these equations quite resemble to those of ordinary linear differential equation with constant coefficients.

1.24.1. Case I. *When* $F(x, y) = e^{ax+by}$ *and* $\phi(a, b) \neq 0$

$$\text{P.I.} = \frac{1}{\phi(D, D')} e^{ax+by}$$

$$= \frac{1}{\phi(a, b)} e^{ax+by} \qquad \text{| Replacing D by } a \text{ and D}' \text{ by } b$$

ILLUSTRATIVE EXAMPLES

Example 1. *Solve:* $s + ap + bq + abz = e^{mx + ny}$.

Sol. The given equation is

$$(DD' + aD + bD' + ab)z = e^{mx + ny}$$

$$\Rightarrow \qquad (D + b)(D' + a)z = e^{mx + ny}$$

Its $\quad \text{C.F.} = e^{-bx} f_1(y) + e^{-ay} f_2(x)$

$$\text{P.I.} = \frac{1}{(D+b)(D'+a)} e^{mx + ny} = \frac{e^{mx + ny}}{(m+b)(n+a)}$$

Hence complete solution is

$$z = \text{C.F.} + \text{P.I.} = e^{-bx} f_1(y) + e^{-ay} f_2(x) + \frac{e^{mx + ny}}{(m+b)(n+a)}$$

where f_1 and f_2 are arbitrary functions.

Example 2. *Solve:* $(D^3 - 3DD' + D' + 4)z = e^{2x + y}$.

Sol. Here $D^3 - 3DD' + D' + 4$ cannot be resolved into linear factors in D and D′.

Let $\qquad z = Ae^{hx + ky}$

$$\therefore \qquad (D^3 - 3DD' + D' + 4)z = A(h^3 - 3hk + k + 4)\, e^{hx+ky}$$

Then $\quad (D^3 - 3DD' + D' + 4)z = 0$ iff $h^3 - 3hk + k + 4 = 0$

$\therefore$ The C.F. is $z = \Sigma\, Ae^{hx + ky}$, where $h^3 - 3hk + k + 4 = 0$

$$\text{P.I.} = \frac{1}{D^3 - 3DD' + D' + 4} e^{2x + y} = \frac{e^{2x + y}}{2^3 - 3(2)(1) + 1 + 4} = \frac{1}{7} e^{2x + y}$$

Hence complete solution is $z = \Sigma\, Ae^{hx + ky} + \frac{1}{7} e^{2x + y}$, where $h^3 - 3hk + k + 4 = 0$.

Example 3. *Solve:* $D(D - 2D' - 3)z = e^{x + 2y}$.

Sol. $\qquad \text{C.F.} = f_1(y) + e^{3x} f_2(y + 2x)$

$$\text{P.I.} = \frac{1}{D(D - 2D' - 3)} e^{x+2y} = \frac{e^{x + 2y}}{1\{1 - 2(2) - 3\}} = -\frac{1}{6} e^{x+2y}$$

Hence the complete solution is

$$z = \text{C.F.} + \text{P.I.} = f_1(y) + e^{3x} f_2(y + 2x) - \frac{1}{6} e^{x + 2y}$$

where f_1 and f_2 are arbitrary functions.

TEST YOUR KNOWLEDGE

Solve:

1. $(D^3 - 3DD' + D + 1)z = e^{2x+3y}$

2. $(D^2 - 4DD' + D - 1)z = e^{3x-2y}$

3. $(D^2 - D'^2 - 3D + 3D')z = e^{x-2y}$

4. $(D - D' - 1)(D + D' - 2)z = e^{2x-y}$

5. $(D^2 - D'^2 + D - D')z = e^{2x+3y}$

6. $(D^2 - 4DD' + 4D'^2 - D + 2D')z = e^{3x+4y}$. *(U.P.T.U. 2013)*

Answers

1. $z = \Sigma\, Ae^{hx+ky} - \frac{1}{7}e^{2x+3y}$, where $h^3 - 3hk + h + 1 = 0$

2. $z = \Sigma\, Ae^{hx+ky} + \frac{1}{35}e^{3x-2y}$, where $h^2 - 4hk + h - 1 = 0$

3. $z = f_1(y + x) + e^{3x} f_2(y - x) - \frac{1}{12}e^{x-2y}$

4. $z = e^x f_1(y + x) + e^{2x} f_2(y - x) - \frac{1}{2}e^{2x-y}$

5. $z = f_1(y + x) + e^{-x} f_2(y - x) - \frac{1}{6}e^{2x+3y}$

6. $z = f_1(y + 2x) + e^x f_2(y + 2x) + \frac{1}{30}e^{3x+4y}$.

1.24.2. Case II. *When* $F(x, y) = \sin(ax + by)$ *or* $\cos(ax + by)$

$$\text{P.I.} = \frac{1}{\phi(D, D')}\{\sin(ax + by) \text{ or } \cos(ax + by)\}$$

$$= \frac{1}{\phi(D^2, DD', D'^2)}\{\sin(ax + by) \text{ or } \cos(ax + by)\}$$

$$= \frac{1}{\phi(-a^2, -ab, -b^2)}\{\sin(ax + by) \text{ or } \cos(ax + by)\}$$

where $\phi(-a^2, -ab, -b^2) \neq 0$. It is to be noted that here D^2 is replaced by $-a^2$, DD' is replaced by $-ab$ and D'^2 is replaced by $-b^2$.

If $\phi(D, D') = \phi(D^2, DD', D'^2, D, D')$, then

$$\text{P.I.} = \frac{1}{\phi(-a^2, -ab, -b^2, D, D')}\{\sin(ax + by) \text{ or } \cos(ax + by)\}$$

which can be evaluated further by operating N^r and D^r by the suitable conjugate operator.

ILLUSTRATIVE EXAMPLES

Example 1. *Solve:* $(D^2 - DD' + D' - 1)z = \sin(x + 2y)$.

Sol. The given equation is $(D^2 - DD' + D' - 1)z = \sin(x + 2y)$

$$\Rightarrow \quad \{(D + 1)(D - 1) - D'(D - 1)\}z = \sin(x + 2y)$$

$$\Rightarrow \quad (D - 1)(D - D' + 1)z = \sin(x + 2y)$$

$$\therefore \quad \text{C.F.} = e^x f_1(y) + e^{-x} f_2(y + x)$$

$$\text{P.I.} = \frac{1}{D^2 - DD' + D' - 1} \sin(x + 2y) = \frac{1}{-1 + 2 + D' - 1} \sin(x + 2y)$$

$$= \frac{1}{D'} \sin(x + 2y) = -\frac{\cos(x + 2y)}{2}$$

Hence complete solution is

$$z = \text{C.F.} + \text{P.I.} = e^{x} f_1(y) + e^{-x} f_2(y + x) - \frac{\cos(x + 2y)}{2}$$

where f_1 and f_2 are arbitrary functions.

Example 2. *Solve : $(D - D' - 1)(D - D' - 2)z = \sin(2x + 3y)$.* (U.K.T.U. 2011)

Sol. C.F. $= e^{x} f_1(y + x) + e^{2x} f_2(y + x)$

$$\text{P.I.} = \frac{1}{(D - D' - 1)(D - D' - 2)} \sin(2x + 3y)$$

$$= \frac{1}{D^2 - 2DD' + D'^2 - 3D + 3D' + 2} \sin(2x + 3y)$$

$$= \frac{1}{-4 + 12 - 9 - 3D + 3D' + 2} \sin(2x + 3y)$$

$$= \frac{1}{-3D + 3D' + 1} \sin(2x + 3y)$$

$$= -\left[\frac{(3D - 3D') + 1}{\{(3D - 3D') + 1\}\{3D - 3D' - 1\}} \sin(2x + 3y)\right]$$

$$= -\left[\frac{(3D - 3D') + 1}{9D^2 + 9D'^2 - 18DD' - 1} \sin(2x + 3y)\right]$$

$$= -\left[\frac{3D - 3D' + 1}{-36 - 81 + 108 - 1} \sin(2x + 3y)\right]$$

$$= \frac{1}{10}(3D - 3D' + 1) \sin(2x + 3y)$$

$$= \frac{1}{10}[6\cos(2x + 3y) - 9\cos(2x + 3y) + \sin(2x + 3y)]$$

$$= \frac{1}{10}[\sin(2x + 3y) - 3\cos(2x + 3y)]$$

Hence the complete solution is

$$z = \text{C.F.} + \text{P.I.} = e^{x} f_1(y + x) + e^{2x} f_2(y + x) + \frac{1}{10}[\sin(2x + 3y) - 3\cos(2x + 3y)]$$

where f_1 and f_2 are arbitrary functions.

Example 3. *Find the particular integral of $2s + t - 3q = 5\cos(3x - 2y)$.*

Sol. The given equation is

$$(2DD' + D'^2 - 3D')z = 5\cos(3x - 2y)$$

$$\text{P.I.} = \frac{1}{2DD' + D'^2 - 3D'} 5\cos(3x - 2y)$$

$$= \frac{1}{2(6) + (-4) - 3D'} 5\cos(3x - 2y) = \frac{1}{(8 - 3D')} 5\cos(3x - 2y)$$

$$= 5\left[\frac{8 + 3D'}{64 - 9D'^2}\cos(3x - 2y)\right] = 5\left[\frac{8 + 3D'}{64 - 9(-4)}\cos(3x - 2y)\right]$$

$$= \frac{1}{20}[8\cos(3x - 2y) + 6\sin(3x - 2y)]$$

$$= \frac{1}{10}[4\cos(3x - 2y) + 3\sin(3x - 2y)].$$

TEST YOUR KNOWLEDGE

Solve the following partial differential equations:

1. $(D^2 - DD' + D' - 1)z = \cos(x + 2y)$
2. $(D^2 - DD' - 2D'^2 + 2D + 2D')z = \sin(2x + y)$ *(A.K.T.U. 2016)*
3. $(D^2 + DD' + D' - 1)z = \sin(x + 2y)$
4. $(D^2 - DD' - 2D)z = \sin(3x + 4y) - e^{2x+y}$
5. $(D - D'^2)z = \cos(x - 3y)$
6. $(D^2 + 2DD' + D'^2 - 2D - 2D')z = \sin(x + 2y)$.

Answers

1. $z = e^x f_1(y) + e^{-x} f_2(y + x) + \frac{1}{2}\sin(x + 2y)$
2. $z = f_1(y - x) + e^{-2x} f_2(2x + y) - \frac{1}{6}\cos(2x + y)$
3. $z = e^{-x} f_1(y) + e^x f_2(y - x) - \frac{1}{10}[2\sin(x + 2y) + \cos(x + 2y)]$
4. $z = f_1(y) + e^{2x} f_2(y + x) + \frac{1}{15}[\sin(3x + 4y) + 2\cos(3x + 4y)] + \frac{1}{2}e^{2x+y}$
5. $z = \Sigma A e^{k^2 x + ky} + \frac{1}{82}[\sin(x - 3y) + 9\cos(x - 3y)]$
6. $z = f_1(y - x) + e^{2x} f_2(y - x) + \frac{1}{39}[2\cos(x + 2y) - 3\sin(x + 2y)]$.

1.24.3. Case III. *When $F(x, y) = x^m y^n$* where m and n being positive integers

$$\text{P.I.} = \frac{1}{\phi(D, D')} x^m y^n = [\phi(D, D')]^{-1}(x^m y^n)$$

which can be evaluated after expanding $[\phi(D, D')]^{-1}$ in ascending powers of $\frac{D'}{D}$ (when $m > n$) or $\frac{D}{D'}$ (when $m < n$) or D or D′ as the case may be.

If a separate constant is present in $\phi(D, D')$ then it should be given preference in taking the term outside the bracket. It is to be noted that if P.I. is obtained by expanding $\phi(D, D')$ in two or more different ways, then difference in P.I. will be immaterial.

ILLUSTRATIVE EXAMPLES

Example 1. *Solve the linear partial differential equation:*

$$(D - D' - 1)(D - D' - 2)z = e^{3x-y} + x.$$

Sol. $\text{C.F.} = e^{x} f_1(y + x) + e^{2x} f_2(y + x)$

$$\text{P.I.} = \frac{1}{(D - D' - 1)(D - D' - 2)} e^{3x-y} + \frac{1}{(D - D' - 1)(D - D' - 2)}(x) = P_1 + P_2$$

where
$$P_1 = \frac{1}{(D - D' - 1)(D - D' - 2)} e^{3x-y} = \frac{1}{(3 + 1 - 1)(3 + 1 - 2)} e^{3x-y} = \frac{1}{6} e^{3x-y},$$

$$P_2 = \frac{1}{(D - D' - 1)(D - D' - 2)}(x) = \frac{1}{(1 - D + D')(2 - D + D')}(x)$$

$$= \frac{1}{\{1 - (D - D')\}2\left\{1 - \left(\frac{D - D'}{2}\right)\right\}}(x)$$

$$= \frac{1}{2}\left[\{1 - (D - D')\}^{-1}\left\{1 - \left(\frac{D - D'}{2}\right)\right\}^{-1}\right](x)$$

$$= \frac{1}{2}\left[(1 + D - D')\left(1 + \frac{D - D'}{2}\right)\right](x) \qquad | \text{ Leaving higher powers}$$

$$= \frac{1}{2}\left[1 + \frac{D}{2} - \frac{D'}{2} + D + \frac{D^2}{2} - \frac{DD'}{2} - D' - \frac{D'D}{2} + \frac{D'^2}{2}\right](x)$$

$$= \frac{1}{2}\left[x + \frac{1}{2} - 0 + 1 + 0 - 0 - 0 - 0 - 0 + 0\right] = \frac{1}{2}\left(x + \frac{3}{2}\right)$$

$\therefore$
$$\text{P.I.} = \frac{1}{6} e^{3x-y} + \frac{1}{2}\left(x + \frac{3}{2}\right).$$

Hence the complete solution is

$$z = \text{C.F.} + \text{P.I.} = e^{x} f_1(y + x) + e^{2x} f_2(y + x) + \frac{1}{6} e^{3x-y} + \frac{1}{2}\left(x + \frac{3}{2}\right)$$

where f_1 and f_2 are arbitrary functions.

Example 2. *Solve the linear partial differential equation*

$$(D + D' - 1)(D + 2D' - 3)z = 4 + 3x + 6y.$$

Sol. C.F. $= e^{x} f_1(y - x) + e^{3x} f_2(y - 2x)$

$$\text{P.I.} = \frac{1}{(D + D' - 1)(D + 2D' - 3)}(4 + 3x + 6y)$$

$$= \frac{1}{(1 - D - D')(3 - D - 2D')}(4 + 3x + 6y)$$

$$= \frac{1}{\{1 - (D + D')\}\, 3\left\{1 - \left(\frac{D + 2D'}{3}\right)\right\}}(4 + 3x + 6y)$$

$$= \frac{1}{3}\left[\{1 - (D + D')\}^{-1}\left\{1 - \left(\frac{D + 2D'}{3}\right)\right\}^{-1}\right](4 + 3x + 6y)$$

$$= \frac{1}{3}\left[(1 + D + D')\left(1 + \frac{D}{3} + \frac{2D'}{3}\right)\right](4 + 3x + 6y)$$

$$= \frac{1}{3}\left[1 + \frac{4D}{3} + \frac{5D'}{3} + \frac{D^2}{3} + DD' + \frac{2D'^2}{3}\right](4 + 3x + 6y)$$

$$= \frac{1}{3}\left[(4 + 3x + 6y) + \frac{4}{3}(3) + \frac{5}{3}(6) + 0 + 0 + 0\right] = 6 + x + 2y.$$

Hence the complete solution is

$$z = \text{C.F.} + \text{P.I.} = e^{x} f_1(y - x) + e^{3x} f_2(y - 2x) + 6 + x + 2y$$

where f_1 and f_2 are arbitrary functions.

Example 3. *Solve the partial differential equation*

$$(D^2 - D'^2 - 3D + 3D')z = xy + e^{x + 2y}.$$ [G.B.T.U. (C.O.) 2011]

Sol. The given equation is

$$(D^2 - D'^2 - 3D + 3D')\,z = 0$$

$$(D - D')(D + D' - 3)z = 0$$

$\therefore$ C.F. $= f_1(y + x) + e^{3x} f_2(y - x)$

P.I. corresponding to $xy = \dfrac{1}{(D - D')(D + D' - 3)}(xy)$

$$= -\frac{1}{3D}\left(1 - \frac{D'}{D}\right)^{-1}\left(1 - \frac{D}{3} - \frac{D'}{3}\right)^{-1}(xy)$$

$$= -\frac{1}{3D}\left(1 + \frac{D'}{D} + \ldots\right)\left(1 + \frac{D}{3} + \frac{D'}{3} + \frac{2DD'}{9} + \ldots\right)(xy)$$

$$= -\frac{1}{3D}\left(1 + \frac{D}{3} + \frac{D'}{3} + \frac{D'}{D} + \frac{D'}{3} + \frac{2DD'}{9} + \ldots\right)(xy)$$

$$= -\frac{1}{3D}\left(xy + \frac{y}{3} + \frac{2}{3}x + \frac{x^2}{2} + \frac{2}{9}\right) \qquad | \text{ Leaving higher powers}$$

$$= -\frac{1}{3}\left[\frac{x^2}{2}y + \frac{xy}{3} + \frac{x^2}{3} + \frac{x^3}{6} + \frac{2}{9}x\right]$$

P.I. corresponding to e^{x+2y}

$$= \frac{1}{(D - D')(D + D' - 3)} e^{x+2y} = \frac{1}{(1-2)(D + D' - 3)} e^{x+2y}$$

$$= -e^{x+2y} . \frac{1}{D + 1 + D' + 2 - 3} (1) = -e^{x+2y} . \frac{1}{D + D'} e^{0x+0y}$$

$$= -e^{x+2y} . x\, e^{0x+0y} = -x\, e^{x+2y}$$

$$\therefore \qquad \text{P.I.} = -\frac{1}{3}\left[\frac{x^2y}{2} + \frac{xy}{3} + \frac{x^2}{3} + \frac{x^3}{6} + \frac{2x}{9}\right] - x\, e^{x+2y}$$

Hence the complete solution is

$$z = \text{C.F.} + \text{P.I.} = f_1(y + x) + e^{3x} f_2(y - x) - \frac{1}{3}\left[\frac{x^2y}{2} + \frac{xy}{3} + \frac{x^2}{3} + \frac{x^3}{6} + \frac{2x}{9}\right] - x\, e^{x+2y}$$

where f_1 and f_2 are arbitrary funtions.

Example 4. *Solve*: $s + p - q = z + xy$.

Sol. The given equation is

$$(DD' + D - D' - 1)z = xy$$

$$\Rightarrow \qquad (D - 1)(D' + 1)z = xy$$

$$\therefore \qquad \text{C.F.} = e^x f_1(y) + e^{-y} f_2(x)$$

$$\text{P.I.} = \frac{1}{(D - 1)(D' + 1)} (xy) = -[(1 - D)^{-1} (1 + D')^{-1}] (xy)$$

$$= -[(1 + D + D^2 + ...) (1 - D' + ...)] (xy)$$

$$= -(1 + D - D' - DD')\, xy \qquad | \text{ Leaving higher powers}$$

$$= -(xy + y - x - 1) = x + 1 - xy - y$$

Hence the complete solution is

$$z = \text{C.F.} + \text{P.I.} = e^x f_1(y) + e^{-y} f_2(x) + x + 1 - xy - y$$

where f_1 and f_2 are arbitrary functions.

Example 5. *Solve:*

(*a*) $r - s + p = 1$ (*b*) $D (D + D' - 1) (D + 3D' - 2)z = x^2 - 4xy + 2y^2$.

Sol. (*a*) The given equation is

$$(D^2 - DD' + D)z = 1$$

$$\Rightarrow \qquad D(D - D' + 1)z = 1$$

$$\therefore \qquad \text{C.F} = f_1(y) + e^{-x} f_2 (y + x)$$

$$\text{P.I.} = \frac{1}{D(D - D' + 1)}(1) = \frac{1}{D}[1 + (D - D')]^{-1}(1) = \frac{1}{D}(1) = x$$

$\therefore$ Complete solution is

$$z = \text{C.F} + \text{P.I.} = f_1(y) + e^{-x} f_2(y + x) + x$$

where f_1 and f_2 are arbitrary functions.

(*b*) $\text{C.F.} = f_1(y) + e^x f_2(y - x) + e^{2x} f_3(y - 3x)$

$$\text{P.I.} = \frac{1}{D(D + D' - 1)(D + 3D' - 2)}(x^2 - 4xy + 2y^2)$$

$$= \frac{1}{2D}\{1 - (D + D')\}^{-1}\left\{1 - \frac{D + 3D'}{2}\right\}^{-1}(x^2 - 4xy + 2y^2)$$

$$= \frac{1}{2D}\{1 + D + D' + (D + D')^2 + \ldots\} . \left\{1 + \frac{D + 3D'}{2} + \left(\frac{D + 3D'}{2}\right)^2 + \ldots\right\}(x^2 - 4xy + 2y^2)$$

$$= \frac{1}{2D}\left[1 + \frac{3D}{2} + \frac{5D'}{2} + \frac{7D^2}{4} + \frac{19D'^2}{4} + \frac{11DD'}{2} + \ldots\right](x^2 - 4xy + 2y^2)$$

$$= \frac{1}{2D}\left[x^2 - 4xy + 2y^2 + 3(x - 2y) + 5(2y - 2x) + \frac{7}{2} + 19 - 22\right]$$

$$= \frac{1}{2D}\left(x^2 - 4xy + 2y^2 - 7x + 4y + \frac{1}{2}\right) = \frac{1}{2}\left(\frac{x^3}{3} - 2x^2y + 2y^2x - \frac{7x^2}{2} + 4yx + \frac{x}{2}\right)$$

Hence the complete solution is

$$z = \text{C.F.} + \text{P.I.} = f_1(y) + e^x f_2(y - x) + e^{2x} f_3(y - 3x) + \frac{1}{6}x^3 - x^2y + xy^2 - \frac{7}{4}x^2 + 2xy + \frac{x}{4}$$

where f_1, f_2 and f_3 are arbitrary functions.

TEST YOUR KNOWLEDGE

Solve:

1. $(D - D' - 1)(D - D' - 2)z = e^{2x-y} + x$
2. $(D + D' - 1)(D + 2D' - 3)z = 2x + 3y$
3. $(D^2 - D' - 1)z = x^2y$
4. $(D + D' - 1)^2 z = xy$ [*G.B.T.U. (AG) 2011*]
5. $(D^2 - DD' - 2D'^2 + 2D + 2D')z = e^{2x + 3y} + \sin(2x + y) + xy.$

Answers

1. $z = e^x f_1(y + x) + e^{2x} f_2(y + x) + \dfrac{e^{2x - y}}{2} + \dfrac{2x + 3}{4}$
2. $z = e^x f_1(y - x) + e^{3x} f_2(y - 2x) + \dfrac{2}{3}x + y + \dfrac{23}{9}$
3. $z = \Sigma A e^{hx + (h^2 - 1)y} + x^2 - x^2y - 2y + 4$

4. $z = e^x f_1(y-x) + xe^x f_2(y-x) + xy + 2y + 2x + 6$

5. $z = f_1(y-x) + e^{-2x} f_2(y+2x) - \frac{1}{10} e^{2x+3y} - \frac{1}{6}\cos(2x+y) + \frac{x^2y}{4} - \frac{xy}{4} - \frac{x^3}{12} + \frac{3x^2}{8} - \frac{x}{2}.$

1.24.4. Case IV. When $F(x, y) = e^{ax+by}$. V, where V is a function of x and y

$$\text{P.I.} = \frac{1}{\phi(D, D')} e^{ax+by} . V = e^{ax+by} . \frac{1}{\phi(D+a, D'+b)} V$$

which can be evaluated further by using any one of the previous cases discussed.

Remark. V can be either

(*i*) $e^{a_1x + b_1y}$ (*ii*) $\sin(ax+by)$ or $\cos(ax+by)$

(*iii*) $x^m y^n$ (*iv*) constant (say 1)

or any other function of x and y in some special cases.

ILLUSTRATIVE EXAMPLES

Example 1. *Solve:* $(D - 3D' - 2)^3 z = 6e^{2x} \sin(3x+y)$.

Sol. Here C.F. $= e^{2x} f_1(y+3x) + xe^{2x} f_2(y+3x) + x^2e^{2x} f_3(y+3x)$

$$\text{P.I.} = \frac{1}{(D-3D'-2)^3} 6e^{2x}\sin(3x+y)$$

$$= 6e^{2x} \frac{1}{(D+2-3D'-2)^3}\sin(3x+y) = 6e^{2x}\frac{1}{(D-3D')^3}\sin(3x+y)$$

$$= 6e^{2x} . \frac{x^3}{6}\sin(3x+y) = x^3e^{2x}\sin(3x+y)$$

Hence complete solution is

$$z = e^{2x} f_1(y+3x) + xe^{2x} f_2(y+3x) + x^2e^{2x} f_3(y+3x) + x^3e^{2x}\sin(3x+y)$$

where f_1, f_2 and f_3 are arbitrary functions.

Example 2. *Solve:* $(D - 3D' - 2)^2 z = 2e^{2x} \tan(y+3x)$.

Sol. C.F. $= e^{2x} f_1(y+3x) + xe^{2x} f_2(y+3x)$

$$\text{P.I.} = \frac{1}{(D-3D'-2)^2} 2e^{2x}\tan(y+3x)$$

$$= 2e^{2x} . \frac{1}{(D+2-3D'-2)^2}\tan(y+3x) = 2e^{2x} . \left[\frac{1}{(D-3D')^2}\tan(y+3x)\right]$$

$$= 2e^{2x} . \left[x . \frac{1}{2(D-3D')}\tan(y+3x)\right] = 2e^{2x} . x^2 . \frac{1}{2}\tan(y+3x)$$

$$= x^2 e^{2x}\tan(y+3x)$$

Hence complete solution is

$$z = \text{C.F.} + \text{P.I.} = e^{2x} f_1(y+3x) + xe^{2x} f_2(y+3x) + x^2 e^{2x}\tan(y+3x)$$

where f_1 and f_2 are arbitrary functions.

Example 3. *Solve the linear partial differential equation*

$$(D^2 - DD' + D' - 1)\, z = \cos (x + 2y) + e^y.$$

Sol. The given equation is

$$(D^2 - DD' + D' - 1)z = \cos (x + 2y) + e^y$$

$$\Rightarrow \quad (D - 1)(D - D' + 1)z = \cos (x + 2y) + e^y$$

$$\text{C.F.} = e^x f_1 (y) + e^{-x} f_2 (y + x)$$

$$\text{P.I.} = \frac{1}{D^2 - DD' + D' - 1} \cos (x + 2y) + \frac{1}{(D - D' + 1)(D - 1)} e^y = P_1 + P_2$$

where
$$P_1 = \frac{1}{D^2 - DD' + D' - 1} \cos (x + 2y)$$

$$= \frac{1}{-(1)^2 - (-2) + D' - 1} \cos (x + 2y) = \frac{1}{D'} \cos (x + 2y) = \frac{\sin (x + 2y)}{2}$$

and
$$P_2 = \frac{1}{D - D' + 1}\left[\frac{1}{D - 1} e^y\right] = \frac{1}{D - D' + 1}\left[\frac{1}{0 - 1} e^y\right]$$

$$= \frac{1}{D - D' + 1}(-e^y) = -e^y \cdot \frac{1}{(D + 0) - (D' + 1) + 1} (1)$$

$$= -e^y \cdot \frac{1}{D - D'} (1) = -e^y \cdot \frac{1}{D - D'} (e^{0x + 0y}) = -e^y \cdot x \cdot e^{0x + 0y} = -xe^y.$$

$$\therefore \quad \text{P.I.} = \frac{1}{2} \sin (x + 2y) - xe^y.$$

Hence the complete solution is

$$z = \text{C.F.} + \text{P.I.} = e^x f_1 (y) + e^{-x} f_2 (y + x) + \frac{1}{2} \sin (x + 2y) - xe^y$$

where f_1 and f_2 are arbitrary functions.

Example 4. *Solve:* $r - 4s + 4t + p - 2q = e^{x+y}$.

Sol. The given equation is

$$(D^2 - 4DD' + 4D'^2 + D - 2D')z = e^{x+y}$$

$$\Rightarrow \quad [(D - 2D')^2 + (D - 2D')]z = e^{x+y}$$

$$\Rightarrow \quad (D - 2D')(D - 2D' + 1)z = e^{x+y}$$

$$\therefore \quad \text{C.F.} = f_1 (y + 2x) + e^{-x} f_2 (y + 2x)$$

$$\text{P.I.} = \frac{1}{(D - 2D')(D - 2D' + 1)} e^{x+y} = \frac{1}{D - 2D' + 1}\left[\frac{1}{D - 2D'} e^{x+y}\right]$$

$$= \frac{1}{D - 2D' + 1}\left[\frac{1}{1 - 2} \int e^u \, du\right] \qquad \text{where } x + y = u$$

$$= -\left[\frac{1}{D - 2D' + 1} e^{x+y}\right] = -e^{x+y} \cdot \frac{1}{D + 1 - 2(D' + 1) + 1}(1)$$

$$= -e^{x+y}\left[\frac{1}{D - 2D'}(1)\right] = -e^{x+y}\left[\frac{1}{D - 2D'}(e^{0x+0y})\right]$$

$$= -xe^{x+y}.$$

Hence the complete solution is

$$z = \text{C.F.} + \text{P.I.} = f_1(y + 2x) + e^{-x} f_2(y + 2x) - xe^{x+y}$$

where f_1 and f_2 are arbitrary functions.

Example 5. *Find the particular integral of* $(D^2 - D')z = xe^{ax+a^2y}$.

Sol. $$\text{P.I.} = \frac{1}{D^2 - D'}(xe^{ax+a^2y}) = e^{ax+a^2y} \cdot \frac{1}{(D + a)^2 - (D' + a^2)}(x)$$

$$= e^{ax+a^2y} \cdot \frac{1}{D^2 + 2aD - D'}(x) = e^{ax+a^2y} \cdot \frac{1}{2aD}\left(1 + \frac{D}{2a} - \frac{D'}{2aD}\right)^{-1}(x)$$

$$= e^{ax+a^2y} \cdot \frac{1}{2aD}\left\{1 - \left(\frac{D}{2a} - \frac{D'}{2aD}\right) + \ldots\right\} x = e^{ax+a^2y} \cdot \frac{1}{2aD}\left(x - \frac{1}{2a}\right)$$

$$= e^{ax+a^2y} \cdot \left(\frac{x^2}{4a} - \frac{x}{4a^2}\right).$$

TEST YOUR KNOWLEDGE

Solve the following partial differential equations:

1. $(3D^2 - 2D'^2 + D - 1)z = 4e^{x+y}\cos(x + y)$
2. $(D^2 + DD' + D + D' - 1)z = e^{-2x}(x^2 + y^2)$
3. $(D + D' - 1)(D + D' - 3)(D + D')z = e^{x+y}\sin(2x + y)$
4. $(D^2 - DD' + D' - 1)z = \cos(x + 2y) + e^x$.

Answers

1. $z = \Sigma Ae^{hx+ky} + \frac{4}{3}e^{x+y}\sin(x + y)$, where $3h^2 - 2k^2 + h - 1 = 0$
2. $z = \Sigma Ae^{hx+ky} + \frac{1}{27}e^{-2x}(9x^2 + 9y^2 + 18x + 6y + 14)$, where $h^2 + hk + h + k + 1 = 0$
3. $z = e^x f_1(y - x) + e^{3x} f_2(y - x) + f_3(y - x) + \frac{1}{130}[3\cos(2x + y) - 2\sin(2x + y)]e^{x+y}$
4. $z = e^x f_1(y) + e^{-x} f_2(y + x) + \frac{1}{2}\sin(x + 2y) + \frac{1}{2}xe^x$.

1.25 EQUATIONS REDUCIBLE TO PARTIAL DIFFERENTIAL EQUATIONS WITH CONSTANT COEFFICIENTS

An equation in which the coefficient of derivative of any order say k is a multiple of the variables of the degree k then it can be reduced to partial differential equation with constant coefficients in the following way.

Let $$x = e^X,\ y = e^Y \text{ so that } X = \log x,\ Y = \log y$$

$$\frac{\partial z}{\partial x} = \frac{\partial z}{\partial X}\cdot\frac{\partial X}{\partial x} = \frac{1}{x}\frac{\partial z}{\partial X}$$

or $$x\frac{\partial z}{\partial x} = \frac{\partial z}{\partial X} \quad \therefore \quad x\frac{\partial}{\partial x} = D\left(\equiv \frac{\partial}{\partial X}\right)$$

Now $$x\frac{\partial}{\partial x}\left(x^{k-1}\frac{\partial^{k-1} z}{\partial x^{k-1}}\right) = x^k\frac{\partial^k z}{\partial x^k} + (k-1)x^{k-1}\frac{\partial^{k-1} z}{\partial x^{k-1}}$$

or $$x^k\frac{\partial^k z}{\partial x^k} = \left(x\frac{\partial}{\partial x} - k + 1\right)x^{k-1}\frac{\partial^{k-1} z}{\partial x^{k-1}}$$

Putting $k = 2, 3, \ldots\ldots$, we get

$$x^2\frac{\partial^2 z}{\partial x^2} = (D-1)x\frac{\partial z}{\partial x} = (D-1)Dz$$

$$x^3\frac{\partial^3 z}{\partial x^3} = (D-2)x^2\frac{\partial^2 z}{\partial x^2} = (D-2)(D-1)Dz \text{ etc.}$$

Similarly, $$y\frac{\partial z}{\partial y} = D'z,\quad y^2\frac{\partial^2 z}{\partial y^2} = (D'-1)D'z,\quad y^3\frac{\partial^3 z}{\partial y^3} = (D'-2)(D'-1)D'z \text{ etc.}$$

and $$xy\frac{\partial^2 z}{\partial x\partial y} = DD'z \ldots\ldots$$

Substituting in the given equation, it reduces to $\psi(D, D')z = V$ which is an equation with constant coefficients.

ILLUSTRATIVE EXAMPLES

Example 1. *Solve the linear partial differential equation*

$$x^2\frac{\partial^2 z}{\partial x^2} - 4xy\frac{\partial^2 z}{\partial x\partial y} + 4y^2\frac{\partial^2 z}{\partial y^2} + 6y\frac{\partial z}{\partial y} = x^3y^4.$$

Sol. Put $x = e^X, y = e^Y$ so that $X = \log x$ and $Y = \log y$ and Let $D \equiv \frac{\partial}{\partial X}, D' \equiv \frac{\partial}{\partial Y}$

and $DD' \equiv \frac{\partial^2}{\partial X\partial Y}$ then the given equation reduces to

$$[D(D-1)-4DD'+4D'(D'-1)+6D']z=e^{3X+4Y}$$

$$\Rightarrow \quad [(D^2-4DD'+4D'^2)-(D-2D')]z=e^{3X+4Y}$$

$$\Rightarrow \quad (D-2D')(D-2D'-1)z=e^{3X+4Y}$$

Its
$$\text{C.F.}=f_1(Y+2X)+e^X f_2(Y+2X)$$
$$=f_1(\log y+2\log x)+xf_2(\log y+2\log x)$$
$$=f_1(\log yx^2)+x f_2(\log yx^2)=g_1(yx^2)+x\,g_2(yx^2)$$

$$\text{P.I.}=\frac{1}{D-2D'-1}\left[\frac{1}{D-2D'}e^{3X+4Y}\right]$$

$$=\frac{1}{D-2D'-1}\left[\frac{1}{3-8}\int e^u\,du\right] \text{ where } 3X+4Y=u$$

$$=\frac{1}{D-2D'-1}\left[-\frac{1}{5}e^{3X+4Y}\right]=-\frac{1}{5}\left[\frac{1}{D-2D'-1}e^{3X+4Y}\right]$$

$$=-\frac{1}{5}\left[\frac{1}{3-8-1}e^{3X+4Y}\right]=\frac{1}{30}e^{3X+4Y}=\frac{1}{30}x^3y^4.$$

Hence the complete solution is

$$z=\text{C.F.}+\text{P.I.}=g_1(yx^2)+x\,g_2(yx^2)+\frac{1}{30}x^3y^4$$

where g_1 and g_2 are arbitrary functions.

Example 2. *Solve:* $x^2r-y^2t+px-qy=\log x$.

Sol. Let $x=e^X$, $y=e^Y$ so that $X=\log x$ and $Y=\log y$ and let $D\equiv\frac{\partial}{\partial X}$ and $D'\equiv\frac{\partial}{\partial Y}$, then the given equation reduces to

$$[D(D-1)-D'(D'-1)+D-D']\,z=X$$

$$\Rightarrow \quad (D^2-D'^2)z=X \qquad \ldots(1)$$

which is a homogeneous linear p.d.e. with constant coefficients.

$$\therefore \quad \text{C.F.}=\phi_1(Y+X)+\phi_2(Y-X)$$

and
$$\text{P.I.}=\frac{1}{D^2-D'^2}(X)=\frac{1}{(1)^2-(0)^2}\iint u\,du\,du \qquad \text{where } X=u$$

$$=\int\frac{u^2}{2}du=\frac{u^3}{6}=\frac{X^3}{6}$$

Hence solution to (1) is

$$z=\phi_1(Y+X)+\phi_2(Y-X)+\frac{X^3}{6}$$

$$=\phi_1(\log y+\log x)+\phi_2(\log y-\log x)+\frac{(\log x)^3}{6}.$$

Therefore, the complete solution to the given differential equation is

$$z = \phi_1(\log xy) + \phi_2\left(\log \frac{y}{x}\right) + \frac{1}{6}(\log x)^3$$

$$z = f_1(xy) + f_2\left(\frac{y}{x}\right) + \frac{1}{6}(\log x)^3$$

where f_1 and f_2 are arbitrary functions.

Example 3. *Solve:*

(*i*) $x^2 \dfrac{\partial^2 z}{\partial x^2} - y^2 \dfrac{\partial^2 z}{\partial y^2} = xy$ (*ii*) $(x^2 D^2 + 2xy\, DD' + y^2 D'^2)z = x^m y^n$.

Sol. (*i*) Let $x = e^X$, $y = e^Y$ so that $X = \log x$ and $Y = \log y$ and Let $D \equiv \dfrac{\partial}{\partial X}$, $D' \equiv \dfrac{\partial}{\partial Y}$ and $DD' \equiv \dfrac{\partial^2}{\partial X \partial Y}$ then the given equation reduces to

$$[D(D-1) - D'(D'-1)]z = e^{X+Y}$$

$$\Rightarrow \quad (D^2 - D'^2 - D + D')z = e^{X+Y}$$

$$\Rightarrow \quad (D - D')(D + D' - 1)z = e^{X+Y}$$

$$\text{C.F.} = f_1(Y + X) + e^X f_2(Y - X)$$

$$= f_1(\log y + \log x) + x f_2(\log y - \log x) = g_1(xy) + x\, g_2\left(\frac{y}{x}\right)$$

$$\text{P.I.} = \frac{1}{(D-D')(D+D'-1)} e^{X+Y} = \frac{1}{D-D'}\left[\frac{1}{1+1-1} e^{X+Y}\right]$$

$$= \frac{1}{D-D'} e^{X+Y} = X.e^{X+Y} = xy \log x$$

Hence the complete solution is

$$z = \text{C.F.} + \text{P.I.} = g_1(xy) + x\, g_2\left(\frac{y}{x}\right) + xy \log x$$

where g_1 and g_2 are arbitrary functions.

(*ii*) Let $x = e^X, y = e^Y$ so that $X = \log x$, $Y = \log y$ and Let $D \equiv \dfrac{\partial}{\partial X}$, $D' \equiv \dfrac{\partial}{\partial Y}$ and $DD' \equiv \dfrac{\partial^2}{\partial X \partial Y}$ then the given equation reduces to

$$[D(D-1) + 2DD' + D'(D'-1)]z = e^{mX+nY}$$

$$\Rightarrow \quad (D^2 + 2DD' + D'^2 - D - D')z = e^{mX+nY}$$

$$\Rightarrow \quad [(D + D')^2 - (D + D')]z = e^{mX+nY}$$

$$\Rightarrow \quad (D + D')(D + D' - 1)z = e^{mX+nY}$$

$$\therefore \quad \text{C.F.} = f_1(Y - X) + e^X f_2(Y - X)$$

$$= f_1(\log y - \log x) + x f_2(\log y - \log x)$$

$$= f_1\left(\log \frac{y}{x}\right) + x f_2\left(\log \frac{y}{x}\right) = g_1\left(\frac{y}{x}\right) + x\, g_2\left(\frac{y}{x}\right)$$

$$\text{P.I.} = \frac{1}{(D+D')(D+D'-1)} e^{mX+nY}$$

$$= \frac{1}{(m+n)(m+n-1)} e^{mX + nY} = \frac{x^m y^n}{(m+n)(m+n-1)}$$

Hence complete solution is

$$z = \text{C.F.} + \text{P.I.} = g_1 (y/x) + x\, g_2 (y/x) + \frac{x^m y^n}{(m+n)(m+n-1)}$$

where g_1 and g_2 are arbitrary functions.

ASSIGNMENT-I

(2 Marks Questions for Section–A)

1. Show that $z = f(x^2 + y^2)$ is a solution of $y\frac{\partial z}{\partial x} - x\frac{\partial z}{\partial y} = 0$. [*G.B.T.U. (AG) 2012*]
2. Find the particular integral of $(D^2 + DD')z = \sin(x + y)$ where $D = \frac{\partial}{\partial x}$, $D' = \frac{\partial}{\partial y}$. (*G.B.T.U. 2012*)
3. Solve:
   $(D - 5D' + 1)^2 z = 0$ (*U.P.T.U. 2015*)
4. Solve:
   $(D - 5D' + 4)^3 z = 0$ (*U.P.T.U. 2015*)
5. Form the partial differential equation by eliminating a and b from $z = (x^2 + a)(y^2 + b)$. [*G.B.T.U. (AG) 2012*]
6. Formulate the PDE by eliminating the arbitrary function from $\phi(x^2 + y^2, y^2 + z^2) = 0$.
7. Find the particular integral of $(D^2 + DD' - 6D'^2)z = \cos(2x + y)$ (*A.K.T.U. 2016*)
8. Solve: $(D^2 - 2DD' + D'^2)z = 0$ (*U.P.T.U. 2015*)
9. Give two examples of non-linear partial differential equation of the first order. (*M.T.U. 2013*)
10. Find the P.I. of $(D^2 - D'^2)z = \cos(x + y)$.
11. Find the P.I. of $(2D^2 - 3DD' + D'^2)z = e^{x + 2y}$. [*M.T.U. (SUM) 2011, G.B.T.U. 2011*]

42. Find the partial differential equation which is satisfied by the relation $z = c_1 xy + c_2$ where c_1 and c_2 are constants. (*M.T.U. 2013*)

13. Solve: $(D^2 + DD')z = 0$ (*U.P.T.U. 2014*)
14. Form a partial differential equation from $z = (a + x)^2 + y$.
15. Find the solution of $xp + yq = z$.
16. Solve: $p + q = z$.
17. Solve: $(y - z)p + (z - x)q = x - y$.
18. Solve: $\frac{\partial^3 z}{\partial x^3} = 0$.
19. Find the partial differential equation of all spheres whose centres lie on z-axis and given by equations $x^2 + y^2 + (z - a)^2 = b^2$; a and b being constants. (*A.K.T.U. 2017*)
20. Write the complementary function of the partial differential equation $\frac{\partial^2 z}{\partial x^2} - \frac{\partial^2 z}{\partial y^2} = \frac{1}{x^2}$. [*G.B.T.U. (AG) 2011*]

21. Write down the auxiliary equations for the differential equation of the type $Pq + Qq = R$.

22. Find the particular integral of $(D^2 - 2DD' + D'^2)z = e^{x+2y}$ [G.B.T.U. (AG) 2011]

23. Solve: $\frac{\partial^2 z}{\partial x \partial y} = 0$.

24. What is the general solution of I order equation $Pq + Qq = R$? [G.B.T.U. (AG) 2011]

25. Solve: $p - q = \log(x + y)$ (M.T.U. 2011)

26. Solve: $(D^3 + D^2D' - DD'^2 - D'^3)z = 0$.

27. Find the order of the partial differential equation obtained by eliminating F from

$$z = F(x^2 + y^2).$$

28. Solve: $(D^3 - 3D^2D' + 2DD'^2)z = 0$.

29. Solve: $r + 6s + 9t = 0$.

30. Form a partial differential equation by eliminating the constants a and b from $z = (x + a)(y + b)$. [M.T.U. (SUM) 2011]

31. Solve: $p - 2q = \sin(x + 2y)$.

Answers

2. $-\frac{1}{2}\sin(x + y)$

3. $z = e^{-x} f_1(y + 5x) + xe^{-x} f_2(y + 5x)$

4. $z = e^{-4x} f_1(y + 5x) + xe^{-4x} f_2(y + 5x) + x^2 e^{-4x} f_3(y + 5x)$

5. $pq = 4xyz$

6. $yzp - xzq = xy$ where $p = \frac{\partial z}{\partial x}$ and $q = \frac{\partial z}{\partial y}$

7. $\frac{x}{5}\sin(2x + y)$

8. $z = f_1(y + x) + x f_2(y + x)$

10. $\frac{x}{2}\sin(x + y)$

11. $-\frac{x}{2}e^{x+2y}$

12. $y\frac{\partial z}{\partial y} = x\frac{\partial z}{\partial x}$

13. $z = f_1(y) + f_2(y - x)$

14. $z = \frac{1}{4}\left(\frac{\partial z}{\partial x}\right)^2 + y$

15. $f\left(\frac{x}{y}, \frac{y}{z}\right) = 0$

16. $f(x - y, y - \log z) = 0$

17. $f(x + y + z) = x^2 + y^2 + z^2$

18. $z = f_1(y) + xf_2(y) + x^2 f_3(y)$

19. $x\frac{\partial z}{\partial y} - y\frac{\partial z}{\partial x} = 0$

20. $z = f_1(y + x) + f_2(y - x)$

21. $\frac{dx}{P} = \frac{dy}{Q} = \frac{dz}{R}$

22. e^{x+2y}

23. $z = \phi(y) + \psi(x)$

24. $\phi(u, v) = 0;\ u = f_1(x, y, z),\ v = f_2(x, y, z)$

25. $\phi[x + y, z - x\log(x + y)] = 0$

26. $z = f_1(y + x) + f_2(y - x) + xf_3(y - x)$

27. First

28. $z = f_1(y) + f_2(y + x) + f_3(y + 2x)$.

29. $z = f_1(y - 3x) + xf_2(y - 3x)$.

30. $z = pq$

31. $3z = \cos(x + 2y)$

MODULE 2

Applications of Partial Differential Equations

2.1 INTRODUCTION

Many physical and engineering problems when formulated in the mathematical language give rise to partial differential equations. Besides these, partial differential equations also play an important role in the theory of Elasticity, Hydraulics etc.

Since the general solution of a partial differential equation in a region R contains arbitrary constants or arbitrary functions, the unique solution of a partial differential equation corresponding to a physical problem will satisfy certain other conditions at the boundary of the region R. These are known as *boundary conditions.* When these conditions are specified for the time $t = 0$, they are known as *initial conditions.* A partial differential equation together with boundary conditions constitutes a *boundary value problem.*

In the applications of ordinary linear differential equations, we first find the general solution and then determine the arbitrary constants from the initial values. But the same method is not applicable to problems involving partial differential equations. Most of the boundary value problems involving linear partial differential equations can be solved by the method of separation of variables. In this method, right from the beginning, we try to find the particular solutions of the partial differential equation which satisfy all or some of the boundary conditions and then adjust them till the remaining conditions are also satisfied. A combination of these particular solutions gives the solution of the problem.

Fourier series is a powerful aid in determining the arbitrary functions.

2.2 CLASSIFICATION OF LINEAR PARTIAL DIFFERENTIAL EQUATIONS OF SECOND ORDER

2.2.1. Consider the Differential Equation of Second Order in Two Independent Variables x and y as

$$A\frac{\partial^2 u}{\partial x^2} + B\frac{\partial^2 u}{\partial x \partial y} + C\frac{\partial^2 u}{\partial y^2} + f\left(x, y, u, \frac{\partial u}{\partial x}, \frac{\partial u}{\partial y}\right) = 0 \quad ...(1)$$

where A, B, C are constants or continuous functions of x and y possessing continuous partial derivatives and A is positive.

Now equation (1) is

(*i*) **elliptic** if $B^2 - 4AC < 0$

(*ii*) **hyperbolic** if $B^2 - 4AC > 0$

(*iii*) **parabolic** if $B^2 - 4AC = 0$

e.g., one dimensional wave equation $z_{xx} = z_{tt}$ is a hyperbolic equation.

One-dimensional heat flow equation $u_t = u_{xx}$ is parabolic type.
while the two-dimensional heat flow equation in steady state given by $z_{xx} + z_{yy} = 0$ is elliptic in nature.

Remark 1. If A, B, C in the given eqn. (1) are constants, then nature of eqn. (1) will be the same in whole region *i.e.,* for all values of x and y.

Remark 2. If A, B, C are functions of x and y in the given eqn. (1), then nature of eqn. (1) will not be the same in the whole region *i.e.,* for all values of x and y.

ILLUSTRATIVE EXAMPLES

Example 1. *Classify the following operators:*

(*i*) $\dfrac{\partial^2 u}{\partial x^2} + \dfrac{\partial^2 u}{\partial x \partial t} + \dfrac{\partial^2 u}{\partial t^2}$ (*ii*) $4\dfrac{\partial^2 u}{\partial x^2} + 4\dfrac{\partial^2 u}{\partial x \partial t} + \dfrac{\partial^2 u}{\partial t^2}$ (*iii*) $5\dfrac{\partial^2 u}{\partial x^2} - 9\dfrac{\partial^2 u}{\partial x \partial t} + 4\dfrac{\partial^2 u}{\partial t^2}$.

Sol. (*i*) Here the dependent variable u is a function of two independent variables x and t only.

Here, $\quad A = 1, B = 1, C = 1$

$\therefore \quad B^2 - 4AC = (1)^2 - 4(1)(1) = 1 - 4 = -3 < 0$

Hence the given operator is elliptic.

(*ii*) Here $\quad A = 4, B = 4, C = 1$

$\therefore \quad B^2 - 4AC = (4)^2 - 4(4)(1) = 16 - 16 = 0$

Hence the given operator is parabolic.

(*iii*) Here $\quad A = 5, B = -9, C = 4$

$\therefore \quad B^2 - 4AC = (-9)^2 - 4(5)(4) = 81 - 80 = 1 > 0$

Hence the given operator is hyperbolic.

Note. Coefficient of $\dfrac{\partial^2 u}{\partial t^2}$ or u_{tt} may be taken as A.

Example 2. *Find whether the following operators are hyperbolic, parabolic and elliptic?*

(*i*) $x^2\dfrac{\partial^2 u}{\partial t^2} - \dfrac{\partial^2 u}{\partial x^2} + u$ (*ii*) $t\dfrac{\partial^2 u}{\partial t^2} + 2\dfrac{\partial^2 u}{\partial x \partial t} + x\dfrac{\partial^2 u}{\partial x^2} + \dfrac{\partial u}{\partial x}$

(*iii*) $x\dfrac{\partial^2 u}{\partial x^2} + t\dfrac{\partial^2 u}{\partial x \partial t} + \dfrac{\partial^2 u}{\partial t^2}$.

Sol. (*i*) Here $\quad A = x^2, B = 0, C = -1$

Now, $\quad B^2 - 4AC = (0)^2 - 4x^2(-1) = 4x^2$

$\therefore$ The operator is hyperbolic if $4x^2 > 0$ *i.e.,* $x > 0$

parabolic if $4x^2 = 0$ *i.e.,* $x = 0$

Since $4x^2$ being a square, cannot be negative hence operator cannot be elliptic.

(*ii*) Here $\quad A = t, B = 2, C = x$

Now, $\quad B^2 - 4AC = 4 - 4tx$

$\therefore$ The operator is hyperbolic if $4 - 4tx > 0$ *i.e.,* $tx < 1$

elliptic if $4 - 4tx < 0$ *i.e.,* $tx > 1$

and parabolic if $4 - 4tx = 0$ *i.e.,* $tx = 1$.

(*iii*) Here $\quad A = x, B = t, C = 1$

Now $\quad B^2 - 4AC = (t)^2 - 4(x)(1) = t^2 - 4x$

∴ The operator is hyperbolic if $t^2 - 4x > 0$ *i.e.*, $t^2 > 4x$

elliptic if $t^2 - 4x < 0$ *i.e.*, $t^2 < 4x$

and parabolic if $t^2 - 4x = 0$ *i.e.*, $t^2 = 4x$.

Example 3. *Classify the following differential equation as to type in the second quadrant of xy-plane*

$$\sqrt{y^2 + x^2}\; u_{xx} + 2(x - y)\, u_{xy} + \sqrt{y^2 + x^2}\; u_{yy} = 0.$$

Sol. Here $A = \sqrt{y^2 + x^2}$, $B = 2(x - y)$, $C = \sqrt{y^2 + x^2}$

Now $B^2 - 4AC = 4(x - y)^2 - 4(y^2 + x^2) = -8xy$

In second quadrant, y is +ve while x is –ve

∴ $B^2 - 4AC = +\text{ve} > 0$

∴ Differential equation is hyperbolic.

Example 4. *Show that the equation* $\dfrac{\partial^2 u}{\partial t^2} = c^2 \dfrac{\partial^2 u}{\partial x^2}$ *is hyperbolic.*

Sol. The given differential equation is

$$\frac{\partial^2 u}{\partial t^2} - c^2 \frac{\partial^2 u}{\partial x^2} = 0$$

Here, $A = 1,\ B = 0,\ C = -c^2$

Now, $B^2 - 4AC = (0)^2 - 4\,(1)\,(-c^2) = 4c^2$

which is always greater than zero.

Hence, $B^2 - 4AC > 0$

∴ The given equation is hyperbolic.

Example 5. *Classify the partial differential equation*

$$\frac{\partial^2 u}{\partial t^2} + t\frac{\partial^2 u}{\partial x\,\partial t} + x\frac{\partial^2 u}{\partial x^2} + 2\frac{\partial u}{\partial t} + \frac{\partial u}{\partial x} + 6u = 0.$$

Sol. Here, $A = 1,\ B = t,\ C = x.$

Now, $B^2 - 4AC = t^2 - 4(1)(x) = t^2 - 4x.$

The equation is elliptic if $t^2 - 4x < 0.$

The equation is parabolic if $t^2 - 4x = 0.$

The equation is hyperbolic if $t^2 - 4x > 0.$

Example 6. *Classify the partial differential equation*

$$x^2\frac{\partial^2 u}{\partial t^2} + 3\frac{\partial^2 u}{\partial x\,\partial t} + x\frac{\partial^2 u}{\partial x^2} + 17\frac{\partial u}{\partial t} = 100u.$$

Sol. Here, $A = x^2,\ B = 3,\ C = x.$

Now, $B^2 - 4AC = (3)^2 - 4x^2.x = 9 - 4x^3.$

The equation is elliptic if $9 - 4x^3 < 0.$

The equation is parabolic if $9 - 4x^3 = 0.$

The equation is hyperbolic if $9 - 4x^3 > 0.$

Example 7. (*i*) *Show that the equation* $u_{xx} + xu_{yy} + u_y = 0$ *is elliptic for* $x > 0$ *and hyperbolic for* $x < 0$.

(*ii*) *Show that the equation* $z_{xx} + 2xz_{xy} + (1 - y^2)\, z_{yy} = 0$ *is elliptic for values of x and y in the region* $x^2 + y^2 < 1$, *parabolic on the boundary and hyperbolic outside this region.*

(A.K.T.U. 2018,U.K.T.U. 2011)

Sol. (*i*) Here, $A = 1,\ B = 0,\ C = x$

Now, $B^2 - 4AC = -4x$

Equation is elliptic if $B^2 - 4AC < 0$

i.e., if $-4x < 0$ or if $x > 0$

and equation is hyperbolic if $B^2 - 4AC > 0$

i.e., if $-4x > 0$ or if $x < 0$

(*ii*) Here, $A = 1,\ B = 2x,\ C = 1 - y^2$

Now, $B^2 - 4AC = 4x^2 - 4(1 - y^2) = 4(x^2 + y^2 - 1)$

Equation is elliptic if $B^2 - 4AC < 0$

i.e., if $4(x^2 + y^2 - 1) < 0$ or if $x^2 + y^2 < 1$

Parabolic if $B^2 - 4AC = 0$

i.e., if $4(x^2 + y^2 - 1) = 0$ or if $x^2 + y^2 = 1$ (on the boundary)

and hyperbolic if $B^2 - 4AC > 0$

i.e., if $4(x^2 + y^2 - 1) > 0$ or if $x^2 + y^2 > 1$

TEST YOUR KNOWLEDGE

1. Classify the following operators:

(*i*) $4\dfrac{\partial^2 u}{\partial x^2} - 4\dfrac{\partial^2 u}{\partial x\,\partial t} + \dfrac{\partial^2 u}{\partial t^2}$ (*ii*) $t\dfrac{\partial^2 u}{\partial t^2} + 3\dfrac{\partial^2 u}{\partial x\,\partial t} + x\dfrac{\partial^2 u}{\partial x^2} + 17\dfrac{\partial u}{\partial x}$.

2. Classify the following differential equations:

(*a*) $z_{xx} = z_{yy}$ (*b*) $z_{xx} + z_{yy} = 0$ (*c*) $z_{xx} = z_y$ (*d*) $u_{xx} + 3u_{xy} + u_{yy} = 0$. (*M.T.U. 2012*)

3. Classify the equation: $(1 - x^2)\dfrac{\partial^2 z}{\partial x^2} - 2xy\dfrac{\partial^2 z}{\partial x\,\partial y} + (1 - y^2)\dfrac{\partial^2 z}{\partial y^2} - 2z = 0.$ (*A.K.T.U. 2016, 2017*)

4. Classify the equation: $y^2 r - 2xys + x^2 t = \dfrac{y^2}{z}\,p + \dfrac{x^2}{y}\,q.$

5. Characterize the following partial differential equations into elliptic, parabolic and hyperbolic equations $A\dfrac{\partial^2 u}{\partial x^2} + 2B\dfrac{\partial^2 u}{\partial x\partial y} + C\dfrac{\partial^2 u}{\partial y^2} = F(x, y, u, u_x, u_y)$. Here A, B, C may be functions of x and y.

6. Give a consize description of the classification of linear partial differential equations of second order with constant coefficients. Give examples of each type of partial differential equations.

Answers

1. (*i*) parabolic (*ii*) hyperbolic if $xt < \dfrac{9}{4}$, parabolic if $xt = \dfrac{9}{4}$, elliptic if $xt > \dfrac{9}{4}$

2. (*a*) hyperbolic (*b*) elliptic (*c*) parabolic (*d*) hyperbolic

3. hyperbolic in the region $x^2 + y^2 > 1$, parabolic at points on $x^2 + y^2 = 1$, and elliptic in the region $x^2 + y^2 < 1$

4. parabolic.

2.2.2. Consider a Linear P.D.E. of II Order in Three Independent Variables *x*, *y*, *z* as

$$a_{11}\frac{\partial^2 u}{\partial x^2}+a_{12}\frac{\partial^2 u}{\partial x\,\partial y}+a_{13}\frac{\partial^2 u}{\partial x\,\partial z}+a_{21}\frac{\partial^2 u}{\partial y\,\partial x}+a_{22}\frac{\partial^2 u}{\partial y^2}+a_{23}\frac{\partial^2 u}{\partial y\,\partial z}+a_{31}\frac{\partial^2 u}{\partial z\,\partial x}+a_{32}\frac{\partial^2 u}{\partial z\,\partial y}$$
$$+a_{33}\frac{\partial^2 u}{\partial z^2}+b_1\frac{\partial u}{\partial x}+b_2\frac{\partial u}{\partial y}+b_3\frac{\partial u}{\partial z}+cu=0 \qquad ...(1)$$

where a_{ij}, b_i and c are constants or some functions of x, y, z.

Eqn. (1) may be rewritten in the form

$$\sum_{i,j=1}^{3} a_{ij}\frac{\partial^2 u}{\partial x_i\,\partial x_j}+\sum_{i=1}^{3} b_i\frac{\partial u}{\partial x_i}+cu=0 \qquad ...(2)$$

In general, when there are n independent variables $x_1, x_2, x_3, ..., x_n$; we may write as

$$\sum_{i,j=1}^{n} a_{ij}\frac{\partial^2 u}{\partial x_i\,\partial x_j}+\sum_{i=1}^{n} b_i\frac{\partial u}{\partial x_i}+cu=0 \qquad ...(3)$$

Now, let $\delta_i \equiv \dfrac{\partial}{\partial x_i}\ (1 \le i \le n)$ and $\delta_i\delta_j \equiv \dfrac{\partial^2}{\partial x_i\,\partial x_j}\ (1 \le i \le n,\ 1 \le j \le n)$

Consider $\phi = \displaystyle\sum_{i=1}^{n}\sum_{j=1}^{n} a_{ij}\,\delta_i\delta_j$ for all real values of δ_i positive or negative.

At any point $x_1, x_2, ..., x_n$; differential equation (3) is said to be

(*i*) parabolic if Δ vanishes, where $\Delta = \begin{vmatrix} a_{11} & a_{12} & \cdots & a_{1n} \\ a_{21} & a_{22} & \cdots & a_{2n} \\ \vdots & \vdots & \vdots & \vdots \\ \vdots & \vdots & \vdots & \vdots \\ a_{n1} & a_{n2} & \cdots & a_{nn} \end{vmatrix}$.

(*ii*) elliptic if ϕ is positive for all real values of δ_i and it reduces to zero only when all δ_i's are zero.

(*iii*) hyperbolic if ϕ can be both positive or negative.

Note. If a_{ij} are functions of $x_1, x_2, ..., x_n$ the same differential equation can be elliptic, hyperbolic and parabolic at different points while if a_{ij} are constant, the equation will have the same nature throughout.

ILLUSTRATIVE EXAMPLES

Example 1. *Classify the following equation:* $\dfrac{\partial^2 u}{\partial x^2}+\dfrac{\partial^2 u}{\partial y^2}+\dfrac{\partial^2 u}{\partial z^2}=\dfrac{1}{c^2}\dfrac{\partial u}{\partial t}$.

Sol. Here, $a_{11}=1, a_{22}=1, a_{33}=1, a_{44}=0$

$a_{13}=a_{23}=a_{31}=a_{24}=a_{34}=0$, etc.

$$\Delta = \begin{vmatrix} 1 & 0 & 0 & 0 \\ 0 & 1 & 0 & 0 \\ 0 & 0 & 1 & 0 \\ 0 & 0 & 0 & 0 \end{vmatrix} = 0$$

Hence the given differential equation is parabolic.

Example 2. *Classify the equation:* $u_{xx} + 2u_{yy} + u_{zz} = 2u_{xy} + 2u_{yz}$.

Sol. Here, $a_{11} = 1, a_{12} = -2, a_{13} = 0, a_{21} = 0, a_{22} = 2,$
$a_{23} = -2, a_{31} = 0, a_{32} = 0, a_{33} = 1$

$$\therefore \quad \Delta = \begin{vmatrix} 1 & -2 & 0 \\ 0 & 2 & -2 \\ 0 & 0 & 1 \end{vmatrix} \neq 0$$

$$\therefore \quad \phi = \delta_1^2 - 2\delta_1\delta_2 + 2\delta_2^2 - 2\delta_2\delta_3 + \delta_3^2 = (\delta_1 - \delta_2)^2 + (\delta_2 - \delta_3)^2$$

which is definite positive for all real values of $\delta_1, \delta_2, \delta_3$.

Hence the equation is elliptic.

TEST YOUR KNOWLEDGE

Classify the following equations:

1. $u_{xx} + u_{yy} + u_{zz} = 0$ **2.** $u_{xx} + u_{yy} = u_z$ **3.** $u_{xx} + u_{yy} = u_{zz}$

4. $u_{xx} + u_{yy} + u_{zx} + 2u_{yz} = 0$.

Answers

1. Elliptic **2.** Parabolic **3.** Hyperbolic **4.** Parabolic.

2.3 METHOD OF SEPARATION OF VARIABLES *(M.T.U. 2012)*

The most important feature of the method of separation of variables is that it successively replaces the partial differential equation by a system of ordinary differential equations that are usually easy to handle. This method requires specific assumptions and transformation formulae in handling partial differential equations. It is well known that this method is applicable if the equation and the boundary conditions are linear and homogeneous. For inhomogeneous boundary conditions, a transformation formula should be employed to transform inhomogeneous boundary conditions to homogeneous boundary conditions.

In this method, we assume the solution to be the product of two functions, each of which involves only one of the variables. The following examples explain the method.

ILLUSTRATIVE EXAMPLES

Example 1. *Use the method of separation of variables to solve the equation*

$$\frac{\partial u}{\partial x} = 2\frac{\partial u}{\partial t} + u, \text{ given that } u(x, 0) = 6e^{-3x}.$$

[G.B.T.U. (AG) 2011]

Sol. The given equation is

$$\frac{\partial u}{\partial x} = 2\frac{\partial u}{\partial t} + u \quad \text{...(1)}$$

Let $u = XT$ ...(2)

where X is a function of x only and T is a function of t only

Then, $$\frac{\partial u}{\partial x} = \frac{\partial}{\partial x}(XT) = T\frac{dX}{dx}$$

$$\frac{\partial u}{\partial t} = \frac{\partial}{\partial t}(\text{XT}) = \text{X}\frac{d\text{T}}{dt}.$$

Substituting in eqn. (1), we get

$$\text{T}\frac{d\text{X}}{dx} = 2\text{X}\frac{d\text{T}}{dt} + \text{XT}$$

$$\Rightarrow \qquad \text{TX}' = 2\text{XT}' + \text{XT}$$

$$\Rightarrow \qquad \text{TX}' = \text{X}(2\text{T}' + \text{T})$$

$$\Rightarrow \qquad \frac{\text{X}'}{\text{X}} = \frac{2\text{T}'}{\text{T}} + 1 = -p^2 \text{ (say)}$$

(*i*)

$$\frac{\text{X}'}{\text{X}} = -p^2$$

$$\frac{d\text{X}}{dx} + p^2\text{X} = 0$$

$$\frac{d\text{X}}{\text{X}} = -p^2\,dx$$

Integration yields

$$\log \text{X} = -p^2x + \log c_1$$

$$\Rightarrow \qquad \text{X} = c_1\,e^{-p^2x} \qquad ...(3)$$

(*ii*)

$$\frac{2\text{T}'}{\text{T}} = -(p^2 + 1)$$

$$\frac{d\text{T}}{\text{T}} = -\left(\frac{p^2+1}{2}\right)dt$$

Integration yields, $\log \text{T} = -\left(\frac{p^2+1}{2}\right)t + \log c_2$

$$\text{T} = c_2\,e^{-\left(\frac{p^2+1}{2}\right)t} \qquad ...(4)$$

From (2), (3) and (4), we get

$$u = \text{XT} = c_1c_2\,e^{-p^2x-\left(\frac{p^2+1}{2}\right)t}$$

or

$$u(x, t) = c_1c_2\,e^{-p^2x-\left(\frac{p^2+1}{2}\right)t} \qquad ...(5)$$

$$u(x, 0) = c_1c_2\,e^{-p^2x}.$$

$$\Rightarrow \qquad 6e^{-3x} = c_1c_2\,e^{-p^2x}$$

Comparison gives, $c_1c_2 = 6$ and $p^2 = 3$.

Hence from (5), we get

$$u(x, t) = 6e^{-3x-2t}.$$

Example 2. *Use the method of separation of variables to solve the equation*

$$\frac{\partial^2 u}{\partial x^2} - 2\frac{\partial u}{\partial x} + \frac{\partial u}{\partial y} = 0.$$ [G.B.T.U. (A.G.) 2012]

Sol. Let $$u = XY \qquad ...(1)$$

where X is a function of x only and Y is a function of y only.

$$\frac{\partial u}{\partial x} = \frac{\partial}{\partial x}(XY) = Y\frac{dX}{dx}$$

$$\frac{\partial u}{\partial y} = \frac{\partial}{\partial y}(XY) = X\frac{dY}{dy}$$

$$\frac{\partial^2 u}{\partial x^2} = \frac{\partial^2}{\partial x^2}(XY) = Y\frac{d^2X}{dx^2}.$$

Substituting these values in (1), we get

$$Y\frac{d^2X}{dx^2} - 2Y\frac{dX}{dx} + X\frac{dY}{dy} = 0$$

$$\Rightarrow \qquad YX'' - 2YX' + XY' = 0$$

$$\Rightarrow \qquad \frac{X'' - 2X'}{X} + \frac{Y'}{Y} = 0$$

$$\Rightarrow \qquad \frac{X'' - 2X'}{X} = -\frac{Y'}{Y} = -p^2 \text{ (say)}$$

(*i*) $$\frac{X'' - 2X'}{X} = -p^2$$

$$\Rightarrow \qquad X'' - 2X' + p^2X = 0.$$

Auxiliary equation is $$m^2 - 2m + p^2 = 0$$

$$\Rightarrow \qquad m = \frac{2 \pm \sqrt{4 - 4p^2}}{2} = 1 \pm \sqrt{1 - p^2}$$

$$\therefore \qquad \text{C.F.} = c_1\, e^{(1+\sqrt{1-p^2})\,x} + c_2\, e^{(1-\sqrt{1-p^2})\,x}$$

$$\text{P.I.} = 0.$$

Hence, $$X = \text{C.F.} + \text{P.I.} = c_1\, e^{(1+\sqrt{1-p^2})\,x} + c_2\, e^{(1-\sqrt{1-p^2})\,x} \qquad ...(2)$$

(*ii*) $$-\frac{Y'}{Y} = -p^2 \;\Rightarrow\; \frac{dY}{dy} = p^2Y \;\Rightarrow\; \frac{dY}{Y} = p^2dy.$$

Integration yields, $\log Y = p^2y + \log c_3$

$$Y = c_3\, e^{p^2y} \qquad ...(3)$$

$$\therefore \qquad u(x, y) = [c_1e^{(1+\sqrt{1-p^2})\,x} + c_2\, e^{(1-\sqrt{1-p^2})\,x}]\; c_3\, e^{p^2y}.$$

Example 3. *Use the method of separation of variables to solve the equation*

$$\frac{\partial^2 u}{\partial x^2} = \frac{\partial u}{\partial y} + 2u.$$ (U.P.T.U. 2014)

Sol. Let $u = XY$ ...(1)

where X is a function of x only and Y is a function of y only.

$$\frac{\partial u}{\partial y} = \frac{\partial}{\partial y}(XY) = X\frac{dY}{dy}$$

$$\frac{\partial^2 u}{\partial x^2} = \frac{\partial^2}{\partial x^2}(XY) = Y\frac{d^2X}{dx^2}.$$

From (1), $$Y\frac{d^2X}{dx^2} = X\frac{dY}{dy} + 2XY$$

$\Rightarrow$ $$YX'' = XY' + 2XY = X(Y' + 2Y)$$

$\Rightarrow$ $$\frac{X''}{X} = \frac{Y'}{Y} + 2 = -p^2 \text{ (say)}$$

(*i*) $$\frac{X''}{X} = -p^2 \Rightarrow \frac{d^2X}{dx^2} + p^2X = 0$$

Auxiliary eqn. is $m^2 + p^2 = 0$

$$m = \pm pi$$

$\therefore$ $$\text{C.F.} = c_1 \cos px + c_2 \sin px$$

$$\text{P.I.} = 0$$

$\therefore$ $$X = c_1 \cos px + c_2 \sin px \quad ...(2)$$

(*ii*) $$\frac{Y'}{Y} + 2 = -p^2$$

$$\frac{Y'}{Y} = -(p^2 + 2)$$

$$\frac{dY}{Y} = -(p^2 + 2)\,dy.$$

Integration yields, $\log Y = -(p^2 + 2)\,y + \log c_3$

$\Rightarrow$ $$Y = c_3\, e^{-(p^2+2)y} \quad ...(3)$$

$\therefore$ $$u(x, y) = (c_1 \cos px + c_2 \sin px)\, c_3\, e^{-(p^2+2)y}.$$

Example 4. *Solve by the method of separation of variables:*

$$4\frac{\partial u}{\partial t} + \frac{\partial u}{\partial x} = 3u, \quad u = 3e^{-x} - e^{-5x}, \quad \textit{when } t = 0.$$

Sol. Let $u = XT$ ...(1)

where X is a function of x only and T is a function of t only.

$\therefore$ $$\frac{\partial u}{\partial t} = \frac{\partial}{\partial t}(XT) = X\frac{dT}{dt} = XT' \text{ (say)}$$

$$\frac{\partial u}{\partial x} = \frac{\partial}{\partial x}(XT) = T\frac{dX}{dx} = TX' \text{ (say)}$$

$\therefore$ From the given equation

$$4XT' + TX' = 3XT$$

$$\frac{4T'}{T} + \frac{X'}{X} = 3$$

$$\frac{4T'}{T} - 3 = -\frac{X'}{X} = p^2 \text{ (say)} \qquad ...(2)$$

(*i*)
$$\frac{4T'}{T} = p^2 + 3$$

$$\frac{dT}{T} = \left(\frac{3 + p^2}{4}\right) dt$$

Integration yields,

$$\log T = \left(\frac{3 + p^2}{4}\right) t + \log c_1 \quad \Rightarrow \quad T = c_1\, e^{\left(\frac{3+p^2}{4}\right)t} \qquad ...(3)$$

(*ii*)
$$\frac{-X'}{X} = p^2 \quad \Rightarrow \quad \frac{X'}{X} = -p^2 \quad \Rightarrow \quad \frac{dX}{X} = -p^2\, dx$$

Integration yields,

$$\log X = -p^2 x + \log c_2$$

$$X = c_2\, e^{-p^2 x} \qquad ...(4)$$

From (1), we get

$$u = XT = c_1 c_2\, e^{-p^2 x + \left(\frac{3+p^2}{4}\right)t}$$

or
$$u(x, t) = b_n\, e^{-p^2 x + \left(\frac{3+p^2}{4}\right)t}$$

Most general solution is

$$u\,(x, t) = \sum_{n=1}^{\infty} b_n\, e^{-p^2 x + \left(\frac{3+p^2}{4}\right)t} \qquad ...(5)$$

when $t = 0$,
$$u(x, 0) = 3e^{-x} - e^{-5x} = \sum_{n=1}^{\infty} b_n\, e^{-p^2 x}$$

Comparing, when $p^2 = 1, b_1 = 3$ and when $p^2 = 5, b_2 = -1$

Hence, from (5), general solution is

$$u(x, t) = 3e^{-x+t} - e^{-5x+2t}.$$

Example 5. *Use the method of separation of variables to solve the equation*

$\frac{\partial^2 v}{\partial x^2} = \frac{\partial v}{\partial t}$ *given that* $v = 0$ *when* $t \to \infty$ *as well as* $v = 0$ *at* $x = 0$ *and* $x = l$.

(G.B.T.U. 2013)

Sol. Let $v = XT$ ...(1)

where X is a function of x only and T is a function of t only.

$$\frac{\partial v}{\partial t} = \frac{\partial}{\partial t}(XT) = X\frac{dT}{dt} = XT' \text{ (say)}$$

$$\frac{\partial^2 v}{\partial x^2} = \frac{\partial^2}{\partial x^2}(XT) = T\frac{d^2X}{dx^2} = TX'' \text{ (say)}$$

From the given equation

$$TX'' = XT'$$

$$\Rightarrow \quad \frac{T'}{T} = \frac{X''}{X} = -p^2 \text{ (say)} \quad ...(2)$$

(*i*) $$\frac{T'}{T} = -p^2 \quad \Rightarrow \quad \frac{dT}{T} = -p^2 dt$$

Integration yields,

$$\log T = -p^2 t + \log c_1 \quad \Rightarrow \quad T = c_1 e^{-p^2 t} \quad ...(3)$$

(*ii*) $$\frac{X''}{X} = -p^2 \quad \Rightarrow \quad X'' + p^2 X = 0$$

Auxiliary eqn. is $m^2 + p^2 = 0 \quad \Rightarrow \quad m = \pm pi$

$$\text{C.F.} = c_2 \cos px + c_3 \sin px$$

$$\text{P.I.} = 0$$

$$\therefore \quad X = c_2 \cos px + c_3 \sin px \quad ...(4)$$

Hence $$v = XT = c_1 e^{-p^2 t}(c_2 \cos px + c_3 \sin px) \quad ...(5)$$

Putting $x = 0$, $v = 0$ in (5), we get

$$0 = c_1 e^{-p^2 t} . c_2$$

$$\therefore \quad c_2 = 0 \quad (\because \ c_1 \neq 0)$$

Hence from (5), $$v = c_1 c_3 e^{-p^2 t} \sin px \quad ...(6)$$

At $x = l$, $v = 0$

$\therefore$ From (6), $$0 = c_1 c_3 e^{-p^2 t} \sin pl$$

$$\Rightarrow \quad \sin pl = 0 = \sin n\pi \quad (\because \ c_3 \neq 0)$$

$$p = \frac{n\pi}{l}, n \in I$$

$\therefore$ From (6), $$v = c_1 c_3 e^{-\frac{n^2\pi^2 t}{l^2}} \sin \frac{n\pi x}{l} = b_n \sin \frac{n\pi x}{l} . e^{-\left(\frac{n^2\pi^2}{l^2}\right)t} \quad ...(7)$$

The above equation (7) satisfies the given conditions for all values of n. Hence the most general solution is

$$v = \sum_{n=1}^{\infty} b_n e^{-\left(\frac{n^2\pi^2}{l^2}\right)t} \sin \frac{n\pi x}{l}.$$

Example 6. *Solve the following equation by the method of separation of variables*

$$\frac{\partial^2 u}{\partial x \partial t} = e^{-t} \cos x$$

given that u = 0 when t = 0 and $\frac{\partial u}{\partial t} = 0$ *when x = 0.* (U.P.T.U. 2013)

Sol. Let $$u = XT \quad ...(1)$$

where X is a function of x only and T is a function of t only.

Then, $$\frac{\partial u}{\partial t} = \frac{\partial}{\partial t}(XT) = X\frac{dT}{dt}$$

$$\therefore \qquad \frac{\partial^2 u}{\partial x \partial t} = \frac{\partial}{\partial x}\left(X\frac{dT}{dt}\right) = \frac{dT}{dt} \cdot \frac{dX}{dx} \qquad ...(2)$$

Substituting (2) in the given equation, we get

$$\frac{dT}{dt}\frac{dX}{dx} = e^{-t} \cos x$$

$$e^t \frac{dT}{dt} = \frac{\cos x}{\left(\frac{dX}{dx}\right)} = -p^2 \text{ (say)} \qquad ...(3)$$

Now, $e^t \frac{dT}{dt} = -p^2$

$\Rightarrow$ $dT = -p^2 e^{-t}\, dt$

Integration yields,

$$T = p^2 e^{-t} + c_1 \qquad ...(4)$$

Also, $\frac{dX}{dx} = -\frac{1}{p^2} \cos x$

$$dX = -\frac{1}{p^2} \cos x\, dx$$

Integration yields,

$$X = -\frac{1}{p^2} \sin x + c_2 \qquad ...(5)$$

Using (4) and (5), we get from (1),

$$u(x, t) = XT = \left(-\frac{1}{p^2} \sin x + c_2\right)(p^2 e^{-t} + c_1) \qquad ...(6)$$

Applying the condition $u = 0$ when $t = 0$ in (6), we get

$$0 = \left(-\frac{1}{p^2} \sin x + c_2\right)(p^2 + c_1)$$

$\Rightarrow$ $p^2 + c_1 = 0 \Rightarrow c_1 = -p^2$

From (6), $$\frac{\partial u}{\partial t} = \left(-\frac{1}{p^2} \sin x + c_2\right)(-p^2 e^{-t}) \qquad ...(7)$$

Applying the condition $\frac{\partial u}{\partial t} = 0$ when $x = 0$ in (7), we get

$$0 = c_2(-p^2 e^{-t})$$

$\Rightarrow$ $c_2 = 0$

Substituting the values of c_1 and c_2 in (6), we get

$$u(x, t) = -\frac{1}{p^2} \sin x\, (p^2 e^{-t} - p^2) = \sin x\, (1 - e^{-t})$$

Example 7. *Solve the P.D.E. by separation of variables method,*

$$u_{xx} = u_y + 2u,\ u(0, y) = 0,\ \frac{\partial}{\partial x} u(0, y) = 1 + e^{-3y}. \qquad \text{(M.T.U. 2013)}$$

Sol. Let $\quad u = XY \qquad ...(1)$

where X is a function of x only and Y is a function of y only.

$$\therefore \quad \frac{\partial u}{\partial y} = \frac{\partial}{\partial y}(XY) = X\frac{dY}{dy} = XY'$$

$$\frac{\partial^2 u}{\partial x^2} = \frac{\partial^2}{\partial x^2}(XY) = Y\frac{d^2X}{dx^2} = YX''$$

From the given equation,

$$YX'' = XY' + 2XY$$

$$\frac{X''}{X} = \frac{Y' + 2Y}{Y}$$

$$\Rightarrow \quad \frac{X''}{X} = \frac{Y'}{Y} + 2 = k(\text{say}) \qquad ...(2)$$

$$(i) \quad \frac{X''}{X} = k$$

$$\Rightarrow \quad X'' - kX = 0$$

Auxiliary equation is

$$m^2 - k = 0$$

$$\Rightarrow \quad m = \pm\sqrt{k}$$

$$\therefore \quad \text{C.F.} = C_1 e^{\sqrt{k}x} + C_2 e^{-\sqrt{k}x}$$

$$\text{P.I.} = 0$$

$$\therefore \quad X = C_1 e^{\sqrt{k}x} + C_2 e^{-\sqrt{k}x} \qquad ...(3)$$

$$(ii) \quad \frac{Y'}{Y} + 2 = k$$

$$\Rightarrow \quad \frac{Y'}{Y} = k - 2$$

$$\Rightarrow \quad \frac{dY}{Y} = (k-2)\,dy$$

Integration yields,

$$\log Y = (k-2)y + \log C_3$$

$$\Rightarrow \quad Y = C_3 e^{(k-2)y} \qquad ...(4)$$

Hence from (1),

$$u(x, y) = (C_1 e^{\sqrt{k}x} + C_2 e^{-\sqrt{k}x})\, C_3 e^{(k-2)y} \qquad ...(5)$$

Applying the condition $u(0, y) = 0$ in (5), we get

$$u(0, y) = 0 = (C_1 + C_2)\, C_3 e^{(k-2)y}$$

$$\Rightarrow \quad C_1 + C_2 = 0 \ \Rightarrow \ C_2 = -C_1 \qquad ...(6)$$

From (5), most general solution is

$$u(x, y) = \Sigma\, C_1C_3(e^{\sqrt{k}x} - e^{-\sqrt{k}x})\, e^{(k-2)y} \qquad ...(7)$$

$$\frac{\partial u}{\partial x} = \Sigma\, C_1C_3\sqrt{k}(e^{\sqrt{k}x} + e^{-\sqrt{k}x})\, e^{(k-2)y}$$

$$\left(\frac{\partial u}{\partial x}\right)_{x=0} = 1 + e^{-3y} = \Sigma C_1C_3\sqrt{k}(2)\, e^{(k-2)y} = \sum_{n=1}^{\infty} b_n\, e^{(k-2)y}$$

Comparing the coefficients, we get

(*i*) $b_1 = 1, \quad k - 2 = 0$

$$2C_1C_3\sqrt{k} = 1, \quad k = 2$$

$$\therefore \qquad C_1C_3 = \frac{1}{2\sqrt{2}}$$

(*ii*) $b_3 = -1, \quad k - 2 = -3$

$$2C_1C_3\sqrt{k} = 1, \quad k = -1$$

$$\therefore \qquad C_1C_3 = \frac{1}{2i}$$

Hence from (7), the particular solution is

$$u(x, y) = \frac{1}{2\sqrt{2}}(e^{\sqrt{2}x} - e^{-\sqrt{2}x}) + \frac{1}{2i}(e^{ix} - e^{-ix})\, e^{-3y}$$

$$\Rightarrow \qquad u(x, y) = \frac{1}{\sqrt{2}} \sinh \sqrt{2}\, x + e^{-3y} \sin x.$$

TEST YOUR KNOWLEDGE

Solve by the method of separation of variables (1–9):

1. $3\frac{\partial u}{\partial x} + 2\frac{\partial u}{\partial y} = 0;\ u(x, 0) = 4e^{-x}$ (*A.K.T.U. 2017*)
2. $\frac{\partial u}{\partial x} = 4\frac{\partial u}{\partial y};\ u(0, y) = 8e^{-3y}$
3. $4\frac{\partial u}{\partial x} + \frac{\partial u}{\partial y} = 3u;\ u(0, y) = 4e^{-y} - e^{-5y}$ (*A.K.T.U. 2018*)
4. $\frac{\partial u}{\partial t} = \frac{\partial u}{\partial x} - 2u;\ u(x, 0) = 10e^{-x} - 6e^{-4x}$ (*A.K.T.U. 2018*)
5. (*i*) $x\frac{\partial u}{\partial x} + y\frac{\partial u}{\partial y} = 0$ (*U.P.T.U. 2015*) (*ii*) $y^3\frac{\partial u}{\partial x} + x^2\frac{\partial u}{\partial y} = 0$ (*G.B.T.U. 2011*)
6. (*i*) $\frac{\partial u}{\partial x} = 2\frac{\partial u}{\partial t} + u;\quad u(x, 0) = 6e^{-5x}$ (*U.P.T.U. 2015*) (*ii*) $x\frac{\partial^2 u}{\partial x \partial y} + 2yu = 0$ (*U.P.T.U. 2015*)
7. (*i*) $\frac{\partial u}{\partial x} = \frac{\partial u}{\partial y}$ (*G.B.T.U. 2012*) (*ii*) $\frac{\partial^2 u}{\partial x^2} - \frac{\partial u}{\partial y} = 0.$ [*U.P.T.U. (SUM) 2007*]
8. Solve the partial differential equation:

   $2\frac{\partial z}{\partial x} + 3\frac{\partial z}{\partial y} + 5z = 0;\qquad z(0, y) = 2e^{-y}$

   by the method of separation of variables. (*A.K.T.U. 2017*)

9. $\frac{\partial z}{\partial x} + \frac{\partial^2 z}{\partial y^2} = 0$; $z(x, 0) = 0$, $z(x, \pi) = 0$, $z(0, y) = 4 \sin 3y$ *(U.P.T.U. 2014)*

10. Solve $\frac{\partial^2 u}{\partial x^2} = 2u + \frac{\partial u}{\partial y}$ using method of separation of variables subject to the conditions $u = 0$ and $\frac{\partial u}{\partial x} = e^{-3y}$ when $x = 0$ for all values of y. *(G.B.T.U. 2013)*

Answers

1. $u(x, y) = 4e^{-x + \frac{3}{2}y}$

2. $u(x, y) = 8e^{-3y-12x}$

3. $u(x, y) = 4e^{x-y} - e^{2x-5y}$

4. $u(x, t) = 10\, e^{-(x+3t)} - 6e^{-2(2x+3t)}$

5. (*i*) $u(x, y) = c_1c_2 \left(\frac{x}{y}\right)^k$ (*ii*) $u(x, y) = c_1c_2\, e^{p^2\left(\frac{y^4}{4} - \frac{x^3}{3}\right)}$

6. (*i*) $u(x, t) = 6e^{-5x-3t}$ (*ii*) $u(x, y) = c_1c_2 x^k e^{-(y^2/k)}$

7. (*i*) $u(x, y) = c_1c_2e^{k(x+y)}$ (*ii*) $u(x, y) = c_1\, e^{-p^2 y}(c_2 \cos px + c_3 \sin px)$

8. $z = 2\, e^{-(x+y)}$

9. $z(x, y) = 4\, e^{9x} \sin 3y$

10. $u(x, y) = e^{-3y} \sin x$.

2.4 VIBRATIONS OF A STRETCHED STRING, ONE DIMENSIONAL WAVE EQUATION $\frac{\partial^2 y}{\partial t^2} = c^2 \frac{\partial^2 y}{\partial x^2}$

Consider a uniform elastic string of length l stretched tightly between two points O and A and displaced slightly from its equilibrium position OA. Taking the end O as the origin, OA as the x-axis and a perpendicular line through O as the y-axis, we shall find the displacement y as a function of the distance x and the time t.

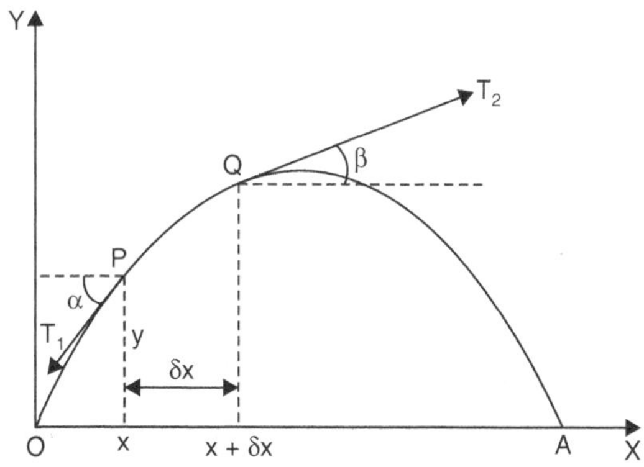

We shall obtain the equation of motion for the string under the following assumptions :

(*i*) The motion takes place entirely in the xy-plane and each particle of the string moves perpendicular to the equilibrium position OA of the string.

(*ii*) The string is perfectly flexible and does not offer resistance to bending.

(*iii*) The tension in the string is so large that the forces due to weight of the string can be neglected.

(*iv*) The displacement y and the slope $\frac{\partial y}{\partial x}$ are small, so that their higher powers can be neglected.

Let m be the mass per unit length of the string. Consider the motion of an element PQ of length δs. Since the string does not offer resistance to bending (by assumption), the tensions T_1 and T_2 at P and Q respectively are tangential to the curve.

Since, there is no motion in the horizontal direction, we have

$$T_1 \cos \alpha = T_2 \cos \beta = T \text{ (constant)} \quad \ldots(1)$$

Mass of element PQ is $m\delta s$. By Newton's second law of motion, the equation of motion in the vertical direction is

$$m\delta s \frac{\partial^2 y}{\partial t^2} = T_2 \sin \beta - T_1 \sin \alpha$$

or
$$\frac{m\delta s}{T} \frac{\partial^2 y}{\partial t^2} = \frac{T_2 \sin \beta}{T_2 \cos \beta} - \frac{T_1 \sin \alpha}{T_1 \cos \alpha} \quad \text{[By using (1)]}$$

or
$$\frac{\partial^2 y}{\partial t^2} = \frac{T}{m\delta s} (\tan \beta - \tan \alpha)$$

or
$$\frac{\partial^2 y}{\partial t^2} = \frac{T}{m\delta x} \left[\left(\frac{\partial y}{\partial x} \right)_{x+\delta x} - \left(\frac{\partial y}{\partial x} \right)_x \right]$$

[Since $\delta s = \delta x$ to a first approximation and $\tan \alpha$ and $\tan \beta$ are the slopes of the curve of the string at x and $x + \delta x$]

or
$$\frac{\partial^2 y}{\partial t^2} = \frac{T}{m} \left[\frac{\left(\frac{\partial y}{\partial x} \right)_{x+\delta x} - \left(\frac{\partial y}{\partial x} \right)_x}{\delta x} \right] = \frac{T}{m} \frac{\partial^2 y}{\partial x^2}, \text{ as } \delta x \to 0$$

or
$$\boxed{\frac{\partial^2 y}{\partial t^2} = c^2 \frac{\partial^2 y}{\partial x^2}} \qquad \text{where } c^2 = \frac{T}{m}$$

This is the partial differential equation giving the transverse vibrations of the string. It is also called the **one-dimensional wave equation.**

The *boundary conditions* which the equation $\frac{\partial^2 y}{\partial t^2} = c^2 \frac{\partial^2 y}{\partial x^2}$ has to satisfy are :

(*i*) $y = 0$ when $x = 0$
(*ii*) $y = 0$ when $x = l$
These should be satisfied for every value of t.

If the string is made to vibrate by pulling it into a curve $y = f(x)$ and then releasing it, the *initial conditions* are

(*i*) $y = f(x)$ when $t = 0$ (*ii*) $\frac{\partial y}{\partial t} = 0$ when $t = 0$.

2.5 SOLUTION OF THE WAVE EQUATION

(*U.P.T.U. 2015*)

The wave equation is

$$\frac{\partial^2 y}{\partial t^2} = c^2 \frac{\partial^2 y}{\partial x^2} \quad \ldots(1)$$

Let $$y = XT \qquad ...(2)$$
where X is a function of x only and T is a function of t only, be a solution of (1)

Then, $$\frac{\partial^2 y}{\partial t^2} = XT'' \quad \text{and} \quad \frac{\partial^2 y}{\partial x^2} = X''T$$

Substituting in (1), we have $$XT'' = c^2X''T$$

Separating the variables, we get $$\frac{X''}{X} = \frac{1}{c^2} \cdot \frac{T''}{T} \qquad ...(3)$$

Now the L.H.S of (3) is a function of x only and the RHS is a function of t only. Since x and t are independent variables, this equation can hold only when both sides reduce to a constant, say k. Then equation (3) leads to the ordinary linear differential equations.

$$X''k - X = 0 \quad \text{and} \quad T'' - kc^2T = 0 \qquad ...(4)$$

Solving equations (4), we get

(*i*) *When k is positive* and $= p^2$, say

$$X = c_1e^{px} + c_2e^{-px},\ T = c_3e^{cpt} + c_4e^{-cpt}$$

(*ii*) *When k is negative* and $= -p^2$, say

$$X = c_1 \cos px + c_2 \sin px, \quad T = c_3 \cos cpt + c_4 \sin cpt$$

(*iii*) *When* $k = 0$

$$X = c_1x + c_2 \qquad T = c_3t + c_4$$

Thus the various possible solutions of the wave equation (1) are:

$$y = (c_1e^{px} + c_2e^{-px})(c_3e^{cpt} + c_4e^{-cpt})$$
$$y = (c_1 \cos px + c_2 \sin px)(c_3 \cos cpt + c_4 \sin cpt)$$
$$y = (c_1x + c_2)(c_3t + c_4)$$

Of these three solutions, we have to choose that solution which is consistent with the physical nature of the problem. Since, we are dealing with a problem on vibrations, y must be a periodic function of x and t. Therefore, the solution must involve trigonometric terms.

Accordingly $$y(x, t) = (c_1 \cos px + c_2 \sin px)(c_3 \cos cpt + c_4 \sin cpt) \qquad ...(5)$$
is the only suitable solution of the wave equation and it corresponds to $k = -p^2$.

Now, applying boundary conditions that

$y = 0$ when $x = 0$ and $y = 0$ when $x = l$, we get

$$0 = c_1(c_3 \cos cpt + c_4 \sin cpt) \qquad ...(6)$$

and $$0 = (c_1 \cos pl + c_2 \sin pl)(c_3 \cos cpt + c_4 \sin cpt) \qquad ...(7)$$

From (6), we have $c_1 = 0$ and equation (7) reduces to

$$c_2 \sin pl\,(c_3 \cos cpt + c_4 \sin cpt) = 0$$

which is satisfied when $\sin pl = 0$ or $pl = n\pi$ or $p = \frac{n\pi}{l}$, where $n = 1, 2, 3, ...$

$\therefore$ A solution of the wave equation satisfying the boundary conditions is

$$y = c_2\left(c_3 \cos \frac{n\pi ct}{l} + c_4 \sin \frac{n\pi ct}{l}\right) \sin \frac{n\pi x}{l}$$
$$= \left(a_n \cos \frac{n\pi ct}{l} + b_n \sin \frac{n\pi ct}{l}\right) \sin \frac{n\pi x}{l}$$

on replacing c_2c_3 by a_n and c_2c_4 by b_n.

Adding up the solutions for different values of n, we get

$$y = \sum_{n=1}^{\infty} \left(a_n \cos \frac{n\pi ct}{l} + b_n \sin \frac{n\pi ct}{l} \right) \sin \frac{n\pi x}{l} \quad \text{...(8)}$$

which is also a solution.

Now applying the initial conditions

$$y = f(x) \quad \text{and} \quad \frac{\partial y}{\partial t} = 0, \text{ when } t = 0, \text{ we have}$$

$$f(x) = \sum_{n=1}^{\infty} a_n \sin \frac{n\pi x}{l} \quad \text{...(9)}$$

and

$$0 = \sum_{n=1}^{\infty} \frac{n\pi c}{l} b_n \sin \frac{n\pi x}{l} \quad \text{...(10)}$$

Since equation (9) represents Fourier series for $f(x)$, we have

$$a_n = \frac{2}{l} \int_0^l f(x) \sin \frac{n\pi x}{l} \, dx \quad \text{...(11)}$$

From (10), $b_n = 0$, for all n

Hence (8) reduces to $$y = \sum_{n=1}^{\infty} a_n \cos \frac{n\pi ct}{l} \sin \frac{n\pi x}{l} \quad \text{...(12)}$$

where a_n is given by (11) when $f(x)$ *i.e.*, $y(x, 0)$ is known.

2.6 INHOMOGENEOUS BOUNDARY CONDITIONS

Here we will discuss wave equations where Dirichlet boundary conditions and Neumann boundary conditions are not homogeneous. It is normal to seek transformation formulae to convert these inhomogeneous boundary conditions to homogeneous conditions.

2.6.1. Dirichlet Boundary Conditions

In this first type of boundary conditions, the displacements $u(0, t) = \alpha$ and $u(L, t) = \beta$ of a vibrating string of length L are given. Consider an initial–boundary value problem (IBVP) as,

$$u_{tt} = c^2 u_{xx}, 0 < x < L, t > 0$$

$$u(0, t) = \alpha, u(L, t) = \beta, \quad t \geq 0$$

and

$$u(x, 0) = f(x), \quad u_t(x, 0) = g(x)$$

To convert the above inhomogeneous boundary conditions to homogeneous boundary conditions, we use the conversion formula

$$u(x, t) = \left[\alpha + \left(\frac{\beta - \alpha}{L} \right) x \right] + v(x, t)$$

Substituting it in above IBVP, $v(x, t)$ is governed by IBVP:

$$v_{tt} = c^2 v_{xx}, 0 < x < L, \quad t > 0$$

$$v(0, t) = 0 = v(L, t)$$

$$v(x, 0) = f(x) - \left[\alpha + \left(\frac{\beta - \alpha}{L} \right) x \right], \; v_t(x, 0) = g(x)$$

Now using method of separation of variables, $v(x, t)$ can be found out. Subsequently, $u(x, t)$ is obtained as final result.

2.6.2 Neumann Boundary Conditions

In this second type of boundary conditions, $u_x(0, t) = \alpha$ and $u_x(L, t) = \beta$ are given. It is noted here that the above transformation formula cannot be used for these boundary conditions although it works effectively for first kind of boundary conditions. Consider the initial boundary value problem (IBVP):

$$\begin{aligned} u_{tt} &= c^2 u_{xx}, & 0 < x < L,\ t > 0 \\ u_x(0, t) &= \alpha, & u_x(L, t) = \beta, \quad t \geq 0 \\ u_x(x, 0) &= f(x), & u_t(x, 0) = g(x) \end{aligned}$$

To convert the inhomogeneous boundary conditions to homogeneous boundary conditions, we use the transformation formula

$$u(x, t) = \alpha x + \left(\frac{\beta - \alpha}{2L}\right)x^2 + c^2\left(\frac{\beta - \alpha}{2L}\right)t^2 + v(x, t)$$

Using above, it is shown that $v(x, t)$ is now governed by the IBVP:

$$\begin{aligned} v_{tt} &= c^2 v_{xx}, & 0 < x < L,\ t > 0 \\ v_x(0, t) &= 0, & v_x(L, t) = 0 \end{aligned}$$

$$v(x, 0) = f(x) - \left[\alpha x + \left(\frac{\beta - \alpha}{2L}\right)x^2\right]$$

$$v_t(x, 0) = g(x)$$

Now, using method of separation of variables, $v(x, t)$ is readily obtained. Consequently $u(x, t)$ is also found as final result.

2.7 WAVE EQUATION IN AN INFINITE DOMAIN: D'ALEMBERT SOLUTION

The physical model that controls the wave motion of a very long string is governed by a PDE and initial conditions only. The method of separation of variables is not applicable in this case.

However, D'Alembert solution allows us to solve the initial value problem on an infinite domain.

Let,
$$y_{tt} = c^2 y_{xx}, \ -\infty < x < \infty, \quad t > 0 \qquad ...(1)$$
$$y(x, 0) = f(x), \quad y_t(x, 0) = g(x)$$

To derive D'Alembert's solution, we consider two new variables ξ and η defined by

$$\left.\begin{aligned} \xi &= x + ct \\ \text{and} \quad \eta &= x - ct \end{aligned}\right\} \qquad ...(2)$$

Using chain rule, we get

$$\left.\begin{aligned} y_{xx} &= y_{\xi\xi} + 2y_{\xi\eta} + y_{\eta\eta} \\ \text{and} \quad y_{tt} &= c^2(y_{\xi\xi} - 2y_{\xi\eta} + y_{\eta\eta}) \end{aligned}\right\} \qquad ...(3)$$

Substituting (3) in (1), we get

$$y_{\xi\eta} = 0 \qquad ...(4)$$

General solution is

$$y(\xi, \eta) = F(\xi) + G(\eta)$$

where F and G are arbitrary functions. Using (2), we write,

$$y(x, t) = F(x + ct) + G(x - ct) \quad ...(5)$$

Using the initial condition $u(x, 0) = f(x)$ in (5), we get

$$F(x) + G(x) = f(x) \quad ...(6)$$

Again, using the initial condition $u_t(x, 0) = g(x)$ in (5), we get

$$cF'(x) - cG'(x) = g(x)$$

Integration yields,

$$F(x) - G(x) = \frac{1}{c}\int_0^x g(r)\,dr + K \quad ...(7)$$

where K is a constant of integration.

Solving (6) and (7), we get

$$F(x) = \frac{1}{2}f(x) + \frac{1}{2c}\int_0^x g(r)\,dr + \frac{1}{2}K$$

$$G(x) = \frac{1}{2}f(x) - \frac{1}{2c}\int_0^x g(r)\,dr - \frac{1}{2}K$$

This means that

$$F(x + ct) = \frac{1}{2}f(x + ct) + \frac{1}{2c}\int_0^{x+ct} g(r)\,dr + \frac{1}{2}K$$

$$G(x - ct) = \frac{1}{2}f(x - ct) - \frac{1}{2c}\int_0^{x-ct} g(r)\,dr - \frac{1}{2}K$$

Hence by (5),

$$\boxed{y(x, t) = \frac{f(x + ct) + f(x - ct)}{2} + \frac{1}{2c}\int_{x-ct}^{x+ct} g(r)\,dr}$$

This completes the formal derivation of D'Alembert's solution.

ILLUSTRATIVE EXAMPLES

Example 1. *A string is stretched and fastened to two points l apart. Motion is started by displacing the string in the form* $y = A\sin\frac{\pi x}{l}$ *from which it is released at time t = 0. Show that the displacement of any point at a distance x from one end at time t is given by*

$y(x, t) = A\sin\frac{\pi x}{l}\cos\frac{\pi ct}{l}$. (A.K.T.U. 2018, 2013, 2017; U.K.T.U. 2011, 2012)

Sol. The equation of the string is

$$\frac{\partial^2 y}{\partial t^2} = c^2\frac{\partial^2 y}{\partial x^2} \quad ...(1)$$

Since, the string is stretched between two fixed points (0, 0) and (l, 0) hence the displacement of the string at these points will be zero

$$\therefore \quad y(0, t) = 0 \quad ...(2)$$

and $$y(l, t) = 0 \quad ...(3)$$

Since, the string is released from rest hence its initial velocity will be zero

$$\therefore \qquad \frac{\partial y}{\partial t} = 0 \quad \text{at} \quad t = 0 \qquad \text{...(4)}$$

Since, the string is displaced from its initial position at time $t = 0$ hence the initial displacement is

$$y(x, 0) = A \sin \frac{\pi x}{l} \qquad \text{...(5)}$$

Conditions (2), (3), (4) and (5) are the boundary conditions.

Let us now proceed to solve equation (1),

Let $\qquad y = XT. \qquad$ ...(6)

where X is a function of x only and T is a function of t only.

$$\frac{\partial y}{\partial t} = \frac{\partial}{\partial t}(XT) = X\frac{dT}{dt}$$

$$\frac{\partial^2 y}{\partial t^2} = \frac{\partial}{\partial t}\left(X\frac{dT}{dt}\right) = X\frac{d^2T}{dt^2}.$$

Similarly, $\qquad \dfrac{\partial^2 y}{\partial x^2} = T\dfrac{d^2X}{dx^2}.$

Substituting the above in equation (1), we get

$$X\frac{d^2T}{dt^2} = c^2 T\frac{d^2X}{dx^2} \qquad \Rightarrow \qquad XT'' = c^2 TX''$$

Case I. $\qquad \dfrac{1}{c^2}\dfrac{T''}{T} = \dfrac{X''}{X} = -p^2$ (say)

(*i*) $\qquad \dfrac{1}{c^2}\dfrac{T''}{T} = -p^2$

$$\frac{d^2T}{dt^2} + c^2p^2T = 0.$$

Auxiliary equation is $\qquad m^2 + c^2p^2 = 0$

$$m^2 = c^2p^2i^2$$

$$m = \pm\, cpi$$

$$\therefore \qquad \text{C.F.} = c_1 \cos cpt + c_2 \sin cpt$$

$$\text{P.I.} = 0$$

$$\therefore \qquad T = \text{C.F.} + \text{P.I.} = c_1 \cos cpt + c_2 \sin cpt \qquad \text{...(7)}$$

(*ii*) $\qquad \dfrac{X''}{X} = -p^2 \quad \Rightarrow \quad \dfrac{d^2X}{dx^2} + p^2X = 0.$

Auxiliary equation is $\qquad m^2 + p^2 = 0$

$$m = \pm\, pi$$

$$\text{C.F.} = c_3 \cos px + c_4 \sin px$$

P.I. = 0

$\therefore$ $X = c_3 \cos px + c_4 \sin px$ ...(8)

Hence, $y(x, t) = (c_1 \cos cpt + c_2 \sin cpt)(c_3 \cos px + c_4 \sin px)$ ...(9)

Case II. $\frac{1}{c^2}\frac{T''}{T} = \frac{X''}{X} = p^2$ (say)

(*i*) $\frac{1}{c^2}\frac{T''}{T} = p^2 \Rightarrow \frac{d^2T}{dt^2} - p^2c^2T = 0$

Auxiliary equation is $m^2 - p^2c^2 = 0 \Rightarrow m = \pm pc$

$\therefore$ C.F. $= c_5\, e^{pct} + c_6\, e^{-pct}$

P.I. = 0

$\therefore$ $T = c_5\, e^{pct} + c_6\, e^{-pct}$.

(*ii*) $\frac{X''}{X} = p^2 \Rightarrow \frac{d^2X}{dx^2} - p^2X = 0$

Auxiliary equation is

$$m^2 - p^2 = 0 \Rightarrow m = \pm p$$

$\therefore$ C.F. $= c_7\, e^{px} + c_8\, e^{-px}$

P.I. = 0

$\therefore$ $X = c_7\, e^{px} + c_8\, e^{-px}$

Hence, $y(x, t) = (c_5\, e^{pct} + c_6\, e^{-pct})(c_7\, e^{px} + c_8\, e^{-px})$ ...(10)

Case III. $\frac{1}{c^2}\frac{T''}{T} = \frac{X''}{X} = 0$ (say)

(*i*) $\frac{1}{c^2}\frac{T''}{T} = 0 \Rightarrow T'' = 0$ or $\frac{d^2T}{dt^2} = 0$

Auxiliary equation is

$$m^2 = 0 \Rightarrow m = 0, 0$$

$\therefore$ C.F. $= c_9 + c_{10}t$

P.I. = 0

$\therefore$ $T = c_9 + c_{10}t$

(*ii*) $\frac{X''}{X} = 0 \Rightarrow X'' = 0$ or $\frac{d^2X}{dx^2} = 0$

Auxiliary equation is

$$m^2 = 0 \Rightarrow m = 0, 0$$

$\therefore$ C.F. $= c_{11} + c_{12}x$

P.I. = 0

$\therefore$ $X = c_{11} + c_{12}x$

Hence, $y(x, t) = (c_9 + c_{10}\, t)\,(c_{11} + c_{12}x)$ ...(11)

Out of these three above solutions (9), (10) and (11), we have to choose the solution which is consistent with the physical nature of the problem. Since, we are dealing with a problem on vibrations, the solution must contain periodic functions. Hence the solution which contains trigonometric terms must be the required solution.

Hence solution (9) is the general solution of one-dimensional wave equation given by equation (1).

Now, $$y(x, t) = (c_1 \cos cpt + c_2 \sin cpt)(c_3 \cos px + c_4 \sin px)$$

Applying the boundary condition,

$$y(0, t) = 0 = (c_1 \cos cpt + c_2 \sin cpt)\, c_3$$

$$\Rightarrow \quad c_3 = 0.$$

$\therefore$ From (9), $$y(x, t) = (c_1 \cos cpt + c_2 \sin cpt)\, c_4 \sin px \qquad \text{...(12)}$$

Again, $$y(l, t) = 0 = (c_1 \cos cpt + c_2 \sin cpt)\, c_4 \sin pl$$

$$\Rightarrow \quad \sin pl = 0 = \sin n\pi \; (n \in \text{I})$$

$$\therefore \quad p = \frac{n\pi}{l}.$$

Hence from (12), $$y(x, t) = \left(c_1 \cos \frac{n\pi ct}{l} + c_2 \sin \frac{n\pi ct}{l}\right) c_4 \sin \frac{n\pi x}{l} \qquad \text{...(13)}$$

$$\frac{\partial y}{\partial t} = \frac{n\pi c}{l}\left[-c_1 \sin \frac{n\pi ct}{l} + c_2 \cos \frac{n\pi ct}{l}\right] c_4 \sin \frac{n\pi x}{l}$$

At $t = 0$,

$$\left(\frac{\partial y}{\partial t}\right)_{t=0} = 0 = \frac{n\pi c}{l}\left[c_2 c_4 \sin \frac{n\pi x}{l}\right]$$

$$\Rightarrow \quad c_2 = 0,$$

$\therefore$ From (13), $$y(x, t) = c_1 c_4 \cos \frac{n\pi ct}{l} \sin \frac{n\pi x}{l} \qquad \text{...(14)}$$

$$y(x, 0) = \text{A} \sin \frac{\pi x}{l} = c_1 c_4 \sin \frac{n\pi x}{l}$$

$$\Rightarrow \quad c_1 c_4 = \text{A},\; n = 1. \qquad | \text{ Comparing}$$

Hence from (14), $$y(x, t) = \text{A} \cos \frac{\pi ct}{l} \sin \frac{\pi x}{l}$$

which is the required solution.

Example 2. *Show how the wave equation* $c^2 \dfrac{\partial^2 y}{\partial x^2} = \dfrac{\partial^2 y}{\partial t^2}$ *can be solved by the method of separation of variables. If the initial displacement and velocity of a string stretched between x = 0 and x = l are given by* $y = f(x)$ *and* $\dfrac{\partial y}{\partial t} = g(x)$*, determine the constants in the series solution.*

Sol. The wave equation is $$\frac{\partial^2 y}{\partial t^2} = c^2 \frac{\partial^2 y}{\partial x^2} \qquad \text{...(1)}$$

Let $$y = \text{XT} \qquad \text{...(2)}$$

where X is a function of x only and T is a function of t only.

$$\frac{\partial y}{\partial t} = \frac{\partial}{\partial t}(\text{XT}) = \text{X} \frac{d\text{T}}{dt}$$

$$\frac{\partial^2 y}{\partial t^2} = X \frac{d^2 T}{dt^2}$$

Similarly, $$\frac{\partial^2 y}{\partial x^2} = T \frac{d^2 X}{dx^2}.$$

Substituting in (1), we get

$$X \frac{d^2 T}{dt^2} = c^2 T \frac{d^2 X}{dx^2} \quad \Rightarrow \quad XT'' = c^2 TX''$$

$$\Rightarrow \quad \frac{1}{c^2} \frac{T''}{T} = \frac{X''}{X} \qquad ...(3)$$

Case I. When $\frac{1}{c^2} \frac{T''}{T} = \frac{X''}{X} = p^2$ (say)

(*i*) $$\frac{1}{c^2} \frac{T''}{T} = p^2 \quad \Rightarrow \quad \frac{d^2 T}{dt^2} - p^2 c^2 T = 0.$$

Auxiliary equation is

$$m^2 - p^2 c^2 = 0$$

$$m = \pm pc$$

$$\text{C.F.} = c_1 e^{pct} + c_2 e^{-pct}$$

$$\text{P.I.} = 0$$

$$\therefore \quad T = \text{C.F.} + \text{P.I.} = c_1 e^{pct} + c_2 e^{-pct}$$

(*ii*) $$\frac{X''}{X} = p^2 \quad \Rightarrow \quad \frac{d^2 X}{dx^2} - p^2 X = 0.$$

Auxiliary equation is

$$m^2 - p^2 = 0$$

$$m = \pm p$$

$$\text{C.F.} = c_3 e^{px} + c_4 e^{-px}$$

$$\text{P.I.} = 0.$$

$$\therefore \quad X = \text{C.F.} + \text{P.I.} = c_3 e^{px} + c_4 e^{-px}.$$

Hence, the solution is

$$y = XT = (c_1 e^{pct} + c_2 e^{-pct})(c_3 e^{px} + c_4 e^{-px}). \qquad ...(4)$$

Case II. When

$$\frac{1}{c^2} \frac{T''}{T} = \frac{X''}{X} = -p^2 \text{ (say)}$$

(*i*) $$\frac{1}{c^2} \frac{T''}{T} = -p^2 \quad \Rightarrow \quad \frac{d^2 T}{dt^2} + p^2 c^2 T = 0.$$

Auxiliary equation is

$$m^2 + p^2 c^2 = 0 \quad \Rightarrow \quad m = \pm pci$$

$$\therefore \quad \text{C.F.} = (c_5 \cos pct + c_6 \sin cpt)$$

$$\text{P.I.} = 0.$$

$$\therefore \quad T = \text{C.F.} + \text{P.I.} = c_5 \cos cpt + c_6 \sin cpt$$

(*ii*) $$\frac{X''}{X} = -p^2 \Rightarrow \frac{d^2X}{dx^2} + p^2X = 0.$$

Auxiliary equation is

$$m^2 + p^2 = 0 \Rightarrow m = \pm pi$$

$\therefore$ $$\text{C.F.} = c_7 \cos px + c_8 \sin px$$

$$\text{P.I.} = 0$$

$\therefore$ $$X = c_7 \cos px + c_8 \sin px.$$

Hence, the solution is

$$y = XT = (c_5 \cos cpt + c_6 \sin cpt)(c_7 \cos px + c_8 \sin px) \quad ...(5)$$

Case III. When, $\frac{1}{c^2}\frac{T''}{T} = \frac{X''}{X} = 0$

(*i*) $$\frac{1}{c^2}\frac{T''}{T} = 0 \Rightarrow \frac{d^2T}{dt^2} = 0$$

$\Rightarrow$ $$T = c_9 + c_{10}t$$

(*ii*) $$\frac{X''}{X} = 0 \Rightarrow \frac{d^2X}{dx^2} = 0$$

$\Rightarrow$ $$X = c_{11} + c_{12}x.$$

Hence, the solution is

$$y(x, t) = (c_9 + c_{10}t)(c_{11} + c_{12}x) \quad ...(6)$$

Of the above three solutions given by (4), (5) and (6), we have to choose the solution which is consistent with the physical nature of the problem. Since, we are dealing with a problem on vibrations, y must be a periodic function of x and t therefore the solution must involve trigonometric terms hence solution (5) is the required solution.

Boundary conditions are

$$y(0, t) = 0, \qquad y(l, t) = 0$$

$$y = f(x) \qquad \text{when } t = 0$$

$$\frac{\partial y}{\partial t} = g(x) \qquad \text{when } t = 0$$

From equation (5), $y(0, t) = (c_5 \cos cpt + c_6 \sin cpt)\, c_7$

$$0 = (c_5 \cos cpt + c_6 \sin cpt)\, c_7$$

$\Rightarrow$ $$c_7 = 0.$$

Hence from (5), $$y(x, t) = (c_5 \cos cpt + c_6 \sin cpt)\, c_8 \sin px \quad ...(7)$$

$$y(l, t) = 0 = (c_5 \cos cpt + c_6 \sin cpt)\, c_8 \sin pl$$

$\Rightarrow$ $$\sin pl = 0 = \sin n\pi \ (n \in I) \Rightarrow p = \frac{n\pi}{l}.$$

$\therefore$ From (7), $$y(x, t) = \left(c_5 \cos \frac{n\pi ct}{l} + c_6 \sin \frac{n\pi ct}{l}\right) c_8 \sin \frac{n\pi x}{l} \quad ...(8)$$

$$= \left(a_n \cos \frac{n\pi ct}{l} + b_n \sin \frac{n\pi ct}{l}\right) \sin \frac{n\pi x}{l}$$

where $c_5c_8 = a_n$ and $c_6c_8 = b_n$

The general solution is

$$y(x, t) = \sum_{n=1}^{\infty} \left(a_n \cos\frac{n\pi ct}{l} + b_n \sin\frac{n\pi ct}{l}\right) \sin\frac{n\pi x}{l} \quad ...(9)$$

$$y(x, 0) = f(x) = \sum_{1}^{\infty} a_n \sin\frac{n\pi x}{l}$$

where $$a_n = \frac{2}{l}\int_0^l f(x) . \sin\frac{n\pi x}{l}\, dx \quad ...(10)$$

From (9), $$\frac{\partial y}{\partial t} = \frac{\pi c}{l}\sum_{1}^{\infty}\left(-n\, a_n \sin\frac{n\pi ct}{l} + n\, b_n \cos\frac{n\pi ct}{l}\right)\sin\frac{n\pi x}{l}$$

At $t = 0$, $$\left(\frac{\partial y}{\partial t}\right)_{t=0} = g(x) = \frac{\pi c}{l}\sum_{1}^{\infty} n b_n \sin\frac{n\pi x}{l}$$

where $$\frac{n\pi c}{l} b_n = \frac{2}{l}\int_0^l g(x) . \sin\frac{n\pi x}{l}\, dx$$

$\Rightarrow$ $$b_n = \frac{2}{n\pi c}\int_0^l g(x) . \sin\frac{n\pi x}{l}\, dx . \quad ...(11)$$

Hence, the required solution is

$$y(x, t) = \sum_{n=1}^{\infty}\left(a_n \cos\frac{n\pi ct}{l} + b_n \sin\frac{n\pi ct}{l}\right)\sin\frac{n\pi x}{l}$$

where $$a_n = \frac{2}{l}\int_0^l f(x) . \sin\frac{n\pi x}{l}\, dx$$

and $$b_n = \frac{2}{n\pi c}\int_0^l g(x) \sin\frac{n\pi x}{l}\, dx.$$

Example 3. *A tightly stretched string with fixed end points $x = 0$ and $x = l$ is initially in a position given by $y = y_0 \sin^3 \frac{\pi x}{l}$. If it is released from rest from this position, find the displacement $y(x, t)$.* [G.B.T.U. (C.O.) 2011]

Sol. The equation of the string is

$$\frac{\partial^2 y}{\partial t^2} = c^2 \frac{\partial^2 y}{\partial x^2} \quad ...(1)$$

The solution of eqn. (1) is

$$y(x, t) = (c_1 \cos cpt + c_2 \sin cpt)(c_3 \cos px + c_4 \sin px) \quad ...(2)$$

| Refer Sol. of Ex. 1

Boundary conditions are

$$y(0, t) = 0 \quad ...(3)$$

$$y(l, t) = 0 \quad ...(4)$$

$$\left(\frac{\partial y}{\partial t}\right)_{t=0} = 0 \quad ...(5)$$

$$y(x, 0) = y_0 \sin^3\frac{\pi x}{l} \quad ...(6)$$

Applying boundary condition in (2),

$$y(0, t) = 0 = (c_1 \cos cpt + c_2 \sin cpt)\, c_3$$

$\Rightarrow$ $c_3 = 0$

$\therefore$ From (2), $y(x, t) = (c_1 \cos cpt + c_2 \sin cpt)\, c_4 \sin px$ ...(7)

Again, $y(l, t) = 0 = (c_1 \cos cpt + c_2 \sin cpt)\, c_4 \sin pl$

$\Rightarrow$ $\sin pl = 0 = \sin n\pi \quad (n \in \text{I})$

$\therefore$ $p = \dfrac{n\pi}{l}$

Hence, from (7),

$$y(x, t) = \left(c_1 \cos \frac{n\pi ct}{l} + c_2 \sin \frac{n\pi ct}{l}\right) c_4 \sin \frac{n\pi x}{l} \quad ...(8)$$

$$\frac{\partial y}{\partial t} = \frac{n\pi c}{l}\left[-c_1 \sin \frac{n\pi ct}{l} + c_2 \cos \frac{n\pi ct}{l}\right] c_4 \sin \frac{n\pi x}{l}$$

At $t = 0$,

$$\left(\frac{\partial y}{\partial t}\right)_{t=0} = 0 = \frac{n\pi c}{l} c_2 c_4 \sin \frac{n\pi x}{l}$$

$\Rightarrow$ $c_2 = 0.$

$\therefore$ From (8),

$$y(x, t) = c_1 c_4 \sin \frac{n\pi x}{l} \cos \frac{n\pi ct}{l} = b_n \sin \frac{n\pi x}{l} \cos \frac{n\pi ct}{l}$$

Most general solution is

$$y(x, t) = \sum_{n=1}^{\infty} b_n \sin \frac{n\pi x}{l} \cos \frac{n\pi ct}{l} \quad ...(9)$$

$$y(x, 0) = y_0 \sin^3 \frac{\pi x}{l} = \sum_{n=1}^{\infty} b_n \sin \frac{n\pi x}{l}$$

$$\Rightarrow \quad y_0 \left(\frac{3 \sin \frac{\pi x}{l} - \sin \frac{3\pi x}{l}}{4}\right) = b_1 \sin \frac{\pi x}{l} + b_2 \sin \frac{2\pi x}{l} + b_3 \sin \frac{3\pi x}{l} + \ldots$$

Comparing, we get

$$b_1 = \frac{3y_0}{4},\ b_2 = 0,\ b_3 = \frac{-y_0}{4},\ b_4 = b_5 = \ldots = 0$$

Hence, from (9),

$$y(x, t) = \frac{3y_0}{4} \sin \frac{\pi x}{l} \cos \frac{\pi ct}{l} - \frac{y_0}{4} \sin \frac{3\pi x}{l} \cos \frac{3\pi ct}{l}.$$

Example 4. *A tightly stretched flexible string has its ends fixed at $x = 0$ and $x = l$. At time $t = 0$, the string is given a shape defined by $F(x) = \mu x(l - x)$, μ is a constant and then released. Find the displacement $y(x, t)$ of any point x of the string at any time $t > 0$.*

Sol. The wave equation is $\dfrac{\partial^2 y}{\partial t^2} = c^2 \dfrac{\partial^2 y}{dx^2}$ ...(1)

The solution of equation (1) is

$y(x, t) = (c_1 \cos cpt + c_2 \sin cpt)(c_3 \cos px + c_4 \sin px)$ ...(2) (Refer sol. of Ex. 1)

Boundary conditions are $y(0, t) = 0$ ...(3)

$$y(l, t) = 0 \quad ...(4)$$

$$\frac{\partial y}{\partial t} = 0 \text{ at } t = 0 \quad ...(5)$$

and $y(x, 0) = \mu x(l - x)$ ...(6)

From (2), $y(0, t) = 0 = (c_1 \cos cpt + c_2 \sin cpt)c_3$

$\Rightarrow$ $c_3 = 0.$

$\therefore$ From (2), $y(x, t) = (c_1 \cos cpt + c_2 \sin cpt)\, c_4 \sin px$ ...(7)

$$y(l, t) = 0 = (c_1 \cos cpt + c_2 \sin cpt)\, c_4 \sin pl$$

$\Rightarrow$ $\sin pl = 0 = \sin n\pi \ (n \in I)$

$$p = \frac{n\pi}{l}.$$

From (7), $$y(x, t) = \left(c_1 \cos \frac{n\pi ct}{l} + c_2 \sin \frac{n\pi ct}{l}\right) c_4 \sin \frac{n\pi x}{l} \quad ...(8)$$

Now from (7), $$\frac{\partial y}{\partial t} = \frac{n\pi c}{l}\left[-c_1 \sin \frac{n\pi ct}{l} + c_2 \cos \frac{n\pi ct}{l}\right] . c_4 \sin \frac{n\pi x}{l}$$

At $t = 0$, $$\left(\frac{\partial y}{\partial t}\right)_{t=0} = 0 = \frac{n\pi c}{l} c_2 c_4 \sin \frac{n\pi x}{l}$$

$\Rightarrow$ $c_2 = 0.$

$\therefore$ From (8), $$y(x, t) = c_1 c_4 \cos \frac{n\pi ct}{l} \sin \frac{n\pi x}{l}$$

$\Rightarrow$ $$y(x, t) = b_n \cos \frac{n\pi ct}{l} \sin \frac{n\pi x}{l} \text{ where } c_1 c_4 = b_n.$$

The most general solution is

$$y(x, t) = \sum_{1}^{\infty} b_n \cos \frac{n\pi ct}{l} \sin \frac{n\pi x}{l} \quad ...(9)$$

$$y(x, 0) = \mu(lx - x^2) = \sum_{1}^{\infty} b_n \sin \frac{n\pi x}{l}$$

where $$b_n = \frac{2}{l}\int_0^l \mu\,(lx - x^2) \sin \frac{n\pi x}{l}\, dx$$

$$= \frac{2\mu}{l}\left[\left\{(lx - x^2)\left(\frac{-\cos \frac{n\pi x}{l}}{\frac{n\pi}{l}}\right)\right\}_0^l - \int_0^l (l - 2x)\left(\frac{-\cos \frac{n\pi x}{l}}{\frac{n\pi}{l}}\right) dx\right]$$

$$= \frac{2\mu}{l}\left[\frac{l}{n\pi}\int_0^l (l - 2x) \cos \frac{n\pi x}{l}\, dx\right]$$

$$= \frac{2\mu}{n\pi}\left[\left\{(l-2x)\frac{\sin\frac{n\pi x}{l}}{\frac{n\pi}{l}}\right\}_0^l - \int_0^l (-2)\frac{\sin\frac{n\pi x}{l}}{\frac{n\pi}{l}}\,dx\right] = \frac{2\mu}{n\pi}\cdot\frac{2l}{n\pi}\int_0^l \sin\frac{n\pi x}{l}\,dx$$

$$= \frac{4\mu l}{n^2\pi^2}\left(\frac{-\cos\frac{n\pi x}{l}}{\frac{n\pi}{l}}\right)_0^l = \frac{4\mu l^2}{n^3\pi^3}(-\cos n\pi + 1) = \frac{4\mu l^2}{n^3\pi^3}[1-(-1)^n].$$

$\therefore$ From (9),

$$y(x, t) = \frac{4\mu l^2}{\pi^3}\sum_1^\infty \frac{[1-(-1)^n]}{n^3}\cos\frac{n\pi ct}{l}\sin\frac{n\pi x}{l}$$

$$= \frac{8\mu l^2}{\pi^3}\sum_{n=1}^\infty \frac{1}{(2n-1)^3}\sin\frac{(2n-1)\pi x}{l}\cos\frac{(2n-1)\pi ct}{l}.$$

Example 5. *A string is stretched between two fixed points (0, 0) and (l, 0) and released at rest from the initial deflection given by*

and
$$f(x) = \begin{cases}\left(\frac{2k}{l}\right)x, & 0 < x < \frac{l}{2}\\ \left(\frac{2k}{l}\right)(l-x), & \frac{l}{2} < x < l\end{cases}$$

Find the deflection of the string at any time.

Sol. The equation for the vibrations of the string is

$$\frac{\partial^2 y}{\partial t^2} = c^2\frac{\partial^2 y}{dx^2} \qquad ...(1)$$

The solution of eqn. (1) is

$$y(x, t) = (c_1 \cos cpt + c_2 \sin cpt)(c_3 \cos px + c_4 \sin px) \qquad ...(2)$$ [Refer Sol. of Ex. 1]

Boundary conditions are, $y(0, t) = 0, y(l, t) = 0$

$$\frac{\partial y}{\partial t} = 0 \quad \text{at} \quad t = 0$$

$$y(x, 0) = \begin{cases}\frac{2k}{l}x, & 0 < x < \frac{l}{2}\\ \frac{2k}{l}(l-x), & \frac{l}{2} < x < l\end{cases}$$

From (2), $y(0, t) = (c_1 \cos cpt + c_2 \sin cpt)\, c_3$

$0 = (c_1 \cos cpt + c_2 \sin cpt)\, c_3$

$\Rightarrow$ $c_3 = 0.$

$\therefore$ From (2), $y(x, t) = (c_1 \cos cpt + c_2 \sin cpt)\, c_4 \sin px$ ...(3)

$y(l, t) = 0 = (c_1 \cos cpt + c_2 \sin cpt)\, c_4 \sin pl$

$\Rightarrow$ $\sin pl = 0 = \sin n\pi\,;\ n \in I$

$$p = \frac{n\pi}{l}.$$

$\therefore$ From (3), $$y(x,t) = \left(c_1 \cos \frac{n\pi ct}{l} + c_2 \sin \frac{n\pi ct}{l}\right) c_4 \sin \frac{n\pi x}{l} \quad ...(4)$$

$$\frac{\partial y}{\partial t} = \frac{n\pi c}{l}\left[-c_1 \sin \frac{n\pi ct}{l} + c_2 \cos \frac{n\pi ct}{l}\right) c_4 \sin \frac{n\pi x}{l}$$

At $t = 0$, $$\left(\frac{\partial y}{\partial t}\right)_{t=0} = 0 = \frac{n\pi c}{l}\left[c_2 c_4 \sin \frac{n\pi x}{l}\right]$$

$\Rightarrow$ $$c_2 = 0.$$

$\therefore$ From (4), $$y(x,t) = c_1 c_4 \cos \frac{n\pi ct}{l} \sin \frac{n\pi x}{l}$$

$$= b_n \cos \frac{n\pi ct}{l} \sin \frac{n\pi x}{l} \qquad \text{(where } c_1 c_4 = b_n) \quad ...(5)$$

The most general solution is

$$y(x,t) = \sum_{n=1}^{\infty} b_n \cos \frac{n\pi ct}{l} \sin \frac{n\pi x}{l} \quad ...(6)$$

$$y(x,0) = \sum_{1}^{\infty} b_n \sin \frac{n\pi x}{l} \qquad \text{[From (6)]}$$

where $$b_n = \frac{2}{l}\int_0^l y(x,0) \sin \frac{n\pi x}{l}\, dx$$

$$= \frac{2}{l}\left[\int_0^{l/2} \frac{2k}{l} x \sin \frac{n\pi x}{l}\, dx + \int_{l/2}^{l} \frac{2k}{l}(l-x) \sin \frac{n\pi x}{l}\, dx\right]$$

$$= \frac{4k}{l^2}\left[\int_0^{l/2} x \sin \frac{n\pi x}{l}\, dx + \int_{l/2}^{l} (l-x) \sin \frac{n\pi x}{l}\, dx\right]$$

$$= \frac{4k}{l^2}\left[x\left(\frac{-\cos \frac{n\pi x}{l}}{\frac{n\pi}{l}}\right)_0^{l/2} - \int_0^{l/2} 1.\left(\frac{-\cos \frac{n\pi x}{l}}{\frac{n\pi}{l}}\right) dx \right.$$
$$\left. + \left\{(l-x)\left(\frac{-\cos \frac{n\pi x}{l}}{\frac{n\pi}{l}}\right)\right\}_{l/2}^{l} - \int_{l/2}^{l} (-1).\left(\frac{-\cos \frac{n\pi x}{l}}{\frac{n\pi}{l}}\right) dx\right]$$

$$= \frac{4k}{l^2}\left[-\frac{l}{n\pi}.\frac{l}{2}\cos\frac{n\pi}{2} + \frac{l}{n\pi}\left(\frac{\sin \frac{n\pi x}{l}}{\frac{n\pi}{l}}\right)_0^{l/2} + \frac{l}{2}.\frac{l}{n\pi}\cos\frac{n\pi}{2} - \frac{l}{n\pi}\left(\frac{\sin \frac{n\pi x}{l}}{\frac{n\pi}{l}}\right)_{l/2}^{l}\right]$$

$$= \frac{4k}{l^2}\left[\frac{l^2}{n^2\pi^2}\sin\frac{n\pi}{2} - \frac{l^2}{n^2\pi^2}\left(\sin n\pi - \sin\frac{n\pi}{2}\right)\right]$$

$$= \frac{4k}{l^2}\left[\frac{2l^2}{n^2\pi^2}\sin\frac{n\pi}{2}\right] = \frac{8k}{n^2\pi^2}\sin\frac{n\pi}{2}.$$

$\therefore$ From (6), $$y(x, t) = \frac{8k}{\pi^2}\sum_{1}^{\infty}\frac{1}{n^2}\sin\frac{n\pi}{2}\cos\frac{n\pi ct}{l}\sin\frac{n\pi x}{l}.$$

Example 6. *A tightly stretched violin string of length l and fixed at both ends is plucked at* $x = \frac{l}{3}$ *and assumes initially the shape of a triangle of height a. Find the displacement y at any distance x and any time t after the string is released from rest.*

Sol. One-Dimensional wave equation is

$$\frac{\partial^2 y}{\partial t^2} = c^2\frac{\partial^2 y}{\partial x^2} \quad ...(1)$$

The solution of eqn. (1) is

$y(x, t) = (c_1 \cos cpt + c_2 \sin cpt)(c_3 \cos px + c_4 \sin px)$ ...(2) (Refer Sol. of Ex. 1)

Eqn. of line OC is $$y - 0 = \frac{a - 0}{\frac{l}{3} - 0}(x - 0)$$

$$y = \frac{3a}{l}x \quad ...(3)$$

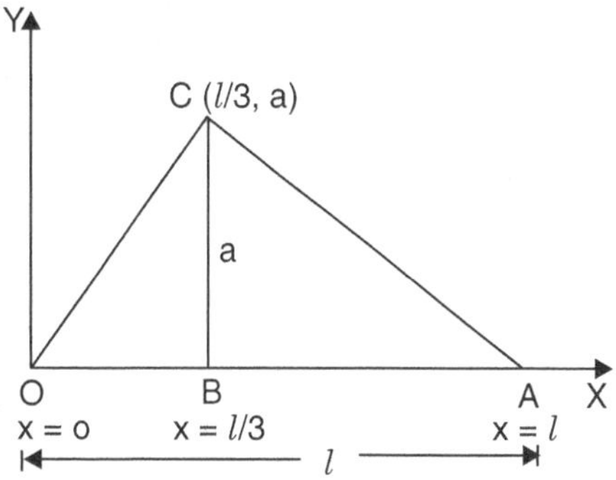

Eqn. of line CA is $$y - a = \frac{0 - a}{l - l/3}\left(x - \frac{l}{3}\right)$$

$$y - a = \frac{-a}{\left(\frac{2l}{3}\right)}\left(x - \frac{l}{3}\right) = -\frac{3a}{2l}\left(x - \frac{l}{3}\right)$$

$$y - a = -\frac{3ax}{2l} + \frac{a}{2}$$

$$y = -\frac{3ax}{2l} + \frac{3a}{2} = \frac{3a}{2}\left(1 - \frac{x}{l}\right) \quad ...(4)$$

Hence the boundary conditions are

$$y(0, t) = 0 \quad ...(5)$$

$$y(l, t) = 0 \quad ...(6)$$

$$\frac{\partial y}{\partial t} = 0 \text{ at } t = 0 \quad ...(7)$$

and $$y(x, 0) = \begin{cases} \frac{3ax}{l}, & 0 < x < l/3 \\ \frac{3a}{2}\left(1 - \frac{x}{l}\right), & \frac{l}{3} < x < l \end{cases} \quad ...(8)$$

From (2), $y(0, t) = 0 = (c_1 \cos cpt + c_2 \sin cpt)\, c_3$

$\Rightarrow$ $c_3 = 0.$

$\therefore$ From (2), $y(x, t) = (c_1 \cos cpt + c_2 \sin cpt)\, c_4 \sin px$ ...(9)

$y(l, t) = 0 = (c_1 \cos cpt + c_2 \sin cpt)\, c_4 \sin pl$

$\Rightarrow \qquad \sin pl = 0 = \sin n\pi (n \in \text{I}).$

$\Rightarrow \qquad p = \dfrac{n\pi}{l}.$

$$\therefore \qquad y(x, t) = \left(c_1 \cos \frac{n\pi ct}{l} + c_2 \sin \frac{n\pi ct}{l}\right) c_4 \sin \frac{n\pi x}{l} \qquad \ldots(10)$$

$$\frac{\partial y}{\partial t} = \frac{n\pi c}{l}\left[- c_1 \sin \frac{n\pi ct}{l} + c_2 \cos \frac{n\pi ct}{l}\right] c_4 \sin \frac{n\pi x}{l}.$$

At $t = 0$,

$$\left(\frac{\partial y}{\partial t}\right)_{t=0} = 0 = \frac{n\pi c}{l}\left[c_2 c_4 \sin \frac{n\pi x}{l}\right]$$

$\Rightarrow \qquad c_2 = 0.$

$$\therefore \qquad y(x, t) = c_1 c_4 \cos \frac{n\pi ct}{l} \sin \frac{n\pi x}{l} = b_n \cos \frac{n\pi ct}{l} \sin \frac{n\pi x}{l}.$$

The most general solution is

$$y(x, t) = \sum_{1}^{\infty} b_n \cos \frac{n\pi ct}{l} \sin \frac{n\pi x}{l} \qquad \ldots(11)$$

From (11), $\qquad y(x, 0) = \sum_{1}^{\infty} b_n \sin \dfrac{n\pi x}{l}$, where

$$b_n = \frac{2}{l}\int_0^l y(x, 0) \sin \frac{n\pi x}{l}\, dx$$

$$= \frac{2}{l}\left[\int_0^{l/3} \frac{3ax}{l} \sin \frac{n\pi x}{l}\, dx + \int_{l/3}^{l} \frac{3a}{2}\left(1 - \frac{x}{l}\right) \sin \frac{n\pi x}{l}\, dx\right]$$

$$= \frac{2}{l}\left[\frac{3a}{l}\int_0^{l/3} x \sin \frac{n\pi x}{l}\, dx + \frac{3a}{2}\int_{l/3}^{l}\left(1 - \frac{x}{l}\right) \sin \frac{n\pi x}{l}\, dx\right]$$

$$= \frac{6a}{l^2}\left[\left(x\left\{\frac{-\cos \frac{n\pi x}{l}}{\frac{n\pi}{l}}\right\}\right)_0^{l/3} - \int_0^{l/3} 1.\left(\frac{-\cos \frac{n\pi x}{l}}{\frac{n\pi}{l}}\right) dx\right]$$

$$+ \frac{3a}{l}\left[\left(\left(1 - \frac{x}{l}\right)\left(\frac{-\cos \frac{n\pi x}{l}}{\frac{n\pi}{l}}\right)\right)_{l/3}^{l} - \int_{l/3}^{l}\left(-\frac{1}{l}\right)\left(\frac{-\cos \frac{n\pi x}{l}}{\frac{n\pi}{l}}\right) dx\right]$$

$$= \frac{6a}{l^2}\left[-\frac{l}{n\pi}.\frac{l}{3}\cos \frac{n\pi}{3} + \frac{l}{n\pi}\left(\frac{\sin \frac{n\pi x}{l}}{\frac{n\pi}{l}}\right)_0^{l/3}\right] + \frac{3a}{l}\left[\frac{l}{n\pi}.\frac{2}{3}\cos \frac{n\pi}{3} - \frac{1}{n\pi}.\left(\frac{\sin \frac{n\pi x}{l}}{\frac{n\pi}{l}}\right)_{l/3}^{l}\right]$$

$$= \frac{6a}{l^2}\left[-\frac{l^2}{3n\pi}\cos\frac{n\pi}{3}+\frac{l^2}{n^2\pi^2}\sin\frac{n\pi}{3}\right]+\frac{3a}{l}\left[\frac{2l}{3n\pi}\cos\frac{n\pi}{3}-\frac{l}{n^2\pi^2}\left(0-\sin\frac{n\pi}{3}\right)\right]$$

$$= \frac{6a}{n\pi}\left[\frac{-1}{3}\cos\frac{n\pi}{3}+\frac{1}{n\pi}\sin\frac{n\pi}{3}\right]+\frac{6a}{n\pi}\left[\frac{1}{3}\cos\frac{n\pi}{3}\right]+\frac{3a}{n^2\pi^2}\sin\frac{n\pi}{3}$$

$$\Rightarrow \qquad b_n = \frac{9a}{n^2\pi^2}\sin\frac{n\pi}{3}$$

$$\therefore \quad \text{From (11),} \quad y(x,t) = \frac{9a}{\pi^2}\sum_{1}^{\infty}\frac{1}{n^2}\sin\frac{n\pi}{3}\cos\frac{n\pi ct}{l}\sin\frac{n\pi x}{l}.$$

Example 7. *The points of trisection of a string are pulled aside through the same distance on opposite sides of the position of equilibrium and the string is released from rest. Derive an expression for the displacement of the string at subsequent time and show that the mid-point of the string always remains at rest.*

Sol. The equation for the vibration of the string is

$$\frac{\partial^2 y}{\partial t^2} = c^2\frac{\partial^2 y}{\partial x^2} \qquad \text{...(1)}$$

The solution of eqn. (1) is

$$y(x,t) = (c_1\cos cpt + c_2\sin cpt)(c_3\cos px + c_4\sin px) \qquad \text{...(2)}$$

Let l be the length of string (Refer Sol. of Ex. 1)

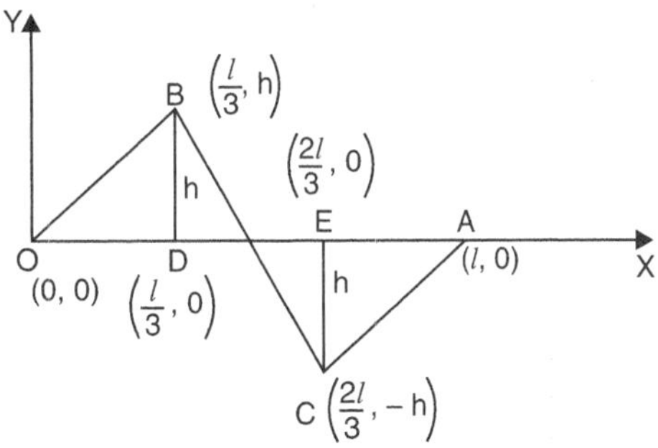

Eqn. of OB is,

$$y - 0 = \frac{h-0}{\frac{l}{3}-0}(x-0)$$

$$\Rightarrow \qquad y = \frac{3h}{l}x \qquad \text{...(3)}$$

Eqn. of BC is,

$$y - h = \frac{-h-h}{\frac{2l}{3}-\frac{l}{3}}\left(x-\frac{l}{3}\right)$$

$$= \frac{-2h}{\left(\frac{l}{3}\right)}\left(x-\frac{l}{3}\right) = -\frac{6h}{l}\left(x-\frac{l}{3}\right)$$

$$y - h = -\frac{6hx}{l}+2h$$

$$y = 3h - \frac{6hx}{l} = 3h\left(1-\frac{2x}{l}\right) \qquad \text{...(4)}$$

Eqn. of CA is,

$$y + h = \frac{0+h}{l-\frac{2l}{3}}\left(x-\frac{2l}{3}\right) = \frac{3h}{l}\left(x-\frac{2l}{3}\right) = \frac{3hx}{l} - 2h$$

$$y = \frac{3hx}{l} - 3h = 3h\left(\frac{x}{l}-1\right) \qquad \text{...(5)}$$

Hence Boundary conditions are

$$y(0, t) = 0, \; y(l, t) = 0$$

$$\frac{\partial y}{\partial t} = 0 \qquad \text{when } t = 0$$

and
$$y(x, 0) = \begin{cases} \dfrac{3h}{l}x, & 0 \le x \le l/3 \\ \dfrac{3h}{l}(l - 2x), & \dfrac{l}{3} \le x \le \dfrac{2l}{3} \\ \dfrac{3h}{l}(x - l), & \dfrac{2l}{3} \le x \le l \end{cases}$$

From (2), $y(0, t) = 0 = (c_1 \cos cpt + c_2 \sin cpt)\, c_3$

$\Rightarrow$ $c_3 = 0.$

$\therefore$ From (2),

$$y\,(x, t) = (c_1 \cos cpt + c_2 \sin cpt)\, c_4 \sin px \qquad ...(6)$$

$$y(l, t) = 0 = (c_1 \cos cpt + c_2 \sin cpt)\, c_4 \sin pl$$

$\Rightarrow$ $\sin pl = 0 = \sin n\pi \; (n \in \text{I})$

$\therefore$ $p = \dfrac{n\pi}{l}.$

$\therefore$ From (6), $$y(x, t) = \left(c_1 \cos \frac{n\pi ct}{l} + c_2 \sin \frac{n\pi ct}{l} \right) c_4 \sin \frac{n\pi x}{l} \qquad ...(7)$$

$$\frac{\partial y}{\partial t} = \frac{n\pi c}{l}\left(-c_1 \sin \frac{n\pi ct}{l} + c_2 \cos \frac{n\pi ct}{l} \right) c_4 \sin \frac{n\pi x}{l}.$$

At $t = 0$, $$\left(\frac{\partial y}{\partial t}\right)_{t=0} = 0 = \frac{n\pi c}{l} c_2 c_4 \sin \frac{n\pi x}{l}$$

$\Rightarrow$ $c_2 = 0$

$\therefore$ From (7), $$y(x, t) = c_1 c_4 \cos \frac{n\pi ct}{l} \sin \frac{n\pi x}{l} = b_n \cos \frac{n\pi ct}{l} \sin \frac{n\pi x}{l}.$$

The most general solution is

$$y(x, t) = \sum_{1}^{\infty} b_n \cos \frac{n\pi ct}{l} \sin \frac{n\pi x}{l} \qquad ...(8)$$

$$y(x, 0) = \sum_{1}^{\infty} b_n \sin \frac{n\pi x}{l}, \text{ where}$$

$$b_n = \frac{2}{l} \int_0^l y(x, 0) \sin \frac{n\pi x}{l}\, dx$$

$$= \frac{2}{l} \left[\int_0^{l/3} \frac{3h}{l} x \sin \frac{n\pi x}{l}\, dx + \int_{l/3}^{2l/3} \frac{3h}{l}(l - 2x) \sin \frac{n\pi x}{l}\, dx + \int_{2l/3}^{l} \frac{3h}{l}(x - l) \sin \frac{n\pi x}{l}\, dx \right]$$

$$= \frac{2}{l}\cdot\frac{3h}{l}\int_0^{l/3} x\sin\frac{n\pi x}{l}\,dx + \frac{2}{l}\cdot\frac{3h}{l}\int_{l/3}^{2l/3}(l-2x)\sin\frac{n\pi x}{l}\,dx + \frac{2}{l}\cdot\frac{3h}{l}\int_{2l/3}^{l}(x-l)\sin\frac{n\pi x}{l}\,dx$$

$$= \frac{6h}{l^2}\left[\left\{x\left(\frac{-\cos\frac{n\pi x}{l}}{\frac{n\pi}{l}}\right)\right\}_0^{l/3} - \int_0^{l/3} 1.\left(\frac{-\cos\frac{n\pi x}{l}}{\frac{n\pi}{l}}\right)dx\right]$$

$$+ \frac{6h}{l^2}\left[\left\{(l-2x)\left(\frac{-\cos\frac{n\pi x}{l}}{\frac{n\pi}{l}}\right)\right\}_{l/3}^{2l/3} - \int_{l/3}^{2l/3}(-2)\left(\frac{-\cos\frac{n\pi x}{l}}{\frac{n\pi}{l}}\right)dx\right]$$

$$+ \frac{6h}{l^2}\left[\left\{(x-l)\left(\frac{-\cos\frac{n\pi x}{l}}{\frac{n\pi}{l}}\right)\right\}_{2l/3}^{l} - \int_{2l/3}^{l} 1.\left(\frac{-\cos\frac{n\pi x}{l}}{\frac{n\pi}{l}}\right)dx\right]$$

$$= \frac{6h}{l^2}\left[\frac{-l}{n\pi}\cdot\frac{l}{3}\cos\frac{n\pi}{3} + \frac{l}{n\pi}\left(\frac{\sin\frac{n\pi x}{l}}{\frac{n\pi}{l}}\right)_0^{l/3}\right]$$

$$+ \frac{6h}{l^2}\left[\frac{-l}{3}\cdot\frac{l}{n\pi}\left(-\cos\frac{2n\pi}{3}\right) + \left(\cos\frac{n\pi}{3}\right)\frac{l}{3}\cdot\frac{l}{n\pi} - \frac{2l}{n\pi}\left(\frac{\sin\frac{n\pi x}{l}}{\frac{n\pi}{l}}\right)_{l/3}^{2l/3}\right]$$

$$+ \frac{6h}{l^2}\left[\frac{-l}{3}\cdot\frac{l}{n\pi}\cos\frac{2n\pi}{3} + \frac{l}{n\pi}\left(\frac{\sin n\pi x/l}{n\pi/l}\right)_{2l/3}^{l}\right]$$

$$= \frac{-2h}{n\pi}\cos\frac{n\pi}{3} + \frac{6h}{n^2\pi^2}\sin\frac{n\pi}{3} + \frac{2h}{n\pi}\cos\frac{2n\pi}{3}$$

$$+ \frac{2h}{n\pi}\cos\frac{n\pi}{3} - \frac{12h}{n^2\pi^2}\left(\sin\frac{2n\pi}{3} - \sin\frac{n\pi}{3}\right) - \frac{2h}{n\pi}\cos\frac{2n\pi}{3} + \frac{6h}{n^2\pi^2}\left(0 - \sin\frac{2n\pi}{3}\right)$$

$$= \frac{18h}{n^2\pi^2}\sin\frac{n\pi}{3} - \frac{18h}{n^2\pi^2}\sin\frac{2n\pi}{3} = \frac{18h}{n^2\pi^2}\sin\frac{n\pi}{3} - \frac{18h}{n^2\pi^2}\sin\left(n\pi - \frac{n\pi}{3}\right)$$

$$= \frac{18h}{n^2\pi^2}\sin\frac{n\pi}{3} + \frac{18h}{n^2\pi^2}\sin\frac{n\pi}{3}\cos n\pi$$

$$= \begin{cases} \frac{36h}{n^2\pi^2}\sin\frac{n\pi}{3}, & \text{when } n \text{ is even} \\ 0, & \text{when } n \text{ is odd} \end{cases}$$

$\therefore$ From (8), $$y(x,t) = \frac{36h}{\pi^2}\sum_{n=2,4,\ldots}^{\infty}\frac{1}{n^2}\sin\frac{n\pi}{3}\cos\frac{n\pi ct}{l}\sin\frac{n\pi x}{l}$$

$$y(x, t) = \frac{9h}{\pi^2} \sum_{m=1,2,\ldots}^{\infty} \frac{1}{m^2} \sin \frac{2m\pi}{3} \cos \frac{2m\pi ct}{l} \sin \frac{2m\pi x}{l} \quad \ldots(9)$$

(where $n = 2m$)

Putting $x = \frac{l}{2}$ in eqn. (6), we get

$$y\left(\frac{l}{2}, t\right) = \frac{9h}{\pi^2} \sum_{m=1}^{\infty} \sin\left(\frac{2m\pi}{3}\right) . \frac{1}{m^2} . \cos \frac{2m\pi ct}{l} . \sin m\pi = 0.$$

Hence, midpoint of the string is always at rest.

Example 8. *If a string of length l is initially at rest in equilibrium position and each of its points is given the velocity* $\left(\frac{\partial y}{\partial t}\right)_{t=0} = b \sin^3 \frac{\pi x}{l}$, *find the displacement y (x, t).*

Sol. The equation for the vibrations of the string is

$$\frac{\partial^2 y}{\partial t^2} = c^2 \frac{\partial^2 y}{\partial x^2} \quad \ldots(1)$$

The solution of equation (1) is

$y(x, t) = (c_1 \cos cpt + c_2 \sin cpt)(c_3 \cos px + c_4 \sin px)$ ...(2) [Refer Sol. of Ex. 1]

Boundary conditions are,

$$y(0, t) = 0 \quad \ldots(3)$$

$$y(l, t) = 0 \quad \ldots(4)$$

$$y(x, 0) = 0 \quad \ldots(5)$$

$$\left(\frac{\partial y}{\partial t}\right) = b \sin^3 \frac{\pi x}{l} \text{ at } t = 0 \quad \ldots(6)$$

From eqn. (2), $y(0, t) = 0 = (c_1 \cos cpt + c_2 \sin cpt)\, c_3$

$\Rightarrow$ $c_3 = 0.$

$\therefore$ From (2), $y(x, t) = (c_1 \cos cpt + c_2 \sin cpt)\, c_4 \sin px$ ...(7)

$y(l, t) = 0 = (c_1 \cos cpt + c_2 \sin cpt)\, c_4 \sin pl$

$\Rightarrow$ $\sin pl = 0 = \sin n\pi \ (n \in \text{I})$

$\therefore$ $p = \frac{n\pi}{l}.$

$\therefore$ From (7), $y(x, t) = \left(c_1 \cos \frac{n\pi ct}{l} + c_2 \sin \frac{n\pi ct}{l}\right) c_4 \sin \frac{n\pi x}{l}$ ...(8)

$y(x, 0) = 0 = c_1 c_4 \sin \frac{n\pi x}{l}$

$\Rightarrow$ $c_1 = 0.$

$\therefore$ From (8), $y(x, t) = c_2 c_4 \sin \frac{n\pi ct}{l} \sin \frac{n\pi x}{l}$

$$= b_n \sin \frac{n\pi ct}{l} \sin \frac{n\pi x}{l} \text{ where } c_2c_4 = b_n$$

The general solution is

$$y(x, t) = \sum_1^\infty b_n \sin \frac{n\pi ct}{l} \sin \frac{n\pi x}{l} \qquad ...(9)$$

$$\frac{\partial y}{\partial t} = \sum_1^\infty b_n \cdot \frac{n\pi c}{l} \cos \frac{n\pi ct}{l} \sin \frac{n\pi x}{l}$$

At $t = 0$,

$$\left(\frac{\partial y}{\partial t}\right)_{t=0} = \sum_1^\infty b_n \cdot \frac{n\pi c}{l} \sin \frac{n\pi x}{l}$$

$$b \sin^3 \frac{\pi x}{l} = \sum_1^\infty b_n \cdot \frac{n\pi c}{l} \sin \frac{n\pi x}{l}$$

$$\frac{b}{4}\left[3 \sin \frac{\pi x}{l} - \sin \frac{3\pi x}{l}\right] = b_1 \frac{\pi c}{l} \sin \frac{\pi x}{l} + \frac{2b_2 \pi c}{l} \sin \frac{2\pi x}{l} + 3b_3 \frac{\pi c}{l} \sin \frac{3\pi x}{l} + \ldots$$

$$\Rightarrow \qquad b_1 \frac{\pi c}{l} = \frac{3b}{4} \quad \Rightarrow \quad b_1 = \frac{3bl}{4\pi c}$$

$$b_2 = 0 \quad \text{and} \quad \frac{3b_3 \pi c}{l} = -\frac{b}{4} \quad \Rightarrow \quad b_3 = -\frac{bl}{12\pi c}$$

Also, $b_4 = 0 = b_5 = \ldots$ etc.

Hence from (9),

$$y(x, t) = \frac{3bl}{4\pi c} \sin \frac{\pi ct}{l} \sin \frac{\pi x}{l} - \frac{bl}{12\pi c} \sin \frac{3\pi ct}{l} \sin \frac{3\pi x}{l}$$

$$= \frac{bl}{12\pi c}\left[9 \sin \frac{\pi x}{l} \sin \frac{\pi ct}{l} - \sin \frac{3\pi x}{l} \sin \frac{3\pi ct}{l}\right].$$

Example 9. *A tightly stretched string with fixed end points $x = 0$ and $x = l$ is initially at rest in its equilibrium position. If it is set vibrating by giving to each of its points an initial velocity $\lambda x(l - x)$, find the displacement of the string at any distance x from one end at any time t.* [M.T.U. (SUM) 2011]

Sol. Here the boundary conditions are $y(0, t) = y(l, t) = 0$

$$y(x, t) = \sum_{n=1}^\infty \left(a_n \cos \frac{n\pi ct}{l} + b_n \sin \frac{n\pi ct}{l}\right) \sin \frac{n\pi x}{l} \qquad ...(1)$$ | Refer Sol. of Ex. 2

Since the string was at rest initially, $y(x, 0) = 0$

$\therefore$ From (1),

$$0 = \sum_{n=1}^\infty a_n \sin \frac{n\pi x}{l} \quad \Rightarrow \quad a_n = 0$$

$$\therefore \qquad y(x, t) = \sum_{n=1}^\infty b_n \sin \frac{n\pi ct}{l} \sin \frac{n\pi x}{l} \qquad ...(2)$$

and

$$\frac{\partial y}{\partial t} = \sum_{n=1}^\infty \frac{n\pi c}{l} b_n \cos \frac{n\pi ct}{l} \sin \frac{n\pi x}{l} = \frac{\pi c}{l} \sum_{n=1}^\infty n b_n \cos \frac{n\pi ct}{l} \sin \frac{n\pi x}{l}$$

But

$$\frac{\partial y}{\partial t} = \lambda x(l - x) \quad \text{when } t = 0$$

$\therefore \qquad \lambda x(l-x) = \dfrac{\pi c}{l} \sum_{n=1}^{\infty} n b_n \sin \dfrac{n\pi x}{l}$

$\Rightarrow \qquad \dfrac{\pi c}{l} n b_n = \dfrac{2}{l} \int_0^l \lambda x(l-x) \sin \dfrac{n\pi x}{l}\, dx$

$$= \frac{2\lambda}{l}\left[x(l-x)\left(-\frac{l}{n\pi}\cos\frac{n\pi x}{l}\right) - (l-2x)\left(-\frac{l^2}{n^2\pi^2}\sin\frac{n\pi x}{l}\right) + (-2)\left(\frac{l^3}{n^3\pi^3}\cos\frac{n\pi x}{l}\right)\right]_0^l$$

$$= \frac{4\lambda l^2}{n^3\pi^3}(1-\cos n\pi) = \frac{4\lambda l^2}{n^3\pi^3}[1-(-1)^n]$$

$$= \begin{cases} 0, & \text{when } n \text{ is even} \\ \dfrac{8\lambda l^2}{n^3\pi^3}, & \text{when } n \text{ is odd} \end{cases} \quad i.e., \ \frac{8\lambda l^2}{\pi^3(2m-1)^3}, \text{ taking } n = 2m-1$$

$$\Rightarrow \ b_n = \frac{8\lambda l^3}{c\pi^4(2m-1)^4}$$

$\therefore$ From (2), the required solution is

$$y(x, t) = \frac{8\lambda l^3}{c\pi^4} \sum_{m=1}^{\infty} \frac{1}{(2m-1)^4} \sin\frac{(2m-1)\pi ct}{l} \sin\frac{(2m-1)\pi x}{l}.$$

Example 10. *Transform the equation* $\dfrac{\partial^2 y}{\partial t^2} = c^2 \dfrac{\partial^2 y}{\partial x^2}$ *to its normal form using the transformation* $u = x + ct$, $v = x - ct$ *and hence solve it. Show that the solution may be put in the form* $y = \dfrac{1}{2}[f(x+ct) + f(x-ct)]$. (M.T.U. 2013)

Assume initial conditions $y = f(x)$ *and* $(\partial y/\partial t) = 0$ *at* $t = 0$.

Sol. One-dimensional wave equation is

$$\frac{\partial^2 y}{\partial t^2} = c^2 \frac{\partial^2 y}{\partial x^2} \qquad ...(1)$$

Let us introduce two new independent variables

$$u = x + ct \qquad ...(2)$$

and

$$v = x - ct \qquad ...(3)$$

so that y becomes a function of u and v.

Then, $\dfrac{\partial y}{\partial x} = \dfrac{\partial y}{\partial u} \cdot \dfrac{\partial u}{\partial x} + \dfrac{\partial y}{\partial v} \cdot \dfrac{\partial v}{\partial x} = \dfrac{\partial y}{\partial u} + \dfrac{\partial y}{\partial v}$ ...(4) [Using (2) and (3)]

Also, $\dfrac{\partial}{\partial x} \equiv \dfrac{\partial}{\partial u} + \dfrac{\partial}{\partial v}$ ...(5)

$$\therefore \quad \frac{\partial^2 y}{\partial x^2} = \frac{\partial}{\partial x}\left(\frac{\partial y}{\partial x}\right) = \left(\frac{\partial}{\partial u} + \frac{\partial}{\partial v}\right)\left(\frac{\partial y}{\partial u} + \frac{\partial y}{\partial v}\right)$$

$$= \frac{\partial^2 y}{\partial u^2} + 2\frac{\partial^2 y}{\partial u \partial v} + \frac{\partial^2 y}{\partial v^2} \qquad ...(6)$$

Also, $$\frac{\partial y}{\partial t} = \frac{\partial y}{\partial u} \cdot \frac{\partial u}{\partial t} + \frac{\partial y}{\partial v} \cdot \frac{\partial v}{\partial t} = c\frac{\partial y}{\partial u} + \frac{\partial y}{\partial v}(-c) = c\left(\frac{\partial y}{\partial u} - \frac{\partial y}{\partial v}\right) \quad ...(7)$$

$$\Rightarrow \quad \frac{\partial}{\partial t} \equiv c\left(\frac{\partial}{\partial u} - \frac{\partial}{\partial v}\right) \quad ...(8)$$

$$\therefore \quad \frac{\partial^2 y}{\partial t^2} = \frac{\partial}{\partial t}\left(\frac{\partial y}{\partial t}\right) = c\left(\frac{\partial}{\partial u} - \frac{\partial}{\partial v}\right) c\left(\frac{\partial y}{\partial u} - \frac{\partial y}{\partial v}\right)$$

$$= c^2\left(\frac{\partial^2 y}{\partial u^2} - 2\frac{\partial^2 y}{\partial u \partial v} + \frac{\partial^2 y}{\partial v^2}\right) \quad ...(9)$$

From (1), (6) and (9), we have

$$c^2\left(\frac{\partial^2 y}{\partial u^2} - 2\frac{\partial^2 y}{\partial u \partial v} + \frac{\partial^2 y}{\partial v^2}\right) = c^2\left(\frac{\partial^2 y}{\partial u^2} + 2\frac{\partial^2 y}{\partial u \partial v} + \frac{\partial^2 y}{\partial v^2}\right)$$

$$\Rightarrow \quad -4c^2 \frac{\partial^2 y}{\partial u \partial v} = 0$$

$$\Rightarrow \quad \frac{\partial^2 y}{\partial u \partial v} = 0 \quad ...(10) \quad (\because \ c^2 \neq 0)$$

Integrating eqn. (10) partially, w.r.t. u, we get

$$\frac{\partial y}{\partial v} = f_1(v).$$

Integrating again w.r.t. v partially, we get

$$y = \int f_1(v)\, \partial v + \psi(u) = \phi(v) + \psi(u)$$

$$\Rightarrow \quad y(x, t) = \phi(x - ct) + \psi(x + ct) \quad ...(11)$$

which is *d'* Alembert's solution of wave equation.

Applying initial conditions $y = f(x)$ and $\frac{\partial y}{\partial t} = 0$ at $t = 0$ in (11), we get

$$f(x) = \phi(x) + \psi(x) \text{ and } 0 = -\phi'(x) + \psi'(x)$$

Hence, $$\phi(x) = \psi(x) = \frac{1}{2} f(x)$$

$$\therefore \quad y = \frac{1}{2}[f(x + ct) + f(x - ct)].$$

Example 11. *A tightly stretched string with fixed end points $x = 0$ and $x = \pi$ is initially at rest in its equilibrium position. If it is set vibrating by giving to each of its points an initial velocity* $\left(\frac{\partial y}{\partial t}\right)_{t=0} = 0.03 \sin x - 0.04 \sin 3x$

then find the displacement $y(x, t)$ at any point of string at any time t.

Sol. The equation for the vibrations of a string is

$$\frac{\partial^2 y}{\partial t^2} = c^2 \frac{\partial^2 y}{\partial x^2} \quad ...(1)$$

Its solution is

$$y(x, t) = (c_1 \cos pct + c_2 \sin pct)(c_3 \cos px + c_4 \sin px) \quad ...(2)$$

Boundary conditions are

$$y(0, t) = 0 = y(\pi, t)$$

$$y(x, 0) = 0$$

and $$\left(\frac{\partial y}{\partial t}\right)_{t=0} = 0.03 \sin x - 0.04 \sin 3x.$$

From (2), $y(0, t) = 0 = (c_1 \cos pct + c_2 \sin pct)\, c_3$

$\Rightarrow$ $c_3 = 0$

From (2), $y(x, t) = (c_1 \cos pct + c_2 \sin pct)\, c_4 \sin px$ ...(3)

$y(\pi, t) = 0 = (c_1 \cos pct + c_2 \sin pct)\, c_4 \sin p\pi$

$\Rightarrow$ $\sin p\pi = 0 = \sin n\pi \; (n \in \text{I})$

$\Rightarrow$ $p = n$

From (3), $y(x, t) = (c_1 \cos nct + c_2 \sin nct)\, c_4 \sin nx$ ...(4)

$y(x, 0) = 0 = c_1 c_4 \sin nx$

$\Rightarrow$ $c_1 = 0.$

$\therefore$ From (4), $y(x, t) = c_2 c_4 \sin nct \sin nx = b_n \sin nct \sin nx$ ...(5)

where $c_2 c_4 = b_n$

The most general solution is

$$y(x, t) = \sum_{n=1}^{\infty} b_n \sin nct \sin nx$$

$$\frac{\partial y}{\partial t} = \sum_{n=1}^{\infty} nc\, b_n \cos nct \sin nx$$

At $t = 0$,

$$\left(\frac{\partial y}{\partial t}\right)_{t=0} = \sum_{n=1}^{\infty} nc\, b_n \sin nx$$

$$0.03 \sin x - 0.04 \sin 3x = cb_1 \sin x + 2cb_2 \sin 2x + 3cb_3 \sin 3x + \ldots.$$

$\Rightarrow$ $cb_1 = 0.03 \;\Rightarrow\; b_1 = \dfrac{0.03}{c}$

$b_2 = 0$

and $3cb_3 = -0.04 \;\Rightarrow\; b_3 = \dfrac{-0.0133}{c}.$

$\therefore$ From (6),

$$y(x, t) = \frac{0.03}{c} \sin ct \sin x - \frac{0.0133}{c} \sin 3ct \sin 3x$$

$$= \frac{1}{c}\,[0.03 \sin x \sin ct - 0.0133 \sin 3x \sin 3ct].$$

TEST YOUR KNOWLEDGE

1. Find the deflection $y(x, t)$ of the vibrating string of length π and ends fixed, corresponding to zero initial velocity and initial deflection $f(x) = k(\sin x - \sin 2x)$ given $c^2 = 1$.

2. Solve: $y_{tt} = 4y_{xx}$; $y(0, t) = 0 = y(5, t), y(x, 0) = 0 \left(\frac{\partial y}{\partial t}\right)_{t=0} = f(x)$

   if (i) $f(x) = 5 \sin \pi x$ (ii) $f(x) = 3 \sin 2\pi x - 2 \sin 5\pi x$.

3. Find the deflection of the vibrating string which is fixed at the ends $x = 0$ and $x = 2$ and the motion is started by displacing the string into the form $\sin^3\left(\frac{\pi x}{2}\right)$ and releasing it with zero initial velocity at $t = 0$. *(U.P.T.U. 2014)*

4. Find the solution of the equation of a vibrating string of length l satisfying the initial conditions :

   $$y = F(x) \qquad \text{when } t = 0$$

   and $$\frac{\partial y}{\partial t} = \phi(x) \qquad \text{when } t = 0$$

   It is assumed that the equation of a vibrating string is $y_{tt} = a^2 y_{xx}$.

5. The vibrations of an elastic string is governed by the partial differential equation

   $$\frac{\partial^2 u}{\partial t^2} = \frac{\partial^2 u}{\partial x^2}$$

   The length of the string is π and ends are fixed. The initial velocity is zero and the initial deflection is $u(x, 0) = 2(\sin x + \sin 3x)$. Find the deflection $u(x, t)$ of the vibrating string at any time t.

6. A tight string of length 20 cms fastened at both ends is displaced from its position of equilibrium by imparting to each of its points an initial velocity given by

   $$v = \left.\begin{matrix} x\,; & 0 \le x \le 10 \\ 20 - x\,; & 10 \le x \le 20 \end{matrix}\right\} ;$$

   x being the distance from one end. Determine the displacement at any subsequent time.

7. Using d' Alembert's method, find the deflection of a vibrating string of unit length having fixed ends, with initial velocity zero and initial deflection $f(x) = a(x - x^3)$.

8. Reduce the equation $u_{xx} - 2u_{xy} + u_{yy} = 0$ to its normal form using the transformation $v = x$, $z = x + y$ and solve it.

9. Solve the equation $\frac{\partial^2 u}{\partial x^2} + \frac{\partial^2 u}{\partial x \partial y} - 2\frac{\partial^2 u}{\partial y^2} = 0$ using the transformation $v = x + y, z = 2x - y$.

10. The ends of a tightly stretched string of length l are fixed at $x = 0$ and $x = l$. The string is at rest with the point $x = b$ drawn aside through a small distance d and released at time $t = 0$. Show that

    $$y(x, t) = \frac{2dl^2}{\pi^2 b(l - b)} \sum_{n=1}^{\infty} \frac{1}{n^2} \sin \frac{n\pi b}{l} \sin \frac{n\pi x}{l} \cos \frac{n\pi ct}{l}.$$

11. Find the deflection of the vibrating string of unit length whose end points are fixed if the initial velocity is zero and the initial deflection is given by $u(x, 0) = \begin{cases} 1, & 0 \le x \le \frac{1}{2} \\ -1, & \frac{1}{2} < x \le 1 \end{cases}$ *(U.P.T.U. 2014)*

12. Find the deflection $u(x, t)$ of a tightly stretched vibrating string of unit length that is initially at rest and whose initial position is given by

$$\sin \pi x + \frac{1}{3} \sin 3\pi x + \frac{1}{5} \sin 5\pi x, \quad 0 \le x \le 1$$ *(G.B.T.U. 2013)*

13. A string is stretched and fastened to two points distance l apart. Find the displacement $y(x, t)$ at any point at a distance x from one end at time t given that:

$$y(x, 0) = A \sin\left(\frac{2\pi x}{l}\right)$$ *(M.T.U. 2013)*

14. Solve the following initial-boundary value problem (IBVP):

$$\frac{\partial^2 u}{\partial t^2} = \frac{\partial^2 u}{\partial x^2}, \qquad 0 < x < 1, \quad t > 0$$

$$u(0, t) = 2, \qquad u(1, t) = 3, \quad t \ge 0$$

$$u(x, 0) = 2 + x + \sin \pi x$$

$$\left(\frac{\partial u}{\partial t}\right)_{t=0} = 0.$$

15. Solve the following initial-boundary value problem (IBVP):

$$\frac{\partial^2 u}{\partial t^2} = \frac{\partial^2 u}{\partial x^2}, \qquad 0 < x < 1, \quad t > 0$$

$$u_x(0, t) = 1, \qquad u_x(1, t) = 3, \quad t \ge 0$$

$$u(x, 0) = x + x^2, \qquad u_t(x, 0) = \pi \cos \pi x.$$

16. Use D'Alembert's formula to solve the IVP:

$$u_t = 4u_{xx}, \qquad -\infty < x < \infty, \quad t > 0$$

$$u(x, 0) = \sin x, \quad u_t(x, 0) = 4.$$

17. If $u(x, t) = \sin x \cos t + 2xt$ is a solution of the initial value problem

$$u_{tt} = u_{xx}, \qquad -\infty < x < \infty, \quad t > 0$$

$$u(x, 0) = f(x), \quad u_t(x, 0) = 2x$$

Use D'Alembert formula to find $f(x)$.

Answers

1. $y(x, t) = k(\cos t \sin x - \cos 2t \sin 2x)$

2. (*i*) $y(x, t) = \frac{5}{2\pi} \sin \pi x \sin 2\pi t$ (*ii*) $y(x, t) = \frac{3}{4\pi} \sin 2\pi x \sin 4\pi t - \frac{1}{5\pi} \sin 5\pi x \sin 10\pi t$

3. $y(x, t) = \frac{3}{4} \sin \frac{\pi x}{2} \cos \frac{\pi ct}{2} - \frac{1}{4} \sin \frac{3\pi x}{2} \cos \frac{3\pi ct}{2}$

4. $y(x, t) = \sum_{n=1}^{\infty} \sin \frac{n\pi x}{l} \left(a_n \cos \frac{n\pi at}{l} + b_n \sin \frac{n\pi at}{l} \right)$

where $a_n = \frac{2}{l} \int_0^l F(x) \sin \frac{n\pi x}{l} \, dx$ and $b_n = \frac{2}{n a \pi} \int_0^l \phi(x) \sin \frac{n\pi x}{l} \, dx$

5. $y(x, t) = 2[\cos t \sin x + \cos 3t \sin 3x]$

6. $y(x, t) = \frac{1600}{a\pi^3} \left[\sin \frac{\pi x}{20} \sin \frac{\pi at}{20} - \frac{1}{3^3} \sin \frac{3\pi x}{20} \sin \frac{3\pi at}{20} + \ldots\ldots \right]$

7. $y(x, t) = ax(1 - x^2 - 3c^2t^2)$

8. $\frac{\partial^2 u}{\partial v^2} = 0$; $u = xf_1(x + y) + f_2(x + y)$.

9. $\frac{\partial^2 u}{\partial v \partial z} = 0$; $u = f_1(x + y) + f_2(2x - y)$

11. $y(x, t) = \frac{2}{\pi} \sum_{n=1}^{\infty} \frac{1}{n} \left(1 + \cos n\pi - 2\cos \frac{n\pi}{2} \right) \sin n\pi x \cos n\pi ct$

12. $u(x, t) = \sin \pi x \cos \pi ct + \frac{1}{3} \sin 3\pi x \cos 3\pi ct + \frac{1}{5} \sin 5\pi x \cos 5\pi ct$

13. $y(x, t) = A \sin \left(\frac{2\pi x}{l} \right) \cos \left(\frac{2\pi ct}{l} \right)$

14. $u(x, t) = 2 + x + \sin \pi x \cos \pi t$

15. $u(x, t) = x + x^2 + t^2 + \cos \pi x \sin \pi t$

16. $u(x, t) = \sin x \cos 2t + 4t$

17. $f(x) = \sin x$

2.8 VIBRATING MEMBRANE—TWO-DIMENSIONAL WAVE EQUATION

We shall now obtain the equation for the vibrations of a tightly stretched membrane (such as the membrane of a drum). We shall assume that the membrane is uniform and the tension in it per unit length is the same at every point in all directions. Let T be the tension per unit length and m be the mass of the membrane per unit area.

Consider the forces on an element $\delta x \delta y$ of the membrane. Due to its displacement u, perpendicular to the xy-plane, the forces $T\delta y$ (tangential to the membrane) on its opposite edges of length δy act at angles α and β to the horizontal. So their vertical component

$= (T\delta y) \sin \beta - (T\delta y) \sin \alpha = T\delta y(\tan \beta - \tan \alpha)$ approximately, since α and β are small

$$= T\delta y \left[\left(\frac{\partial u}{\partial x} \right)_{x+\delta x} - \left(\frac{\partial u}{\partial x} \right)_x \right] = T\delta y \, \delta x \left[\frac{\left(\frac{\partial u}{\partial x} \right)_{x+\delta x} - \left(\frac{\partial u}{\partial x} \right)_x}{\delta x} \right]$$

$= T\delta x \delta y \frac{\partial^2 u}{\partial x^2}$ upto a first order of approximation.

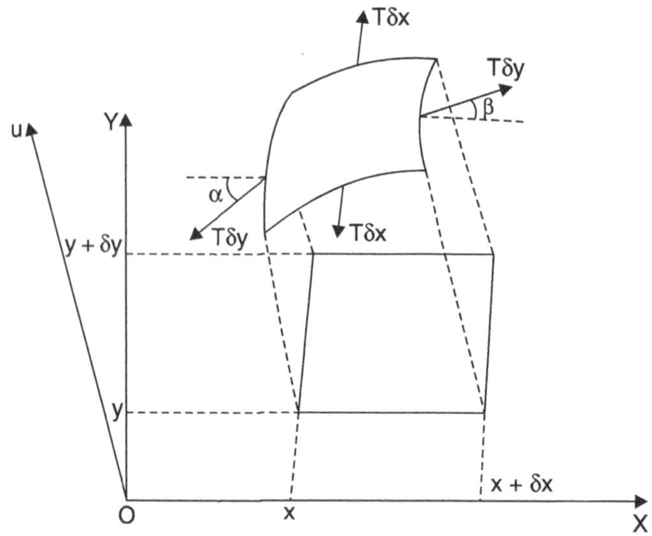

Similarly, the forces $T\delta x$ acting on the edges of length δx have the vertical component $T\delta x\delta y \dfrac{\partial^2 u}{\partial y^2}$.

Hence the equation of motion of the element is

$$(m\delta x\delta y)\frac{\partial^2 u}{\partial t^2} = T\left(\frac{\partial^2 u}{\partial x^2} + \frac{\partial^2 u}{\partial y^2}\right)\delta x\delta y$$

or

$$\boxed{\frac{\partial^2 u}{\partial t^2} = c^2\left(\frac{\partial^2 u}{\partial x^2} + \frac{\partial^2 u}{\partial y^2}\right)} \quad \text{where } c^2 = \frac{T}{m}.$$

This is the wave equation in two dimensions.

2.9 SOLUTION OF TWO-DIMENSIONAL WAVE EQUATION

Two-dimensional wave equation is

$$\frac{\partial^2 u}{\partial t^2} = c^2\left(\frac{\partial^2 u}{\partial x^2} + \frac{\partial^2 u}{\partial y^2}\right) \quad ...(1)$$

Let $u = XYT$ ...(2)

where X is a function of x only, Y is a function of y only and T is a function of t only, be a solution of (1).

Then, $\dfrac{\partial^2 u}{\partial t^2} = XYT''$, $\dfrac{\partial^2 u}{\partial x^2} = X''YT$ and $\dfrac{\partial^2 u}{\partial y^2} = XY''T$

Substituting in (1), we have $\dfrac{1}{c^2} XYT'' = X''YT + XY''T$

Dividing by XYT, we have $$\frac{1}{c^2}\frac{T''}{T} = \frac{X''}{X} + \frac{Y''}{Y} \quad ...(3)$$

This will be true only when *each member is a constant.* Choosing the constants suitably, we have

$$\frac{d^2X}{dx^2} + k^2X = 0, \quad \frac{d^2Y}{dy^2} + l^2Y = 0 \quad \text{and} \quad \frac{d^2T}{dt^2} + (k^2 + l^2)\, c^2T = 0$$

The solutions of these equations are respectively

$$X = c_1 \cos kx + c_2 \sin kx$$
$$Y = c_3 \cos ly + c_4 \sin ly$$

and $$T = c_5 \cos \sqrt{(k^2 + l^2)}\, ct + c_6 \sin \sqrt{(k^2 + l^2)}\, ct$$

Hence from (2), a solution of (1) is

$$u(x, y, t) = (c_1 \cos kx + c_2 \sin kx)(c_3 \cos ly + c_4 \sin ly)$$
$$[c_5 \cos \sqrt{(k^2 + l^2)}\, ct + c_6 \sin \sqrt{(k^2 + l^2)}\, ct] \quad \text{...(4)}$$

Now let us suppose that the membrane is rectangular and is stretched between the lines

$$x = 0, x = a, y = 0, y = b.$$

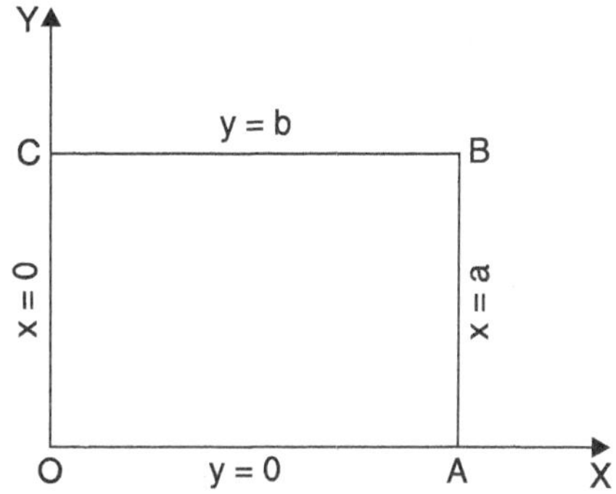

Then the boundary conditions are:

(*i*) $u = 0$, when $x = 0$

(*ii*) $u = 0$, when $x = a$

(*iii*) $u = 0$, when $y = 0$

(*iv*) $u = 0$, when $y = b$ for all t.

Applying the condition (*i*), we have

$$0 = c_1(c_3 \cos ly + c_4 \sin ly)[c_5 \cos \sqrt{(k^2 + l^2)}\, ct + c_6 \sin \sqrt{(k^2 + l^2)}\, ct]$$

i.e., $$c_1 = 0$$

Putting $c_1 = 0$ in (3) and applying the condition (*ii*), we have $\sin ka = 0$ or $k = \dfrac{m\pi}{a}$, where m is an integer.

Similarly, applying the conditions (*iii*) and (*iv*), we get

$$c_3 = 0 \quad \text{or} \quad l = \frac{n\pi}{b}, \text{ where } n \text{ is an integer.}$$

Therefore, the solution (3) becomes

$$u(x, y, t) = c_2c_4 \sin \frac{m\pi x}{a} \sin \frac{n\pi y}{b} (c_5 \cos pt + c_6 \sin pt)$$

where $p = \pi c \sqrt{\dfrac{m^2}{a^2} + \dfrac{n^2}{b^2}}$

Replacing the arbitrary constants suitably, we can write the general solution as

$$u(x, y, t) = \sum_{m=1}^{\infty} \sum_{n=1}^{\infty} \sin \frac{m\pi x}{a} \sin \frac{n\pi y}{b} (A_{mn} \cos pt + B_{mn} \sin pt) \quad \text{...(5)}$$

Now suppose the membrane starts from rest from the initial position $u = f(x, y)$ *i.e.*, $u(x, y, 0) = f(x, y)$.

Then applying the condition: $\frac{\partial u}{\partial t} = 0$ when $t = 0$, we get $B_{mn} = 0$.

Also using the condition: $u = f(x, y)$ when $t = 0$, we get

$$f(x, y) = \sum_{m=1}^{\infty} \sum_{n=1}^{\infty} A_{mn} \sin \frac{m\pi x}{a} \sin \frac{n\pi y}{b} \qquad ...(6)$$

This is a double Fourier series. Multiplying both sides by $\sin \frac{m\pi x}{a} \sin \frac{n\pi y}{b}$ and integrating from $x = 0$ to $x = a$ and $y = 0$ to $y = b$, every term on the right except one becomes zero. Thus, we get

$$\int_0^a \int_0^b f(x, y) \sin \frac{m\pi x}{a} \sin \frac{n\pi y}{b}\, dy\, dx = \frac{ab}{4} A_{mn}$$

i.e.,

$$A_{mn} = \frac{4}{ab} \int_0^a \int_0^b f(x, y) \sin \frac{m\pi x}{b} \sin \frac{n\pi y}{a}\, dy\, dx \qquad ...(7)$$

Hence, from (5), the required solution is

$$u(x, y, t) = \sum_{m=1}^{\infty} \sum_{n=1}^{\infty} A_{mn} \sin \frac{m\pi x}{a} \sin \frac{n\pi y}{b} \cos pt$$

where A_{mn} is given by (7) and $p = \pi c \sqrt{\frac{m^2}{a^2} + \frac{n^2}{b^2}}$.

ILLUSTRATIVE EXAMPLES

Example 1. *Find the deflection u(x, y, t) of the square membrane with a = b = c = 1, if the initial velocity is zero and the initial deflection f(x, y) = A sin πx sin 2πy.*

Sol. The vibrations of the square membrane are governed by two-dimensional wave equation

$$\frac{\partial^2 u}{\partial t^2} = c^2 \left(\frac{\partial^2 u}{\partial x^2} + \frac{\partial^2 u}{\partial y^2} \right) \qquad ...(1)$$

Here the boundary conditions are,

$$u(0, y, t) = 0,\ u(1, y, t) = 0, \quad u(x, 0, t) = 0,\ u(x, 1, t) = 0$$

and the initial conditions are

$$u(x, y, 0) = f(x, y) = A \sin \pi x \sin 2\pi y \quad \text{and} \quad \left(\frac{\partial u}{\partial t} \right)_{t=0} = 0$$

To solve eqn. (1), let $u = XYT$ ...(2)

where X is function of x only, Y is a function of y only, and T is a function of t only.

$$\frac{\partial^2 u}{\partial t^2} = \frac{\partial^2}{\partial t^2}(XYT) = XY \frac{d^2 T}{dt^2}$$

$$\frac{\partial^2 u}{\partial x^2} = \frac{\partial^2}{\partial x^2}(XYT) = YT \frac{d^2 X}{dx^2}$$

$$\frac{\partial^2 u}{\partial y^2} = \frac{\partial^2}{\partial y^2}(\text{XYT}) = \text{XT}\frac{d^2\text{Y}}{dy^2},$$

From (1), $\quad \text{XYT}'' = (\text{YTX}'' + \text{XTY}'')c^2$

$$\Rightarrow \quad \frac{1}{c^2}\frac{\text{T}''}{\text{T}} = \frac{\text{X}''}{\text{X}} + \frac{\text{Y}''}{\text{Y}}$$

This will be true only when each member is a constant. Choosing the constant suitably, we have

$$\frac{d^2\text{X}}{dx^2} + k^2\text{X} = 0, \ \frac{d^2\text{Y}}{dy^2} + l^2\text{Y} = 0 \ \text{ and } \ \frac{d^2\text{T}}{dt^2} + (k^2 + l^2)\,c^2\text{T} = 0.$$

The solutions of these equations are respectively,

$$\text{X} = c_1 \cos kx + c_2 \sin kx$$

$$\text{Y} = c_3 \cos ly + c_4 \sin ly$$

and
$$\text{T} = c_5 \cos \sqrt{k^2 + l^2}\,ct + c_6 \sin \sqrt{k^2 + l^2}\,ct.$$

$$\therefore \quad u(x, y, t) = (c_1 \cos kx + c_2 \sin kx)(c_3 \cos ly + c_4 \sin y)$$

$$[c_5 \cos \sqrt{k^2 + l^2}\,ct + c_6 \sin \sqrt{k^2 + l^2}\,ct] \quad \ldots(3)$$

From (3), $\quad u(0, y, t) = 0 = c_1(c_3 \cos ly + c_4 \sin ly)(c_5 \cos \sqrt{k^2 + l^2}\,ct + c_6 \sin \sqrt{k^2 + l^2}\,ct)$

$\Rightarrow \quad c_1 = 0.$

$\therefore$ From (3),

$$u(x, y, t) = c_2 \sin kx(c_3 \cos ly + c_4 \sin ly)(c_5 \cos \sqrt{k^2 + l^2}\,ct + c_6 \sin \sqrt{k^2 + l^2}\,ct) \quad \ldots(4)$$

$$u(1, y, t) = 0 = c_2 \sin k(c_3 \cos ly + c_4 \sin ly)(c_5 \cos \sqrt{k^2 + l^2}\,ct + c_6 \sin \sqrt{k^2 + l^2}\,ct)$$

$\Rightarrow \quad \sin k = 0 = \sin m\pi \ (m \in \text{I})$

$k = m\pi.$

$\therefore$ From (4), $\quad u(x, y, t) = c_2 \sin m\pi x(c_3 \cos ly + c_4 \sin ly)$

$$(c_5 \cos \sqrt{k^2 + l^2}\,ct + c_6 \sin \sqrt{k^2 + l^2}\,ct) \quad \ldots(5)$$

$$u(x, 0, t) = 0 = c_2 \sin m\pi x \,.\, c_3(c_5 \cos \sqrt{k^2 + l^2}\,ct + c_6 \sin \sqrt{k^2 + l^2}\,ct]$$

$\Rightarrow \quad c_3 = 0.$

$\therefore$ From (5), $\quad u(x, y, t) = c_2c_4 \sin m\pi x \sin ly\ (c_5 \cos \sqrt{k^2 + l^2}\,ct + c_6 \sin \sqrt{k^2 + l^2}\,ct) \quad \ldots(6)$

$$u(x, 1, t) = 0 = c_2c_4 \sin m\pi x \sin l\ (c_5 \cos \sqrt{k^2 + l^2}\,ct + c_6 \sin \sqrt{k^2 + l^2}\,ct)$$

$\Rightarrow \quad \sin l = 0 = \sin n\pi \ (n \in \text{I})$

$\Rightarrow \quad l = n\pi.$

$\therefore$ From (6), $u(x, y, t) = c_2c_4 \sin m\pi x \sin n\pi y\,(c_5 \cos \sqrt{k^2 + l^2}\, ct + c_6 \sin \sqrt{k^2 + l^2}\, ct)$

$$u(x, y, t) = \sin m\pi x \sin n\pi y\,(A_{mn} \cos pt + B_{mn} \sin pt) \qquad ...(7)$$

where $p = \pi c\sqrt{m^2 + n^2}$.

The most general solution is

$$u(x, y, t) = \sum_{m=1}^{\infty}\sum_{n=1}^{\infty} \sin m\pi x \sin n\pi y\,(A_{mn} \cos pt + B_{mn} \sin pt) \qquad ...(8)$$

Applying the condition,

$$\frac{\partial u}{\partial t} = 0 \text{ at } t = 0$$

we get $B_{mn} = 0$ ...(9)

Also, using the condition $u = f(x, y) = A \sin \pi x \sin 2\pi y$ when $t = 0$, we get

$$A \sin \pi x \sin 2\pi y = \sum_{m=1}^{\infty}\sum_{n=1}^{\infty} A_{mn} \sin m\pi x \sin n\pi y$$

This is a double Fourier Series.

$$A_{mn} = \frac{2}{1}\cdot\frac{2}{1}\int_0^1\int_0^1 A \sin \pi x \sin 2\pi y \sin m\pi x \sin n\pi y\, dx\, dy$$

Obviously, $A_{m1} = A_{m3} = A_{m4} = \ldots\ldots = 0$

But,

$$A_{m2} = 4A\int_0^1\int_0^1 \sin \pi x \sin m\pi x \sin^2 2\pi y\, dx\, dy$$

$$= 2A\int_0^1\int_0^1 \sin \pi x \sin m\pi x\,(1 - \cos 4\pi y)\, dx\, dy$$

$$= 2A\int_0^1 \sin \pi x \sin m\pi x \left(y - \frac{\sin 4\pi y}{4\pi}\right)_0^1 dx$$

$$= 2A\int_0^1 \sin \pi x \sin m\pi x\, dx$$

Again, obviously,

$$A_{22} = A_{32} = A_{42} = \ldots = 0$$

But,

$$A_{12} = 2A\int_0^1 \sin^2 \pi x\, dx = A\int_0^1 (1 - \cos 2\pi x)\, dx$$

$$= A\left(x - \frac{\sin 2\pi x}{2\pi}\right)_0^1 = A \qquad ...(10)$$

$\therefore$ From (8), (9) and (10), we get

$$u(x, y, t) = A \sin \pi x \sin 2\pi y \cos pt \qquad ...(11)$$

where $p = \pi c\sqrt{m^2 + n^2} = \pi(1)\sqrt{1 + 4} = \pi\sqrt{5}$.

$\therefore$ From (11), $u(x, y, t) = A \cos \pi\sqrt{5}\,t \sin \pi x \sin 2\pi y$.

Example 2. *Find the deflection $u(x, y, t)$ of a rectangular membrane $(0 \le x \le a, 0 \le y \le b)$ whose boundary is fixed ; given that it starts from rest and $u(x, y, 0) = xy(a-x)(b-y)$. Show that the deflection u is given by*

$$u(x, y, t) = \sum_{m=1}^{\infty} \sum_{n=1}^{\infty} A_{mn} \cos ckt \sin \frac{m\pi x}{a} \sin \frac{n\pi y}{b}$$

where $A_{mn} = \dfrac{16a^2b^2}{m^3n^3\pi^6}(1 - \cos m\pi)(1 - \cos n\pi)$ *and* $k^2 = \pi^2 \left(\dfrac{m^2}{a^2} + \dfrac{n^2}{b^2}\right)$.

Sol. Proceeding as in Art. 2.7, we have

$$u(x, y, t) = \sum_{m=1}^{\infty} \sum_{n=1}^{\infty} A_{mn} \sin \frac{m\pi x}{a} \sin \frac{n\pi y}{b} \cos pt$$

where $$p = \pi c \sqrt{\frac{m^2}{a^2} + \frac{n^2}{b^2}} \qquad ...(1)$$

From (1), $$u(x, y, 0) = \sum_{m=1}^{\infty} \sum_{n=1}^{\infty} A_{mn} \sin \frac{m\pi x}{a} \sin \frac{n\pi y}{b}$$

$$xy(a-x)(b-y) = \sum_{m=1}^{\infty} \sum_{n=1}^{\infty} A_{mn} \sin \frac{m\pi x}{a} \sin \frac{n\pi y}{b}$$

where
$$A_{mn} = \frac{2}{a} \cdot \frac{2}{b} \int_0^a \int_0^b xy(a-x)(b-y) \sin \frac{m\pi x}{a} \sin \frac{n\pi y}{b}\, dy\, dx$$

$$= \frac{4}{ab} \int_0^a x(a-x) \sin \frac{m\pi x}{a}\, dx \,.\int_0^b y(b-y) \sin \frac{n\pi y}{b}\, dy$$

$$= \frac{4}{ab} \left[x(a-x) \left(\frac{-\cos \frac{m\pi x}{a}}{\frac{m\pi}{a}} \right) - (a-2x) \left(\frac{-\sin \frac{m\pi x}{a}}{\frac{m^2\pi^2}{a^2}} \right) + (-2) \left(\frac{\cos \frac{m\pi x}{a}}{\frac{m^3\pi^3}{a^3}} \right) \right]_0^a$$

$$\times \left[y(b-y) \left(\frac{-\cos \frac{n\pi y}{b}}{\frac{n\pi}{b}} \right) - (b-2y) \left(\frac{-\sin \frac{n\pi y}{b}}{\frac{n^2\pi^2}{b^2}} \right) + (-2) \left(\frac{\cos \frac{n\pi y}{b}}{\frac{n^3\pi^3}{b^3}} \right) \right]_0^b$$

$$= \frac{4}{ab} \left[\frac{-2a^3}{m^3\pi^3} \cos m\pi + \frac{2a^3}{m^3\pi^3} \right] \left[\frac{-2b^3}{n^3\pi^3} \cos n\pi + \frac{2b^3}{n^3\pi^3} \right]$$

$$= \frac{4}{ab} \cdot \frac{2a^3}{m^3\pi^3} \cdot \frac{2b^3}{n^3\pi^3} [1 - (-1)^n] [1 - (-1)^m]$$

$$A_{mn} = \frac{16a^2b^2}{m^3n^3\pi^6} [1 - (-1)^n] [1 - (-1)^m].$$

Hence from (1),

$$u(x, y, t) = \frac{16a^2b^2}{\pi^6} \sum_{m=1}^{\infty} \sum_{n=1}^{\infty} \frac{[1-(-1)^n][1-(-1)^m]}{m^3n^3} \sin \frac{m\pi x}{a} \sin \frac{n\pi y}{b} \cos pt$$

where $$p = \pi c \sqrt{\frac{m^2}{a^2} + \frac{n^2}{b^2}} \qquad ...(2)$$

Expression (2) may also be put in as,

$$u(x, y, t) = \sum_{m=1}^{\infty} \sum_{n=1}^{\infty} A_{mn} \cos ckt \sin \frac{m\pi x}{a} \sin \frac{n\pi y}{b}$$

where $$A_{mn} = \frac{16a^2b^2}{m^3\pi^6n^3}(1 - \cos m\pi)(1 - \cos n\pi) \quad \text{and} \quad k^2 = \pi^2\left(\frac{m^2}{a^2} + \frac{n^2}{b^2}\right).$$

Example 3. *A tightly stretched unit square membrane starts vibrating from rest and its initial displacement is k sin 2πx sin πy. Show that the deflection at any instant is*

$$k \sin 2\pi x \sin \pi y \cos (\sqrt{5}\, \pi ct).$$

Sol. Here we have to solve the equation $\dfrac{\partial^2 u}{\partial t^2} = c^2\left(\dfrac{\partial^2 u}{\partial x^2} + \dfrac{\partial^2 u}{\partial y^2}\right)$

with boundary conditions $u(0, y, t) = u(1, y, t) = u(x, 0, t) = u(x, 1, t) = 0$

and the initial conditions $u(x, y, 0) = f(x, y) = k \sin 2\pi x \sin \pi y$

$$\frac{\partial u}{\partial t} = 0 \quad \text{when } t = 0$$

Proceeding as in Ex. 1, we have

$$u(x, y, t) = \sum_{m=1}^{\infty} \sum_{n=1}^{\infty} A_{mn} \sin m\pi x \sin n\pi y \cos pt \qquad ...(1)$$

Since, $a = b = 1$, where $p = \pi c \sqrt{m^2 + n^2}$

and $$A_{mn} = \frac{4}{1 \times 1} \int_0^1 \int_0^1 k \sin 2\pi x \sin \pi y \sin m\pi x \sin n\pi y \, dy dx$$

$$= 4k \int_0^1 \sin m\pi x \sin 2\pi x \, dx \int_0^1 \sin n\pi y \sin \pi y \, dy$$

$$= 0 \quad \text{for } m \neq 2 \quad \text{or} \quad n \neq 1$$

$\therefore$ $$A_{21} = 4k(\tfrac{1}{2})(\tfrac{1}{2}) = k \quad \text{and} \quad p = \pi c \sqrt{(2)^2 + (1)^2} = \sqrt{5}\, \pi c$$

Hence solution (1) reduces to $u(x, y, t) = k \sin 2\pi x \sin \pi y \cos (\sqrt{5}\, \pi ct)$.

TEST YOUR KNOWLEDGE

1. Find the deflecting $u(x, y, t)$ of a rectangular membrane $(0 < x < 1, 0 < y < 2)$ whose boundary is fixed, given that it starts from rest and $u(x, y, 0) = xy(1-x)(2-y)$.
2. Find the deflection $u(x, y, t)$ of a rectangular membrane $(0 < x < a, 0 < y < b)$ whose boundary is fixed, given that it starts from rest and $u(x, y, 0) = xy(a^2 - x^2)(b^2 - y^2)$.
3. Find the deflection $u(x, y, t)$ of the tightly stretched rectangular membrane with sides a and b having wave velocity $c = 1$ if the initial velocity is zero and the initial deflection is

$$f(x, y) = \sin\frac{2\pi x}{a}\sin\frac{3\pi y}{b}. \qquad \text{(G.B.T.U. 2011)}$$

Answers

1. $u(x, y, t) = \sum_{m=1}^{\infty}\sum_{n=1}^{\infty} A_{mn}\sin m\pi x\sin\frac{n\pi y}{2}\cos pt$, where $A_{mn} = \frac{256}{m^3n^3\pi^6}$, both m, n odd

and $p = \pi c\sqrt{m^2 + \frac{n^2}{4}}$

2. $u(x, y, t) = \sum_{m=1}^{\infty}\sum_{n=1}^{\infty} A_{mn}\sin\frac{m\pi x}{a}\sin\frac{n\pi y}{b}\cos pt$, where $A_{mn} = \frac{144a^3b^3}{m^3n^3\pi^6}$ and $p = \pi c\sqrt{\frac{m^2}{a^2} + \frac{n^2}{b^2}}$.
3. $u(x, y, t) = \sin\frac{2\pi x}{a}\sin\frac{3\pi y}{b}\cos pt$ where $p = \pi\sqrt{\frac{4}{a^2} + \frac{9}{b^2}}$.

2.10 ONE-DIMENSIONAL HEAT FLOW $\frac{\partial u}{\partial t} = c^2\frac{\partial^2 u}{\partial x^2}$

Consider the flow of heat by conducting in a uniform bar. It is assumed that the sides of the bar are insulated and the loss of heat from the sides by conduction or radiation is negligible. Take one end of the bar as origin and the direction of flow as the positive x-axis. The temperature u at any point of the bar depends on the distance x of the point from one end and the time t. Also, the temperature of all points of any cross-section is the same.

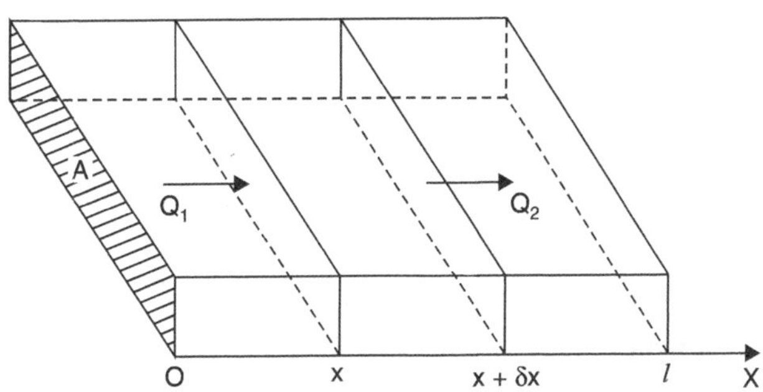

The amount of heat crossing any section of the bar per second depends on the area A of the cross-section, the conductivity K of the material of the bar and the temperature gradient $\frac{\partial u}{\partial x}$ *i.e.*, rate of change of temperature w.r.t. distance normal to the area.

$\therefore$ Q_1, the quantity of heat flowing into the section at a distance x

$$= -KA \left(\frac{\partial u}{\partial x}\right)_x \text{ per sec.}$$

(the negative sign on the right is attached because as x increases, u decreases).

Q_2, the quantity of heat flowing out of the section at a distance $x + \delta x$

$$= -KA \left(\frac{\partial u}{\partial x}\right)_{x+\delta x} \text{ per sec.}$$

Hence the amount of heat retained by the slab with thickness δx is

$$Q_1 - Q_2 = KA \left[\left(\frac{\partial u}{\partial x}\right)_{x+\delta x} - \left(\frac{\partial u}{\partial x}\right)_x\right] \text{ per sec.} \quad \text{...(1)}$$

But the rate of increase of heat in the slab $= S\rho A \, \delta x \, \dfrac{\partial u}{\partial t}$ ...(2)

where S is the specific heat and ρ, the density of the material.

$\therefore$ From (1) and (2), $S\rho A \, \delta x \, \dfrac{\partial u}{\partial t} = KA \left[\left(\dfrac{\partial u}{\partial x}\right)_{x+\delta x} - \left(\dfrac{\partial u}{\partial x}\right)_x\right]$

or
$$S\rho \frac{\partial u}{\partial t} = K \left[\frac{\left(\frac{\partial u}{\partial x}\right)_{x+\delta x} - \left(\frac{\partial u}{\partial x}\right)_x}{\delta x}\right]$$

Taking the limit as $\delta x \to 0$, we have

$$S\rho \frac{\partial u}{\partial t} = K \frac{\partial^2 u}{\partial x^2} \quad \text{or} \quad \frac{\partial u}{\partial t} = \frac{K}{S\rho} \frac{\partial^2 u}{\partial x^2}$$

or
$$\boxed{\frac{\partial u}{\partial t} = c^2 \frac{\partial^2 u}{\partial x^2}}, \quad \text{where } c^2 = \frac{K}{S\rho} \text{ is known as diffusivity of the material of the bar.}$$

2.11 SOLUTION OF THE HEAT EQUATION [G.B.T.U. (C.O.) 2011]

The heat equation is
$$\frac{\partial u}{\partial t} = c^2 \frac{\partial^2 u}{\partial x^2} \quad \text{...(1)}$$

Let
$$u = XT \quad \text{...(2)}$$

where X is a function of x only and T is a function of t only, be a solution of (1).

Then
$$\frac{\partial u}{\partial t} = XT' \quad \text{and} \quad \frac{\partial^2 u}{\partial x^2} = X''T$$

Substituting in (1), we have $XT' = c^2 X''T$

Separating the variables, we get
$$\frac{X''}{X} = \frac{1}{c^2} \frac{T'}{T} \quad \text{...(3)}$$

Now the LHS of (3) is a function of x only and the RHS is a function of t only. Since x and t are independent variables, this equation can hold only when both sides reduce to a constant, say k. The equation (3) leads to the ordinary differential equations

$$\frac{d^2X}{dx^2} - kX = 0 \quad \text{and} \quad \frac{dT}{dt} - kc^2T = 0 \qquad ...(4)$$

Solving equations (4), we get

(*i*) *When k is positive* and $= p^2$, say

$$X = c_1 e^{px} + c_2 e^{-px}, \; T = c_3 \, e^{c^2 p^2 t}$$

(*ii*) *When k is negative* and $= -p^2$, say

$$X = c_1 \cos px + c_2 \sin px, \; T = c_3 \, e^{-c^2 p^2 t}$$

(*iii*) *When k = 0*

$$X = c_1 x + c_2, \; T = c_3.$$

Thus the various possible solutions of the heat equation (1) are:

$$u = (c_1 e^{px} + c_2 e^{-px}) \, . \, c_3 \, e^{c^2 p^2 t} \qquad ...(5)$$

$$u = (c_1 \cos px + c_2 \sin px) \, . \, c_3 e^{-c^2 p^2 t} \qquad ...(6)$$

$$u = (c_1 x + c_2) \, c_3 \qquad ...(7)$$

Of these three solutions, we have to choose that solution which is consistent with the physical nature of the problem. Since u decreases as time t increases, the only suitable solution of the heat equation is

$$u = (c_1 \cos px + c_2 \sin px) \, c_3 \, e^{-c^2 p^2 t}.$$

Solution (5) is rejected since $u \to \infty$ as $t \to \infty$.

Also, solution (7) is rejected as it gives a non-zero temperature at all times.

2.12 INHOMOGENEOUS BOUNDARY CONDITIONS

Now consider the case where the ends of a rod are kept at constant temperatures different from zero.

Consider the IBVP, $u_t = u_{xx}, \; 0 < x < L, \; t > 0$

$$u(0, t) = \alpha, \; u(L, t) = \beta; \; t \geq 0$$

$$u(x, 0) = f(x)$$

To convert the inhomogeneous boundary conditions to homogeneous boundary conditions, we use the following transformation formula:

$$u(x, t) = \left[\alpha + \left(\frac{\beta - \alpha}{L}\right) x\right] + v(x, t)$$

We can easily show that now $v(x, t)$ will be governed by the IBVP:

$$v_t = v_{xx}, \quad 0 < x < L, \qquad t > 0$$

$$v(0, t) = 0 = v(L, t), \quad t \geq 0$$

$$v(x, 0) = f(x) - \left[\alpha + \left(\frac{\beta - \alpha}{L}\right) x\right]$$

$v(x, t)$ can now easily be found using method of separation of variables. Consequently $u(x, t)$ can readily be obtained as a final result.

ILLUSTRATIVE EXAMPLES

Example 1. *A rod of length l with insulated sides is initially at a uniform temperature u_0. Its ends are suddenly cooled to 0°C and are kept at that temperature. Find the temperature function u(x, t).* (U.P.T.U. 2015)

Sol. The temperature function $u(x, t)$ satisfies the differential equation

$$\frac{\partial u}{\partial t} = c^2 \frac{\partial^2 u}{\partial x^2}$$

As proved in Art. 2.11, we have

$$u(x, t) = (c_1 \cos px + c_2 \sin px)\, c_3 e^{-c^2 p^2 t} \quad ...(1)$$

Since the ends $x = 0$ and $x = l$ are cooled to 0°C and kept at that temperature throughout, the boundary conditions are $u(0, t) = u(l, t) = 0$ for all t

Also $u(x, 0) = u_0$ is the initial condition.

Since $u(0, t) = 0$, we have from (1),

$$0 = c_1 c_3 e^{-c^2 p^2 t} \quad \Rightarrow \quad c_1 = 0$$

$\therefore$ From (1), $\quad u(x, t) = c_2 c_3 \sin px .\, e^{-c^2 p^2 t}$ ...(2)

Since $u(l, t) = 0$, we have from (2),

$$0 = c_2 c_3 \sin pl \,.\, e^{-c^2 p^2 t}$$

$$\Rightarrow \quad \sin pl = 0 \quad \Rightarrow \quad pl = n\pi$$

$$\therefore \quad p = \frac{n\pi}{l}, \; n \text{ being an integer}$$

Solution (2) reduces to $u(x, t) = b_n \sin \frac{n\pi x}{l} . e^{\frac{-c^2 n^2 \pi^2 t}{l^2}}$ on replacing $c_2 c_3$ by b_n.

The most general solution is obtained by adding all such solutions for $n = 1, 2, 3,......$

$$\therefore \quad u(x, t) = \sum_{n=1}^{\infty} b_n \sin \frac{n\pi x}{l} . e^{\frac{-c^2 n^2 \pi^2 t}{l^2}} \quad ...(3)$$

Since $u(x, 0) = u_0$, we have $\quad u_0 = \sum_{n=1}^{\infty} b_n \sin \frac{n\pi x}{l}$

which is half-range sine series for u_0.

$$b_n = \frac{2}{l} \int_0^l u_0 \sin \frac{n\pi x}{l} dx = \begin{cases} 0 \;, & \text{when } n \text{ is even} \\ \frac{4u_0}{n\pi} \;, & \text{when } n \text{ is odd} \end{cases}$$

Hence the temperature function

$$u(x, t) = \frac{4u_0}{\pi} \sum_{n=1, 3, 5, ...}^{\infty} \frac{1}{n} \sin \frac{n\pi x}{l} e^{-\frac{c^2 n^2 \pi^2 t}{l^2}}$$

$$= \frac{4u_0}{\pi} \sum_{n=1}^{\infty} \frac{1}{2n-1} \sin \frac{(2n-1)\pi x}{l} e^{-\frac{c^2 (2n-1)^2 \pi^2 t}{l^2}}.$$

Example 2. *Find the temperature in a bar of length 2 whose ends are kept at zero and lateral surface insulated if the initial temperature is* $\sin \frac{\pi x}{2} + 3 \sin \frac{5\pi x}{2}$.

(U.P.T.U. 2015)

Sol. Let $u(x, t)$ be the temperature in the bar. The boundary conditions are

$$u(0, t) = 0 = u(2, t) \text{ for any } t. \quad \text{...(1)}$$

The initial condition is

$$u(x, 0) = \sin \frac{\pi x}{2} + 3 \sin \frac{5\pi x}{2} \quad \text{...(2)}$$

One-dimensional heat flow equation is

$$\frac{\partial u}{\partial t} = c^2 \frac{\partial^2 u}{\partial x^2} \quad \text{...(3)}$$

Its solution is

$$u(x, t) = (c_1 \cos px + c_2 \sin px)\, c_3\, e^{-c^2 p^2 t} \quad \text{...(4)}$$

$$u(0, t) = 0 = c_1 c_3 e^{-c^2 p^2 t} \quad | \text{ Using (1)}$$

$$\Rightarrow \quad c_1 = 0$$

$\therefore$ From (4),

$$u(x, t) = c_2 c_3 \sin px\ e^{-c^2 p^2 t} \quad \text{...(5)}$$

$$u(2, t) = 0 = c_2 c_3 \sin 2p\, e^{-c^2 p^2 t} \quad | \text{ Using (1)}$$

$$\Rightarrow \quad \sin 2p = 0 = \sin n\pi$$

$$\therefore \quad p = \frac{n\pi}{2}, n \in \text{I}$$

Hence from (5),

$$u(x, t) = b_n \sin \frac{n\pi x}{2}\ e^{-\frac{n^2\pi^2c^2t}{4}} \quad | \because c_2 c_3 = b_n$$

The most general solution is

$$u(x, t) = \sum_{n=1}^{\infty} b_n \sin \frac{n\pi x}{2}\ e^{-\frac{n^2\pi^2c^2t}{4}} \quad \text{...(6)}$$

$$u(x, 0) = \sin \left(\frac{\pi x}{2}\right) + 3 \sin \left(\frac{5\pi x}{2}\right) = \sum_{n=1}^{\infty} b_n \sin \frac{n\pi x}{2}$$

$$= b_1 \sin \left(\frac{\pi x}{2}\right) + b_2 \sin \left(\frac{2\pi x}{2}\right) + \ldots + b_5 \sin \left(\frac{5\pi x}{2}\right) + \ldots$$

Comparing, we get

$$b_1 = 1 \text{ and } b_5 = 3$$

Hence from (6),

$$u(x, t) = \sin\left(\frac{\pi x}{2}\right) e^{-\pi^2 c^2 t/4} + 3 \sin\left(\frac{5\pi x}{2}\right) e^{-25\pi^2 c^2 t/4}.$$

Example 3. *An insulated rod of length l has its ends A and B maintained at 0°C and 100°C respectively until steady state conditions prevail. If B is suddenly reduced to 0°C and maintained at 0°C, find the temperature at a distance x from A at time t.*

[G.B.T.U. (A.G.) 2011, U.K.T.U. 2011]

Find also the temperature if the change consists of raising the temperature of A to 20°C and reducing that of B to 80°C.

Sol. Initial temperature distribution in the rod is

$$u_1 = 0 + \left(\frac{100 - 0}{l}\right) x = \frac{100}{l} x$$

Final temperature distribution (*i.e.*, in steady state) is

$$u_2 = 0 + \left(\frac{0 - 0}{l}\right) x = 0$$

To get u in the intermediate period,

$$u = u_2(x) + u_1(x, t)$$

where $u_2(x)$ is the steady state temperature distribution in the rod. $u_1(x, t)$ is the transient temperature distribution which tends to zero as t increases.

$u_1(x, t)$ satisfies one-dimensional heat flow equation

$$\therefore \quad u(x, t) = \sum_{n=1}^{\infty} (a_n \cos px + b_n \sin px)\, e^{-c^2 p^2 t} \quad \text{...(1)}$$

In steady state, $u(0, t) = 0 = u(l, t)$ ...(2)

$\therefore$ From (1), $$u(0, t) = 0 = \sum_{n=1}^{\infty} a_n e^{-c^2 p^2 t} \Rightarrow a_n = 0 \quad \text{...(3)}$$

Also, $$u(l, t) = 0 = \sum_{n=1}^{\infty} b_n \sin pl\, e^{-c^2 p^2 t} \quad | \text{ using (3)}$$

$\Rightarrow$ $\sin pl = 0 = \sin n\pi,\ n \in I$

or $$p = \frac{n\pi}{l} \quad \text{...(4)}$$

$\therefore$ From (1), (3) and (4), $$u(x, t) = \sum_{n=1}^{\infty} b_n \sin\frac{n\pi x}{l}\, e^{-\left(\frac{n\pi}{l}\right)^2 c^2 t} \quad \text{...(5)}$$

Using initial condition,

$$u(x, 0) = \frac{100}{l} x = \sum_{n=1}^{\infty} b_n \sin\frac{n\pi x}{l}$$

which is half-range sine series for $\frac{100}{l} x$.

$$\therefore \qquad b_n = \frac{2}{l}\int_0^l \frac{100}{l} x \sin\frac{n\pi x}{l}\, dx$$

$$= \frac{200}{l^2}\left[\left\{x.\left(\frac{-\cos\frac{n\pi x}{l}}{\frac{n\pi}{l}}\right)\right\}_0^l - \int_0^l 1.\left(\frac{-\cos\frac{n\pi x}{l}}{\frac{n\pi}{l}}\right)dx\right]$$

$$= \frac{200}{l^2}\left[\frac{-l^2}{n\pi}\cos n\pi + \frac{l}{n\pi}\left(\frac{\sin\frac{n\pi x}{l}}{\frac{n\pi}{l}}\right)_0^l\right] = \frac{-200}{n\pi}(-1)^n$$

Hence the temperature function

$$u(x, t) = -\frac{200}{\pi}\sum_{n=1}^{\infty}\frac{(-1)^n}{n}\sin\frac{n\pi x}{l}\, e^{-\frac{n^2\pi^2c^2t}{l^2}}$$

In the second part, the initial condition remains the same as in first part *i.e.,*

$$u(x, 0) = \frac{100}{l}x.$$

Boundary conditions are $u(0, t) = 20$ and $u(l, t) = 80$ for all values of t then, final temperature distribution is

$$u_2 = 20 + \left(\frac{80-20}{l}\right)x = 20 + \frac{60}{l}x$$

Then, $$u = u_2(x) + u_1(x, t)$$

$$u = 20 + \frac{60}{l}x + \sum_{n=1}^{\infty}(a_n\cos px + b_n\sin px)\, e^{-c^2p^2t} \qquad ...(6)$$

$$u(0, t) = 20 = 20 + \sum_{n=1}^{\infty} a_n e^{-c^2p^2t} \qquad | \text{ From (6)}$$

$$\Rightarrow \qquad a_n = 0$$

$\therefore$ From (6), $$u = 20 + \frac{60}{l}x + \sum_{n=1}^{\infty} b_n \sin px\, e^{-c^2p^2t} \qquad ...(7)$$

$$u(l, t) = 80 = 20 + \frac{60}{l}l + \sum_{n=1}^{\infty} b_n \sin pl\, e^{-c^2p^2t} \qquad | \text{ From (7)}$$

$$\Rightarrow \qquad 0 = \sum_{n=1}^{\infty} b_n \sin pl\, e^{-c^2p^2t}$$

$$\sin pl = 0 = \sin n\pi, n \in I$$

$$\therefore \qquad p = \frac{n\pi}{l} \qquad ...(8)$$

From (7) and (8), $$u = 20 + \frac{60}{l}x + \sum_{n=1}^{\infty} b_n \sin\frac{n\pi x}{l}\, e^{-\left(\frac{n\pi c}{l}\right)^2 t} \qquad ...(9)$$

Using initial condition,

$$u(x, 0) = \frac{100}{l}x = 20 + \frac{60}{l}x + \sum_{n=1}^{\infty} b_n \sin\frac{n\pi x}{l}$$

$$\Rightarrow \qquad \frac{40}{l}x - 20 = \sum_{n=1}^{\infty} b_n \sin\frac{n\pi x}{l}$$

where

$$b_n = \frac{2}{l}\int_0^l \left(\frac{40}{l}x - 20\right)\sin\frac{n\pi x}{l}\,dx$$

$$= \frac{2}{l}\left[\left\{\left(\frac{40}{l}x - 20\right)\left(\frac{-\cos\frac{n\pi x}{l}}{\frac{n\pi}{l}}\right)\right\}_0^l - \int_0^l \frac{40}{l}.\left(\frac{-\cos\frac{n\pi x}{l}}{\frac{n\pi}{l}}\right)dx\right]$$

$$= \frac{2}{l}\left[\frac{-20l}{n\pi}\cos n\pi - \frac{20l}{n\pi} + \frac{40}{n\pi}\left(\frac{\sin\frac{n\pi x}{l}}{\frac{n\pi}{l}}\right)_0^l\right]$$

$$= \frac{-40}{n\pi}(1 + \cos n\pi) = \begin{cases} 0, & \text{when } n \text{ is odd} \\ \dfrac{-80}{n\pi}, & \text{when } n \text{ is even} \end{cases}$$

Hence equation (9) becomes,

$$u(x, t) = 20 + \frac{60}{l}x - \frac{80}{\pi}\sum_{\substack{n=2,4,\ldots \\ (n \text{ is even})}}^{\infty} \frac{1}{n}\sin\frac{n\pi x}{l}\, e^{-\left(\frac{n\pi c}{l}\right)^2 t}$$

$$= 20 + \frac{60}{l}x - \frac{40}{\pi}\sum_{m=1}^{\infty}\frac{1}{m}\sin\frac{2m\pi x}{l}\, e^{-\frac{4c^2m^2\pi^2 t}{l^2}}. \qquad \text{(taking } n = 2m)$$

Example 4. *The ends A and B of a rod of length 20 cm are at temperatures 30°C and 80°C until steady state prevails. Then the temperature of the rod ends are changed to 40°C and 60°C respectively. Find the temperature distribution function u(x, t). The specific heat, density and the thermal conductivity of the material of the rod are such that the combination* $\frac{k}{\rho\sigma} = c^2 = 1$.

Sol. Initial temperature distribution in the rod is

$$u_1 = 30 + \left(\frac{80 - 30}{20}\right)x = 30 + \frac{5}{2}x$$

Final temperature distribution (*i.e.*, in steady state) is

$$u_2 = 40 + \left(\frac{60 - 40}{20}\right)x = 40 + x$$

To get u in the intermediate period,

$$u = u_1(x, t) + u_2(x)$$

where $u_2(x)$ is the steady state temperature distribution in the rod $u_1(x, t)$ is the transient temperature distribution which tends to zero as t increases.

$\because$ $u_1(x, t)$ satisfies one-dimensional heat flow equation.

$$\therefore \qquad u = 40 + x + \sum_{n=1}^{\infty} (a_n \cos px + b_n \sin px)\, e^{-p^2 t} \qquad ...(1)$$

In steady state,

$$u(0, t) = 40 \qquad ...(2)$$

$$u(20, t) = 60 \qquad ...(3)$$

From (1), and (2), $u(0, t) = 40 = 40 + \sum_{1}^{\infty} a_n\, e^{-p^2 t}$ | From (2)

$$0 = \sum_{1}^{\infty} a_n\, e^{-p^2 t} \quad \Rightarrow \quad a_n = 0 \qquad ...(4)$$

$\therefore$ From (1),

$$u = 40 + x + \sum_{n=1}^{\infty} b_n \sin px\, e^{-p^2 t}$$

Again, $\quad u(20, t) = 60 = 60 + \sum_{n=1}^{\infty} b_n \sin 20p\, e^{-p^2 t}$

$$\Rightarrow \qquad 0 = \sum_{1}^{\infty} b_n \sin 20p\, e^{-p^2 t}$$

$$\sin 20p = 0 = \sin n\pi,\ n \in \text{I}$$

$$\Rightarrow \qquad p = \frac{n\pi}{20}$$

$$\therefore \qquad u = 40 + x + \sum_{n=1}^{\infty} b_n \sin \frac{n\pi x}{20}\, e^{-\left(\frac{n\pi}{20}\right)^2 t} \qquad ...(5)$$

Using initial condition,

$$u(x, 0) = 30 + \frac{5}{2}x \quad \text{in eqn. (5), we get}$$

$$\Rightarrow \qquad 30 + \frac{5}{2}x = 40 + x + \sum_{n=1}^{\infty} b_n \sin \frac{n\pi x}{20}$$

$$\Rightarrow \qquad \frac{3}{2}x - 10 = \sum_{n=1}^{\infty} b_n \sin \frac{n\pi x}{20}$$

where $\quad b_n = \frac{2}{20}\int_0^{20} \left(\frac{3}{2}x - 10\right) \sin \frac{n\pi x}{20}\, dx = -\frac{20}{n\pi}\, [2(-1)^n + 1]$

From (5), $\quad u(x, t) = 40 + x - \frac{20}{\pi} \sum_{n=1}^{\infty} \left\{\frac{2(-1)^n + 1}{n}\right\} \sin \frac{n\pi x}{20}\, e^{-\left(\frac{n\pi}{20}\right)^2 t}.$

Example 5. *The temperature distribution in a bar of length π which is perfectly insulated at ends $x = 0$ and $x = \pi$ is governed by partial differential equation*

$$\frac{\partial u}{\partial t} = \frac{\partial^2 u}{\partial x^2}.$$

Assuming the initial temperature distribution as u(x, 0) = f(x) = cos 2x, find the temperature distribution at any instant of time. (M.T.U. 2011)

Sol. $$\frac{\partial u}{\partial t} = \frac{\partial^2 u}{\partial x^2} \quad ...(1)$$

Its solution is $$u(x, t) = c_1 e^{-p^2 t} (c_2 \cos px + c_3 \sin px) \quad ...(2)$$

Since ends of bar are insulated, no heat can pass from either sides and boundary conditions are

$$\frac{\partial u}{\partial x} = 0 \quad \text{at } x = 0 \quad ...(3)$$

and $$\frac{\partial u}{\partial x} = 0 \quad \text{at } x = \pi \quad ...(4)$$

From (2), $$\frac{\partial u}{\partial x} = c_1 e^{-p^2 t} (-pc_2 \sin px + pc_3 \cos px)$$

At $x = 0$,

$$0 = c_1 e^{-p^2 t} pc_3 \Rightarrow c_3 = 0$$

$\therefore$ From (2), $$u(x, t) = c_1 c_2 \, e^{-p^2 t} \cos px \quad ...(5)$$

Again $$\frac{\partial u}{\partial x} = -pc_1 c_2 \, e^{-p^2 t} \sin px$$

At $x = \pi$,

$$0 = -pc_1 c_2 e^{-p^2 t} \sin p\pi$$

$\Rightarrow$ $$\sin p\pi = 0 = \sin n\pi \ (n \in I)$$

$$p\pi = n\pi \Rightarrow p = n$$

$\therefore$ From (5), $u(x, t) = b_n \, e^{-n^2 t} \cos nx,$ where $c_1 c_2 = b_n$

Most general solution is

$$u(x, t) = \sum_{n=1}^{\infty} b_n \, e^{-n^2 t} \cos nx \quad ...(6)$$

$$u(x, 0) = \cos 2x = \sum_{n=1}^{\infty} b_n \cos nx$$

Comparing, we get $b_2 = 1$ and $n = 2$. All other b_i's are zero.

$\therefore$ From (6), $u(x, t) = e^{-4t} \cos 2x$.

Example 6. *Solve the equation* $\frac{\partial u}{\partial t} = \frac{\partial^2 u}{\partial x^2}$ *with boundary condition u(x, 0) = 3 sin nπx, u(0, t) = 0, u (l, t) = 0, where 0 < x < l.*

Sol. The solution to the equation

$$\frac{\partial u}{\partial t} = \frac{\partial^2 u}{\partial x^2} \quad ...(1)$$

is given by $u(x, t) = c_1 e^{-p^2 t} (c_2 \cos px + c_3 \sin px)$ ...(2)

From (2), $u(0, t) = c_1 c_2 e^{-p^2 t}$

$\Rightarrow$ $0 = c_1 c_2 e^{-p^2 t}$

$\Rightarrow$ $c_2 = 0.$

$\therefore$ From (2), $u(x, t) = c_1 c_3 e^{-p^2 t} \sin px$ ...(3)

$u(l, t) = 0 = c_1 c_3 e^{-p^2 t} \sin pl$

$\Rightarrow$ $\sin pl = 0 = \sin n\pi (n \in \text{I})$

$\therefore$ $p = \dfrac{n\pi}{l}.$

From (3), $u(x, t) = c_1 c_3 e^{-\frac{n^2\pi^2}{l^2} t} \sin \dfrac{n\pi x}{l} = b_n e^{-\frac{n^2\pi^2 t}{l^2}} \sin \dfrac{n\pi x}{l}$

The most general solution is

$$u(x, t) = \sum_{n=1}^{\infty} b_n e^{-(n^2\pi^2 t/l^2)} \sin \frac{n\pi x}{l} \quad ...(4)$$

$\therefore$ From (4), $u(x, 0) = \displaystyle\sum_{1}^{\infty} b_n \sin \frac{n\pi x}{l}$

$\Rightarrow$ $3 \sin n\pi x = \displaystyle\sum_{1}^{\infty} b_n \sin \frac{n\pi x}{l}.$

Comparison gives, $b_n = 3, l = 1.$

Hence from (4), the required solution is

$$u(x, t) = 3 \sum_{n=1}^{\infty} e^{-n^2\pi^2 t} \sin n\pi x.$$

Example 7. *A bar with insulated sides is initially at a temperature 0°C throughout. The end x = 0 is kept at 0°C, and heat is suddenly applied at the end x = l so that* $\dfrac{\partial u}{\partial x} = A$ *for x = l, where A is a constant. Find the temperature function u(x, t).*

Sol. One-dimensional heat flow equation is

$$\frac{\partial u}{\partial t} = c^2 \frac{\partial^2 u}{\partial x^2} \quad ...(1)$$

Its solution is

$$u(x, t) = c_1 e^{-p^2 c^2 t} (c_2 \cos px + c_3 \sin px)$$

or $u(x, t) = (\text{A}_1 \cos px + \text{B} \sin px)\, e^{-p^2 c^2 t}$ ...(2)

Applying the zero end conditions as,

$u(0, t) = 0 = \text{A}_1 e^{-p^2 c^2 t}$

$\Rightarrow$ $\text{A}_1 = 0.$

$\therefore$ From (2), $u(x, t) = \text{B} \sin px e^{-p^2c^2t}$ ...(3)

From (3), $\dfrac{\partial u}{\partial x} = p\text{B} \cos px e^{-p^2c^2t}$.

At $x = l$, $\left(\dfrac{\partial u}{\partial x}\right)_{x=l} = 0 = p\text{B} \cos pl e^{-p^2c^2t}$

$\Rightarrow$ $\cos pl = 0 = \cos\left(n\pi - \dfrac{\pi}{2}\right)$; $n \in \text{I}$ or $pl = (2n-1)\dfrac{\pi}{2}$

$\Rightarrow$ $p = (2n-1)\dfrac{\pi}{2l}$.

$\therefore$ From (3), $u(x, t) = \text{B} \sin px e^{-p^2c^2t}$ ...(4) where $p = (2n-1)\dfrac{\pi}{2l}$

The most general solution is

$$u(x, t) = \sum_{n=1}^{\infty} \text{B}_n \sin px\, e^{-p^2c^2t} \quad \text{...(5) where } p = (2n-1)\frac{\pi}{2l}.$$

The end conditions given for this problem are

(*i*) $u = 0$ at $x = 0$ (*ii*) $\dfrac{\partial u}{\partial x} = \text{A}$ at $x = l$ ...(6)

These conditions are different from the zero end conditions. So we add to (5) the solution

$$u = \text{A}_1 x + \text{B}$$

Choosing A_1 and B so that (6) is satisfied.
This gives, $0 = \text{B}$ and $\text{A}_1 = \text{A}$

$$\therefore \quad u(x, t) = \text{A}x + \sum_{n=1}^{\infty} \text{B}_n \sin px\, e^{-p^2c^2t} \quad \text{...(7) where } p = (2n-1)\frac{\pi}{2l}.$$

Applying the condition that $u = 0$ at $t = 0$, we have

$$0 = \text{A}x + \sum_{n=1}^{\infty} \text{B}_n \sin px$$

or
$$-\text{A}x = \sum_{n=1}^{\infty} \text{B}_n \sin px$$

where
$$\text{B}_n = \frac{2}{l}\int_0^l (-\text{A}x) \sin px\, dx, \text{ where } p = (2n-1)\frac{\pi}{2l}$$

$$= \frac{-2\text{A}}{l}\left[\left\{x\left(\frac{-\cos px}{p}\right)\right\}_0^l - \int_0^l 1.\left(\frac{-\cos px}{p}\right)dx\right]$$

$$= -\frac{2\text{A}}{l}\left[\frac{-l\cos pl}{p} + \frac{1}{p}\left(\frac{\sin px}{p}\right)_0^l\right]$$

$$= \frac{-2A}{l}\left[\frac{-l\cos pl}{p} + \frac{1}{p^2}\sin pl\right] = -\frac{2A(2l)^2}{l(2n-1)^2\pi^2}\sin(2n-1)\frac{\pi}{2}$$

$(\because \quad \cos pl = 0)$

$$= \frac{-8Al}{\pi^2(2n-1)^2}\sin\left(n\pi - \frac{\pi}{2}\right) = \frac{8Al}{\pi^2(2n-1)^2}(-1)^n \quad \text{...(8)}$$

$\therefore$ From (7),
$$u(x, t) = Ax + \frac{8Al}{\pi^2}\sum_{n=1}^{\infty}\left[\frac{(-1)^n}{(2n-1)^2}\right]\sin(2n-1)\frac{\pi x}{2l}\, e^{-\left[\frac{(2n-1)^2\pi^2c^2t}{4l^2}\right]}$$

Example 8. *Solve:* $\frac{\partial u}{\partial t} = k\frac{\partial^2 u}{\partial x^2}$ *under the conditions*

(i) $u \neq \infty$ *if* $t \to \infty$ (ii) $\frac{\partial u}{\partial x} = 0$ *for* $x = 0$ *and* $x = l$

(iii) $u = lx - x^2$ *for* $t = 0$ *between* $x = 0$ *and* $x = l$. (U.P.T.U. 2015)

Sol. Solution to $\frac{\partial u}{\partial t} = k\frac{\partial^2 u}{\partial x^2}$ is

$$u(x, t) = c_1 e^{-c^2kt}(c_2\cos cx + c_3\sin cx) \quad \text{...(1)}$$

Eqn. (1) satisfies the condition $u \neq \infty$ if $t \to \infty$

Applying $\frac{\partial u}{\partial x} = 0$ for $x = 0$ and $x = l$ to (1), we get

$$c_3 = 0$$

and
$$c = \frac{n\pi}{l}, n \in I$$

$\therefore$
$$u = c_1c_2\, e^{-\left(\frac{n^2\pi^2kt}{l^2}\right)}\cos\frac{n\pi x}{l} = a_n\cos\frac{n\pi x}{l}\, e^{-\left(\frac{n^2\pi^2kt}{l^2}\right)} \quad \text{...(2)}$$

Again, the second possible solution is

$$u = c_1(c_2x + c_3) \quad \text{...(3)} \quad | \text{ if } c^2 = 0$$

Applying $\frac{\partial u}{\partial x} = 0$ for $x = 0$ and $x = l$ to (3), we get $c_2 = 0$

$\therefore$
$$u = c_1c_3 = \frac{a_0}{2} \text{ (say)} \quad \text{...(4)} \quad | \text{ From (3)}$$

$\therefore$ The general solution is the sum of solutions (2) and (4) for various n.

$\therefore$
$$u(x, t) = \frac{a_0}{2} + \sum_{n=1}^{\infty} a_n\cos\frac{n\pi x}{l}\, e^{-\left(\frac{n^2\pi^2kt}{l^2}\right)} \quad \text{...(5)}$$

Now applying $u = lx - x^2$ for $t = 0$ to eqn. (5), we get

$$lx - x^2 = \frac{a_0}{2} + \sum_{n=1}^{\infty} a_n\cos\frac{n\pi x}{l}$$

Here,
$$a_0 = \frac{2}{l}\int_0^l (lx - x^2)\,dx = \frac{l^2}{3}$$

$$a_n = \frac{2}{l}\int_0^l (lx - x^2)\cos\frac{n\pi x}{l}\,dx$$

$$= \begin{cases} -\dfrac{4l^2}{n^2\pi^2}; & \text{when } n \text{ is even} \\ 0; & \text{when } n \text{ is odd} \end{cases} \qquad \text{| On simplification}$$

$$\therefore \qquad u = \frac{l^2}{6} - \frac{4l^2}{\pi^2}\sum_{n=2,4,...}^{\infty}\frac{1}{n^2}\cos\frac{n\pi x}{l}\, e^{-\left(\frac{n^2\pi^2 kt}{l^2}\right)}$$

Put $n = 2m$, we get

$$u(x, t) = \frac{l^2}{6} - \frac{l^2}{\pi^2}\sum_{m=1}^{\infty}\frac{1}{m^2}\cos\frac{2m\pi x}{l}\, e^{-\left(\frac{4m^2\pi^2 kt}{l^2}\right)}$$

TEST YOUR KNOWLEDGE

1. (*i*) Solve: $\frac{\partial u}{\partial t} = \alpha\frac{\partial^2 u}{\partial x^2}$; α constant, subject to the boundary conditions $u(0, t) = 0$, $u(\pi, t) = 0$ and the initial condition $u(x, 0) = \sin 2x$. *(M.T.U. 2012)*

(*ii*) Solve: $\frac{\partial u}{\partial t} = a^2\frac{\partial^2 u}{\partial x^2}$ given that

(*a*) $u = 0$ when $x = 0$ and $x = l$ for all t (*b*) $u = 3\sin\frac{\pi x}{l}$ when $t = 0$ for all x.

(*iii*) Solve: $u_t = a^2 u_{xx}$ under the conditions $u_x(0, t) = 0 = u_x(\pi, t)$ and $u(x, 0) = x^2 (0 < x < \pi)$.

2. (*i*) Determine the solution of one-dimensional heat equation $\frac{\partial u}{\partial t} = c^2\frac{\partial^2 u}{\partial x^2}$ where the boundary conditions are $u(0, t) = 0$, $u(l, t) = 0$ $(t > 0)$ and the initial condition $u(x, 0) = x$: l being the length of the bar. *(U.P.T.U. 2015)*

(*ii*) Find the temperature distribution in a rod of length 2 m whose end points are fixed at temperature zero and the initial temperature distribution is $f(x) = 100x$. *(G.B.T.U. 2012)*

3. The heat flow in a bar of length 10 cm of homogeneous material is governed by partial diff. eqn. $u_t = c^2 u_{xx}$. The ends of the bar are kept at temp. 0°C and initial temp. is $f(x) = x(10 - x)$. Find the temp. in the bar at any instant of time. *(U.P.T.U. 2014)*

4. Find the temperature $u(x, t)$ in a homogeneous bar of heat conducting material of length L cm. with its ends kept at zero temperature and initial temperature given by $\frac{x(\text{L} - x)\,d}{\text{L}^2}$.

5. A homogeneous rod of conducting material of length 100 cm has its ends kept at zero temperature and the temperature initially is $u(x, 0) = \begin{cases} x, & 0 \le x \le 50 \\ 100 - x, & 50 \le x \le 100 \end{cases}$

Find the temperature $u(x, t)$ at any time.

6. Find the temperature $u(x, t)$ in a slab whose ends $x = 0$ and $x = L$ are kept at zero temperature and whose initial temperature $f(x)$ is given by

$$f(x) = \begin{cases} k, & \text{when } 0 < x < \frac{1}{2}L \\ 0, & \text{when } \frac{1}{2}L < x < L \end{cases}.$$

7. Find the temperature distribution in a rod of length π which is totally insulated including the ends and the initial temperature distribution is $100 \cos x$. *(U.P.T.U. 2015)*

8. Find the temperature in a thin metal rod of length L with both ends insulated (so that there is no passage of heat through the ends) and with initial temperature $\sin \frac{\pi x}{L}$ in the rod.

9. (*i*) The temperature of a bar 50 cm long with insulated sides is kept at 0° at one end and 100° at the other end until steady conditions prevail. The two ends are then suddenly insulated so that the temperature gradient is zero at each end thereafter. Find the temperature distribution.

(*ii*) A bar 10 cm long, with insulated sides, has its ends A and B maintained at temperatures 50°C and 100°C respectively, until steady-state conditions prevail. The temperature at A is suddenly raised to 90°C and at the same time that at B is lowered to 60°C. Find the temperature distribution in the bar at time t.

(*iii*) A bar of 10 cm length with insulated sides A and B are kept and 20°C and 40°C respectively until steady state conditions prevail. The temperature at A is then suddenly varied to 50°C and the same instant at B, lowered at 10°C. Find the subsequent temperature at any point of the bar at any time. *(A.K.T.U. 2017)*

10. A homogeneous rod of conducting material of length '1' has its ends kept at zero temperature. The temperature at the centre is T and falls uniformly to zero at the two ends. Find the temperature distribution. *(U.K.T.U. 2012)*

$$\left[\textbf{Hint: } u(x, 0) = \begin{cases} 2Tx, & 0 \le x \le \frac{1}{2} \\ 2T(1 - x), & \frac{1}{2} \le x \le 1 \end{cases}\right]$$

11. Solve $\frac{\partial \theta}{\partial t} = k \frac{\partial^2 \theta}{\partial x^2}$, such that

(*i*) θ is finite when $t \to \infty$, (*ii*) $\frac{\partial \theta}{\partial x} = 0$ when $x = 0$ and $\theta = 0$ when $x = l$ for all t,

(*iii*) $\theta = \theta_0$ when $t = 0$ for all values of x between 0 and l.

12. Find a solution of the heat conduction equation $\frac{\partial u}{\partial t} = \alpha \frac{\partial^2 u}{\partial x^2}$ such that

(*i*) u is finite when $t \to \infty$, (*ii*) $u = 100$ when $x = 0$ or π for all values of t,

(*iii*) $u = 0$ when $t = 0$ for all values of x between 0 and π.

(Here, the initially ice-cold rod has its ends in boiling water.)

13. Solve the following IBVP: $u_t = u_{xx}$, $0 < x < 1$, $t > 0$

$u(0, t) = 1$, $u(1, t) = 2$, $t \ge 0$

$u(x, 0) = 1 + x + 2 \sin \pi x$

14. Solve the following IBVP: $u_t = u_{xx}$, $0 < x < 1$, $t > 0$

$u(0, t) = 1$, $u(1, t) = 3$

$u(x, 0) = 1 + 2x + 3 \sin \pi x$

Answers

1. (i) $u(x, t) = \sin 2x\, e^{-4\alpha t}$ (ii) $u(x, t) = 3 \sin \frac{\pi x}{l} e^{-(a^2\pi^2 t/l^2)}$

(iii) $u(x, t) = \frac{\pi^3}{3} + 4 \sum_{n=1}^{\infty} \frac{(-1)^n}{n^2} \cos nx\, e^{-a^2 n^2 t}$

2. (i) $u(x, t) = -\frac{2l}{\pi} \sum_{n=1}^{\infty} \frac{\cos n\pi}{n} \sin \frac{n\pi x}{l} e^{-\left(\frac{c^2 n^2 \pi^2 t}{l^2}\right)}$

(ii) $u(x, t) = -\frac{400}{\pi} \sum_{n=1}^{\infty} \frac{\cos n\pi}{n} \sin \frac{n\pi x}{2} e^{-\left(\frac{c^2 n^2 \pi^2 t}{4}\right)}$

3. $u(x, t) = \frac{800}{\pi^3} \sum_{n=1}^{\infty} \frac{1}{(2n-1)^3} \sin \frac{(2n-1)\pi x}{10} e^{-\left[\frac{(2n-1)^2 \pi^2 c^2 t}{100}\right]}$.

4. $u(x, t) = \frac{8d}{\pi^3} \sum_{n=1}^{\infty} \frac{1}{(2n-1)^3} \sin \frac{(2n-1)\pi x}{L} e^{-\frac{(2n-1)^2 \pi^2 c^2 t}{L^2}}$.

5. $u(x, t) = \frac{400}{\pi^2} \sum_{m=0}^{\infty} \frac{(-1)^m}{(2m+1)^2} \sin \frac{(2m+1)\pi x}{100} e^{-\left[\frac{(2m+1)c\pi}{100}\right]^2 t}$

6. $u(x, t) = \frac{4k}{\pi} \sum_{n=1}^{\infty} \frac{1}{n} \sin^2 \frac{n\pi}{4} \sin \frac{n\pi x}{L} e^{-\left(\frac{c^2 n^2 \pi^2 t}{L^2}\right)}$

7. $100\, e^{-t} \cos x$

8. $u(x, t) = \frac{2}{\pi} - \frac{4}{\pi} \sum_{m=1}^{\infty} \frac{1}{(4m^2-1)} \cos\left(\frac{2m\pi x}{L}\right) e^{-\frac{4m^2\pi^2 c^2 t}{L^2}}$

9. (i) $u(x, t) = \frac{200}{\pi} \sum_{n=1}^{\infty} \frac{(-1)^{n+1}}{n} \sin \frac{n\pi x}{50} e^{-\left(\frac{n^2\pi^2 kt}{2500}\right)}$

(ii) $u(x, t) = 90 - 3x - \frac{80}{\pi} \sum_{n=1}^{\infty} \frac{1}{n} \sin \frac{n\pi x}{5} e^{-\left(\frac{c^2 n^2 \pi^2 t}{25}\right)}$

(iii) $u(x, t) = 50 - 4x - \frac{120}{\pi} \sum_{n=1}^{\infty} \frac{(-1)^n}{n} \sin \frac{n\pi x}{10} e^{-\frac{n^2\pi^2 t}{100}}$

10. $u(x, t) = \frac{8T}{\pi^2} \sum_{m=1}^{\infty} \frac{(-1)^{m+1}}{(2m-1)^2} \sin (2m-1)\pi x\, e^{-[(2m-1)^2 \pi^2 c^2 t]}$

11. $\theta = \frac{4\theta_0}{\pi}\left[e^{-(\pi/2l)^2 kt} \cos \frac{\pi x}{2l} - \frac{1}{3} e^{-\left(\frac{3\pi}{2l}\right)^2 kt} \cos \frac{3\pi x}{2l} + \frac{1}{5} e^{-\left(\frac{5\pi}{2l}\right)^2 kt} \cos \frac{5\pi x}{2l} - \ldots\right]$

12. $u(x, t) = 100 - \dfrac{400}{\pi} \displaystyle\sum_{m=1}^{\infty} \dfrac{\sin (2m-1)x}{2m-1} e^{-(2m-1)^2 \alpha t}$

13. $u(x, t) = 1 + x + 2e^{-\pi^2 t} \sin \pi x$

14. $u(x, t) = 1 + 2x + 3e^{-\pi^2 t} \sin \pi x.$

2.13 TWO-DIMENSIONAL HEAT FLOW

Consider the flow of heat in a metal plate, in the XOY plane. If the temperature at any point is independent of the z-coordinate and depends on x, y and t only, then the flow is called two dimensional and the heat-flow lies in the plane XOY only and is zero along the normal to the plane XOY.

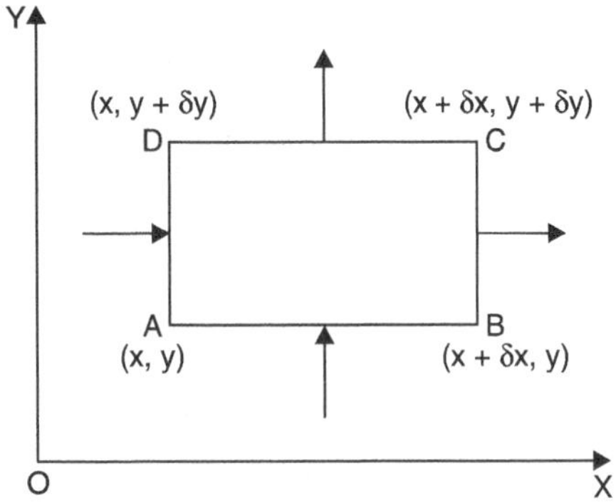

Take a rectangular element of the plate with sides δx and δy and thickness α. As discussed in the one-dimensional heat flow along a bar, the quantity of heat that enters the plate per second from the sides AB and AD is given by

$$-k\alpha\, \delta x \left(\frac{\partial u}{\partial y}\right)_y \quad \text{and} \quad -k\alpha\, \delta y \left(\frac{\partial u}{\partial x}\right)_x$$

respectively and that which flows out through the sides CD and BC per second is

$$-k\alpha\, \delta x \left(\frac{\partial u}{\partial y}\right)_{y+\delta y} \quad \text{and} \quad -k\alpha\, \delta y \left(\frac{\partial u}{\partial x}\right)_{x+\delta x} \quad \text{respectively.}$$

Therefore, the total gain of heat by the rectangular plate ABCD per second

$$= -k\alpha\delta x \left(\frac{\partial u}{\partial y}\right)_y - k\alpha\delta y \left(\frac{\partial u}{\partial x}\right)_x + k\alpha\delta x \left(\frac{\partial u}{\partial y}\right)_{y+\delta y} + k\alpha\delta y \left(\frac{\partial u}{\partial x}\right)_{x+\delta x}$$

$$= k\alpha\delta x\delta y \left[\frac{\left(\frac{\partial u}{\partial x}\right)_{x+\delta x} - \left(\frac{\partial u}{\partial x}\right)_x}{\delta x} + \frac{\left(\frac{\partial u}{\partial y}\right)_{y+\delta y} - \left(\frac{\partial u}{\partial y}\right)_y}{\delta y} \right] \quad \text{...(1)}$$

The rate of gain of heat by the plate is also given by

$$s\rho\delta x\delta y \frac{\partial u}{\partial t} \quad \text{...(2)}$$

where s = specific heat and ρ = density of the metal plate.

Equating (1) and (2), we obtain

$$k\alpha\delta x\delta y \left[\frac{\left(\frac{\partial u}{\partial x}\right)_{x+\delta x} - \left(\frac{\partial u}{\partial x}\right)_x}{\delta x} + \frac{\left(\frac{\partial u}{\partial y}\right)_{y+\delta y} - \left(\frac{\partial u}{\partial y}\right)_y}{\delta y} \right] = s\rho\delta x\delta y \frac{\partial u}{\partial t}$$

Dividing both sides by $\alpha\delta x\delta y$ and taking the limit as $\delta x \to 0$, $\delta y \to 0$, we get

$$k\left(\frac{\partial^2 u}{\partial x^2} + \frac{\partial^2 u}{\partial y^2}\right) = s\rho \frac{\partial u}{\partial t}$$

or $$c^2\left(\frac{\partial^2 u}{\partial x^2}+\frac{\partial^2 u}{\partial y^2}\right)=\frac{\partial u}{\partial t} \quad \text{where } c^2=\frac{k}{s\rho} \qquad ...(3)$$

Equation (3) gives the temperature distribution of the plate in the transient state.

Note 1. In steady state, u is independent of t, so that $\frac{\partial u}{\partial t}=0$ and the above equation reduces to

$$\frac{\partial^2 u}{\partial x^2}+\frac{\partial^2 u}{\partial y^2}=0 \qquad ...(4)$$

which is known as **Laplace's Equation in two dimensions.**

Note 2. The equation of heat flow in a solid (Three-dimensional heat flow) can similarly be derived as

$$c^2\left(\frac{\partial^2 u}{\partial x^2}+\frac{\partial^2 u}{\partial y^2}+\frac{\partial^2 u}{\partial z^2}\right)=\frac{\partial u}{\partial t}$$

In the steady state, it reduces to $$\frac{\partial^2 u}{\partial x^2}+\frac{\partial^2 u}{\partial y^2}+\frac{\partial^2 u}{\partial z^2}=0$$

which is **Laplace's Equation in three dimensions.**

2.14 SOLUTION OF LAPLACE'S EQUATION IN TWO DIMENSIONS *(G.B.T.U. 2011)*

Laplace's equation in two dimensions is

$$\frac{\partial^2 u}{\partial x^2}+\frac{\partial^2 u}{\partial y^2}=0 \qquad ...(1)$$

Let $$u = XY \qquad ...(2)$$

where X is a function of x only and Y is a function of y only, be a solution of (1).

Then $$\frac{\partial^2 u}{\partial x^2}=X''Y \text{ and } \frac{\partial^2 u}{\partial y^2}=XY''$$

Substituting in (1), we have $X''Y + XY'' = 0$ or $$\frac{X''}{X}=-\frac{Y''}{Y} \qquad ...(3)$$

Now the LHS of (3) is a function of x only and the R.H.S is a function of y only. Since x and y are independent variables, this equation can hold only when both sides reduce to a constant, say k. Then equation (3) leads to the ordinary differential equations

$$\frac{d^2X}{dx^2}-kX=0 \quad \text{and} \quad \frac{d^2Y}{dy^2}+kY=0 \qquad ...(4)$$

Solving equations (4), we get

(*i*) *When k is positive* and $= p^2$, say

$$X = c_1e^{px}+c_2e^{-px},\ Y = c_3\cos py + c_4 \sin py$$

(*ii*) *When k is negative* and $= -p^2$, say

$$X = c_1\cos px + c_2\sin px,\ Y = c_3e^{py}+c_4e^{-py}$$

(*iii*) *When k = 0*

$$X = c_1x + c_2,\ Y = c_3y + c_4$$

Thus, the various possible solutions of Laplace's equation (1) are:

$$u = (c_1 e^{px} + c_2 e^{-px})(c_3 \cos py + c_4 \sin py) \quad ...(5)$$

$$u = (c_1 \cos px + c_2 \sin px)(c_3 e^{py} + c_4 e^{-py}) \quad ...(6)$$

$$u = (c_1 x + c_2)(c_3 y + c_4) \quad ...(7)$$

Of these three solutions, we have to choose that solution which is consistent with the physical nature of the problem and the given boundary conditions. Solution (6) is the required solution.

$$\boxed{u(x, y) = (c_1 \cos px + c_2 \sin px)(c_3 e^{py} + c_4 e^{-py}).}$$

ILLUSTRATIVE EXAMPLES

Example 1. *Use separation of variables method to solve the equation*

$$\frac{\partial^2 u}{\partial x^2} + \frac{\partial^2 u}{\partial y^2} = 0$$

subject to the boundary conditions $u(0, y) = u(l, y) = u(x, 0) = 0$ *and* $u(x, a) = \sin \frac{n\pi x}{l}$.

[A.K.T.U. 2017]

Sol. The given equation is

$$\frac{\partial^2 u}{\partial x^2} + \frac{\partial^2 u}{\partial y^2} = 0 \quad ...(1)$$

Let $\quad u = XY \quad ...(2)$

where X is a function of x only and Y is a function of y only then,

$$\frac{\partial^2 u}{\partial x^2} = \frac{\partial^2}{\partial x^2}(XY) = Y\frac{d^2X}{dx^2} \quad \text{and} \quad \frac{\partial^2 u}{\partial y^2} = \frac{\partial^2}{\partial y^2}(XY) = X\frac{d^2Y}{dy^2}$$

$\therefore$ From (1), $\quad YX'' + XY'' = 0$

$$\Rightarrow \quad \frac{X''}{X} + \frac{Y''}{Y} = 0$$

Case I. When $\quad \frac{X''}{X} = -\frac{Y''}{Y} = p^2$ (say)

(*i*) $\quad \frac{X''}{X} = p^2$

$$X'' - p^2 X = 0$$

Auxiliary equation is $\quad m^2 - p^2 = 0$

$$m = \pm p$$

$\therefore \quad$ C.F. $= c_1 e^{px} + c_2 e^{-px}$

P.I. $= 0$

$\therefore \quad X = c_1 e^{px} + c_2 e^{-px}$

(*ii*) $\quad \frac{-Y''}{Y} = p^2 \quad \Rightarrow \quad Y'' + p^2 Y = 0$

Auxiliary equation is $\quad m^2 + p^2 = 0 \quad \Rightarrow \quad m = \pm pi$

$\therefore \quad$ C.F. $= c_3 \cos py + c_4 \sin py$

$$\text{P.I.} = 0$$

$\therefore$ $$y = c_3 \cos py + c_4 \sin py$$

Now, $$X(0) = 0$$

$\Rightarrow$ $$c_1 + c_2 = 0 \Rightarrow c_2 = -c_1$$

$$X(l) = 0$$

$\Rightarrow$ $$c_1 e^{pl} + c_2 e^{-pl} = 0 \Rightarrow c_1 (e^{pl} - e^{-pl}) = 0$$

$\Rightarrow$ $$c_1 = 0 \qquad | \text{ Since } e^{pl} - e^{-pl} \neq 0 \text{ (as } p \neq 0 \neq l)$$

$\therefore$ $$c_2 = 0$$

$\therefore$ $$X = 0 \Rightarrow u = XY = 0 \text{ which is impossible}$$

Hence we reject case I.

Case II. When $$\frac{X''}{X} = -\frac{Y''}{Y} = 0 \text{ (say)}$$

(*i*) $$\frac{X''}{X} = 0$$

$\Rightarrow$ $$X'' = 0 \Rightarrow X = c_5 x + c_6$$

(*ii*) $$\frac{-Y''}{Y} = 0$$

$\Rightarrow$ $$Y'' = 0 \Rightarrow Y = c_7 y + c_8$$

Now, $$X(0) = 0 \Rightarrow c_6 = 0$$

$$X(l) = 0$$

$\Rightarrow$ $$c_5 l + c_6 = 0 \Rightarrow c_5 l = 0$$

$\Rightarrow$ $$c_5 = 0 \qquad \text{(Since } l \neq 0)$$

$\therefore$ $$X = 0$$

$\therefore$ $$u = XY = 0 \quad \text{which is impossible}$$

Hence we also reject case II.

Case III. When $$\frac{X''}{X} = -\frac{Y''}{Y} = -p^2 \text{ (say)}$$

(*i*) $$\frac{X''}{X} = -p^2$$

$\Rightarrow$ $$X'' + p^2 X = 0 \Rightarrow \frac{d^2X}{dx^2} + p^2 X = 0.$$

Auxiliary equation is $m^2 + p^2 = 0 \Rightarrow m = \pm pi$

$$\text{C.F.} = c_9 \cos px + c_{10} \sin px$$

$$\text{P.I.} = 0$$

$$X = c_9 \cos px + c_{10} \sin px$$

(*ii*) $$-\frac{Y''}{Y} = -p^2$$

$\Rightarrow$ $$\frac{Y''}{Y} = p^2 \Rightarrow \frac{d^2Y}{dy^2} - p^2 Y = 0.$$

Auxiliary equation is

$$m^2 - p^2 = 0$$

$$m = \pm p.$$

$\therefore$ $$\text{C.F.} = c_{11}e^{py} + c_{12}e^{-py}$$

$$\text{P.I.} = 0$$

Hence, $$Y = c_{11}\, e^{py} + c_{12}\, e^{-py}.$$

Now, $$X(0) = 0 \;\Rightarrow\; c_9 = 0$$

$\therefore$ $$X = c_{10} \sin px$$

$$X(l) = 0$$

$$c_{10} \sin pl = 0$$

$\Rightarrow$ $$\sin pl = 0 = \sin n\pi,\ n \in I$$

$\therefore$ $$p = \frac{n\pi}{l}$$

$\therefore$ $$X = c_{10} \sin \frac{n\pi x}{l} \qquad \text{...(3)}$$

Again, $$Y(0) = 0$$

$\Rightarrow$ $$c_{11} + c_{12} = 0 \;\Rightarrow\; c_{12} = -c_{11}$$

$$Y = c_{11}(e^{py} - e^{-py}) = c_{11}\left(e^{\frac{n\pi y}{l}} - e^{-\frac{n\pi y}{l}}\right) \qquad \text{...(4)}$$

$\therefore$ $$u = XY = c_{10}c_{11} \sin \frac{n\pi x}{l}\,[e^{(n\pi y/l)} - e^{(-n\pi y/l)}]$$

or $$u(x, y) = b_n \sin \frac{n\pi x}{l}\,[e^{(n\pi y/l)} - e^{(-n\pi y/l)}] \qquad \text{...(5)}$$

Now, $$u(x, a) = \sin \frac{n\pi x}{l} = b_n \sin \frac{n\pi x}{l}\,[e^{(n\pi a/l)} - e^{-(n\pi a/l)}]$$

$\Rightarrow$ $$b_n = \frac{1}{e^{\frac{n\pi a}{l}} - e^{-\frac{n\pi a}{l}}} = \frac{1}{2 \sin h\left(\frac{n\pi a}{l}\right)}$$

$\therefore$ $$u(x, y) = \frac{e^{(n\pi y/l)} - e^{-(n\pi y/l)}}{2 \sinh\left(\frac{n\pi a}{l}\right)} \sin \frac{n\pi x}{l} = \frac{\sinh\left(\frac{n\pi y}{l}\right)}{\sinh\left(\frac{n\pi a}{l}\right)} \sin \frac{n\pi x}{l}.$$

Example 2. *A rectangular plate with insulated surfaces is 8 cm wide and so long compared to its width that it may be considered infinite in length without introducing an appreciable error. If the temperature along one short edge y = 0 is given by*

$$u(x, 0) = 100 \sin \frac{\pi x}{8},\ 0 < x < 8$$

while the two long edges x = 0 and x = 8 as well as the other short edge are kept at 0°C, show that the steady state temperature at any point of the plate is given by

$$u(x, y) = 100e^{-\frac{\pi y}{8}} \sin \frac{\pi x}{8}. \qquad \text{(A. K. T. U. 2018)}$$

Sol. Let $u(x, y)$ be the temperature at any point P of the plate.

Two dimensional heat flow equation in steady state is given by

$$\frac{\partial^2 u}{\partial x^2} + \frac{\partial^2 u}{\partial y^2} = 0 \qquad ...(1)$$

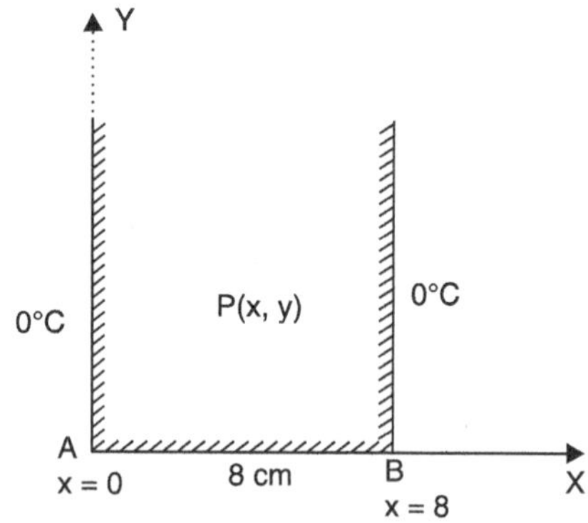

Its solution is

$$u(x, y) = (c_1 \cos px + c_2 \sin px)(c_3 e^{py} + c_4 e^{-py}) \qquad ...(2)$$

Boundary conditions are

$$u(0, y) = 0 = u(8, y)$$

$$\underset{y \to \infty}{\text{Lt}}\ u(x, y) = 0$$

$$u(x, 0) = 100 \sin \frac{\pi x}{8}, \quad 0 < x < 8$$

From (2), $\quad u(0, y) = 0 = c_1 (c_3 e^{py} + c_4 e^{-py})$

$\Rightarrow \quad c_1 = 0.$

$\therefore$ From (2), $\quad u(x, y) = c_2 \sin px\, (c_3\, e^{py} + c_4 e^{-py}) \qquad ...(3)$

$$u(8, y) = 0 = c_2 \sin 8p(c_3\, e^{py} + c_4\, e^{-py})$$

$\Rightarrow \quad \sin 8p = 0 = \sin n\pi$

$\Rightarrow \quad p = \frac{n\pi}{8} \quad (n \in \text{I})$

$\therefore$ From (3), $\quad u(x, y) = c_2 \sin \frac{n\pi x}{8} (c_3 e^{\frac{n\pi y}{8}} + c_4 e^{-\frac{n\pi y}{8}}) \qquad ...(4)$

$$\underset{y \to \infty}{\text{Lt}}\ u(x, y) = 0 = c_2 \sin \frac{n\pi x}{8} \lim_{y \to \infty} (c_3 e^{\frac{n\pi y}{8}} + c_4 e^{-\frac{n\pi y}{8}})$$

which is satisfied only when

$$c_3 = 0.$$

$\therefore$ From (4), $\quad u(x, y) = c_2 c_4 \sin \frac{n\pi x}{8} e^{-\frac{n\pi y}{8}} = b_n \sin \frac{n\pi x}{8} e^{-\frac{n\pi y}{8}} \qquad ...(5)$

From (5), $\quad u(x, 0) = 100 \sin \frac{\pi x}{8} = b_n \sin \frac{n\pi x}{8}$

$\Rightarrow \quad b_n = 100, n = 1.$

$\therefore$ From (5), $\quad u(x, y) = 100 \sin \left(\frac{\pi x}{8}\right) e^{-(\pi y/8)}$

which is the required steady state temperature at any point of the plate.

Example 3. *An infinitely long plane uniform plate is bounded by two parallel edges and an end at right angles to them. The breadth is π. This end is maintained at temperature u_0 at all points and the other edges are at zero temperature. Determine the temperature at any point of the plate in the steady state.* (U.P.T.U. 2013)

Sol. In steady state, two dimensional heat flow equation is

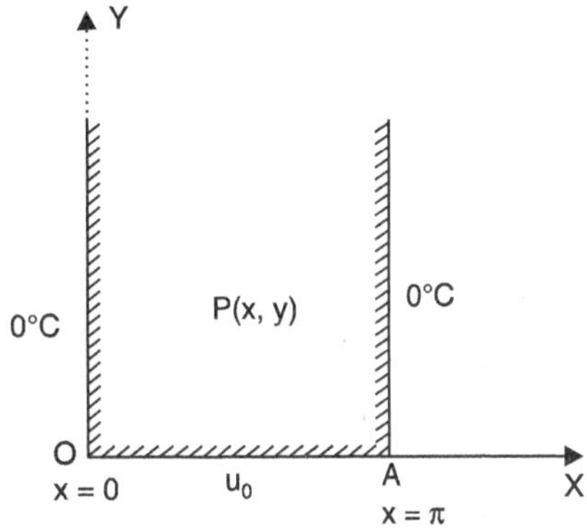

$$\frac{\partial^2 u}{\partial x^2} + \frac{\partial^2 u}{\partial y^2} = 0 \quad ...(1)$$

Boundary conditions are,

$$u(0, y) = 0 = u(\pi, y)$$

$$\underset{y \to \infty}{\text{Lt}}\ u(x, y) = 0\ (0 < x < \pi)$$

and $$u(x, 0) = u_0\ (0 < x < \pi)$$

Solution to equation (1) is

$$u(x, t) = (c_1 \cos px + c_2 \sin px)(c_3 e^{py} + c_4 e^{-py}) \quad ...(2)$$

From (2), $u(0, y) = 0 = c_1(c_3 e^{py} + c_4 e^{-py})$

$\Rightarrow$ $c_1 = 0.$

From (2), $u(x, y) = c_2 \sin px(c_3 e^{py} + c_4 e^{-py})$

$u(\pi, y) = 0 = c_2 \sin p\pi(c_3 e^{py} + c_4 e^{-py})$

$\Rightarrow$ $\sin p\pi = 0 = \sin n\pi\ (n \in I)$ ...(3)

$\therefore$ $p = n.$

$\therefore$ From (3), $u(x, y) = c_2 \sin nx\ (c_3 e^{ny} + c_4 e^{-ny})$ ...(4)

$$\lim_{y \to \infty} u(x, y) = 0 = c_2 \sin nx \lim_{y \to \infty} (c_3 e^{ny} + c_4 e^{-ny})$$

which is satisfied only when $c_3 = 0$.

$\therefore$ From (4), $u(x, y) = c_2 c_4 e^{-ny} \sin nx = b_n e^{-ny} \sin nx,$ where $c_2 c_4 = b_n$

The most general solution is

$$u(x, y) = \sum_{n=1}^{\infty} b_n e^{-ny} \sin nx \quad ...(5)$$

$$u(x, 0) = u_0 = \sum_{n=1}^{\infty} b_n \sin nx$$

where $$b_n = \frac{2}{\pi} \int_0^{\pi} u_0 \sin nx\, dx$$

$$= \frac{2u_0}{\pi} \left(\frac{-\cos nx}{n} \right)_0^{\pi} = \frac{2u_0}{n\pi} \{1 - (-1)^n\} = \begin{cases} \dfrac{4u_0}{n\pi}; & \text{if } n \text{ is odd} \\ 0; & \text{if } n \text{ is even} \end{cases}$$

$\therefore$ From (5), $$u(x, y) = \frac{4u_0}{\pi} \sum_{n=1,3,5,...}^{\infty} \frac{\sin nx}{n} e^{-ny} \quad (n \text{ is odd})$$

or $$u(x, y) = \frac{4u_0}{\pi} \sum_{n=1}^{\infty} \frac{1}{(2n-1)} \sin (2n-1)\, x\, e^{-(2n-1)y}.$$

Example 4. *A rectangular plate with insulated surfaces is 10 cm wide and so long compared to its width that it may be considered infinite in length without introducing an appreciable error. If the temperature along the short edge y = 0 is given by*

and $$u(x, y) = \begin{cases} 20x, & 0 < x \le 5 \\ 20(10 - x), & 5 < x < 10 \end{cases}$$

and the two long edges x = 0 and x = 10 as well as other short edge are kept at 0°C. Find the temperature u at any point P(x, y).

Sol. In steady state, two-dimensional heat flow equation is

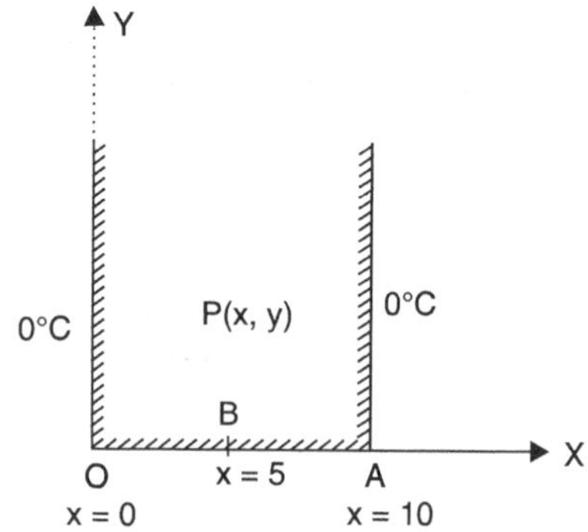

$$\frac{\partial^2 u}{\partial x^2} + \frac{\partial^2 u}{\partial y^2} = 0 \qquad ...(1)$$

Its solution is

$$u(x, y) = (c_1 \cos px + c_2 \sin px)(c_3 e^{py} + c_4 e^{-py}) \qquad ...(2)$$

Boundary conditions are $u(0, y) = 0$

$$u(10, y) = 0$$

$$\lim_{y \to \infty} u(x, y) = u(x, \infty) = 0$$

and $$u(x, 0) = \begin{cases} 20x, & 0 < x \le 5 \\ 20(10 - x), & 5 < x \le 10 \end{cases}$$

From (2), $$u(x, y) = 0 = c_1(c_3 e^{py} + c_4 e^{-py})$$

$\Rightarrow$ $$c_1 = 0$$

From (2), $$u(x, y) = c_2 \sin px \, (c_3 e^{py} + c_4 e^{-py}) \qquad ...(3)$$

$$u(10, y) = 0 = c_2 \sin 10p \, (c_3 e^{py} + c_4 e^{-py})$$

$\Rightarrow$ $$\sin 10p = 0 = \sin n\pi$$

or $$10p = n\pi \ (n \in \text{I})$$

$\Rightarrow$ $$p = \frac{n\pi}{10}.$$

$\therefore$ From (3), $$u(x, y) = c_2 \sin \frac{n\pi x}{10} (c_3 e^{\frac{n\pi y}{10}} + c_4 e^{-\frac{n\pi y}{10}}) \qquad ...(4)$$

$$\lim_{y \to \infty} u(x, y) = c_2 \sin \frac{n\pi x}{10} \lim_{y \to \infty} (c_3 e^{\frac{n\pi y}{10}} + c_4 e^{-\frac{n\pi y}{10}})$$

which is satisfied only when $c_3 = 0$.

Hence from (4), $$u(x, y) = c_2 c_4 \sin \frac{n\pi x}{10} e^{-\frac{n\pi y}{10}} = b_n \sin \frac{n\pi x}{10} e^{-\frac{n\pi y}{10}} \qquad ...(5)$$

The most general solution is

$$u(x, y) = \sum_{n=1}^{\infty} b_n \sin \frac{n\pi x}{10} e^{-\frac{n\pi y}{10}} \qquad ...(6)$$

$$u(x, 0) = \sum_{n=1}^{\infty} b_n \sin \frac{n\pi x}{10},$$

where $b_n = \dfrac{2}{10}\int_0^{10} u(x,0)\sin\dfrac{n\pi x}{10}\,dx$

$$= \frac{1}{5}\left[\int_0^5 20x\sin\frac{n\pi x}{10}\,dx + \int_5^{10} 20(10-x)\sin\frac{n\pi x}{10}\,dx\right]$$

$$= 4\left[\left\{x\left(\frac{-\cos\frac{n\pi x}{10}}{\frac{n\pi}{10}}\right)\right\}_0^5 - \int_0^5 1\left(\frac{-\cos\frac{n\pi x}{10}}{\frac{n\pi}{10}}\right)dx + \left\{(10-x)\left(\frac{-\cos\frac{n\pi x}{10}}{\frac{n\pi}{10}}\right)\right\}_5^{10} - \int_5^{10}(-1)\left(\frac{-\cos\frac{n\pi x}{10}}{\frac{n\pi}{10}}\right)dx\right]$$

$$= 4\left[\frac{10}{n\pi}(-5)\cos\frac{n\pi}{2} + \frac{10}{n\pi}\left(\frac{\sin\frac{n\pi x}{10}}{\frac{n\pi}{10}}\right)_0^5 + \frac{50}{n\pi}\cos\frac{n\pi}{2} - \frac{10}{n\pi}\left(\frac{\sin\frac{n\pi x}{10}}{\frac{n\pi}{10}}\right)_5^{10}\right]$$

$$= 4\left[\frac{-50}{n\pi}\cos\frac{n\pi}{2} + \frac{100}{n^2\pi^2}\sin\frac{n\pi}{2} + \frac{50}{n\pi}\cos\frac{n\pi}{2} - \frac{100}{n^2\pi^2}\left(0-\sin\frac{n\pi}{2}\right)\right]$$

$$= \frac{800}{n^2\pi^2}\sin\frac{n\pi}{2}.$$

From (6), $$u(x, y) = \frac{800}{\pi^2}\sum_1^{\infty}\frac{\sin n\pi/2}{n^2}\sin\frac{n\pi x}{10}\,e^{-\frac{n\pi y}{10}}.$$

Example 5. *Solve* $\dfrac{\partial^2 u}{\partial x^2} + \dfrac{\partial^2 u}{\partial y^2} = 0,\ 0 < x < \pi,\ 0 < y < \pi$, *which satisfies the conditions:*

$u(0, y) = u(\pi, y) = u(x, \pi) = 0$ *and* $u(x, 0) = \sin^2 x$. (U.K.T.U. 2011)

Sol. The given equation is $\dfrac{\partial^2 u}{\partial x^2} + \dfrac{\partial^2 u}{\partial y^2} = 0$ ...(1)

Its solution consistent with boundary condition is

$$u(x, y) = (c_1\cos px + c_2\sin px)(c_3 e^{py} + c_4 e^{-py}) \quad ...(2)$$

From (2), $u(0, y) = 0 = c_1(c_3 e^{py} + c_4 e^{-py})$

$\Rightarrow$ $c_1 = 0.$

$\therefore$ From (2), $u(x, y) = c_2\sin px\,(c_3 e^{py} + c_4 e^{-py})$ ...(3)

$u(\pi, y) = 0 = c_2\sin p\pi(c_3 e^{py} + c_4 e^{-py})$

$\Rightarrow$ $\sin p\pi = 0 = \sin n\pi\ (n \in I)$

$\therefore$ $p = n.$

Hence from (3), $u(x, y) = c_2\sin nx\,(c_3 e^{ny} + c_4 e^{-ny}) = \sin nx\,(Ae^{ny} + Be^{-ny})$ ...(4)

where $c_2 c_3 = A$ and $c_2 c_4 = B$.

From (4), $u(x, \pi) = \sin nx(Ae^{n\pi} + Be^{-n\pi})$

$$0 = \sin nx(Ae^{n\pi} + Be^{-n\pi})$$

$\Rightarrow$ $$0 = Ae^{n\pi} + Be^{-n\pi}$$

$\Rightarrow$ $$Ae^{n\pi} = -Be^{-n\pi} = -\frac{1}{2}B_n \text{ (say)}$$

then (4) becomes, $$u(x, y) = \sin nx\left[-\frac{1}{2}B_n e^{-n\pi} e^{ny} + \frac{1}{2}B_n e^{n\pi} e^{-ny}\right]$$

$$= \frac{1}{2}B_n[e^{n(\pi-y)} - e^{-n(\pi-y)}] \sin nx = B_n \sinh n(\pi - y) \sin nx.$$

The most general solution is

$$u(x, y) = \sum_{n=1}^{\infty} B_n \sinh n(\pi - y) \sin nx \qquad ...(5)$$

$$u(x, 0) = \sin^2 x = \sum_{n=1}^{\infty} B_n \sinh n\pi \sin nx$$

where $$B_n \sinh n\pi = \frac{2}{\pi}\int_0^{\pi} \sin^2 x \sin nx\, dx$$

$$= \frac{1}{\pi}\int_0^{\pi} (1 - \cos 2x) \sin nx\, dx$$

$$= \frac{1}{\pi}\int_0^{\pi}\left[\sin nx - \frac{1}{2}\{\sin (n+2)x + \sin (n-2)x\}\right] dx$$

$$= \frac{1}{\pi}\left[\frac{-\cos nx}{n} + \frac{\cos (n+2)x}{2(n+2)} + \frac{\cos (n-2)x}{2(n-2)}\right]_0^{\pi}$$

$$= \frac{1}{2\pi}\left[\left(\frac{1}{n+2} + \frac{1}{n-2} - \frac{2}{n}\right)\{(-1)^n - 1\}\right], \text{ when } n \neq 2$$

$$B_n \sinh n\pi = \begin{cases} \dfrac{-8}{\pi n(n^2 - 4)}, & \text{when } n \text{ is odd} \\ 0, & \text{when } n \text{ is even and } \neq 2 \end{cases}$$

when $n = 2$,

$$B_2 \sinh 2\pi = \frac{2}{\pi}\int_0^{\pi} \sin^2 x \sin 2x\, dx$$

$$= \frac{1}{\pi}\int_0^{\pi} (1 - \cos 2x) \sin 2x\, dx = \frac{1}{\pi}\int_0^{\pi}\left(\sin 2x - \frac{1}{2}\sin 4x\right) dx$$

$$= \frac{1}{\pi}\left(\frac{-\cos 2x}{2} + \frac{1}{8}\cos 4x\right)_0^{\pi} = 0$$

$\therefore$ $$B_2 = 0.$$

Hence the solution (5) becomes,

$$u(x, y) = \frac{-8}{\pi} \sum_{n=1,3,5,\ldots}^{\infty} \frac{\sin nx \sinh n(\pi - y)}{n(n^2 - 4)\sinh n\pi}$$

or
$$u(x, y) = -\frac{8}{\pi} \sum_{m=1,2,3,\ldots}^{\infty} \frac{\sin (2m-1)x \sinh (2m-1)(\pi - y)}{(2m-1)\{(2m-1)^2 - 4\}\sinh (2m-1)\pi}.$$

Example 6. *Solve* $\frac{\partial^2 u}{\partial x^2} + \frac{\partial^2 u}{\partial y^2} = 0$, *with the rectangle* $0 \le x \le a$, $0 \le y \le b$; *given that* $u(x, b) = u(0, y) = u(a, y) = 0$ *and* $u(x, 0) = x(a - x)$.

Sol. The equation is

$$\frac{\partial^2 u}{\partial x^2} + \frac{\partial^2 u}{\partial y^2} = 0 \qquad \ldots(1)$$

Its solution is

$$u(x, y) = (c_1 \cos px + c_2 \sin px)(c_3 e^{py} + c_4 e^{-py}) \qquad \ldots(2)$$

$$u(0, y) = 0 = c_1(c_3 e^{py} + c_4 e^{-py})$$

$\Rightarrow$ $c_1 = 0.$

$\therefore$ From (2), $u(x, y) = c_2 \sin px\, (c_3\, e^{py} + c_4 e^{-py}) \qquad \ldots(3)$

$$u(a, y) = 0 = c_2 \sin ap\, (c_3\, e^{py} + c_4 e^{-py})$$

$\Rightarrow$ $\sin ap = 0 = \sin n\pi \; (n \in \text{I})$

$\Rightarrow$ $ap = n\pi$ or $p = \frac{n\pi}{a}.$

$\therefore$ From (3), $u(x, y) = c_2 \sin \frac{n\pi x}{a}\left(c_3 e^{\frac{n\pi y}{a}} + c_4 e^{-\frac{n\pi y}{a}}\right)$

$$u(x, y) = \sin \frac{n\pi x}{a}\left(\text{A}\, e^{\frac{n\pi y}{a}} + \text{B}\, e^{-\frac{n\pi y}{a}}\right) \qquad \ldots(4)$$

where $c_2 c_3 = \text{A}$ and $c_2 c_4 = \text{B}$

$$u(x, b) = \sin \frac{n\pi x}{a}\left(\text{A} e^{\frac{n\pi b}{a}} + \text{B} e^{-\frac{n\pi b}{a}}\right)$$

$$0 = \sin \frac{n\pi x}{a}\left(\text{A} e^{\frac{n\pi b}{a}} + \text{B} e^{-\frac{n\pi b}{a}}\right)$$

$\Rightarrow$ $\text{A}\, e^{\frac{n\pi b}{a}} + \text{B} e^{-\frac{n\pi b}{a}} = 0$

$$\text{A}\, e^{\frac{n\pi b}{a}} = -\text{B} e^{-\frac{n\pi b}{a}} = -\frac{1}{2}\text{B}_n \text{ (say).}$$

Then (4) becomes,

$$u(x, y) = \sin \frac{n\pi x}{a}\left[\frac{-1}{2}\text{B}_n e^{-\frac{n\pi b}{a}} e^{\frac{n\pi y}{a}} + \frac{1}{2}\text{B}_n e^{\frac{n\pi b}{a}} e^{-\frac{n\pi y}{a}}\right]$$

$$= \frac{1}{2} B_n \sin \frac{n\pi x}{a} [e^{\frac{n\pi}{a}(b-y)} - e^{-\frac{n\pi}{a}(b-y)}]$$

$$= \frac{1}{2} B_n \sin \frac{n\pi x}{a} . 2 \sinh \frac{n\pi}{a} (b-y) = B_n \sin \frac{n\pi x}{a} \sinh \frac{n\pi}{a} (b-y).$$

The most general solution is

$$u(x, y) = \sum_{n=1}^{\infty} B_n \sin \frac{n\pi x}{a} \sinh \frac{n\pi}{a} (b-y) \qquad ...(5)$$

Applying to this the condition $u(x, 0) = x(a-x)$, we get

From (5), $$u(x, 0) = \sum_{n=1}^{\infty} B_n \sinh \frac{n\pi b}{a} \sin \frac{n\pi x}{a}$$

$$\Rightarrow \qquad x(a-x) = \sum_{n=1}^{\infty} B_n \sinh \frac{n\pi b}{a} \sin \frac{n\pi x}{a}$$

where $$B_n \sinh \frac{n\pi}{a} b = \frac{2}{a} \int_0^a x(a-x) \sin \frac{n\pi}{a} x \, dx$$

$$= \frac{2}{a} \left[\left\{ (ax - x^2) \left(\frac{-\cos \frac{n\pi}{a} x}{\frac{n\pi}{a}} \right) \right\}_0^a - \int_0^a (a-2x) . \left(\frac{-\cos \frac{n\pi}{a} x}{\frac{n\pi}{a}} \right) dx \right]$$

$$= \frac{2}{a} . \frac{a}{n\pi} \int_0^a (a-2x) . \cos \frac{n\pi}{a} x \, dx$$

$$= \frac{2}{n\pi} \left[\left\{ (a-2x) \frac{\sin \frac{n\pi}{a} x}{\frac{n\pi}{a}} \right\}_0^a - \int_0^a (-2) \left(\frac{\sin \frac{n\pi}{a} x}{\frac{n\pi}{a}} \right) dx \right]$$

$$= \frac{4}{n\pi} . \frac{a}{n\pi} \left(\frac{-\cos \frac{n\pi}{a} x}{\frac{n\pi}{a}} \right)_0^a = \frac{4a}{n^2\pi^2} . \frac{a}{n\pi} (1 - \cos n\pi)$$

$$= \frac{4a^2}{n^3\pi^3} [1 - (-1)^n] = \begin{cases} \dfrac{8a^2}{n^3\pi^3} & \text{when } n \text{ is odd} \\ 0, & \text{when } n \text{ is even} \end{cases}$$

$$\therefore \qquad B_n = \begin{cases} \dfrac{8a^2}{\sinh\left(\frac{n\pi}{a} b\right)(n^3\pi^3)}, & \text{when } n \text{ is odd} \\ 0, & \text{when } n \text{ is even} \end{cases}$$

$\therefore$ From (5), $$u(x, y) = \frac{8a^2}{\pi^3} \sum_{n=1, 3, 5, \ldots}^{\infty} \frac{\sin \frac{n\pi x}{a}}{n^3 \sinh \frac{n\pi}{a} b} \cdot \sinh \frac{n\pi}{a}(b - y)$$

(n is odd)

or $$u(x, y) = \frac{8a^2}{\pi^3} \sum_{n=0}^{\infty} \frac{1}{(2n+1)^3} \sin (2n+1)\frac{\pi x}{a} \cdot \frac{\sinh \frac{(2n+1)\pi}{a}(b - y)}{\sinh \frac{(2n+1)\pi}{a} b}$$

Example 7. *A thin rectangular plate whose surface is impervious to heat flow has at $t = 0$ an arbitrary distribution of temperature $f(x, y)$. Its four edges $x = 0$, $x = a$, $y = 0$, $y = b$ are kept at zero temperature. Determine the temperature at a point of a plate as t increases.*

Sol. Two dimensional heat flow equation is

$$\frac{\partial^2 u}{\partial x^2} + \frac{\partial^2 u}{\partial y^2} = \frac{1}{c^2} \frac{\partial u}{\partial t}. \qquad \ldots(1)$$

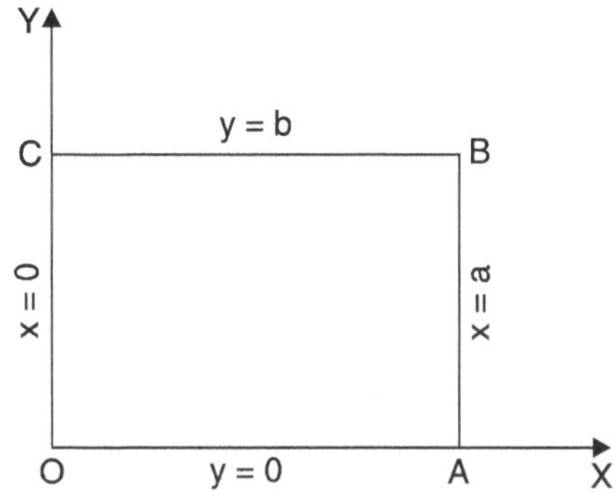

Boundary conditions are

$$u(0, y, t) = 0 = u(a, y, t)$$
$$u(x, 0, t) = 0 = u(x, b, t)$$

and $$u(x, y, t) = f(x, y) \text{ at } t = 0.$$

Let the solution be $u = XYT$

where X is a function of x only, Y is a function of y only and T is a function of t only.

$$\frac{\partial u}{\partial t} = \frac{\partial}{\partial t}(XYT) = XY \frac{dT}{dt}$$

$$\frac{\partial^2 u}{\partial x^2} = \frac{\partial^2}{\partial x^2}(XYT) = YT \frac{d^2 X}{dx^2}$$

$$\frac{\partial^2 u}{\partial y^2} = \frac{\partial^2}{\partial y^2}(XYT) = XT \frac{d^2 Y}{dy^2}.$$

From (1), $YT X'' + XTY'' = \frac{1}{c^2}(XYT')$

$$\Rightarrow \qquad \frac{X''}{X} + \frac{Y''}{Y} = \frac{T'}{c^2 T} \qquad \ldots(2)$$

There are three possibilities :

(*i*) $\frac{X''}{X} = 0, \quad \frac{Y''}{Y} = 0, \quad \frac{T'}{c^2 T} = 0$

(*ii*) $\frac{X''}{X} = K_1^2, \quad \frac{Y''}{Y} = K_2^2, \quad \frac{T'}{c^2 T} = K^2$

(*iii*) $\frac{X''}{X} = -K_1^2, \quad \frac{Y''}{Y} = -K_2^2, \quad \frac{T'}{c^2 T} = -K^2$

where $K^2 = K_1^2 + K_2^2$.

Of these three solutions, we have to select the solution which is consistent with the physical nature of the problem.

The solution satisfying the given boundary conditions will be given by (*iii*).

Then, $X = c_1 \cos K_1x + c_2 \sin K_1x$

$Y = c_3 \cos K_2y + c_4 \sin K_2y$

$T = c_5 e^{-c^2K^2t}$

$\therefore$ $u = XYT$

$$\Rightarrow \quad u(x, y, t) = (c_1 \cos K_1x + c_2 \sin K_1x)(c_3 \cos K_2y + c_4 \sin K_2y)(c_5 e^{-c^2K^2t}) \quad ...(3)$$

$$u(0, y, t) = 0 = c_1(c_3 \cos K_2y + c_4 \sin K_2y)c_5 e^{-c^2K^2t}$$

$\Rightarrow$ $c_1 = 0.$

$\therefore$ From (3), $u(x, y, t) = c_2 \sin K_1x(c_3 \cos K_2y + c_4 \sin K_2y)(c_5 e^{-c^2K^2t})$

$$= c_6 \sin K_1x\,(c_3 \cos K_2y + c_4 \sin K_2y)(e^{-c^2K^2t}) \quad ...(4)$$

where $c_2c_5 = c_6$

From (4), $u(a, y, t) = 0 = c_6 \sin K_1a(c_3 \cos K_2y + c_4 \sin K_2y)e^{-c^2K^2t}$

$\Rightarrow$ $\sin K_1a = 0 = \sin n\pi \; (n \in I)$

$\therefore$ $K_1 = \dfrac{n\pi}{a}.$

From (4), $$u(x, y, t) = c_6 \sin \frac{n\pi x}{a}\,(c_3 \cos K_2y + c_4 \sin K_2y)\,(e^{-c^2K^2t}) \quad ...(5)$$

$$u(x, 0, t) = 0 = c_6 \sin \frac{n\pi x}{a} \cdot c_3\, e^{-c^2K^2t}$$

$\Rightarrow$ $c_3 = 0.$

$\therefore$ From (5), $$u(x, y, t) = c_6c_4 \sin \frac{n\pi x}{a} \sin K_2y\, e^{-c^2K^2t} \quad ...(6)$$

$$u(x, b, t) = 0 = c_6c_4 \sin \frac{n\pi x}{a} \sin K_2b\, e^{-c^2K^2t}$$

$\Rightarrow$ $\sin K_2b = 0 = \sin m\pi \; (m \in I)$

$K_2b = m\pi$

$\Rightarrow$ $K_2 = \dfrac{m\pi}{b}.$

$\therefore$ From (6), $u(x, y, t) = c_6c_4 \sin \dfrac{n\pi x}{a} \sin \dfrac{m\pi y}{b}\, e^{-c^2K^2t}$

$$= A_{mn} \sin \frac{n\pi x}{a} \sin \frac{m\pi y}{b}\, e^{-c^2K^2t} \quad ...(7) \quad | \text{ where } c_6c_4 = A_{mn}$$

But, $$K^2 = K_1^2 + K_2^2 = \frac{n^2\pi^2}{a^2} + \frac{m^2\pi^2}{b^2}$$

or $$K^2_{mn} = \pi^2 \left(\frac{n^2}{a^2} + \frac{m^2}{b^2}\right).$$

By using K_{mn}, equation (7) becomes,

$$u(x, y, t) = \sum_{m=1}^{\infty} \sum_{n=1}^{\infty} A_{mn} \sin \frac{n\pi x}{a} \sin \frac{m\pi y}{b} e^{-c^2 K^2_{mn} t} \quad ...(8)$$

which is the most general solution.

$$u(x, y, 0) = f(x, y) = \sum_{m=1}^{\infty} \sum_{n=1}^{\infty} A_{mn} \sin \frac{n\pi x}{a} \sin \frac{m\pi y}{b}$$

which is the double Fourier half-range sine series for $f(x, y)$.

where $$A_{mn} = \frac{2}{a} \cdot \frac{2}{b} \int_{x=0}^{a} \int_{y=0}^{b} \sin \frac{n\pi x}{a} \sin \frac{m\pi y}{b} f(x, y) dx\, dy.$$

Example 8. *Solve the Laplace equation* $\frac{\partial^2 u}{\partial x^2} + \frac{\partial^2 u}{\partial y^2} = 0$ *in a rectangle in the xy-plane with* $u(x, 0) = 0$, $u(x, b) = 0$, $u(0, y) = 0$ *and* $u(a, y) = f(y)$ *parallel to y-axis.*

(A.K.T.U. 2016)

Sol. The given equation is

$$\frac{\partial^2 u}{\partial x^2} + \frac{\partial^2 u}{\partial y^2} = 0 \quad ...(1)$$

Let $$u = XY \quad ...(2)$$

where X is a function of x only and Y is a function of y only. Then,

$$\frac{\partial^2 u}{\partial x^2} = \frac{\partial^2}{\partial x^2}(XY) = YX''$$

and $$\frac{\partial^2 u}{\partial y^2} = \frac{\partial^2}{\partial y^2}(XY) = XY''$$

$\therefore$ From (1), $YX'' + XY'' = 0 \Rightarrow \frac{Y''}{Y} = -\frac{X''}{X}$ ...(3)

Case I. When $\frac{Y''}{Y} = -\frac{X''}{X} = p^2$ (say)

(*i*) $$\frac{Y''}{Y} = p^2$$

$$\Rightarrow \quad Y'' - p^2 Y = 0$$

Auxiliary equation is

$$m^2 - p^2 = 0 \Rightarrow m = \pm p$$

$$\therefore \quad \text{C.F.} = c_1 e^{py} + c_2 e^{-py}$$

$$\text{P.I.} = 0$$

$$\therefore \quad Y = c_1 e^{py} + c_2 e^{-py}$$

(ii) $$-\frac{X''}{X} = p^2$$

$\Rightarrow$ $$X'' + p^2X = 0$$

Auxiliary equation is

$$m^2 + p^2 = 0 \quad \Rightarrow \quad m = \pm pi$$

$\therefore$ $$\text{C.F.} = c_3 \cos px + c_4 \sin px$$

$$\text{P.I.} = 0$$

$\therefore$ $$X = c_3 \cos px + c_4 \sin px$$

Now, $$Y(0) = 0$$

$\Rightarrow$ $$c_1 + c_2 = 0 \quad \Rightarrow \quad c_2 = -c_1$$

$$Y(b) = 0$$

$\Rightarrow$ $$c_1e^{pb} + c_2 e^{-pb} = 0$$

$\Rightarrow$ $$c_1(e^{pb} - e^{-pb}) = 0$$

$\Rightarrow$ $$c_1 = 0$$ | Since $e^{pb} - e^{-pb} \neq 0$ (as $p \neq 0 \neq b$)

$\therefore$ $$Y = 0 \quad \Rightarrow \quad u = XY = 0$$ which is impossible.

Hence, we reject case I.

Case II. When $$\frac{Y''}{Y} = -\frac{X''}{X} = 0 \text{ (say)}$$

(i) $$\frac{Y''}{Y} = 0$$

$\Rightarrow$ $$Y'' = 0 \quad \Rightarrow \quad Y = c_5 + c_6y$$

(ii) $$-\frac{X''}{X} = 0$$

$\Rightarrow$ $$X'' = 0 \quad \Rightarrow \quad X = c_7 + c_8x$$

Now, $$Y(0) = 0 \quad \Rightarrow \quad c_5 = 0$$

$$Y(b) = 0 \quad \Rightarrow \quad c_6b = 0 \quad \Rightarrow \quad c_6 = 0$$ | $\because b \neq 0$

$\therefore$ $$Y = 0$$

$\therefore$ $u = XY = 0$ which is impossible.

Hence, we also reject case II.

Case III. When $$\frac{Y''}{Y} = -\frac{X''}{X} = -p^2 \text{ (say)}$$

(i) $$\frac{Y''}{Y} = -p^2$$

$\Rightarrow$ $$Y'' + p^2Y = 0$$

Auxiliary equation is

$$m^2 + p^2 = 0 \quad \Rightarrow \quad m = \pm pi$$

$\therefore$ $$\text{C.F.} = c_9 \cos py + c_{10} \sin py$$

$$\text{P.I.} = 0$$

$\therefore$ $$Y = c_9 \cos py + c_{10} \sin py$$

(*ii*) $$-\frac{X''}{X} = -p^2 \Rightarrow X'' - p^2X = 0$$

Auxiliary equation is

$$m^2 - p^2 = 0 \Rightarrow m = \pm p$$

$$\therefore \quad \text{C.F.} = c_{11}\, e^{px} + c_{12}\, e^{-px}$$

$$\text{P.I.} = 0$$

$$\therefore \quad X = c_{11}\, e^{px} + c_{12}\, e^{-px}$$

Now, $$Y(0) = 0 \Rightarrow c_9 = 0$$

$$Y(b) = 0 \Rightarrow c_{10} \sin bp = 0$$

$$\therefore \quad \sin bp = 0 = \sin n\pi,\ n \in I$$

$$p = \frac{n\pi}{b}$$

Hence, $$u = XY = c_{10} \sin \frac{n\pi y}{b} \left(c_{11}\, e^{\frac{n\pi x}{b}} + c_{12} e^{-\frac{n\pi x}{b}} \right) \qquad ...(4)$$

Now, $$u(0, y) = 0 = c_{10} \sin \frac{n\pi y}{b} (c_{11} + c_{12})$$

$$\Rightarrow \quad c_{11} + c_{12} = 0 \Rightarrow c_{12} = -c_{11}$$

$\therefore$ From (4), $$u(x, y) = c_{10} c_{11} \sin \frac{n\pi y}{b} \left(e^{\frac{n\pi x}{b}} - e^{-\frac{n\pi x}{b}} \right)$$

$$= b_n \sin \frac{n\pi y}{b} \sinh \frac{n\pi x}{b} \qquad ...(5)$$

| where $b_n = 2\, c_{10}\, c_{11}$

Most general solution is

$$u(x, y) = \sum_{n=1}^{\infty} b_n \sin \frac{n\pi y}{b} \sinh \frac{n\pi x}{b} \qquad ...(6)$$

Now, $$u(a, y) = f(y) = \sum_{n=1}^{\infty} b_n \sin \frac{n\pi y}{b} \sinh \frac{n\pi a}{b}$$

where $$\left(\sinh \frac{n\pi a}{b} \right) b_n = \frac{2}{b} \int_0^b f(y) \sin \frac{n\pi y}{b}\, dy$$

$$\Rightarrow \quad b_n = \frac{2}{b \sinh\left(\frac{n\pi a}{b}\right)} \int_0^b f(y) \sin \frac{n\pi y}{b}\, dy. \qquad ...(7)$$

Example 9. *Find the steady state temperature distribution in a rectangular thin plate with its two surfaces insulated and with the conditions $u(0, y) = 0$, $u(x, 0) = 0$, $u(a, y) = g(y)$, $u(x, b) = f(x)$.*

Sol. Superposition applied to boundary conditions dismantles the given problem to solution of two simpler problems each of which can easily be solved by the method of separation of variables.

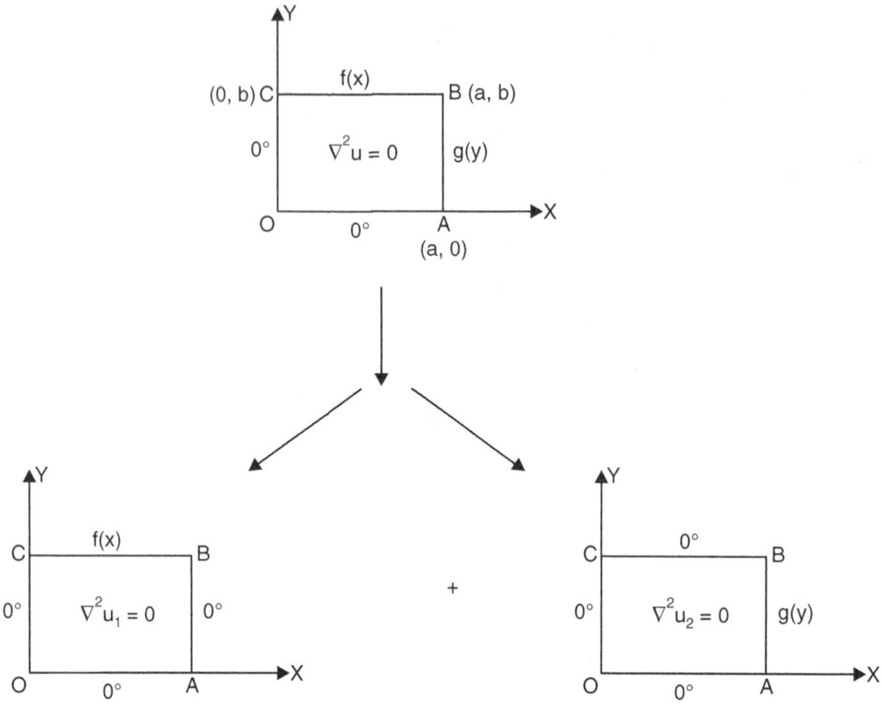

Now, the following two problems are required to be solved:

Problem 1. $\dfrac{\partial^2 u_1}{\partial x^2} + \dfrac{\partial^2 u_1}{\partial y^2} = 0$ subject to conditions

$$u_1(0, y) = 0,\ u_1(a, y) = 0,\ u_1(x, 0) = 0,\ u_1(x, b) = f(x)$$

Problem 2. $\dfrac{\partial^2 u_2}{\partial x^2} + \dfrac{\partial^2 u_2}{\partial y^2} = 0$ subject to conditions

$$u_2(0, y) = 0,\ u_2(x, 0) = 0,\ u_2(x, b) = 0,\ u_2(a, y) = g(y)$$

Let us now proceed to solve problem 1.

$$u_1(x, y) = (c_1 \cos px + c_2 \sin px)\,(c_3 e^{py} + c_4\, e^{-py}) \quad \text{...(1)}$$

$$u_1(0, y) = 0 = c_1\,(c_3 e^{py} + c_4 e^{-py})$$

$$\Rightarrow \quad c_1 = 0$$

$$\therefore \quad u_1(x, y) = c_2 \sin px\,(c_3 e^{py} + c_4 e^{-py}) \quad \text{...(2)}$$

$$u_1(a, y) = 0 = c_2 \sin pa\,(c_3 e^{py} + c_4 e^{-py})$$

$$\Rightarrow \quad \sin ap = 0 = \sin n\pi \quad (n \in I)$$

$$p = \frac{n\pi}{a}$$

Hence, from (2), $$u_1(x, y) = c_2 \sin \frac{n\pi x}{a}\left(c_3\, e^{\frac{n\pi y}{a}} + c_4\, e^{-\frac{n\pi y}{a}}\right) \quad \text{...(3)}$$

$$u_1(x, 0) = 0 = c_2 \sin \frac{n\pi x}{a} (c_3 + c_4)$$

$\Rightarrow$ $$c_3 + c_4 = 0 \quad \Rightarrow \quad c_4 = -c_3$$

$\therefore$ $$u_1(x, y) = c_2 c_3 \sin \frac{n\pi x}{a} \left(e^{\frac{n\pi y}{a}} - e^{-\frac{n\pi y}{a}} \right) = 2\, c_2 c_3 \sin \frac{n\pi x}{a} \sinh \frac{n\pi y}{a}$$

$$= b_n \sin \frac{n\pi x}{a} \sinh \frac{n\pi y}{a} \text{ where } b_n = 2c_2c_3$$

Most general solution to problem 1 is

$$u_1(x, y) = \sum_{n=1}^{\infty} b_n \sin \frac{n\pi x}{a} \sinh \frac{n\pi y}{a} \qquad ...(4)$$

Now, $$u_1(x, b) = f(x) = \sum_{n=1}^{\infty} b_n \sin \frac{n\pi x}{a} \sinh \frac{n\pi b}{a}$$

where $$b_n \sinh \frac{n\pi b}{a} = \frac{2}{a} \int_0^a f(x) \sin \frac{n\pi x}{a}\, dx$$

$\Rightarrow$ $$b_n = \frac{2}{a \sinh\left(\frac{n\pi b}{a}\right)} \int_0^a f(x) \sin \frac{n\pi x}{a}\, dx \qquad ...(5)$$

Similarly, the most general solution to problem 2 is

$$u_2(x, y) = \sum_{n=1}^{\infty} B_n \sin\left(\frac{n\pi y}{b}\right) \sinh\left(\frac{n\pi x}{b}\right) \qquad ...(6)$$

where $$B_n = \frac{2}{b \sinh\left(\frac{n\pi a}{b}\right)} \int_0^b g(y) \sin\left(\frac{n\pi y}{b}\right) dy \qquad ...(7)$$

| Interchanging x by y, a by b and $f(x)$ by $g(y)$

Hence, the required solution to the given original problem is

$$u(x, y) = u_1(x, y) + u_2(x, y)$$

$$= \sum_{n=1}^{\infty} \left[b_n \sin\left(\frac{n\pi x}{a}\right) \sinh\left(\frac{n\pi y}{a}\right) + B_n \sin\left(\frac{n\pi y}{b}\right) \sinh\left(\frac{n\pi x}{b}\right) \right]$$

where $$b_n = \frac{2}{a \sinh\left(\frac{n\pi b}{a}\right)} \int_0^a f(x) \sin \frac{n\pi x}{a}\, dx$$

and $$B_n = \frac{2}{b \sinh\left(\frac{n\pi a}{b}\right)} \int_0^b g(y) \sin \frac{n\pi y}{b}\, dy.$$

TEST YOUR KNOWLEDGE

1. A long rectangular plate of width a cm with insulated surface has its temperature v equal to zero on both the long sides and one of the short sides so that $v(0, y) = 0$, $v(a, y) = 0$,

$$\lim_{y \to \infty} v(x, y) = 0 \quad \text{and} \quad v(x, 0) = kx$$

Show that the steady-state temperature within the plate is

$$v(x, y) = \frac{2ak}{\pi} \sum_{n=1}^{\infty} \frac{(-1)^{n+1}}{n} e^{-\left(\frac{n\pi y}{a}\right)} \sin \frac{n\pi x}{a}.$$

2. A square plate is bounded by the lines $x = 0, y = 0, x = 20$ and $y = 20$. Its faces are insulated. The temperature along the upper horizontal edge is given by $u(x, 20) = x(20 - x)$ when $0 < x < 20$ while other three edges are kept at 0°C. Find the steady state temperature in the plate. *(A.K.T.U. 2014, 2017)*

3. A rectangular plate has sides a and b. Let the side of length a be taken along OX and that of length b along OY and the other sides along $x = a$ and $y = b$. The sides $x = 0, x = a$ and $y = b$ are insulated and the edge $y = 0$ is kept at temperature $u_0 \cos \frac{\pi x}{a}$. Find the steady-state temperature at any point (x, y).

[**Hint:** Boundary conditions are $(u_x)_{x=0} = 0$, $(u_x)_{x=a} = 0$, $(u_y)_{y=b} = 0$ and $u(x, 0) = u_0 \cos(\pi x/a)$]

4. The temperature u is maintained at 0° along three edges of a square plate of length 100 cm and the fourth edge is maintained at 100° until steady-state conditions prevail. Find an expression for the temperature u at any point (x, y).

Hence, show that the temperature at the centre of the plate

$$= \frac{200}{\pi}\left[\frac{1}{\cosh \frac{\pi}{2}} - \frac{1}{3\cosh \frac{3\pi}{2}} + \frac{1}{5\cosh \frac{5\pi}{2}} - \dots\right].$$

5. A rectangular plate is bounded by the lines $x = 0, y = 0, x = a, y = b$. Its surfaces are insulated and the temperature along the upper horizontal edge is 100°C while the other three edges are kept at 0°C. Find the steady state temperature function $u(x, y)$ and also the temperature at the point $\left(\frac{1}{2}a, \frac{1}{2}b\right)$.

6. Solve the following Laplace equation $\frac{\partial^2 u}{\partial x^2} + \frac{\partial^2 u}{\partial y^2} = 0$ in a rectangle with $u(0, y) = 0$, $u(a, y) = 0$, $u(x, b) = 0$ and $u(x, 0) = f(x)$ along x-axis.

7. Solve the boundary value problem:

$$\frac{\partial^2 u}{\partial x^2} + \frac{\partial^2 u}{\partial y^2} = 0, \; 0 \le x \le a, \; 0 \le y \le b$$

with the boundary conditions:

$$u_x(0, y) = u_x(a, y) = u_y(x, 0) = 0 \text{ and } u_y(x, b) = f(x)$$ *[M.T.U. (SUM) 2011]*

8. Solve $\frac{\partial^2 u}{\partial x^2} + \frac{\partial^2 u}{\partial y^2} = 0$ subject to the conditions $u(0, y) = 0$, $u(x, 0) = 0$, $u(1, y) = 0$ and $u(x, 1) = 100 \sin \pi x$. *(G.B.T.U. 2013)*

9. Solve $\dfrac{\partial^2 u}{\partial x^2} + \dfrac{\partial^2 u}{\partial y^2} = 0$ subject to the conditions $u(0, y) = 0$, $u(a, y) = 0$, $u(x, 0) = 0$ and $u(x, b) = x$

(U.P.T.U. 2015)

10. Solve $\dfrac{\partial^2 u}{\partial x^2} + \dfrac{\partial^2 u}{\partial y^2} = 0$ subject to the conditions $u(x, 0) = 0$, $u(x, 1) = 0$, $u(\infty, y) = 0$ and $u(0, y) = u_0$.

(U.P.T.U. 2015)

11. The initial temperature distribution in a square plate of unit length is 100°C. Find the temperature distribution $u(x, t)$ if all the sides are maintained at zero degree temperature. *(U.P.T.U. 2014)*

Answers

2. $$u(x, y) = \frac{3200}{\pi^3} \sum_{n=1}^{\infty} \frac{\sin \dfrac{(2n-1)\pi x}{20} \sin \dfrac{(2n-1)\pi y}{20}}{(2n-1)^3 \sinh (2n-1)\pi}$$

3. $$u(x, y) = u_0 \cos \frac{\pi x}{a} \cosh \frac{\pi}{a}(b - y) \operatorname{sech} \frac{\pi b}{a}$$

5. $$u(x, y) = \frac{400}{\pi} \sum_{n=1}^{\infty} \frac{\sin \left\{(2m-1)\dfrac{\pi x}{a}\right\} \sinh \left\{(2m-1)\dfrac{\pi y}{a}\right\}}{(2m-1) \sinh \left\{(2m-1)\dfrac{\pi b}{a}\right\}};$$

$$u\left(\frac{1}{2}a, \frac{1}{2}a\right) = \frac{200}{\pi}\left[\frac{1}{\cosh\left(\dfrac{\pi b}{2a}\right)} - \frac{1}{3\cosh\left(\dfrac{3\pi b}{2a}\right)} + \ldots\right]$$

6. $$u(x, y) = \sum_{n=1}^{\infty} B_n \sin\left(\frac{n\pi x}{a}\right) \sinh \frac{n\pi}{a}(b - y), \text{ where } B_n = \frac{2}{a \sinh\left(\dfrac{n\pi b}{a}\right)} \int_0^a f(x) \sin \frac{n\pi x}{a}\, dx$$

7. $$u(x, y) = \sum_{n=1}^{\infty} b_n \cos \frac{n\pi x}{a}\left(e^{\frac{n\pi y}{a}} - e^{-\frac{n\pi y}{a}}\right) \text{ where, } b_n = \frac{1}{n\pi \cosh \dfrac{n\pi}{a} b} \int_0^a f(x) \cos \frac{n\pi x}{a}\, dx$$

8. $$u(x, y) = 100 \sin \pi x \left(\frac{\sinh \pi y}{\sinh \pi}\right)$$

9. $$u(x, y) = \frac{-2a}{\pi} \sum_{n=1}^{\infty} \frac{\cos n\pi \sin\left(\dfrac{n\pi x}{a}\right) \sinh\left(\dfrac{n\pi y}{a}\right)}{n \sinh\left(\dfrac{n\pi b}{a}\right)}$$

10. $$u(x, y) = \frac{2u_0}{\pi} \sum_{n=1}^{\infty} \left\{\frac{1 - (-1)^n}{n}\right\} e^{-n\pi x} \sin n\pi y$$

11. $$u(x, y, t) = \frac{400}{\pi^2} \sum_{n=1}^{\infty} \sum_{m=1}^{\infty} \left(\frac{1 - \cos n\pi}{n}\right)\left(\frac{1 - \cos m\pi}{m}\right) \sin n\pi x \sin m\pi y\, e^{-c^2 k_{mn}^2 t}$$

where $k_{mn}^2 = \pi^2(n^2 + m^2)$.

2.15 LAPLACE EQUATION

Laplace's equation has wide applications in Physics and engineering. The theory of its solutions is called the *potential theory* and its solutions are called *harmonic functions.* The solution of Laplace's equation, subject to certain boundary conditions, is simplified by a proper choice of coordinate system.

Note 1. If the problem involves *rectangular boundaries*, we prefer to take Laplace's equation in *cartesian coordinates given by* $\frac{\partial^2 u}{\partial x^2}+\frac{\partial^2 u}{\partial y^2}=0$ and $\frac{\partial^2 u}{\partial x^2}+\frac{\partial^2 u}{\partial y^2}+\frac{\partial^2 u}{\partial z^2}=0$.

Note 2. If the problem involves circular boundaries, we prefer to take Laplace's equation in polar coordinates given by

$$\frac{\partial^2 u}{\partial r^2}+\frac{1}{r}\frac{\partial u}{\partial r}+\frac{1}{r^2}\frac{\partial^2 u}{\partial \theta^2}=0.$$

This equations can be obtained from $u_{xx}+u_{yy}=0$ by putting $x=r\cos\theta, y=r\sin\theta$, thus changing the independent variables from (x, y) to (r, θ).

Note 3. If the problem involves *cylindrical boundaries*, we prefer to take Laplace's equation in *cylindrical coordinates* given by

$$\frac{\partial^2 u}{\partial r^2}+\frac{1}{r}\frac{\partial u}{\partial r}+\frac{1}{r^2}\frac{\partial^2 u}{\partial \theta^2}+\frac{\partial^2 u}{\partial z^2}=0.$$

This equation can be obtained from $u_{xx}+u_{yy}+u_{zz}=0$ by putting $x=r\cos\theta, y=r\sin\theta, z=z$, thus changing the independent variables (x, y, z) to (r, θ, z).

Note 4. If the problem involves *spherical boundaries,* we prefer to take Laplace's equation in *spherical polar coordinates* given by

$$\frac{\partial^2 u}{\partial r^2}+\frac{2}{r}\frac{\partial u}{\partial r}+\frac{1}{r^2}\frac{\partial^2 u}{\partial \theta^2}+\frac{\cot\theta}{r^2}\frac{\partial u}{\partial \theta}+\frac{1}{r^2\sin^2\theta}\frac{\partial^2 u}{\partial \phi^2}=0.$$

This equation can be obtained from $u_{xx}+u_{yy}+u_{zz}=0$ by putting,

$$x=r\sin\theta\cos\phi, y=r\sin\theta\sin\phi, z=r\cos\theta,$$

thus changing the independent variables (x, y, z) to (r, θ, ϕ).

2.16 SOLUTIONS OF LAPLACE'S EQUATION

(*a*) Solution of Laplace's Equation in Two-dimensional Cartesian Form

We have already discussed the solution of $\frac{\partial^2 u}{\partial x^2}+\frac{\partial^2 u}{\partial y^2}=0$ in Art. 2.14.

(*b*) Solution of Laplace's Equation in Polar Coordinates

Laplace's equation in polar coordinates is $r^2\frac{\partial^2 u}{\partial r^2}+r\frac{\partial u}{\partial r}+\frac{\partial^2 u}{\partial \theta^2}=0$ ...(1)

Let $u(r, \theta)=R(r)F(\theta)$ or simply $u=RF$ ...(2)

where R is a function of r only and F is a function of θ only, be a solution of (1).

Substituting it in (1), we get

$$r^2R''F+rR'F+RF''=0 \quad \text{or} \quad (r^2R''+rR')F+RF''=0$$

Separating the variables, $\frac{r^2R''+rR'}{R}=-\frac{F''}{F}=\text{constant}=k$ (say)

Thus, we get ordinary differential equations

$$r^2\frac{d^2R}{dr^2} + r\frac{dR}{dr} - kR = 0 \quad ...(3)$$

and
$$\frac{d^2F}{d\theta^2} + kF = 0 \quad ...(4)$$

Now (3) is a homogeneous linear differential equation.

Putting $r = e^z$, (3) reduces to $\frac{d^2R}{dr^2} - kR = 0$ ...(5)

Solving (5) and (4), we get

(*i*) *When k is positive and* $= p^2$, *say*

$$R = c_1e^{pz} + c_2e^{-pz} = c_1r^p + c_2r^{-p},$$
$$F = c_3 \cos p\theta + c_4 \sin p\theta.$$

(*ii*) *When k is negative and* $= -p^2$, *say*

$$R = c_1 \cos pz + c_2 \sin pz = c_1 \cos (p \log r) + c_2 \sin (p \log r)$$
$$F = c_3e^{p\theta} + c_4e^{-p\theta}$$

(*iii*) *When* $k = 0$
$$R = c_1z + c_2 = c_1 \log r + c_2$$
$$F = c_3\theta + c_4$$

Thus the three possible solution of (1) are

$$u = (c_1r^p + c_2r^{-p})(c_3 \cos p\theta + c_4 \sin p\theta) \quad ...(6)$$
$$u = [c_1 \cos (p \log r) + c_2 \sin (p \log r)] (c_3e^{p\theta} + c_4e^{-p\theta}) \quad ...(7)$$
$$u = (c_1 \log r + c_2)(c_3\theta + c_4) \quad ...(8)$$

Of these solutions, we choose the one which is consistent with the physical nature of the problem.

Note. Usually we require a solution extending up to the origin.

Since u must be finite at the origin, we reject solutions (7) and (8). Also from (6), $c_2 = 0$.

$\therefore$ In this case, the solution may be written as

$$u = (A \cos p\theta + B \sin p\theta)r^p$$

The general solution will consist of a sum of similar terms with different (arbitrary) values of A, B and p.

(*c*) Solution of Laplace's Equation in Three-dimensional Cartesian Form

Laplace's equation in three-dimensional cartesian form is

$$\frac{\partial^2 u}{\partial x^2} + \frac{\partial^2 u}{\partial y^2} + \frac{\partial^2 u}{\partial z^2} = 0 \quad ...(1)$$

Let $u(x, y, z) = X(x)\, Y(y)\, Z(z)$ or simply $u = XYZ$ ...(2) be a solution of (1).

Substituting it in (1), we get $X''YZ + XY''Z + XYZ'' = 0$

Dividing to XYZ, we get
$$\frac{X''}{X} + \frac{Y''}{Y} + \frac{Z''}{Z} = 0$$

or
$$\frac{1}{X}\cdot\frac{d^2X}{dx^2} + \frac{1}{Y}\cdot\frac{d^2Y}{dy^2} + \frac{1}{Z}\cdot\frac{d^2Z}{dz^2} = 0 \quad ...(3)$$

which is of the form $F_1(x) + F_2(y) + F_3(z) = 0$.

Since x, y, z are independent, this is possible only when F_1, F_2, F_3 are constants. Assuming these constants to be k^2, l^2 and $-(k^2 + l^2)$ respectively, (3) gives rise to the following equations:

$$\frac{d^2X}{dx^2} - k^2X = 0, \frac{d^2Y}{dy^2} - l^2Y = 0, \frac{d^2Z}{dz^2} + (k^2 + l^2)Z = 0$$

Their solutions are $X = c_1e^{kx} + c_2e^{-kx}, Y = c_3e^{ly} + c_4e^{-ly}$

$$Z = c_5 \cos \sqrt{(k^2 + l^2)}\, z + c_6 \sin \sqrt{(k^2 + l^2)}\, z$$

Hence a solution of (1) is

$$u = (c_1e^{kx} + c_2e^{-kx})(c_3e^{ly} + c_4e^{-ly})[c_5 \cos \sqrt{(k^2 + l^2)}\, z + c_6 \sin \sqrt{(k^2 + l^2)}\, z]$$

Since, the three constants could have been taken as $-k^2, -l^2$ and $k^2 + l^2$, and alternative solution of (1) is

$$u = (c_1 \cos kx + c_2 \sin kx)(c_3 \cos ly + c_4 \sin ly)\ [c_5e^{\sqrt{(k^2 + l^2)}\, z} + c_6e^{-\sqrt{(k^2 + l^2)}\, z}]$$

The choice of the constants and hence the general solution depends on the given initial and boundary conditions.

(*d*) **Solution of Laplace's Equation in Cylindrical Coordinates**

Laplace's equation in cylindrical coordinates is

$$\frac{\partial^2 u}{\partial r^2} + \frac{1}{r}\frac{\partial u}{\partial r} + \frac{1}{r^2}\frac{\partial^2 u}{\partial \theta^2} + \frac{\partial^2 u}{\partial z^2} = 0 \qquad ...(1)$$

Let $u(r, \theta, z) = R(r)F(\theta)Z(z)$

or simply $u = RFZ$ ...(2) be a solution of (1).

Substituting it in (1), we get $R''FZ + \frac{1}{r} R'FZ + \frac{1}{r^2} RF''Z + RFZ'' = 0$

Dividing by RFZ, we get

$$\frac{1}{R}\left(\frac{d^2R}{dr^2} + \frac{1}{r}\frac{dR}{dr}\right) + \frac{1}{r^2F}\frac{d^2F}{d\theta^2} + \frac{1}{z}\frac{d^2Z}{dz^2} = 0 \qquad ...(3)$$

Assuming $\frac{d^2F}{d\theta^2} = -n^2F$ and $\frac{d^2Z}{dz^2} = k^2Z$ ...(4)

Equation (3) reduces to $\frac{1}{R}\left(\frac{d^2R}{dr^2} + \frac{1}{r}\frac{dR}{dz}\right) - \frac{n^2}{r^2} + k^2 = 0$

or

$$r^2 \frac{d^2R}{dr^2} + r\frac{dR}{dz} + (k^2r^2 - n^2)R = 0$$

This is Bessel's equation. Its solution is $R = c_1J_n(kr) + c_2Y_n(kr)$

The solutions of equations (3) are $F = c_3 \cos n\theta + c_4 \sin n\theta, Z = c_5e^{kz} + c_6e^{-kz}$

Hence a solution of (1) is $u = [c_1J_n(kr) + c_2Y_n(kr)]\ (c_3 \cos n\theta + c_4 \sin n\theta)(c_5e^{kz} + c_6e^{-kz})$ which is known as a *cylindrical harmonic.*

(e) Solution of Laplace's Equation in Spherical Coordinates

Laplace's equation is spherical coordinates is

$$\frac{\partial^2 u}{\partial r^2}+\frac{2}{r}\frac{\partial u}{\partial r}+\frac{1}{r^2}\frac{\partial^2 u}{\partial \theta^2}+\frac{\cot\theta}{r^2}\frac{\partial u}{\partial \theta}+\frac{1}{r^2\sin^2\theta}\frac{\partial^2 u}{\partial \phi^2}=0 \qquad ...(1)$$

Let $u(r, \theta, \phi) = R(r)G(\theta)H(\phi)$.

or simply $u = RGH$ ...(2) be a solution of (1).

Substituting it in (1), we get

$$R''GH+\frac{2}{r}R'GH+\frac{1}{r^2}RG''H+\frac{\cot\theta}{r^2}RG'H+\frac{1}{r^2\sin^2\theta}RGH''=0$$

Dividing by RGH, we get

$$\frac{1}{R}\left(r^2\frac{d^2R}{dr^2}+2r\frac{dR}{dr}\right)+\frac{1}{G}\left(\frac{d^2G}{d\theta^2}+\cot\theta\frac{dG}{d\theta}\right)+\frac{1}{H\sin^2\theta}\frac{d^2H}{d\phi^2}=0 \qquad ...(3)$$

Putting

$$\frac{1}{R}\left(r^2\frac{d^2R}{dr^2}+2r\frac{dR}{dr}\right)=n(n+1) \qquad ...(4)$$

and

$$\frac{1}{H}\frac{d^2H}{d\phi^2}=-m^2 \qquad ...(5)$$

Equation (3) reduces to $\dfrac{d^2G}{d\theta^2}+\cot\theta\dfrac{dG}{d\theta}+[n(n+1)-m^2\operatorname{cosec}^2\theta]\,G=0$

This is associated Legendre's equation and its solution is

$$G = c_1 P_n^m(\cos\theta) + c_2 Q_n^m(\cos\theta)$$

The solution of (5) is $H = c_3\cos m\phi + c_4\sin m\phi$

To solve (4), assume that $R = r^k$ so that

$$k(k-1)+2k = n(n+1) \quad \text{or} \quad (k^2-n^2)+(k-n)=0$$

or

$$(k-n)(k+n+1)=0 \quad \therefore \quad k = n \text{ or } -n-1$$

Thus

$$R = c_5 r^n + c_6 r^{-n-1}$$

Hence, the general solution of (1) is

$$u=\sum_{n=0}^{\infty}\sum_{m=0}^{\infty}[c_1P_n^m(\cos\theta)+c_2Q_n^m(\cos\theta)](c_3\cos m\phi+c_4\sin m\phi)(c_5r^n+c_6r^{-n-1})$$

Any solution of (1) is known as a *spherical harmonic.*

ILLUSTRATIVE EXAMPLES

Example 1. *The diameter of a semi-circular plate of radius a is kept at 0°C and the temperature at the semi-circular boundary is T°C. Show that the steady state temperature in the plate is given by*

$$u(r,\theta)=\frac{4T}{\pi}\sum_{n=1}^{\infty}\frac{1}{2n-1}\left(\frac{r}{a}\right)^{2n-1}\sin(2n-1)\theta. \qquad \text{[G.B.T.U. (C.O.) 2011]}$$

Sol. Take the centre of the circle as the pole and the bounding diameter as the initial line. Let the steady state temperature at any point P(r, θ) be $u(r, \theta)$, so that u satisfies the equation

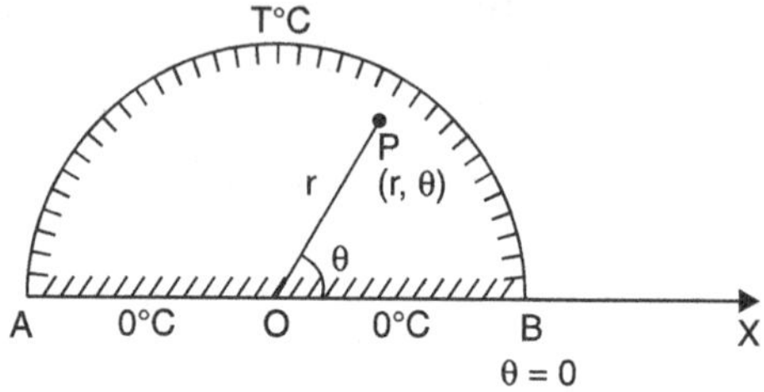

$$r^2 \frac{\partial^2 u}{\partial r^2} + r \frac{\partial u}{\partial r} + \frac{\partial^2 u}{\partial \theta^2} = 0 \qquad ...(1)$$

Let $$u = \text{RT} \qquad ...(2)$$

where R is a function of r only and T is a function of θ only.

$$\frac{\partial u}{\partial r} = \frac{\partial}{\partial r}(\text{RT}) = \text{T}\frac{d\text{R}}{dr}$$

$$\frac{\partial^2 u}{\partial r^2} = \frac{\partial}{\partial r}\left(\text{T}\frac{d\text{R}}{dr}\right) = \text{T}\frac{d^2\text{R}}{dr^2}$$

$$\frac{\partial^2 u}{\partial \theta^2} = \frac{\partial^2}{\partial \theta^2}(\text{RT}) = \text{R}\frac{d^2\text{T}}{d\theta^2}.$$

$\therefore$ From (1), $$r^2\text{TR}'' + r\text{TR}' + \text{RT}'' = 0$$

$$\frac{r^2\text{R}'' + r\text{R}'}{\text{R}} + \frac{\text{T}''}{\text{T}} = 0 \Rightarrow \frac{r^2\text{R}'' + r\text{R}'}{\text{R}} = -\frac{\text{T}''}{\text{T}}$$

Case I. When $$\frac{r^2\text{R}'' + r\text{R}'}{\text{R}} = -\frac{\text{T}''}{\text{T}} = p^2 \text{ (say)}$$

(*i*) $$\frac{r^2\text{R}'' + r\text{R}'}{\text{R}} = p^2 \Rightarrow r^2\text{R}'' + r\text{R}' - p^2\text{R} = 0.$$

Put $r = e^z$ so that $z = \log r$ and let $\text{D} \equiv \frac{d}{dz}$ then above equation reduces to

$$\{\text{D}(\text{D} - 1) + \text{D} - p^2\}\,\text{R} = 0$$

$$(\text{D}^2 - p^2)\text{R} = 0$$

Auxiliary equation is $m^2 - p^2 = 0 \Rightarrow m = \pm p$

$$\text{C.F.} = c_1e^{pz} + c_2e^{-pz} = c_1e^{p\log r} + c_2e^{-p\log r} = c_1r^p + c_2r^{-p}$$

$$\text{P.I.} = 0.$$

$\therefore$ $$\text{R} = c_1r^p + c_2r^{-p}$$

(*ii*) $$-\frac{\text{T}''}{\text{T}} = p^2$$

$$\frac{d^2\text{T}}{d\theta^2} + p^2\text{T} = 0.$$

Auxiliary equation is $m^2 + p^2 = 0$

$\Rightarrow$ $$m = \pm pi$$

$\therefore$ $$\text{C.F.} = c_3\cos p\theta + c_4\sin p\theta$$

$$\text{P.I.} = 0$$

$\therefore$ $$\text{T} = c_3\cos p\theta + c_4\sin p\theta$$

$\therefore$ $$u(r, \theta) = (c_1r^p + c_2r^{-p})(c_3\cos p\theta + c_4\sin p\theta) \qquad ...(3)$$

Case II. When $$\frac{r^2R'' + rR'}{R} = -\frac{T''}{T} = -p^2 \text{ (say)}$$

(*i*) $$\frac{r^2R'' + rR'}{R} = -p^2$$

$$\Rightarrow \quad r^2R'' + rR' + p^2R = 0$$

Put $r = e^z$ so that $z = \log r$ and let $D \equiv \frac{d}{dz}$ then above equation reduces to

$$[D(D-1) + D + p^2]\, R = 0$$

$$(D^2 + p^2)R = 0.$$

Auxiliary equation is $m^2 + p^2 = 0$

$$m = \pm pi$$

$$\text{C.F.} = (c_5 \cos pz + c_6 \sin pz)$$

$$\text{P.I.} = 0$$

$$\therefore \quad R = c_5 \cos pz + c_6 \sin pz$$

$$= c_5 \cos (p \log r) + c_6 \sin (p \log r).$$

(*ii*) $$-\frac{T''}{T} = -p^2$$

$$\Rightarrow \quad \frac{T''}{T} = p^2 \quad \text{or} \quad \frac{d^2T}{d\theta^2} - p^2T = 0.$$

Auxiliary equation is $m^2 - p^2 = 0$

$$m = \pm p$$

$$\therefore \quad \text{C.F.} = c_7e^{p\theta} + c_8e^{-p\theta}$$

$$\text{P.I.} = 0.$$

$$\therefore \quad T = c_7e^{p\theta} + c_8e^{-p\theta}.$$

Hence, $$u(r, \theta) = [c_5 \cos (p \log r) + c_6 \sin (p \log r)](c_7e^{p\theta} + c_8e^{-p\theta}) \qquad ...(4)$$

Case III. When $$\frac{r^2R'' + rR'}{R} = -\frac{T''}{T} = 0 \text{ (say)}$$

(*i*) $$r^2R'' + rR' = 0.$$

Put $r = e^z$ so that $z = \log r$ and let $D \equiv \frac{d}{dz}$, then above equation reduces to

$$[D(D-1) + D]R = 0$$

$$D^2R = 0$$

Auxiliary equation is $m^2 = 0$

$$\Rightarrow \quad m = 0, 0$$

$$\therefore \quad \text{C.F.} = (c_9 + c_{10}z)\, e^{0z} = c_9 + c_{10} \log r$$

$$\text{P.I.} = 0$$

$$\therefore \quad R = c_9 + c_{10} \log r.$$

(*ii*) $$-\frac{T''}{T} = 0$$

$\Rightarrow$ $$T'' = 0$$

$\Rightarrow$ $$T = c_{11} + c_{12}\theta.$$

$\therefore$ $$u(r, \theta) = (c_9 + c_{10}\log r)(c_{11} + c_{12}\theta) \quad ...(5)$$

Of these three solutions (3), (4) and (5), we choose the solution consistent with the given boundary conditions.

Boundary conditions are

$$u(r, 0) = 0 \quad ...(6)$$

$$u(r, \pi) = 0 \quad ...(7)$$

$$u(a, \theta) = T \quad ...(8)$$

and $$u \to 0 \text{ as } r \to 0 \quad ...(9)$$

Solutions (4) and (5) do not satisfy boundary condition (9).

Hence, the consistent solution is

$$u(r, \theta) = (c_1 r^p + c_2 r^{-p})(c_3 \cos p\theta + c_4 \sin p\theta) \quad ...(10)$$

From (10), $$u(r, 0) = 0 = (c_1 r^p + c_2 r^{-p})\, c_3$$

$\Rightarrow$ $$c_3 = 0.$$

From (10),

$$u(r, \theta) = (c_1 r^p + c_2 r^{-p})\, c_4 \sin p\theta \quad ...(11)$$

$$u(r, \pi) = 0 = (c_1 r^p + c_2 r^{-p})\, c_4 \sin p\pi$$

$\Rightarrow$ $$\sin p\pi = 0 = \sin n\pi \ (n \in I)$$

$\therefore$ $$p = n.$$

$\therefore$ From (11), $$u(r, \theta) = (c_1 r^n + c_1 r^{-n})\, c_4 \sin n\theta \quad ...(12)$$

Condition $u \to 0$ as $r \to 0$ is satisfied only and only when $c_2 = 0$.

Hence from (12), $u(r, \theta) = c_1 c_4 r^n \sin n\theta = b_n r^n \sin n\theta.$

The most general solution is

$$u(r, \theta) = \sum_{n=1}^{\infty} b_n r^n \sin n\theta \quad ...(13)$$

$$u(a, \theta) = T = \sum_{n=1}^{\infty} b_n a^n \sin n\theta$$

where $$b_n a^n = \frac{2}{\pi}\int_0^{\pi} \{T \sin n\theta \, d\theta\}$$

$$= \frac{2T}{\pi}\left(\frac{-\cos n\theta}{n}\right)_0^{\pi} = \frac{2T}{n\pi}(1 - \cos n\pi)$$

$$= \frac{2T}{n\pi}\{1 - (-1)^n\} = \begin{cases} \dfrac{4T}{n\pi}; & n \text{ is odd} \\ 0 \ ; & n \text{ is even} \end{cases}$$

$\therefore$ $$b_n = \begin{cases} \dfrac{4T}{n\pi a^n}; & n \text{ is odd} \\ 0 \ ; & n \text{ is even} \end{cases}$$

From (13),
$$u(r, \theta) = \frac{4T}{\pi} \sum_{n=1,3,5,...}^{\infty} \frac{1}{n} \cdot \frac{r^n}{a^n} \sin n\theta$$
$$u(r, \theta) = \frac{4T}{\pi} \sum_{n=1}^{\infty} \frac{1}{(2n-1)} \left(\frac{r}{a}\right)^{2n-1} \sin (2n-1)\theta$$
which is the required steady state temperature in the plate.

Example 2. *Solve:* $\frac{\partial^2 V}{\partial r^2} + \frac{1}{r}\frac{\partial V}{\partial r} + \frac{1}{r^2}\frac{\partial^2 V}{\partial \theta^2} = 0$ *with boundary conditions*

(i) V is finite when $r \to 0$ *(ii)* $V = \Sigma C_n \cos n\theta$ *on* $r = a$.

Sol. Solution to given differential equation is
$$V = \Sigma (A_n r^n + B_n r^{-n}) \cos (n\theta + \alpha)$$
When $r = a$,
$$V = \Sigma C_n \cos n\theta$$
$$\therefore \quad \Sigma C_n \cos n\theta = \Sigma(A_n a^n + B_n a^{-n}) \cos (n\theta + \alpha)$$
$$\therefore \quad C_n = A_n a^n + B_n a^{-n}, \alpha = 0$$
When $r \to 0$, V is finite.
$$\therefore \quad B_n = 0 \qquad \text{(otherwise V becomes } \infty\text{)}$$
$$\therefore \quad A_n = \frac{C_n}{a^n}$$
$$\therefore \quad V = \Sigma C_n \left(\frac{r}{a}\right)^n \cos n\theta.$$

Example 3. *The edge r = a of a circular plate is kept at temperature f(θ). The plate is insulated so that there is no loss of heat from either surface. Find the temperature distribution in steady state.*

Sol. Here, we have to take the solution in polar coordinates.

The solution is
$$u = (c_1 \cos p\theta + c_2 \sin p\theta)(c_3 r^p + c_4 r^{-p}) \qquad ...(1)$$
Since, the temperature remains finite at $r = 0$
$$\therefore \quad c_4 = 0 \qquad ...(2)$$
Also, if we increase θ by 2π, we arrive at the same point. So the solution (1) should be periodic with period 2π.

Therefore $p = n$, an integer. Hence, we may write the general solution as
$$u = \sum_{n=0}^{\infty} (c_1 \cos n\theta + c_2 \sin n\theta) c_3 r^n$$
$$= \sum_{n=0}^{\infty} (A_n \cos n\theta + B_n \sin n\theta) r^n \qquad | \; c_1c_3 = A_n, c_2c_3 = B_n \text{ (say)}$$
Applying to this, the condition
$$u = f(\theta) \quad \text{for} \quad r = a, \text{ we get}$$
$$f(\theta) = \sum_{n=0}^{\infty} (A_n \cos n\theta + B_n \sin n\theta) a^n$$

where $$a^n A_n = \frac{1}{\pi}\int_0^{2\pi} f(\theta)\cos n\theta \, d\theta$$

and $$a^n B_n = \frac{1}{\pi}\int_0^{2\pi} f(\theta)\sin n\theta \, d\theta$$

Example 4. *Find a harmonic function* ϕ *in the semi-circle* $r < a$, $0 < \theta < \pi$ *which vanishes on* $\theta = 0$ *and takes the value c on* $\theta = \pi$ *and on the curved portion* $r = a$.

Sol. A harmonic function is a function satisfying Laplace's equation.

Solution to $r^2 \dfrac{\partial^2\phi}{\partial r^2} + r\dfrac{\partial\phi}{\partial r} + \dfrac{\partial^2\phi}{\partial\theta^2} = 0$ is

$$\phi = (A\cos p\theta + B\sin p\theta)(Cr^p + Dr^{-p}) \quad ...(1)$$

Since $\phi \to 0$ as $r \to 0$ $\therefore$ $D = 0$

$$\therefore \quad \phi = (A\cos p\theta + B\sin p\theta)\, Cr^p \quad ...(2)$$

Imposing on (2), the zero boundary conditions

$\phi(r, 0) = 0$ and $\phi(r, \pi) = 0$, we get

$A = 0$ and $p = n \ (n \in I)$

Putting in (2) and adding up various solutions for $n = 1, 2, 3,...,$ we get

$$\phi = \sum_{n=1}^{\infty} B_n r^n \sin n\theta \quad ...(3)$$

This solution will not satisfy both the given boundary conditions of the problem namely

$$\left.\begin{array}{ll} & (i)\ \phi = 0 \text{ when } \theta = 0 \\ \text{and} & (ii)\ \phi = c \text{ when } \theta = \pi \end{array}\right\} \quad ...(4)$$

So we add to (3), the solution $\phi = A_0\theta + B_0$ which satisfies Laplace's equation. We choose A_0, B_0 so that (4) is satisfied.

Then, $$B_0 = 0 \quad \text{and} \quad c = A_0\pi$$

Hence, $$\phi = \frac{c\theta}{\pi} + \sum_{n=1}^{\infty} B_n r^n \sin n\theta \quad ...(5) \quad \text{satisfies (4)}$$

Applying to (5), the condition that $\phi = c$ when $r = a$, we get

$$c = \frac{c\theta}{\pi} + \sum_{n=1}^{\infty} B_n a^n \sin n\theta$$

$$\Rightarrow \quad c\left(1 - \frac{\theta}{\pi}\right) = \sum_{n=1}^{\infty} B_n a^n \sin n\theta$$

$$\therefore \quad B_n a^n = \frac{2}{\pi}\int_0^{\pi} c\left(1 - \frac{\theta}{\pi}\right)\sin n\theta \, d\theta$$

$$B_n a^n = \frac{2c}{n\pi} \Rightarrow B_n = \frac{2c}{n\pi a^n} \quad ...(6)$$

∴ From (5) and (6),

$$\phi = \frac{c\theta}{\pi} + \frac{2c}{\pi} \sum_{n=1}^{\infty} \frac{1}{n}\left(\frac{r}{a}\right)^n \sin n\theta$$

TEST YOUR KNOWLEDGE

1. Show that the steady state temperature distribution in a semi-circular plate of radius a whose bounding diameter is kept at 0°C, while the circumference is kept at 60°C is given by

$$u(r, \theta) = \frac{240}{\pi} \sum_{n=1}^{\infty} \frac{1}{(2n-1)}\left(\frac{r}{a}\right)^{2n-1} . \sin (2n-1)\theta.$$

2. A semi-circular plate of radius a has its circumference kept at temperature $u(a, \theta) = k\theta(\pi - \theta)$ while the bounding diameter is kept at zero temperature. Assuming the surfaces of the plate to be insulated, show that the steady-state temperature distribution of the plate is given by

$$u(r, \theta) = \frac{8k}{\pi} \sum_{n=1}^{\infty} \left(\frac{r}{a}\right)^{2n-1} . \frac{\sin (2n-1)\theta}{(2n-1)^3}.$$

3. The bounding diameter of a semi-circular plate of radius a is kept at 0°C and the temperature along the semi-circular boundary is given by

$$u(a, \theta) = \begin{cases} 50\theta, & \text{when } 0 < \theta < \frac{\pi}{2} \\ 50(\pi - \theta), & \text{when } \frac{\pi}{2} < \theta < \pi \end{cases}.$$

Show that the steady-state temperature distribution given by

$$u(r, \theta) = \frac{200}{\pi} \sum_{n=1}^{\infty} \frac{(-1)^{n+1}}{(2n-1)^2}\left(\frac{r}{a}\right)^{2n-1} . \sin (2n-1)\theta.$$

2.17 TRANSMISSION LINE EQUATIONS

Consider the flow of electricity in an insulated cable. Let V be the potential and I the current at time t at a point P of the cable at a distance x from a given point O. Let V + δV and I + δI be the potential and the current at the point Q(x + δx) at the same time.

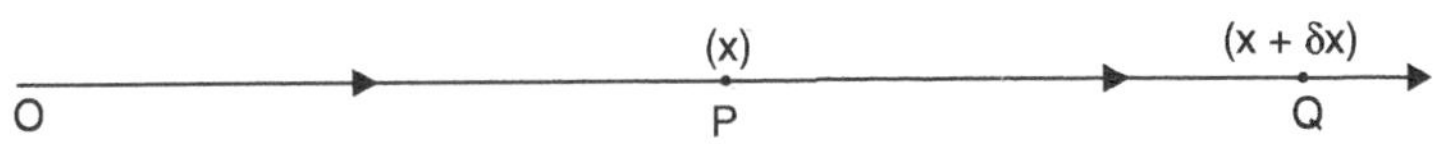

Let R, L, C, G be respectively the resistance, inductance, capacitance and leakance to the ground per unit length of the cable, each assumed to be constant.

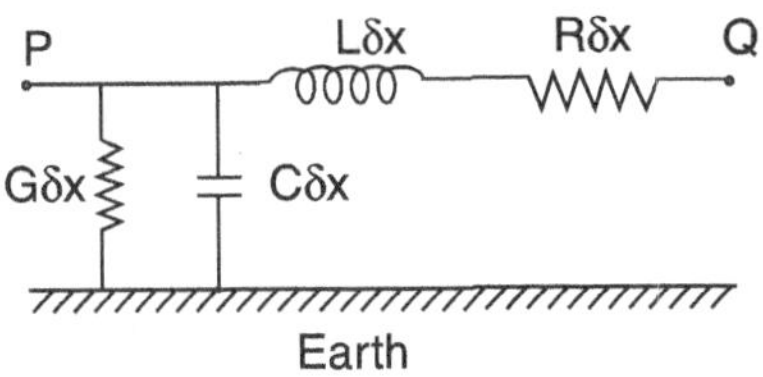

Since, the potential drop across the segment PQ = potential drop due to resistance + potential drop due to inductance

$$\therefore \qquad - \delta V = (R\delta x)I + (L\delta x)\frac{\partial I}{\partial t}$$

Dividing by δx and proceeding to the limits as $\delta x \to 0$, we have

$$-\frac{\partial V}{\partial x} = RI + L\frac{\partial I}{\partial t} \qquad \text{...(1)}$$

Similarly, the decrease in current in crossing the segment PQ = decrease in current due to capacitance and leakance

$$\therefore \qquad - \delta I = (G\delta x)V + (C\delta x)\frac{\partial V}{\partial t}$$

Dividing by δx and proceeding to the limits as $\delta x \to 0$, we have

$$-\frac{\partial I}{\partial x} = GV + C\frac{\partial V}{\partial t} \qquad \text{...(2)}$$

Differentiating (1) and (2) partially w.r.t. x and t respectively, we have

$$-\frac{\partial^2 V}{\partial x^2} = R\frac{\partial I}{\partial x} + L\frac{\partial^2 I}{\partial x \partial t} \qquad \text{...(3)}$$

and

$$-\frac{\partial^2 I}{\partial x \partial t} = G\frac{\partial V}{\partial t} + C\frac{\partial^2 V}{\partial t^2} \qquad \text{...(4)}$$

Eliminating I from (2), (3) and (4), we get

$$-\frac{\partial^2 V}{\partial x^2} = R\left(-GV - C\frac{\partial V}{\partial t}\right) + L\left(-G\frac{\partial V}{\partial t} - C\frac{\partial^2 V}{\partial t^2}\right)$$

or

$$\boxed{\frac{\partial^2 V}{\partial x^2} = LC\frac{\partial^2 V}{\partial t^2} + (RC + LG)\frac{\partial V}{\partial t} + RGV} \qquad \text{...(5)}$$

Now, differentiating (1) and (2) w.r.t. t and x respectively, we have

$$-\frac{\partial^2 V}{\partial x \partial t} = R\frac{\partial I}{\partial t} + L\frac{\partial^2 I}{\partial t^2} \qquad \text{...(6)}$$

and

$$-\frac{\partial^2 I}{\partial x^2} = G\frac{\partial V}{\partial x} + C\frac{\partial^2 V}{\partial x \partial t} \qquad \text{...(7)}$$

Eliminating V from (1), (6), and (7), we get

$$-\frac{\partial^2 I}{\partial x^2} = G\left(-RI - L\frac{\partial I}{\partial t}\right) + C\left(-R\frac{\partial I}{\partial t} - L\frac{\partial^2 I}{\partial t^2}\right)$$

or

$$\boxed{\frac{\partial^2 I}{\partial x^2} = LC\frac{\partial^2 I}{\partial t^2} + (LG + RC)\frac{\partial I}{\partial t} + RGI} \qquad \text{...(8)}$$

The equations (5) and (8) are called the **telephone equations.**

2.17.1. If L = G = 0, the equations (5) and (8) become

$$\boxed{\frac{\partial^2 V}{\partial x^2} = RC\frac{\partial V}{\partial t}} \quad \text{and} \quad \boxed{\frac{\partial^2 I}{\partial x^2} = RC\frac{\partial I}{\partial t}}$$

which are known as the *telegraph equations*. They are similar to the equations in one dimensional heat flow. [*M.T.U. 2013*]

2.17.2. If R = G = 0, the equations (5) and (8) become

$$\boxed{\frac{\partial^2 V}{\partial x^2} = LC\frac{\partial^2 V}{\partial t^2}} \quad \text{and} \quad \boxed{\frac{\partial^2 I}{\partial x^2} = LC\frac{\partial^2 I}{\partial t^2}}$$

which are known as the *radio equations*. They are similar to the wave equation.

2.17.3. If R and G are negligible, the transmission lines become

$$\frac{\partial v}{\partial x} = -L\frac{\partial i}{\partial t} \quad \text{and} \quad \frac{\partial i}{\partial x} = -C\frac{\partial v}{\partial t}$$

2.17.4. If L = C = 0, the equations (5) and (8) become

$$\frac{\partial^2 V}{\partial x^2} = RGV \quad \text{and} \quad \frac{\partial^2 I}{\partial x^2} = RGI$$

which are *equations for submarine cable*.

Note. We have sofar considered the transient state solutions only. The **steady-state solutions** of transmission lines are however obtained by assuming $v = V e^{iwt}$ and $i = I e^{iwt}$, where V and I are complex functions of x only. Substituting these in (5) and (6), we get two ordinary linear diff. eqns. of II order which can be solved at once.

ILLUSTRATIVE EXAMPLES

Example 1. *Find the current i and voltage e in a line of length l, t seconds after the ends are suddenly grounded, given that*

$$i(x, 0) = i_0,\ e(x, 0) = e_0 \sin\frac{\pi x}{l}$$

Also R and G are negligible.

Sol. Since R and G are negligible, transmission line equations become

$$\frac{\partial e}{\partial x} = -L\frac{\partial i}{\partial t} \qquad \text{...(1)}$$

and

$$\frac{\partial i}{\partial x} = -C\frac{\partial e}{\partial t} \qquad \text{...(2)}$$

For elimination of i, differentiating (1) partially w.r.t. x and (2) partially w.r.t. t, we have

$$\frac{\partial^2 e}{\partial x^2} = -L\frac{\partial^2 i}{\partial x \partial t} \quad \text{and} \quad \frac{\partial^2 i}{\partial t \partial x} = -C\frac{\partial^2 e}{\partial t^2}$$

whence $$\frac{\partial^2 e}{\partial x^2} = LC\frac{\partial^2 e}{\partial t^2} \qquad ...(3)$$

The initial conditions are $i(x, 0) = i_0,\ e(x, 0) = e_0 \sin \frac{\pi x}{l}$ ...(4)

Since, the ends are suddenly grounded, the boundary conditions are

$$e(0, t) = e(l, t) = 0 \qquad ...(5)$$

Also $i = i_0$ (constant) when $t = 0$

$\therefore$ $$\frac{\partial i}{\partial x} = 0 \text{ which gives } \frac{\partial e}{\partial t} = 0 \text{ when } t = 0 \qquad ...(6)$$ | Using (2)

Now let $e = XT$ be a solution of (3) where X is a function of x only and T is a function of t only.

$$\frac{\partial^2 e}{\partial x^2} = X''T \text{ and } \frac{\partial^2 e}{\partial t^2} = XT''$$

$\therefore$ From (3), $X''T = LCXT''$

Separating the variables $\frac{X''}{X} = LC\frac{T''}{T} = -p^2$ (say)

This leads to the ordinary differential equations

$$\frac{d^2X}{dx^2} + p^2X = 0 \quad \text{and} \quad \frac{d^2T}{dt^2} + \frac{p^2}{LC}T = 0$$

$\therefore$ $$X = c_1 \cos px + c_2 \sin px$$

$$T = c_3 \cos \frac{pt}{\sqrt{LC}} + c_4 \sin \frac{pt}{\sqrt{LC}}$$

$\Rightarrow$ $$e = XT = (c_1 \cos px + c_2 \sin px)\left(c_3 \cos \frac{pt}{\sqrt{LC}} + c_4 \sin \frac{pt}{\sqrt{LC}}\right) \qquad ...(7)$$

Applying the boundary conditions (5), we have

$$c_1 = 0 \quad \text{and} \quad p = \frac{n\pi}{l}, \ n \text{ being an integer.}$$

$\therefore$ Equation (7) becomes

$$e = c_2 \sin \frac{n\pi x}{l}\left(c_3 \cos \frac{n\pi t}{l\sqrt{LC}} + c_4 \sin \frac{n\pi t}{l\sqrt{LC}}\right)$$

or $$e = \sin \frac{n\pi x}{l}\left(A \cos \frac{n\pi t}{l\sqrt{LC}} + B \sin \frac{n\pi t}{l\sqrt{LC}}\right) \qquad ...(8)$$ | where $A = c_2c_3$ and $B = c_2c_4$

$$\frac{\partial e}{\partial t} = \sin \frac{n\pi x}{l}\left(-\frac{An\pi}{l\sqrt{LC}} \sin \frac{n\pi t}{l\sqrt{LC}} + \frac{Bn\pi}{l\sqrt{LC}} \cos \frac{n\pi t}{l\sqrt{LC}}\right)$$

Since $\frac{\partial e}{\partial t} = 0$ when $t = 0$, we get [From (6)]

$$B = 0$$

$\therefore$ From (8), $e = A \sin \frac{n\pi x}{l} \cos \frac{n\pi t}{l\sqrt{LC}}$

By superposition, $e = \sum_{n=1}^{\infty} A_n \sin \frac{n\pi x}{l} \cos \frac{n\pi t}{l\sqrt{LC}}$ is also a solution.

But $e = e_0 \sin \frac{\pi x}{l}$ when $t = 0$ [From (4)]

$\therefore$ $$e_0 \sin \frac{\pi x}{l} = \sum_{n=1}^{\infty} A_n \sin \frac{n\pi x}{l}$$

$\Rightarrow$ $$A_1 = e_0 \quad \text{and} \quad A_2 = A_3 = \ldots\ldots = 0$$

Hence, $$e = e_0 \sin \frac{\pi x}{l} \cos \frac{\pi t}{l\sqrt{LC}}$$

Now, $$-L \frac{\partial i}{\partial t} = \frac{\partial e}{\partial x}$$ [From (1)]

$\therefore$ $$\frac{\partial i}{\partial t} = -\frac{1}{L} \cdot \frac{e_0 \pi}{l} \cos \frac{\pi x}{l} \cos \frac{\pi t}{l\sqrt{LC}}$$

Integrating w.r.t. t, regarding x as constant

$$i = -\frac{e_0 \pi}{Ll} \cos \frac{\pi x}{l} \cdot \frac{l\sqrt{LC}}{\pi} \sin \frac{\pi t}{l\sqrt{LC}} + f(x) \qquad \ldots(9)$$

where $f(x)$ is an arbitrary constant function.

Since $i = i_0$ when $t = 0$, we have $i_0 = 0 + f(x)$ or $f(x) = i_0$

$\therefore$ From (9), we have $i = i_0 - e_0 \sqrt{\frac{C}{L}} \cos \frac{\pi x}{l} \sin \frac{\pi t}{l\sqrt{LC}}$.

Example 2. *A transmission line 1000 km long is initially under steady-state conditions with potential 1300 volts at the sending end (x = 0) and 1200 volts at the receiving end (x = 1000). The terminal end of the line is suddenly grounded, but the potential at the source is kept at 1300 volts. Assuming the inductance and leakance to be negligible, find the potential E(x, t).*

Sol. Since L and G are negligible, we use the telegraph equation

$$\frac{\partial^2 E}{\partial x^2} = RC \frac{\partial E}{\partial t} \quad \text{or} \quad \frac{\partial E}{\partial t} = \frac{1}{RC} \cdot \frac{\partial^2 E}{\partial x^2} \qquad \ldots(1)$$

Here, E_s = initial steady voltage satisfying $\frac{\partial^2 E}{\partial x^2} = 0$

$$= 1300 - \frac{1300 - 1200}{1000} x = 1300 - 0.1x = E(x, 0) \qquad \ldots(2)$$

E_s' = steady voltage (after grounding the terminal end) when steady conditions are ultimately reached

$$= 1300 - \frac{1300 - 0}{1000} x = 1300 - 1.3x$$

$\therefore$ $\quad E(x, t) = E_s' + E_r(x, t)$, where $E_r(x, t)$ is the transient part

$$= 1300 - 1.3x + \sum_{n=1}^{\infty} b_n \sin \frac{n\pi x}{l} e^{-\frac{n^2\pi^2 t}{l^2 RC}} \quad \text{...(3) where } l = 1000 \text{ km}$$

Putting $t = 0$ in (3), we have

$$E(x, 0) = 1300 - 1.3x + \sum_{n=1}^{\infty} b_n \sin \frac{n\pi x}{l}$$

or
$$1300 - 0.1x = 1300 - 1.3x + \sum_{n=1}^{\infty} b_n \sin \frac{n\pi x}{l} \quad \text{[Using (2)]}$$

or
$$1.2x = \sum_{n=1}^{\infty} b_n \sin \frac{n\pi x}{l}$$

where
$$b_n = \frac{2}{l}\int_0^l 1.2x \sin \frac{n\pi x}{l} dx$$

$$= \frac{2.4}{l}\left[x\left(-\frac{\cos \frac{n\pi x}{l}}{\frac{n\pi}{l}}\right) - (1)\left(-\frac{\sin \frac{n\pi x}{l}}{\left(\frac{n\pi}{l}\right)^2}\right)\right]_0^l = \frac{2.4}{l}\cdot\frac{l}{n\pi}[-l \cos n\pi]$$

$$= -\frac{2.4 \times 1000}{n\pi} \times (-1)^n \quad [\because \ l = 1000]$$

$$= \frac{2400}{\pi}\cdot\frac{(-1)^{n+1}}{n}$$

Hence,
$$E(x, t) = 1300 - 1.3x + \frac{2400}{\pi}\sum_{n=1}^{\infty} \frac{(-1)^{n+1}}{n} \sin \frac{n\pi x}{l} e^{-\frac{n^2\pi^2 t}{l^2 RC}}.$$

Example 3. *Neglecting R and G, find the e.m.f. $v(x, t)$ in a line of length l, t seconds after the ends were suddenly grounded, given that $i(x, 0) = i_0$ and $v(x, 0) = e_1 \sin \frac{\pi x}{l} + e_5 \sin \frac{5\pi x}{l}$.*

(M.T.U 2012)

Sol. Since R and G are negligible, we use the radio equation

$$\frac{\partial^2 v}{\partial x^2} = LC \frac{\partial^2 v}{\partial t^2} \quad \text{...(1)}$$

Since, the ends are suddenly grounded, we have the boundary conditions

$$v(0, t) = 0, \ v(l, t) = 0 \quad \text{...(2)}$$

Also, the initial conditions are

$$i(x, 0) = i_0$$

and
$$\left.\begin{aligned} & i(x,0) = i_0 \\ & v(x, 0) = e_1 \sin \frac{\pi x}{l} + e_5 \sin \frac{5\pi x}{l} \end{aligned}\right\} \quad \text{...(3)}$$

$\therefore$
$$\frac{\partial i}{\partial x} = -C \frac{\partial v}{\partial t} \text{ gives}$$

$$\left(\frac{\partial v}{\partial t}\right)_{t=0} = 0 \qquad ...(4)$$

Let v = XT be the solution of (1) where X is a function of x only and T is a function of t only.

$\therefore$ From (1), $\quad TX'' = LCXT''$

$$\Rightarrow \quad \frac{X''}{X} = LC\frac{T''}{T} = -p^2 \quad \text{(say)}$$

$$\therefore \quad X'' + p^2X = 0 \Rightarrow X = c_1 \cos px + c_2 \sin px$$

and
$$T'' + \left(\frac{p^2}{LC}\right)T = 0$$

$$\Rightarrow \quad T = c_3 \cos\frac{p}{\sqrt{LC}}t + c_4 \sin\frac{p}{\sqrt{LC}}t$$

$$\therefore \quad v = XT = (c_1 \cos px + c_2 \sin px)\left(c_3 \cos\frac{pt}{\sqrt{LC}} + c_4 \sin\frac{pt}{\sqrt{LC}}\right) \qquad ...(5)$$

Using the boundary conditions (2), we get

$$c_1 = 0 \text{ and } p = \frac{n\pi}{l}, n \in I$$

$$\therefore \quad v = \sin\frac{n\pi x}{l}\left(a_n \cos\frac{n\pi t}{l\sqrt{LC}} + b_n \sin\frac{n\pi t}{l\sqrt{LC}}\right) \qquad ...(6)$$

Using initial condition (4), we get $b_n = 0$

$\therefore$ From (6),
$$v = a_n \sin\frac{n\pi x}{l}\cos\frac{n\pi t}{l\sqrt{LC}}$$

Thus the most general solution of (1) is

$$v = \sum_{n=1}^{\infty} a_n \sin\frac{n\pi x}{l}\cos\frac{n\pi t}{l\sqrt{LC}} \qquad ...(7)$$

Finally by the initial condition (3), we have

$$e_1 \sin\frac{\pi x}{l} + e_5 \sin\frac{5\pi x}{l} = \sum_{n=1}^{\infty} a_n \sin\frac{n\pi x}{l}$$

$\therefore \quad a_1 = e_1$ and $a_5 = e_5$ while all other a's are zero.

Hence,
$$v = e_1 \sin\frac{\pi x}{l}\cos\frac{\pi t}{l\sqrt{LC}} + e_5 \sin\frac{5\pi x}{l}\cos\frac{5\pi t}{l\sqrt{LC}}$$

Example 4. *Solve* $\frac{\partial^2 V}{\partial x^2} = LC\frac{\partial^2 V}{\partial t^2}$ *assuming that the initial voltage is* $V_0 \sin\frac{\pi x}{l}$; $V_t(x_0) = 0$ *and* $V = 0$ *at the ends* $x = 0$ *and* $x = l$ *for all* t.

Sol. Let $\quad V = XT \qquad ...(1)$

where X is a function of x only and T is a function of t only.

$$\frac{\partial^2 V}{\partial x^2} = TX'' \quad \text{and} \quad \frac{\partial^2 V}{\partial t^2} = XT''$$

Substituting in the given equations, we get

$$TX'' = LCXT''$$

$$\frac{X''}{X} = LC\frac{T''}{T} = -p^2 \text{ (say)}$$

(*i*) $$\frac{X''}{X} = -p^2 \Rightarrow X'' + p^2X = 0$$

$\therefore$ $$X = c_1 \cos px + c_2 \sin px$$

(*ii*) $$LC\frac{T''}{T} = -p^2 \Rightarrow T'' + \frac{p^2}{LC}T = 0$$

$\therefore$ $$T = c_3 \cos \frac{pt}{\sqrt{LC}} + c_4 \sin \frac{pt}{\sqrt{LC}}$$

Hence, $$V = XT = (c_1 \cos px + c_2 \sin px)\left(c_3 \cos\frac{pt}{\sqrt{LC}} + c_4 \sin\frac{pt}{\sqrt{LC}}\right) \quad ...(2)$$

Boundary conditions are

$$V(0, t) = 0 = V(l, t) \quad \text{and} \quad \frac{\partial V}{\partial t} = 0 \quad \text{when } t = 0$$

Applying conditions on (2), we get

$$0 = c_1\left(c_3 \cos\frac{pt}{\sqrt{LC}} + c_4 \sin\frac{pt}{\sqrt{LC}}\right)$$

$\Rightarrow$ $$c_1 = 0$$

From (2), $$V = c_2 \sin px\left(c_3 \cos\frac{pt}{\sqrt{LC}} + c_4 \sin\frac{pt}{\sqrt{LC}}\right) \quad ...(3)$$

$$V(l, t) = 0 = c_2 \sin pl\left(c_3 \cos\frac{pt}{\sqrt{LC}} + c_4 \sin\frac{pt}{\sqrt{LC}}\right)$$

$\Rightarrow$ $$\sin pl = 0 = \sin n\pi \ (n \in I)$$

$\therefore$ $$p = \frac{n\pi}{l}$$

Hence, from (3),

$$V = c_2 \sin\frac{n\pi x}{l}\left[c_3 \cos\frac{n\pi t}{l\sqrt{LC}} + c_4 \sin\frac{n\pi t}{l\sqrt{LC}}\right] \quad ...(4)$$

$$\frac{\partial V}{\partial t} = c_2 \frac{n\pi}{l\sqrt{LC}} \sin\frac{n\pi x}{l}\left[-c_3 \sin\frac{n\pi t}{l\sqrt{LC}} + c_4 \cos\frac{n\pi t}{l\sqrt{LC}}\right]$$

At $t = 0$,

$$0 = c_2 \frac{n\pi}{l\sqrt{LC}} \sin \frac{n\pi x}{l} \cdot c_4$$

$\Rightarrow$ $c_4 = 0$

Hence from (4),

$$V = c_2 c_3 \sin \frac{n\pi x}{l} \cos \frac{n\pi t}{l\sqrt{LC}} \quad \text{...(5)}$$

Now, $V(x, 0) = V_0 \sin \frac{\pi x}{l} = c_2 c_3 \sin \frac{n\pi x}{l}$

Comparing, we get

$$c_2 c_3 = V_0 \quad \text{and} \quad n = 1$$

Hence, the required solution is

$$V(x, t) = V_0 \sin \frac{\pi x}{l} \cos \frac{\pi t}{l\sqrt{LC}}.$$

TEST YOUR KNOWLEDGE

1. A steady voltage distribution of 20 volts at the sending end and 12 volts at the receiving end is maintained in a telephone wire of length l. A time $t = 0$, the receiving end is grounded. Find the voltage and current t sec later. Neglect leakance and inductance. *[M.T.U. 2011]*

2. A line of length l is initially uncharged so that $i(x, 0) = 0$, $v(x, 0) = 0$, $\frac{\partial v}{\partial t} = 0$ when $t = 0$. At $t = 0$, one end $x = l$ is suddenly connected with a constant potential E while the other end is grounded. Neglecting R and G, show that

$$v(x, t) = \frac{Ex}{l} + \frac{2E}{\pi} \sum_{n=1}^{\infty} \frac{(-1)^n}{n} \sin \frac{n\pi x}{l} \cos\left(\frac{n\pi t}{l\sqrt{LC}}\right).$$

3. Show that a transmission line with negligible resistance and leakage propagates waves of current and potential with velocity equal to $\frac{i}{\sqrt{LC}}$ where L is self-inductance and C is the capacitance.

4. Derive the partial differential equation for transmission line and also derive the telegraph and radio equations under suitable assumptions.

Answer

1. $$v(x, t) = \frac{20(l - x)}{l} + \frac{24}{\pi} \sum_{n=1}^{\infty} \frac{(-1)^{n+1}}{n} \sin \frac{n\pi x}{l} e^{-n^2\pi^2 t/RCl^2}$$

$$i(x, t) = \frac{20}{lR} + \frac{24}{lR} \sum_{n=1}^{\infty} (-1)^n \cos \frac{n\pi x}{l} e^{-n^2\pi^2 t/RCl^2}.$$

ASSIGNMENT-II

(2 Marks Questions for Section–A)

1. Classify the differential equation: $\frac{\partial^2 z}{\partial x^2} + x^2 \frac{\partial^2 z}{\partial y^2} = 0$.

2. Classify the partial differential equation: $\frac{\partial^2 u}{\partial x^2} + \frac{\partial^2 u}{\partial y^2} = \frac{1}{c^2} \frac{\partial u}{\partial t}$. *(M.T.U. 2011)*

3. Classify the partial differential equation: $\frac{\partial^2 u}{\partial x^2} + \frac{\partial^2 u}{\partial y^2} = 0$. Also explain your answer. *(A.K.T.U. 2016)*

4. Classify: $f_{xx} + 2 f_{xy} + 4 f_{yy} = 0$. *(U.P.T.U. 2014)*

5. Classify: $\frac{\partial^2 z}{\partial x^2} = \frac{\partial^2 z}{\partial y^2}$. *[G.B.T.U. (SUM) 2010]*

6. Classify the partial differential equation: $\frac{\partial^2 u}{\partial x^2} + 3 \frac{\partial^2 u}{\partial x \partial y} + \frac{\partial^2 u}{\partial y^2} = 0$. *(U.P.T.U. 2015)*

7. Explain briefly the method of separation of variables in solving a given partial differential equation. *(M.T.U. 2012)*

8. Write down the two-dimensional wave equation. *(A..K.T.U. 2017)*

9. Mention two applications of partial differential equations in engineering. *[G.B.T.U. (AG) 2012]*

10. Name the following equations: *[G.B.T.U. (AG) 2012]*

    (*i*) $\frac{\partial^2 u}{\partial t^2} = c \frac{\partial^2 u}{\partial x^2}$ (*ii*) $\frac{\partial u}{\partial t} = c^2 \frac{\partial^2 u}{\partial x^2}$.

11. Apply the method of separation of variables to find the most appropriate solution of $\frac{\partial^2 u}{\partial x^2} + \frac{\partial^2 u}{\partial y^2} = \frac{1}{c^2} \frac{\partial u}{\partial t}$. *[G.B.T.U. (A.G.) 2012]*

12. What does the two-dimensional wave equation represent?

13. Write down the partial differential equation for one-dimensional wave equation. *(A.K.T.U. 2015, 2017)*

14. Solve: $4u_x + u_y = 3u$; $u(0, y) = e^{-5y}$.

15. Solve: $\frac{\partial u}{\partial t} = c^2 \frac{\partial^2 u}{\partial x^2}$ in the steady state. *[G.B.T.U. (A.G.) 2011]*

16. Classify: $\frac{\partial u}{\partial t} = c^2 \frac{\partial^2 u}{\partial x^2}$

17. Name the equation $\frac{\partial^2 u}{\partial x^2} + \frac{\partial^2 u}{\partial y^2} = 0$

18. Write down the equation of steady state heat conduction in the rectangular plate. *(G.B.T.U. 2011)*

19. Write down the two-dimensional steady state heat flow equation in polar coordinates.

20. Solve: $3u_x + 2u_y = 0$ where $u_x = \frac{\partial u}{\partial x}$, $u_y = \frac{\partial u}{\partial y}$.

21. Classify the following differential equation in the first quadrant:

$$y^2 u_{xx} - x^2 u_{yy} = 0$$ *(G.B.T.U. 2013)*

22. Solve $\frac{\partial u}{\partial x} = 3\frac{\partial u}{\partial t}$ using method of separation of variables. *(G.B.T.U. 2013)*

23. Write the boundary conditions and initial conditions for the displacement of a finite string of length L that is fixed at both ends and is released from rest with an initial displacement $f(x)$. *(G.B.T.U. 2013)*

24. (*i*) Write telegraph equations. *(U.P.T.U. 2014)*

(*ii*) Write two-dimensional heat equation. *(A.K.T.U. 2016)*

25. (*i*) Characterize the following partial differential equation into elliptic, parabolic and hyperbolic equations: $a\frac{\partial^2 z}{\partial x^2} + 2h\frac{\partial^2 z}{\partial x \partial y} + b\frac{\partial^2 z}{\partial y^2} + 2f\frac{\partial z}{\partial x} + 2g\frac{\partial z}{\partial x} + cz = f(x, y)$ where a, b, c, h, f, g are constants. *(M.T.U. 2013)*

(*ii*) Specify with suitable example, the classification of partial differential (PDE) for elliptic, parabolic and hyperbolic differential equations. *(A.K.T.U. 2017)*

26. Find the condition for which the following partial differential equation is parabolic:

$$yu_{xx} + (x + y)u_{xy} + xu_{yy} = 0$$ *(U.P.T.U. 2013)*

27. Classify the partial differential equation:

$$2\frac{\partial^2 z}{\partial x^2} - 3\frac{\partial^2 z}{\partial x \partial y} + \frac{\partial^2 z}{\partial y^2} - 3\frac{\partial z}{\partial x} + \frac{\partial z}{\partial y} = 0.$$ *(U.P.T.U. 2015)*

28. Classify the following partial differential equation along the line $y = x$:

$$yu_{xx} + (x + y)u_{xy} + xu_{yy} = 0.$$ *(U.P.T.U. 2014)*

29. (*i*) Find the steady state temperature distribution in a rod of length L when its one end is kept at 0°C and the other end is kept at 100°C. *(U.P.T.U. 2013)*

(*ii*) Find the steady state temperature distribution in a plate of length of 20 whose ends are kept at 40°C and 100°C respectively. *(U.P.T.U. 2015)*

30. (*i*) Find the steady state temperature distribution in a rod of 2 m whose ends are kept at 30°C and 70°C respectively. *(U.P.T.U. 2015)*

(*ii*) Find the steady state temperature distribution in a rod of length 20 cm, whose ends are kept at 0°C and 60°C. *(U.P.T.U. 2014)*

Answers

1. Elliptic **2.** Parabolic **3.** Elliptic

4. Elliptic **5.** Hyperbolic **6.** Hyperbolic

8. $\frac{\partial^2 u}{\partial t^2} = c^2 \left(\frac{\partial^2 u}{\partial x^2} + \frac{\partial^2 u}{\partial y^2}\right)$

10. (*i*) One-dimensional wave equation (*ii*) One-dimensional heat equation

11. $u(x, y, t) = (c_1 \cos k_1 x + c_2 \sin k_1 x)(c_3 \cos k_2 y + c_4 \sin k_2 y)\, c_5 e^{-c^2 k^2 t}$

12. Vibrations of a tightly stretched membrane

13. $\frac{\partial y}{\partial t} = c^2 \frac{\partial^2 y}{\partial x^2}$

14. $u(x, y) = e^{2x - 5y}$

15. $u = c_1 x + c_2$

16. Parabolic

17. Laplace equation in two dimensions

18. $\frac{\partial^2 u}{\partial x^2} + \frac{\partial^2 u}{\partial y^2} = 0$

19. $r^2 \frac{\partial^2 u}{\partial r^2} + r \frac{\partial u}{\partial r} + \frac{\partial^2 u}{\partial \theta^2} = 0$

20. $u(x, y) = c\, e^{\frac{k}{6}(2x - 3y)}$

21. hyperbolic

22. $u(x, t) = c_1 c_2\, e^{-p^2(3x+t)}$

23. $y(0, t) = 0 = y(L, t)\}$ boundary conditions

$$\left.\begin{array}{l} \left(\frac{\partial y}{\partial t}\right)_{t=0} = 0 \\ y(x, 0) = f(x) \end{array}\right\} \text{initial conditions}$$

24. (*i*) $\frac{\partial^2 V}{\partial x^2} = RC \frac{\partial V}{\partial t}$ and $\frac{\partial^2 I}{\partial x^2} = RC \frac{\partial I}{\partial t}$

(*ii*) $\frac{\partial u}{\partial t} = c^2 \left(\frac{\partial^2 u}{\partial x^2} + \frac{\partial^2 u}{\partial y^2}\right)$

25. $h^2 < ab \to$ elliptic
$h^2 = ab \to$ parabolic
$h^2 > ab \to$ hyperbolic

26. $y = x$

27. hyperbolic

28. parabolic

29. (*i*) $u(x, 0) = \frac{100}{L} x$

(*ii*) $u(x, 0) = 40 + 3x$

30. (*i*) $u(x, 0) = 30 + 20\,x$

(*ii*) $u(x, 0) = 3x$

MODULE 3

Statistical Techniques – I

3.1 INTRODUCTION

Statistics is as old as the human society itself.

It is difficult to imagine any face of our life untouched by numerical data. The modern society is essentially data-oriented. It is, therefore, essential to know how to extract useful information from such data. This is the primary objective of statistics. Statistics concerns itself with the collection, presentation and drawing of inferences from numerical data which vary.

In singular sense, Statistics is used to describe the principles and methods which are employed in collection, presentation, analysis and interpretation of data. These devices help to simplify the complex data and make it possible for a common man to understand it without much difficulty. Human mind is unable to assimilate complicated data at a stretch. Statistical methods make these figures intelligible and readily understandable.

In plural sense, Statistics is considered as a numerical description of quantitative aspect of things.

Definition *Statistics is the science which deals with methods of collecting, classifying, presenting, comparing and interpreting numerical data collected to throw light on any sphere of enquiry.*

3.2 VARIABLE (OR VARIATE)

A quantity which can vary from one individual to another is called a **variable** or **variate,** *e.g.*, heights, weights, ages, wages of persons, rainfall records of cities etc.

Quantities which can take any numerical value within a certain range are called **continuous variables** *e.g.*, as the child grows, his/her height takes all possible values from 50 cm to 100 cm.

Quantities which are incapable of taking all possible values are called **discrete** or **discontinuous variables** *e.g.*, the number of children a man can have are positive integers 1, 2, 3 etc. (no value between any two consecutive integers).

3.3 COMPARISON OF FREQUENCY DISTRIBUTIONS

When two or more different series of *the same type* are compared, tabulation of observations is not sufficient. It is often desirable to define quantitatively the characteristics of frequency distribution.

There are two fundamental characteristics in which similar frequency distributions may differ :

(*i*) They may differ *in measures of location or central tendency i.e.*, in the value of the variate x round which they centre.

(*ii*) They may differ in the extent to which observations are scattered about the central value. Measures of this kind are called measures of dispersion.

We will only study here 'measures of central tedency'.

3.4 MEASURES OF CENTRAL TENDENCY

Tabulation arranges facts in a logical order and helps their understanding and comparison. But often, the groups tabulated are still too large for their characteristics to be readily grasped. What is desired is a numerical expression which summarises the characteristic of the group. Measures of central tendency or measures of location (also popularly called Averages) serve this purpose.

A figure which is used to represent a whole series should neither have the lowest value nor the highest in the series, but a value somewhere between these two limits, possibly in the centre, where most of the items of the series cluster. Such figures are called *Measures of Central Tendency* (or Averages).

According to Professor Bowley, "Averages are Statistical Constants which enable us to comprehend in a single effort the significance of the whole."

There are five types of averages in common use :

1. Arithmetic Average or Mean 2. Median 3. Mode
4. Geometric Mean 5. Harmonic Mean

3.4.1 Arithmetic Mean

In the case of Individual Observations (*i.e.*, where frequency is not given) :

1. Direct Method. If $x : x_1, x_2, ..., x_n$ then A.M. $\bar{x}$ is given by

$$\bar{x} = \frac{x_1 + x_2 + ... + x_n}{n} = \frac{1}{n}\,\Sigma x.$$

2. Short Cut Method. (Shift of origin) Shifting the origin to an *arbitrary point a*, the formula $\bar{x} = \frac{1}{n}\,\Sigma x$ becomes

$$\bar{x} - a = \frac{1}{n}\,\Sigma(x - a)$$

or,
$$\bar{x} = a + \frac{1}{n}\,\Sigma d_x \quad \text{where} \quad d_x = x - a$$

Here, a = arbitrary number, called *Assumed Mean*

$\Sigma d_x = \Sigma(x - a) = (x_1 - a) + (x_2 - a) + \cdots\cdots + (x_n - a)$

= sum of the deviations of the variate x from a

n = number of observations.

In the case of Discrete Series :

1. Direct Method. If the frequency distribution is

$x : x_1, \quad x_2, ..., x_n$

$f : f_1, \; f_2, ..., f_n,$ then

$$\bar{x} = \frac{f_1x_1 + f_2x_2 + ... + f_nx_n}{f_1 + f_2 + ... + f_n} = \frac{\Sigma fx}{N} \quad \text{where } N = f_1 + f_2 + ... + f_n = \Sigma f$$

2. Short Cut Method. (Shift of origin) Shifting the origin to an arbitrary point 'a', the formula $\bar{x} = \frac{1}{N} \Sigma fx$ becomes

$$\bar{x} - a = \frac{1}{N} \Sigma f(x - a)$$

or $$\bar{x} = a + \frac{1}{N} \Sigma fd_x, \quad \text{where } d_x = x - a \text{ and } = \text{assumed mean}$$

$$\Sigma fd_x = \Sigma f(x - a) = f_1(x_1 - a) + f_2(x_2 - a) + ... + f_n(x_n - a)$$

= sum of the products of f and the deviation of the corresponding variate x from a.

$N = f_1 + f_2 + ... + f_n = \Sigma f.$

Note. If the frequencies are given in terms of class intervals, the mid-values of class intervals are considered as x and then the above formulae are applied.

In the case of Continuous Series having equal class intervals, say of width h, we use a different formula (Shift of origin and change of scale ; Step Deviation Method).

Let $$u = \frac{x - a}{h} \text{ then } x = a + hu$$

$\therefore$ $$\Sigma fx = \Sigma f(a + hu) = a\Sigma f + h\Sigma fu$$

Dividing both sides by $N = \Sigma f$, we get

$$\frac{\Sigma fx}{N} = a + \frac{h\Sigma fu}{N} \quad \text{or} \quad \bar{x} = a + h\frac{\Sigma fu}{N}, \quad \text{where } u = \frac{x - a}{h}.$$

Weighted Arithmetic Mean. If the variate-values are not of equal importance, we may attach to them 'weights' $w_1, w_2, ..., w_n$ as measures of their importance.

The weighted mean $\bar{x}_w$ is defined as $\bar{x}_w = \frac{w_1x_1 + w_2x_2 + ... + w_nx_n}{w_1 + w_2 + ... + w_n} = \frac{\Sigma wx}{\Sigma w}.$

ILLUSTRATIVE EXAMPLES

Example 1. *Find the mean from the following data:*

Marks	*No. of students*	*Marks*	*No. of students*
Below 10	*5*	*Below 60*	*60*
Below 20	*9*	*Below 70*	*70*
Below 30	*17*	*Below 80*	*78*
Below 40	*29*	*Below 90*	*83*
Below 50	*45*	*Below 100*	*85*

Sol. The frequency distribution table can be written as:

Marks	*Mid values* (x)	f	$x-55$	$u=\frac{x-55}{10}$	fu
0—10	5	5	– 50	– 5	– 25
10—20	15	4	– 40	– 4	– 16
20—30	25	8	– 30	– 3	– 24
30—40	35	12	– 20	– 2	– 24
40—50	45	16	– 10	– 1	– 16
50—60	55	15	0	0	0
60—70	65	10	10	1	10
70—80	75	8	20	2	16
80—90	85	5	30	3	15
90—100	95	2	40	4	8
		$N=\Sigma f=85$			$\Sigma fu=-56$

$$\text{Mean } \bar{x} = a + h\,\frac{\Sigma fu}{N} = 55 + 10 \times \left(\frac{-56}{85}\right) \qquad [\text{Here } a = 55, h = 10]$$

$$= 55 - \frac{112}{17} = 48.41 \text{ marks.}$$

Example 2. *The mean of 200 items was 50. Later on it was discovered that two items were misread as 92 and 8 instead of 192 and 88. Find out the correct mean.*

Sol. Here incorrect value of $\bar{x} = 50, \quad n = 200$

Since $\bar{x} = \frac{\Sigma x}{n} \quad \therefore \quad \Sigma x = n\bar{x}$

Using incorrect value of $\bar{x}$,

Incorrect $\Sigma x = 200 \times 50 = 10000$

$\therefore$ Corrected value of $\Sigma x = 10000 - (92 + 8) + (192 + 88) = 10180$

$$\text{Correct mean} = \frac{\text{Corrected } \Sigma x}{n} = \frac{10180}{200} = 50.9.$$

Properties of Arithmetic Mean

Property I. *The algebraic sum of the deviations of all the variates from their arithmetic mean is zero.*

Proof. Let d_x be the deviation of the variate x from the mean $\bar{x}$, then $d_x = x - \bar{x}$

$$\therefore \quad \Sigma fd_x = \Sigma f(x - \bar{x}) = \Sigma fx - \bar{x}\,\Sigma f$$

$$= N\bar{x} - N\bar{x} = 0 \qquad \left|\ \because \bar{x} = \frac{\Sigma fx}{N}, \text{ where } N = \Sigma f.\right.$$

Property II. *The sum of the squares of the deviations of a set of values is minimum when taken about mean.*

Proof. Let the frequency distribution be x_i/f_i, $i = 1, 2, \ldots, n$. Let z be the sum of the squares of the deviations of the given values from an arbitrary point a (say).

$\Rightarrow$ Let $$z = \sum_{i=1}^{n} f(x-a)^2.$$

We have to show that z is minimum when $a = \bar{x}$.

z will be minimum when $\dfrac{dz}{da} = 0$ and $\dfrac{d^2z}{da^2} > 0$

Now, $$\frac{dz}{da} = \sum_{i=1}^{n} 2f(x-a).(-1) = -2\sum_{i=1}^{n} f(x-a)$$

$$\therefore \quad \frac{dz}{da} = 0 \quad \Rightarrow \quad -2\Sigma f(x-a) = 0$$

$$\Rightarrow \quad \Sigma fx - a\Sigma f = 0$$

$$\Rightarrow \quad N\bar{x} - aN = 0 \qquad \left[\because \bar{x} = \frac{\Sigma fx}{N}, \Sigma f = N\right]$$

$$\Rightarrow \quad \bar{x} - a = 0 \qquad (\because \ N = \Sigma f \neq 0)$$

$$\Rightarrow \quad a = \bar{x}$$

Also, $$\frac{d^2z}{da^2} = -2\sum_{i=1}^{n} f(-1) = 2\Sigma f = 2N > 0.$$

Hence z is minimum when $a = \bar{x}$.

Property III. (*Mean of the composite series*)

If $\bar{x}_i$ (i = 1, 2, ..., k) be the arithmetic means of k distributions with respective frequencies n_i (i = 1, 2, ..., k), then mean $\bar{x}$ of the whole distribution obtained by combining the k distributions is given by

$$\bar{x} = \frac{n_1\bar{x}_1 + n_2\bar{x}_2 + \cdots + n_k\bar{x}_k}{n_1 + n_2 + \cdots + n_k} = \frac{\sum_i n_i\bar{x}_i}{\sum_i n_i}$$

Proof. Let $x_{11}, x_{12}, x_{13}, \ldots, x_{1n_1}$ be the variables of the first distribution, $x_{21}, x_{22}, \ldots,$ x_{2n_2} be the variables of the second distribution, and so on. Then by definition

$$\left.\begin{aligned} \bar{x}_1 &= \frac{1}{n_1}(x_{11} + x_{12} + \cdots\cdots + x_{1n_1}) \\ \bar{x}_2 &= \frac{1}{n_2}(x_{21} + x_{22} + \cdots\cdots + x_{2n_2}) \\ &\cdots\cdots\cdots\cdots\cdots\cdots\cdots\cdots \\ \bar{x}_k &= \frac{1}{n_k}(x_{k_1} + x_{k_2} + \cdots\cdots + x_{kn_k}) \end{aligned}\right\} \quad \ldots(A)$$

The mean $\bar{x}$ of the whole distribution of size $(n_1 + n_2 + \ldots\ldots + n_k)$ is given by

$$\bar{x} = \frac{(x_{11} + x_{12} + \cdots\cdots + x_{1n_1}) + (x_{21} + x_{22} + \cdots\cdots + x_{2n_2}) + \cdots\cdots + (x_{k_1} + x_{k_2} + \cdots\cdots + x_{kn_k})}{n_1 + n_2 + \cdots\cdots + n_k}$$

$$= \frac{n_1\bar{x}_1 + n_2\bar{x}_2 + \cdots\cdots + n_k\bar{x}_k}{n_1 + n_2 + \cdots\cdots + n_k} = \frac{\sum_i n_i\bar{x}_i}{\sum_i n_i}.$$

Example 3. *The mean annual salary paid to all employees of a company was ₹ 5000. The mean annual salaries paid to male and female employees were ₹ 5200 and ₹ 4200 respectively. Determine the percentage of males and females employed by the company.*

Sol. Let p_1 and p_2 represent percentage of males and females respectively.

Then $p_1 + p_2 = 100$ ...(1)

Mean annual salary of all employees ($\overline{x}$) = ₹ 5000

Mean annual salary of all males ($\overline{x}_1$) = ₹ 5200

Mean annual salary of all females ($\overline{x}_2$) = ₹ 4200

Using $\overline{x} = \dfrac{p_1\overline{x}_1 + p_2\overline{x}_2}{p_1 + p_2}$, we get $5000 = \dfrac{5200p_1 + 4200p_2}{100}$

or $52p_1 + 42p_2 = 5000$ or $26p_1 + 21(100 - p_1) = 2500$ [Using (1)]

or $5p_1 = 2500 - 2100 = 400$ $\therefore$ $p_1 = 80$ and $p_2 = 100 - 80 = 20$

Hence the percentage of males and females is 80 and 20 respectively.

3.4.2 Median

1. Median is the central value of the variable **when the values are arranged in ascending or descending order of magnitude.** When the observations are arranged in the order of their size, the median is the value of that item which has equal number of observations on either side. Median divides the distribution into two equal parts. Median is, thus, a potential average.

For the computation of median, it is necessary that the items be arranged in ascending or descending order.

2. For an ungrouped frequency distribution, if the n values of the variate are arranged in ascending or descending order of magnitude.

(*a*) *When n is odd,* the middle value *i.e.*, $\left(\dfrac{n+1}{2}\right)^{\text{th}}$ value gives the median.

(*b*) *When n is even,* there are two middle values $\left(\dfrac{n}{2}\right)^{\text{th}}$ and $\left(\dfrac{n}{2}+1\right)^{\text{th}}$.

The arithmetic mean of these two values gives the median.

3. For discrete frequency distribution, median is obtained by considering cumulative frequencies. Find $\dfrac{N+1}{2}$ where $N = \Sigma f_i$. Find cumulative frequency just $\geq \dfrac{N+1}{2}$. The corresponding value of x is the median.

4. For a grouped frequency distribution, the median is given by the formula,

$$\textbf{Median} = l + \frac{h}{f}\left(\frac{N}{2} - C\right)$$

where, l = lower limit of median class, where median class is the class corresponding to cumulative frequency just $\geq \dfrac{N}{2}$

h = width of the median class ; f = frequency of the median class

$N = \Sigma f$; C = cumulative frequency of the class **preceding** the median class.

5. Partition values. These are the values of the variate which divide the total frequency into a number of equal parts. Median being that value of the variate which divides the total frequency into two equal parts.

(*a*) **Quartiles.** Quartiles are those values of the variate which divide the total frequency into four equal parts. When the lower half before the median is divided into two equal parts, the value of the dividing variate is called Lower Quartile and is denoted by Q_1. The value of the variate dividing the upper half into two equal parts is called the Upper Quartile and is denoted by Q_3. (Q_2 being the median.) The formulae for computation are

$$Q_1 = l + \frac{h}{f}\left(\frac{N}{4} - C\right); \quad Q_3 = l + \frac{h}{f}\left(\frac{3N}{4} - C\right)$$

(*b*) **Deciles.** Deciles are those values of the variate which divide the total frequency into 10 equal parts. D_1, D_2, ... denote respectively the first, second, ... deciles.

$$D_1 = l + \frac{h}{f}\left(\frac{N}{10} - C\right), \quad D_4 = l + \frac{h}{f}\left(\frac{4N}{10} - C\right), \quad D_7 = l + \frac{h}{f}\left(\frac{7N}{10} - C\right),$$

(The fifth decile D_5 is the median)

(*c*) **Percentiles.** Percentiles are those values of the variate which divide the total frequency into 100 equal parts. If P_1, P_2, ... denote respectively the first, second, ... percentiles, then

$$P_9 = l + \frac{h}{f}\left(\frac{9N}{100} - C\right), \quad P_{72} = l + \frac{h}{f}\left(\frac{72N}{100} - C\right) \text{ etc.}$$

(The 50th percentile P_{50} is the median.)

In the above formulae for Quartiles, Deciles and Percentiles, the letters l, i, f, N, C have been used in the same sense in which they have been used in the formula for the median.

Example 4. *Below are given the marks obtained by a batch of 20 students in a certain class in Mathematics and Physics :*

Roll Nos.	:	*1*	*2*	*3*	*4*	*5*	*6*	*7*	*8*	*9*	*10*
Marks in Maths.	:	*53*	*54*	*52*	*32*	*30*	*60*	*47*	*46*	*35*	*28*
Marks in Physics	:	*58*	*55*	*25*	*32*	*26*	*85*	*44*	*80*	*33*	*72*
Roll Nos.	:	*11*	*12*	*13*	*14*	*15*	*16*	*17*	*18*	*19*	*20*
Marks in Maths.	:	*25*	*42*	*33*	*48*	*72*	*51*	*45*	*33*	*65*	*29*
Marks in Physics	:	*10*	*42*	*15*	*46*	*50*	*64*	*39*	*38*	*30*	*36*

In which subject is the level of knowledge of the students higher?

Sol. To find out the subject in which the level of knowledge of the students is higher, we find out the medians of both the series. The subject for which the median value is higher will be the subject in which the level of knowledge of the students is higher. Let us arrange the marks in ascending order of magnitude.

S. No.	*Marks in Maths*	*Marks in Physics*	*S. No.*	*Marks in Maths*	*Marks in Physics*
1	25	10	11	46	42
2	28	15	12	47	44
3	29	25	13	48	46
4	30	26	14	51	50
5	32	30	15	52	55
6	33	32	16	53	58
7	33	33	17	54	64
8	35	36	18	60	72
9	42	38	19	65	80
10	45	39	20	72	85

Number of items in each case = 20 (even)

Median marks in Mathematics

$$= \text{A.M. of sizes of } \left(\frac{20}{2}\right) \text{th and } \left(\frac{20}{2}+1\right) \text{th items}$$

$$= \text{A.M. of sizes of 10th and 11th items} = \frac{45+46}{2} = 45.5.$$

Median marks in Physics $= \text{A.M. of sizes of 10th and 11th items} = \dfrac{39+42}{2} = 40.5.$

Since the median marks in Mathematics are greater than the median marks in Physics, the level of knowledge in Mathematics is higher.

Example 5. *Obtain the median for the following frequency distribution :*

x :	1	2	3	4	5	6	7	8	9
f :	8	10	11	16	20	25	15	9	6

Sol. The cumulative frequency distribution table is given below :

x	f	$c.f.$
1	8	8
2	10	18
3	11	29
4	16	45
5	20	65
6	25	90
7	15	105
8	9	114
9	6	120

Here $N = 120 \quad \therefore \quad \frac{N+1}{2} = 60.5$

Cumulative frequency just greater than $\frac{N+1}{2}$ is 65 and the value of x corresponding to *c.f.* 65 is 5. Hence median is 5.

Example 6. *Find the median, lower and upper quartiles from the following table :*

Marks	*No. of students*	*Marks*	*No. of students*
Below 10	*15*	*Below 50*	*94*
Below 20	*35*	*Below 60*	*127*
Below 30	*60*	*Below 70*	*198*
Below 40	*84*	*Below 80*	*249*

Sol. From the above table, we reconstruct the *c.f.* table with class intervals.

Marks	*No. of students (f)*	*c.f.*
0—10	15	15
10—20	20	35
20—30	25	60
30—40	24	84
40—50	10	94
50—60	33	127
60—70	71	198
70—80	51	249

Here $N = 249$

(*i*) Calculation of Median

$$\frac{N}{2} = 124.5 \quad \therefore \quad \text{median class is 50—60, } l = 50\text{ ; } \quad h = 10, f = 33, C = 94$$

$$\therefore \quad \text{Median} = l + \frac{h}{f}\left(\frac{N}{2} - C\right) = 50 + \frac{10}{33}(124.5 - 94) = 59.24 \text{ marks}$$

(*ii*) Calculation of lower quartile Q_1

$$\frac{N}{4} = 62.25 \quad \therefore \quad \text{lower quartile class is 30—40, } l = 30$$

$$h = 10, f = 24, C = 60$$

$$\therefore \quad Q_1 = l + \frac{h}{f}\left(\frac{N}{4} - C\right) = 30 + \frac{10}{24}(62.25 - 60) = 30.94 \text{ marks}$$

(*iii*) Calculation of upper quartile Q_3

$$\frac{3N}{4} = \frac{747}{4} = 186.75 \quad \therefore \quad \text{upper quartile class is } 60\text{—}70$$

$$l = 60,\ h = 10,\ f = 71,\ C = 127$$

$$\therefore \qquad Q_3 = l + \frac{h}{f}\left(\frac{3N}{4} - C\right) = 60 + \frac{10}{71}(186.75 - 127) = 68.41 \text{ marks}$$

3.4.3 Mode

1. Mode. Mode is the value which occurs most frequently in a set of observations and around which the other items of the set cluster densely. It is the point of maximum frequency or the point of greatest density. In other words, the mode or modal value of the distribution is that value of the variate for which frequency is maximum.

2. Calculation of Mode.

(*a*) In the case of discrete frequency distribution, mode is the value of x corresponding to maximum frequency.

But in any one (or more) of the following cases:

(*i*) if the maximum frequency is repeated

(*ii*) if the maximum frequency occurs in the very beginning or at the end of the distribution

(*iii*) if there are irregularities in the distribution, the value of mode is determined by the *method of grouping* (illustrated in examples below).

(*b*) In the case of continuous frequency distribution, mode is given by the formula:

$$\text{Mode} = l + \frac{f_m - f_1}{2f_m - f_1 - f_2} \times h$$

where l is the lower limit, h the width and f_m the frequency of the model class, f_1 and f_2 are the frequencies of the classes preceding and succeeding the modal class respectively.

While applying the above formula, it is necessary to see that the class-intervals are of the same size. **If they are unequal, they should first be made equal** on the assumption that the frequencies are equally distributed throughout the class.

In case $f_m - f_1 < 0$ or $2f_m - f_1 - f_2 = 0$, use the formula

$$\text{Mode} = l + \frac{\Delta_1}{\Delta_1 + \Delta_2} \times h$$

where $\qquad \Delta_1 = |\, f_m - f_1 \,|$ and $\Delta_2 = |\, f_m - f_2 \,|$.

(*c*) **For a symmetrical distribution, mean, median and mode coincide.**

(*d*) Where mode is **ill-defined** *i.e.,* where the method of grouping also, fails, its value can be ascertained by the formula

Mode = 3 Median – 2 Mean

This measure is called the **empirical mode.**

Example 7. *Calculate the mode from the following frequency distribution :*

Size (x) :	4	5	6	7	8	9	10	11	12	13
Frequency (f) :	2	5	8	9	12	14	14	15	11	13

Sol. Method of Grouping :

Size (x)	*Frequency*					
	I	*II*	*III*	*IV*	*V*	*VI*
4	2					
		7				
5	5			15		
			13			
6	8				22	
		17				
7	9					29
			21			
8	12			35		
		26				
9	14				**40**	
			28			
10	14					**43**
		29				
11	15			**40**		
			26			
12	11				39	
		24				
13	13					

Explanation :

In column I, *original* frequencies are written.

In column II, frequencies of column I are combined *two by two.*

In column III, *leave the first frequency of column I* and combine the others *two by two.*

In column IV, frequencies of column I are combined *three by three.*

In column V, *leave the first frequency of column I* and combine the others *three by three.*

In column VI, *leave the first two frequencies in column I* and combine the others *three by three.*

In all these columns, the maximum frequency is written in bold black type.

Note. All operations are done on column I.

Now we frame another table in which against every maximum item of columns I to VI, we write down the corresponding size or sizes. The size (x) which occurs maximum number of times is the mode.

Columns	*Size of item having max. frequency*
I	11
II	10, 11
III	9, 10
IV	10, 11, 12
V	8, 9, 10
VI	9, 10, 11

Since the item 10 occurs maximum number of times (*i.e.*, 5 times), hence the mode is 10.

Example 8. *Find the mode of the following :*

Marks	:	*1—5*	*6—10*	*11—15*	*16—20*	*21—25*
No. of candidates	:	*7*	*10*	*16*	*32*	*24*
Marks	:	*26—30*	*31—35*	*36—40*	*41—45*	
No. of candidates	:	*18*	*10*	*5*	*1*	

Sol. Here the greatest frequency 32 lies in the class 16—20. Hence modal class is 16—20. But the actual limits of this class are 15.5—20.5.

$$l = 15.5, f_m = 32, f_1 = 16, f_2 = 24, h = 5$$

$$\therefore \quad \text{Mode} = l + \frac{f_m - f_1}{2f_m - f_1 - f_2} \times h = 15.5 + \frac{32 - 16}{64 - 16 - 24} \times 5$$

$$= 15.5 + \frac{16}{24} \times 5 = 15.5 + \frac{10}{3} = 18.83 \text{ marks.}$$

3.4.4 Geometric Mean

Geometric Mean. (*a*) Geometric Mean (G.M.) of n individual observations $x_1, x_2, \ldots\ldots, x_n$ $(x_i \neq 0)$ is the nth root of their product.

Thus $$G = (x_1, x_2, \ldots, x_n)^{1/n}$$

Taking logarithms of both sides $\log G = \frac{1}{n}(\log x_1 + \log x_2 + \ldots + \log x_n) = \frac{1}{n}\sum_{i=1}^{n} \log x_i$

$$\therefore \quad G = \text{antilog}\left[\frac{1}{n}\sum_{i=1}^{n} \log x_i\right]$$

(*b*) If $x_1, x_2, \ldots\ldots, x_n$ occur $f_1, f_2, \ldots\ldots, f_n$ times respectively and $N = \sum_{i=1}^{n} f_i$, then G.M. is given by

$$G = (x_1^{f_1} x_2^{f_2} \ldots\ldots x_n^{f_n})^{1/N}$$

Taking logarithms of both sides

$$\log G = \frac{1}{N}(f_1 \log x_1 + f_2 \log x_2 + \ldots + f_n \log x_n) = \frac{1}{N}\sum_{i=1}^{n} f_i \log x_i$$

$$G = \text{antilog}\left[\frac{1}{N}\sum_{i=1}^{n} f_i \log x_i\right]$$

(*c*) In the case of continuous frequency distribution, x is taken to be the value corresponding to the mid-points of the class-intervals.

Example 9. *Compute the geometric mean from the following data :*

Marks	*No. of students*
0—10	*10*
10—20	*5*
20—30	*8*
30—40	*7*
40—50	*20*

Sol. Table for the calculation of mean:

Marks	*Mid-values* (x)	*No. of Students* (f)	*log x*	*f log x*
0—10	5	10	0.6990	6.9900
10—20	15	5	1.1761	5.8805
20—30	25	8	1.3979	11.1832
30—40	35	7	1.5441	10.8087
40—50	45	20	1.6532	33.0640
		50		67.9264

$$\log G = \frac{1}{N} \Sigma f \log x = \frac{67.9264}{50} = 1.3585$$

$$G = \text{antilog } 1.3585 = 22.83.$$

3.4.5 Harmonic Mean

Harmonic Mean. Harmonic mean of a number of observations is the reciprocal of the arithmetic mean of the reciprocals of the given values. Thus, the harmonic mean H of n observations $x_1, x_2, ..., x_n$ is

$$H = \frac{1}{\frac{1}{n} \sum_{i=1}^{n} \frac{1}{x_i}} = \frac{n}{\frac{1}{x_1} + \frac{1}{x_2} + ... + \frac{1}{x_n}}.$$

If $x_1, x_2, ..., x_n$ (none of them being zero) have the frequencies $f_1, f_2, ..., f_n$ respectively, then harmonic mean is given by

$$H = \frac{1}{\frac{1}{N} \sum_{i=1}^{n} \frac{f_i}{x_i}} = \frac{N}{\frac{f_1}{x_1} + \frac{f_2}{x_2} + ... + \frac{f_n}{x_n}}, \; N = \sum_{i=1}^{n} f_i$$

In the case of class-intervals, x is taken to be the mid-value of the class-interval.

Example 10. *Find out the harmonic mean of the following data :*

Marks (out of 150)	*No. of students*
10	*2*
20	*3*
40	*6*
60	*5*
120	*4*

Sol.

x	f	$\frac{1}{x}$	$\frac{f}{x}$
10	2	0.100	0.200
20	3	0.050	0.150
40	6	0.025	0.150
60	5	0.017	0.085
120	4	0.008	0.032
	20		0.617

$$\text{H.M.} = \frac{N}{\Sigma \frac{f}{x}} = \frac{20}{0.617} = 32.4.$$

Example 11. *An aeroplane flies along the four sides of a square at speeds of 100, 200, 300 and 400 km/hr respectively. What is the average speed of the aeroplane in its flight around the square ?*

Sol. When *equal distances* are covered with unequal speeds, the harmonic mean is the proper average.

$$\therefore \quad \text{Average speed} = \frac{4}{\frac{1}{100}+\frac{1}{200}+\frac{1}{300}+\frac{1}{400}} = 192 \text{ km/hr.}$$

TEST YOUR KNOWLEDGE

1. The minimum temperature in (°C) for Delhi for the month of July, 1983 as reported by the Meterological Department is given below. Construct a frequency distribution table for it.

   30.3, 30.0, 25.8, 26.5, 24.2, 25.2, 28.0, 28.0, 29.5, 27.8, 30.0, 31.1, 27.2, 25.9, 27.6, 24.5, 24.4, 27.0, 28.1, 26.0, 25.4, 28.0, 26.9, 25.7, 27.2, 25.5, 26.6, 28.5, 28.0, 27.7, 24.0.

2. The following are the monthly rents (in rupees) of 40 shops. Tabulate the data by grouping in intervals of ₹ 8.

   38, 42, 49, 37, 82, 37, 75, 62, 54, 79, 84, 75, 63, 44, 74, 44, 36, 69, 54, 48, 74, 47, 52, 57, 62, 67, 72, 77, 82, 51, 31, 38, 43, 75, 67, 77, 47, 64, 84, 81.

3. Draw a histogram representing the following frequency distribution:

Monthly Wages (in ₹)	*Number of Workers*
15	2
20	20
25	26
30	16
35	9
40	4
45	3

[**Hint.** Mid-values of class intervals of size 5 are given.]

4. Represent the following distribution by a (*i*) histogram and (*ii*) frequency polygon.

Scores	*Frequency*
90—99	2
80—89	12
70—79	22
60—69	20
50—59	14
40—49	3
30—39	1

5. Represent the following distribution by an ogive :

Marks	*No. of students*	*Marks*	*No. of students*
0—10	5	50—60	4
10—20	13	60—70	1
20—30	12	70—80	3
30—40	11	80—90	1
40—50	8	90—100	2

6. Compute the arithmetic mean for the following data :

Height (in cm) :	219	216	213	210	207	204	201	198	195
No. of persons :	2	4	6	10	11	7	5	4	1

7. Find the average marks of students from the following data :

Marks	*No. of students*	*Marks*	*No. of students*
Above 0	80	Above 60	28
Above 10	77	Above 70	16
Above 20	72	Above 80	10
Above 30	65	Above 90	8
Above 40	55	Above 100	0
Above 50	43		

8. Two hundred persons were interviewed by a public opinion polling agency. The frequency distribution gives the ages of persons interviewed.

Age Group (years)	*Frequency*	*Age Group (years)*	*Frequency*
80—89	2	40—49	56
70—79	2	30—39	40
60—69	6	20—29	42
50—59	20	10—19	32

Calculate the arithmetic mean of the data.

9. Calculate the arithmetic mean from the following data :

Class interval	*Frequency*	*Class interval*	*Frequency*
0—1	8	15—25	11
1—3	8	25—28	10
3—5	10	28—30	9
5—10	12	30—45	8
10—15	18	45—60	6

10. Find the class intervals if the arithmetic mean of the following distribution is 33 and assumed mean is 35.

Step deviation (u) :	– 3	– 2	– 1	0	1	2
Frequency (f) :	5	10	25	30	20	10

11. The average height of a group of 25 children was calculated to be 78.4 cm. It was later discovered that one value was misread as 69 cm instead of the correct value 96 cm. Calculate the correct average.

12. A candidate obtains the following percentage in an examination : English 60, Hindi 75, Mathematics 63, Physics 59 and Chemistry 55. Find the weighted mean if weights 2, 1, 5, 5, 3 are allotted to the subjects.

13. From the following data calculate the missing frequency :

No. of tablets	*No. of persons cured*	*No. of tablets*	*No. of persons cured*
4—8	11	24—28	9
8—12	13	28—32	17
12—16	16	32—36	6
16—20	14	36—40	4
20—24	?		

The average number of tablets to cure a person is 20.

14. The frequencies of values 0, 1, 2, ..., n of a variable are given by

$q^n, {}^nC_1q^{n-1}p, {}^nC_2q^{n-2}p^2, ..., p^n$ where $p + q = 1$. Show that the mean is np.

15. The mean marks obtained by 300 students in the subject of Statistics are 45. The mean of the top 100 of them was found to be 70 and the mean of the last 100 was known to be 20. What is the mean of the remaining 100 students ?

16. In a certain examination, the average grade of all students in class A is 68.4 and that of all students in class B is 71.2. If the average of both classes combined is 70, find the ratio of the number of students in class A to the number in class B.

17. The following are the monthly salaries in rupees of 30 employees of a firm :

91	139	126	119	100	87	65	77	99	95	108	127	86	148	116
76	69	88	112	118	89	116	97	105	95	80	86	106	93	135

The firm gave bonuses of 10, 15, 20, 25, 30, 35, 40, 45 and 50 to employees in the respective salary groups : exceeding 60 but not exceeding 70, exceeding 70 but not exceeding 80, and so on upto exceeding 140 but not exceeding 150. Find the average bonus paid per employee.

18. According to the census of 1981, following are the population figures in thousand, of 10 cities :

2000, 1180, 1785, 1500, 560, 782, 1200, 385, 1123, 222.

Find the median.

19. Find the median from the following table:

x :	5	7	9	11	13	15	17	19
f :	1	2	7	9	11	8	5	4

20. Calculate the mean and median from the following table:

Class interval	*Frequency*
6.5—7.5	5
7.5—8.5	12
8.5—9.5	25
9.5—10.5	48
10.5—11.5	32
11.5—12.5	6
12.5—13.5	1

21. Compute the median from the following data:

Mid-value	*Frequency*	*Mid-value*	*Frequency*
115	6	165	60
125	25	175	38
135	48	185	22
145	72	195	3
155	116		

22. Find the median, quartiles, 7th decile and 85th percentile from the following data:

Monthly Rent (₹)	*No. of families*	*Monthly Rent* (₹)	*No. of families*
20—40	6	120—140	15
40—60	9	140—160	10
60—80	11	160—180	8
80—100	14	180—200	7
100—120	20		

23. An incomplete frequency distribution is given as follows:

Variable	*Frequency*	*Variable*	*Frequency*
10—20	12	50—60	?
20—30	30	60—70	25
30—40	?	70—80	18
40—50	65	Total	229

Given that the median value is 46, determine the missing frequencies using the median formula.

24. Find the median, lower and upper quartiles, 4th decile and 60th percentile for the following distribution:

Marks	*No.of students*	*Marks*	*No. of students*
0—4	10	14—18	5
4—8	12	18—20	8
8—12	18	20—25	4
12—14	7	25 and above	6

[**Hint.** *Here the class-intervals are not all equal. To find any partition value, there is no need to make them equal.*]

25. Find the mode of the following frequency distribution:

Size :	1	2	3	4	5	6	7	8	9	10	11	12
Frequency :	3	8	15	23	35	40	32	28	20	45	14	6

26. Find the mode and median from the following table:

Marks	*No. of students*	*Marks*	*No. of students*
0—10	2	40—50	35
10—20	18	50—60	20
20—30	30	60—70	6
30—40	45	70—80	3

27. Calculate the mode of the following distribution:

Monthly wages (in ₹)	*No. of workers*	*Monthly wages (in ₹)*	*No. of workers*
50—70	4	150—170	8
70—90	44	170—190	12
90—110	38	190—210	2
110—130	28	210—230	2
130—150	6		

[**Hint.** Use the method of grouping for finding the modal class.]

28. An incomplete distribution of families according to their expenditure per week is given below. The median and mode for the distribution are ₹ 25 and ₹ 24 respectively. Calculate the missing frequencies.

Expenditure :	0—10	10—20	20—30	30—40	40—50
No. of families :	14	?	27	?	15

29. Compute the geometric mean of the following data :

x :	10	15	18	20	25
f :	2	3	5	6	4

30. If n_1 and n_2 are the sizes, G_1 and G_2 the geometric means of two series respectively, then the geometric mean G of the combined series is given by $\log G = \dfrac{n_1 \log G_1 + n_2 \log G_2}{n_1 + n_2}$.

31. The marks obtained by 25 students in a test are given below:

Marks :	11	12	13	14	15
No. of students :	3	7	8	5	2

Find the harmonic mean.

32. Compute the harmonic mean of the following data:

Class	*Frequency*
0—10	4
10—20	6
20—30	10
30—40	7
40—50	3

33. Three cities A, B, C are equidistant from each other. A motorist travels from A to B at 30 km/hr, from B to C at 40 km/hr and from C to A at 50 km/hr. Determine the average speed.

34. Show that in finding the arithmetic mean of a set of readings on a thermometer, it does not matter whether we measure temperature in Centigrade or Fahrenheit, but that in finding the geometric mean, it does matter which scale we use.

Answers

6. 207.54 cm **7.** 51.75 marks **8.** 35.8 years **9.** 17.36
10. 0—10, 10—20, 20—30, 30—40, 40—50, 50—60 **11.** 79.48 cm **12.** 60.63% marks
13. 14 **15.** 45 **16.** 3 : 4 **17.** ₹ 27.5
18. 1151.5 thousands **19.** 13 **20.** Mean = 9.87, Median = 9.97
21. 153.8 **22.** (₹) 110, 78.18, 140, 133.33, 160 **23.** 34, 45
24. 10.89, 6.5, 18.125, 9.33, 12.57 **25.** 6 **26.** 36, 36.6
27. ₹ 97.50 **28.** 25, 24 **29.** 18.20 **31.** 12.7 marks
32. 16.03 **33.** 38.3 km/hr.

3.5 MOMENTS

Moments are statistical tools, used in statistical investigations. The moments of a distribution are the arithmetic means of the various powers of the deviations of items from some given number.

3.6 MOMENTS ABOUT MEAN (Central Moments)

3.6.1 For an Individual Series

If $x_1, x_2, ..., x_n$ are the values of the variable under consideration, the rth moment μ_r about mean $\bar{x}$ is defined as

$$\mu_r = \frac{\sum_{i=1}^{n} (x_i - \bar{x})^r}{n} \quad ; r = 0, 1, 2, ...$$

3.6.2 For a Frequency Distribution

If $x_1, x_2, ..., x_n$ are the values of a variable x with the corresponding frequencies $f_1, f_2, ..., f_n$ respectively then r^{th} moment μ_r about the mean $\bar{x}$ is defined as

$$\mu_r = \frac{\sum_{i=1}^{n} f_i(x_i - \bar{x})^r}{N} \; ; r = 0, 1, 2, ... \quad \text{where} \quad N = \sum_{i=1}^{n} f_i$$

In particular, $\mu_0 = \frac{1}{N}\sum_{i=1}^{n} f_i(x_i - \bar{x})^0 = \frac{1}{N}\sum_{i=1}^{n} f_i = \frac{N}{N} = 1$

∴ For any distribution, $\boxed{\mu_0 = 1}$

For $r = 1$,

$$\mu_1 = \frac{1}{N}\sum_{i=1}^{n} f_i(x_i - \bar{x}) = \frac{1}{N}\sum_{i=1}^{n} f_i x_i - \bar{x}\left(\frac{1}{N}\sum_{i=1}^{n} f_i\right) = \bar{x} - \bar{x} = 0$$

∴ For any distribution, $\boxed{\mu_1 = 0}$

For $r = 2$,

$$\mu_2 = \frac{1}{N}\sum_{i=1}^{n} f_i(x_i - \bar{x})^2 = (\text{S.D.})^2 = \text{Variance}$$

∴ For any distribution, μ_2 coincides with the variance of the distribution.

Similarly, $\mu_3 = \frac{1}{N}\sum_{i=1}^{n} f_i(x_i - \bar{x})^3, \; \mu_4 = \frac{1}{N}\sum_{i=1}^{n} f_i(x_i - \bar{x})^4$

and so on.

Note. In case of a frequency distribution with class intervals, the values of x are the midpoints of the intervals.

ILLUSTRATIVE EXAMPLES

Example 1. *Find the first four moments for the following individual series:*

x	*3*	*6*	*8*	*10*	*18*

Sol. **Calculation of Moments**

S. No.	*x*	$x - \bar{x}$	$(x - \bar{x})^2$	$(x - \bar{x})^3$	$(x - \bar{x})^4$
1	3	– 6	36	– 216	1296
2	6	– 3	9	– 27	81
3	8	– 1	1	– 1	1
4	10	1	1	1	1
5	18	9	81	729	6561
$n = 5$	$\Sigma x = 45$	$\Sigma(x - \bar{x}) = 0$	$\Sigma(x - \bar{x})^2 = 128$	$\Sigma(x - \bar{x})^3 = 486$	$\Sigma(x - \bar{x})^4 = 7940$

Now, $\bar{x} = \frac{\Sigma x}{n} = \frac{45}{5} = 9$

$$\therefore \quad \mu_1 = \frac{\Sigma(x - \bar{x})}{n} = \frac{0}{5} = 0,$$

$$\mu_2 = \frac{\Sigma(x - \bar{x})^2}{n} = \frac{128}{5} = 25.6,$$

$$\mu_3 = \frac{\Sigma(x - \bar{x})^3}{n} = \frac{486}{5} = 97.2,$$

$$\mu_4 = \frac{\Sigma(x - \bar{x})^4}{n} = \frac{7940}{5} = 1588.$$

Example 2. *Calculate μ_1, μ_2, μ_3, μ_4 for the following frequency distribution:*

Marks	*5–15*	*15–25*	*25–35*	*35–45*	*45–55*	*55–65*
No. of students	*10*	*20*	*25*	*20*	*15*	*10*

Sol. **Calculation of Moments**

Marks	*No. of students (f)*	*Mid-point (x)*	*fx*	$x - \bar{x}$	$f(x - \bar{x})$	$f(x - \bar{x})^2$	$f(x - \bar{x})^3$	$f(x - \bar{x})^4$
5–15	10	10	100	– 24	– 240	5760	– 138240	3317760
15–25	20	20	400	– 14	– 280	3920	– 54880	768320
25–35	25	30	750	– 4	– 100	400	– 1600	6400
35–45	20	40	800	6	120	720	4320	25920
45–55	15	50	750	16	240	3840	61440	983040
55–65	10	60	600	26	260	6760	175760	4569760
	N = 100		Σfx = 3400		$\Sigma f(x - \bar{x})$ = 0	$\Sigma f(x - \bar{x})^2$ = 21400	$\Sigma f(x - \bar{x})^3$ = 46800	$\Sigma f(x - \bar{x})^4$ = 9671200

Now, $\bar{x} = \frac{\Sigma fx}{N} = \frac{3400}{100} = 34$

$$\therefore \quad \mu_1 = \frac{\Sigma f(x - \bar{x})}{N} = \frac{0}{100} = 0, \qquad \mu_2 = \frac{\Sigma f(x - \bar{x})^2}{N} = \frac{21400}{100} = 214$$

$$\mu_3 = \frac{\Sigma f(x - \bar{x})^3}{N} = \frac{46800}{100} = 468, \qquad \mu_4 = \frac{\Sigma f(x - \bar{x})^4}{N} = \frac{9671200}{100} = 96712.$$

3.7 SHEPPARD'S CORRECTIONS FOR MOMENTS

While computing moments for frequency distribution with class intervals, we take variable x as the midpoint of class-intervals which means that we have assumed the frequencies concentrated at the midpoints of class-intervals.

The above assumption is true when the distribution is symmetrical and the no. of class-intervals is not greater than $\frac{1}{20}^{\text{th}}$ of the range, otherwise the computation of moments will have certain error called **grouping error.**

This error is corrected by the following formulae given by **W.F. Sheppard**.

$$\mu_2 \text{ (corrected)} = \mu_2 - \frac{h^2}{12}$$

$$\mu_4 \text{ (corrected)} = \mu_4 - \frac{1}{2}h^2\mu_2 + \frac{7}{240}h^4$$

where h is the width of the class-interval while μ_1 and μ_3 require no correction.

These formulae are known as **Sheppard's corrections.**

Example 3. *Find the corrected values of the following moments using Sheppard's correction. The width of classes in the distribution is 10:*

$\mu_2 = 214,\quad \mu_3 = 468,\quad \mu_4 = 96712.$

Sol. We have $\mu_2 = 214,\quad \mu_3 = 468,\quad \mu_4 = 96712,\quad h = 10.$

Now, $$\mu_2 \text{ (corrected)} = \mu_2 - \frac{h^2}{12} = 214 - \frac{(10)^2}{12} = 214 - 8.333 = 205.667.$$

$$\mu_3 \text{ (corrected)} = \mu_3 = 468$$

$$\mu_4 \text{ (corrected)} = \mu_4 - \frac{1}{2}h^2\mu_2 + \frac{7}{240}h^4 = 96712 - \frac{(10)^2}{2}(214) + \frac{7}{240}(10)^4$$

$$= 96712 - 10700 - 291.667 = 86303.667.$$

3.8 MOMENTS ABOUT AN ARBITRARY NUMBER (Raw Moments)

If $x_1, x_2, x_3, ..., x_n$ are the values of a variable x with the corresponding frequencies $f_1, f_2, f_3, ..., f_n$ respectively then r^{th} moment μ_r' about the number $x = \text{A}$ is defined as

$$\mu_r' = \frac{1}{\text{N}}\sum_{i=1}^{n} f_i(x_i - \text{A})^r \,;\, r = 0, 1, 2,... \quad \text{where,} \quad \text{N} = \sum_{i=1}^{n} f_i$$

For $r = 0$, $$\mu_0' = \frac{1}{\text{N}}\sum_{i=1}^{n} f_i(x_i - \text{A})^0 = 1$$

For $r = 1$, $$\mu_1' = \frac{1}{\text{N}}\sum_{i=1}^{n} f_i(x_i - \text{A}) = \frac{1}{\text{N}}\sum_{i=1}^{n} f_i x_i - \frac{\text{A}}{\text{N}}\sum_{i=1}^{n} f_i = \bar{x} - \text{A}$$

For $r = 2$, $$\mu_2' = \frac{1}{\text{N}}\sum_{i=1}^{n} f_i(x_i - \text{A})^2$$

For $r = 3$, $$\mu_3' = \frac{1}{\text{N}}\sum_{i=1}^{n} f_i(x_i - \text{A})^3 \text{ and so on.}$$

In calculation work, if we find that there is some common factor h (> 1) in values of $x - \text{A}$, we can ease our calculation work by defining $u = \dfrac{x - \text{A}}{h}$. In that case, we have

$$\mu_r' = \frac{1}{\text{N}}\left(\sum_{i=1}^{n} f_i u_i^r\right) h^r \;;\; r = 0, 1, 2, \ldots$$

Note. For an individual series,

1. $\mu_r' = \dfrac{1}{n}\displaystyle\sum_{i=1}^{n} (x_i - \text{A})^r \;;\; r = 0, 1, 2, \ldots$

2. $\mu_r' = \dfrac{1}{\text{N}}\left(\displaystyle\sum_{i=1}^{n} u_i^r\right) h^r \;;\; r = 0, 1, 2, \ldots$ $\quad\Big|$ for $u = \dfrac{x - \text{A}}{h}$

3.9 MOMENTS ABOUT THE ORIGIN

If $x_1, x_2, \ldots, x_n$ be the values of a variable x with corresponding frequencies $f_1, f_2, \ldots, f_n$ respectively then r^{th} moment about the origin ν_r is defined as

$$\nu_r = \frac{1}{\text{N}}\sum_{i=1}^{n} f_i x_i^r \;;\; r = 0, 1, 2, \ldots \quad \text{where,} \quad \text{N} = \sum_{i=1}^{n} f_i$$

For $r = 0$, $\quad \nu_0 = \dfrac{1}{\text{N}}\displaystyle\sum_{i=1}^{n} f_i x_i^0 = \dfrac{\text{N}}{\text{N}} = 1$

For $r = 1$, $\quad \nu_1 = \dfrac{1}{\text{N}}\displaystyle\sum_{i=1}^{n} f_i x_i = \bar{x}$

For $r = 2$, $\quad \nu_2 = \dfrac{1}{\text{N}}\displaystyle\sum_{i=1}^{n} f_i x_i^2$ and so on.

3.10 RELATION BETWEEN μ_r AND μ_r'

We know that,

$$\mu_r = \frac{\sum_{i=1}^{n} f_i (x_i - \bar{x})^r}{\text{N}} = \frac{1}{\text{N}}\sum_{i=1}^{n} f_i\,[(x_i - \text{A}) - (\bar{x} - \text{A})]^r$$

$$= \frac{1}{\text{N}}\sum_{i=1}^{n} f_i\,[(x_i - \text{A}) - \mu'_1]^r \qquad \Big|\ \because\ \mu_1' = \bar{x} - \text{A}$$

$$= \frac{1}{\text{N}}\sum_{i=1}^{n} f_i\,[(x_i - \text{A})^r - {}^rC_1 (x_i - \text{A})^{r-1}\mu_1' + {}^rC_2 (x_i - \text{A})^{r-2}\mu_1'^2 - \ldots\ldots + (-1)^r \mu_1'^r]$$

| Using binomial theorem

$$\Rightarrow \qquad \mu_r = \mu_r' - {}^rC_1\,\mu_{r-1}'\mu_1' + {}^rC_2\,\mu_{r-2}'\mu_1'^2 - \ldots\ldots + (-1)^r \mu_1'^r$$

Putting $r = 2, 3, 4$, we get

$$\mu_2 = \mu_2' - 2\mu_1'^2 + \mu_1'^2 = \mu_2' - \mu_1'^2 \qquad | \because \mu_0' = 1$$
$$\mu_3 = \mu_3' - 3\mu_2'\mu_1' + 3\mu_1'^3 - \mu_1'^3 = \mu_3' - 3\mu_2'\mu_1' + 2\mu_1'^3$$
$$\mu_4 = \mu_4' - 4\mu_3'\mu_1' + 6\mu_2'\mu_1'^2 - 3\mu_1'^4$$

Hence, we have the following relations:

$$\boxed{\mu_1 = 0}$$

$$\boxed{\mu_2 = \mu_2' - \mu_1'^2}$$

$$\boxed{\mu_3 = \mu_3' - 3\mu_2'\mu_1' + 2\mu_1'^3}$$

and $$\boxed{\mu_4 = \mu_4' - 4\mu'_3\mu_1' + 6\mu_2'\mu_1'^2 - 3\mu_1'^4}$$

3.11 RELATION BETWEEN ν_r AND μ_r

We know that,

$$\nu_r = \frac{1}{N}\sum_{i=1}^{n} f_i x_i^r \; ; r = 0, 1, 2, \ldots$$

$$= \frac{1}{N}\sum_{i=1}^{n} f_i (x_i - A + A)^r$$

$$= \frac{1}{N}\sum_{i=1}^{n} f_i [(x_i - A)^r + {}^rC_1 (x_i - A)^{r-1} \,.\, A + \ldots + A^r]$$

$$= \mu_r' + {}^rC_1 \mu_{r-1} A + \ldots + A^r$$

If we take, $A = \bar{x}$ (for μ_r) then

$$\nu_r = \mu_r + {}^rC_1\mu_{r-1}\bar{x} + {}^rc_2\mu_{r-2}\bar{x}^2 + \ldots + \bar{x}^r \qquad \ldots(1)$$

Putting, $r = 1, 2, 3, 4$ in (1), we get

$$\nu_1 = \mu_1 + \mu_0\bar{x} = \bar{x} \qquad | \because \mu_1 = 0, \mu_0 = 1$$
$$\nu_2 = \mu_2 + {}^2c_1\mu_1\bar{x} + {}^2c_2\,\mu_0\bar{x}^2 = \mu_2 + \bar{x}^2$$
$$\nu_3 = \mu_3 + {}^3c_1\mu_2\bar{x} + {}^3c_2\mu_1\bar{x}^2 + {}^3c_3\mu_0\bar{x}^3 = \mu_3 + 3\mu_2\bar{x} + \bar{x}^3$$
$$\nu_4 = \mu_4 + {}^4c_1\mu_3\bar{x} + {}^4c_2\mu_2\bar{x}^2 + {}^4c_3\mu_1\bar{x}^3 + {}^4c_4\mu_0\bar{x}^4$$
$$= \mu_4 + 4\mu_3\bar{x} + 6\mu_2\bar{x}^2 + \bar{x}^4$$

Hence, we have the following relations:

$$\boxed{\nu_1 = \bar{x}} \qquad \boxed{\nu_2 = \mu_2 + \bar{x}^2}$$

$$\boxed{\nu_3 = \mu_3 + 3\mu_2\,\bar{x} + \bar{x}^3} \quad \text{and} \quad \boxed{\nu_4 = \mu_4 + 4\mu_3\bar{x} + 6\mu_2\bar{x}^2 + \bar{x}^4.}$$

3.12 KARL PEARSON'S β AND γ COEFFICIENTS

Karl Pearson defined the following four coefficients based upon the first four moments of a frequency distribution about its mean:

$$\left.\begin{aligned}\beta_1 &= \frac{\mu_3^2}{\mu_2^3}\\ \beta_2 &= \frac{\mu_4}{\mu_2^2}\end{aligned}\right\} \qquad (\beta\text{-coefficients})$$

$$\left.\begin{aligned}\gamma_1 &= +\sqrt{\beta_1}\\ \gamma_2 &= \beta_2 - 3\end{aligned}\right\} \qquad (\gamma\text{-coefficients})$$

The practical use of these coefficients is to measure the skewness and kurtosis of a frequency distribution. These coefficients are pure numbers independent of units of measurement.

Example 4. *The first three moments of a distribution, about the value '2' of the variable are 1, 16 and – 40. Show that the mean is 3, variance is 15 and $\mu_3 = -86$.*

Sol. We have $A = 2, \quad \mu_1' = 1, \quad \mu_2' = 16, \quad \text{and} \quad \mu_3' = -40$

We know that $\mu_1' = \bar{x} - A \Rightarrow \bar{x} = \mu_1' + A = 1 + 2 = 3$

$$\text{Variance} = \mu_2 = \mu_2' - \mu_1'^2 = 16 - (1)^2 = 15$$

$$\mu_3 = \mu_3' - 3\mu_2'\mu_1' + 2\mu_1'^3 = -40 - 3(16)(1) + 2(1)^3 = -40 - 48 + 2 = -86.$$

Example 5. *The first four moments of a distribution, about the value '35' are – 1.8, 240, – 1020 and 144000. Find the values of $\mu_1, \mu_2, \mu_3, \mu_4$.*

Sol. $\mu_1 = 0.$

$$\mu_2 = \mu_2' - \mu_1'^2 = 240 - (-1.8)^2 = 236.76$$

$$\mu_3 = \mu_3' - 3\mu_2'\mu_1' + 2\mu_1'^3 = -1020 - 3(240)(-1.8) + 2(-1.8)^3 = 264.36$$

$$\mu_4 = \mu_4' - 4\mu_3'\mu_1' + 6\mu_2'\mu_1'^2 - 3\mu_1'^4$$

$$= 144000 - 4(-1020)(-1.8) + 6(240)(-1.8)^2 - 3(-1.8)^4 = 141290.11.$$

Example 6. *Calculate the variance and third central moment from the following data:*

x_i	0	1	2	3	4	5	6	7	8
f_i	1	9	26	59	72	52	29	7	1

Sol. **Calculation of Moments**

x	f	$u = \frac{x - A}{h}$ $A = 4, h = 1$	fu	fu^2	fu^3
0	1	– 4	– 4	16	– 64
1	9	– 3	– 27	81	– 243
2	26	– 2	– 52	104	– 208
3	59	– 1	– 59	59	– 59
4	72	0	0	0	0

5	52	1	52	52	52
6	29	2	58	116	232
7	7	3	21	63	189
8	1	4	4	16	64
	$N = \Sigma f = 256$		$\Sigma fu = -7$	$\Sigma fu^2 = 507$	$\Sigma fu^3 = -37$

Now, moments about the point $x = A = 4$ are

$$\mu_1' = \left(\frac{\Sigma fu}{N}\right)h = \frac{-7}{256} = -0.02734$$

$$\mu_2' = \left(\frac{\Sigma fu^2}{N}\right)h^2 = \frac{507}{256} = 1.9805$$

$$\mu_3' = \left(\frac{\Sigma fu^3}{N}\right)h^3 = \frac{-37}{256} = -0.1445$$

Moments about mean

$$\mu_1 = 0$$

$$\mu_2 = \mu_2' - \mu_1'^2 = 1.9805 - (-.02734)^2 = 1.97975$$

$\therefore$ Variance = 1.97975

Also, $\mu_3 = \mu_3' - 3\mu_2'\mu_1' + 2\mu_1'^3$

$$= (-0.1445) - 3(1.9805)(-0.02734) + 2(-0.02734)^3$$

$$= 0.0178997$$

$\therefore$ Third central moment = 0.0178997.

Example 7. *The first three moments of a distribution about the value 2 of the variable are 1, 16 and – 40 respectively. Find the values of the first three moments about the origin.*

Sol. We have $A = 2, \quad \mu_1' = 1, \quad \mu_2' = 16, \; \mu_3' = -40$

$\therefore$ $\nu_1 = \bar{x} = A + \mu_1' = 2 + 1 = 3$

$$\nu_2 = \mu_2 + \bar{x}^2 = 15 + (3)^2 = 24$$

$$\nu_3 = \mu_3 + 3\mu_2\bar{x} + \bar{x}^3 = -86 + 3(15)(3) + (3)^3 = 76.$$

Example 8. *The first four moments of a distribution about x = 2 are 1, 2.5, 5.5 and 16. Calculate the first four moments about the mean and about origin.*

Sol. We have $A = 2, \quad \mu_1' = 1, \quad \mu_2' = 2.5, \quad \mu_3' = 5.5, \quad \mu_4' = 16.$

Moments about mean

$$\mu_1 = 0$$

$$\mu_2 = \mu_2' - (\mu_1')^2 = 2.5 - (1)^2 = 1.5$$

$$\mu_3 = \mu_3' - 3\mu_2'\mu_1' + 2(\mu_1')^3 = 5.5 - 3(2.5)(1) + 2(1)^3 = 0$$

$$\mu_4 = \mu_4' - 4\mu_3'\mu_1' + 6\mu_2'\mu_1'^2 - 3\mu_1'^4 = 16 - 4(5.5)(1) + 6(2.5)(1)^2 - 3(1)^4 = 6.$$

Moments about origin

$$\nu_1 = \bar{x} = A + \mu_1', \qquad \nu_2 = \mu_2 + \bar{x}^2$$

$$\nu_3 = \mu_3 + 3\mu_2\bar{x} + \bar{x}^3, \qquad \nu_4 = \mu_4 + 4\mu_3\bar{x} + 6\mu_2\bar{x}^2 + \bar{x}^4$$

$\therefore$ $\nu_1 = \bar{x} = 2 + 1 = 3, \qquad \nu_2 = 1.5 + (3)^2 = 10.5$

$$\nu_3 = 0 + 3(1.5)(3) + (3)^3 = 40.5, \qquad \nu_4 = 6 + 4(0)(3) + 6(1.5)(3)^2 + (3)^4 = 168.$$

Example 9. *For a distribution, the mean is 10, variance is 16, γ_1 is 1, and β_2 is 4. Find the first four moments about the origin.*

Sol. $\bar{x} = 10, \quad \mu_2 = 16, \quad \gamma_1 = 1, \quad \beta_2 = 4$ | given

Now, $\gamma_1 = 1$

$\Rightarrow \quad \beta_1 = 1$ $\quad | \because \gamma_1 = \sqrt{\beta_1}$

$$\Rightarrow \quad \frac{\mu_3^2}{\mu_2^3} = 1 \Rightarrow \mu_3^2 = \mu_2^3 = (16)^3 = (64)^2$$

$\Rightarrow \quad \mu_3 = 64$

and $\quad \beta_2 = 4$

$$\Rightarrow \quad \frac{\mu_4}{\mu_2^2} = 4 \quad \Rightarrow \mu_4 = 4(16)^2 = 1024 \qquad | \because \mu_2 = 16$$

Moments about the origin

$$\nu_1 = \bar{x} = 10$$
$$\nu_2 = \mu_2 + \bar{x}^2 = 16 + 100 = 116$$
$$\nu_3 = \mu_3 + 3\mu_2\bar{x} + \bar{x}^3 = 64 + 480 + 1000 = 1544$$
$$\nu_4 = \mu_4 + 4\mu_3\bar{x} + 6\mu_2\bar{x}^2 + \bar{x}^4 = 1024 + 4(64)(10) + 6(16)(100) + (10)^4$$
$$= 22184$$

Example 10. *In a certain distribution, the first four moments about the point $x = 4$ are $-1.5, 17, -30$ and 308. Find the moments about mean and about origin. Also, calculate β_1 and β_2.*

(U.P.T.U. 2014)

Sol. We have, $A = 4, \mu'_1 = -1.5, \mu'_2 = 17, \mu'_3 = -30, \mu'_4 = 308$

Moments about mean

$$\mu_1 = 0$$
$$\mu_2 = \mu'_2 - \mu_1'^2 = 17 - (-1.5)^2 = 14.75$$
$$\mu_3 = \mu'_3 - 3\mu'_2\mu'_1 + 2\mu_1'^3 = -30 - 3(17)(-1.5) + 2(-1.5)^3 = 39.75$$
$$\mu_4 = \mu'_4 - 4\mu'_3\mu'_1 + 6\mu'_2\mu_1'^2 - 3\mu_1'^4$$
$$= 308 - 4(-30)(-1.5) + 6(17)(-1.5)^2 - 3(-1.5)^4 = 342.3125$$

Moments about origin

$$\nu_1 = \bar{x} = \mu'_1 + A = -1.5 + 4 = 2.5$$

$$\nu_2 = \mu_2 + \bar{x}^2 = 14.75 + (2.5)^2 = 21$$

$$\nu_3 = \mu_3 + 3\mu_2\bar{x} + \bar{x}^3 = 166$$

$$\nu_4 = \mu_4 + 4\mu_3\bar{x} + 6\mu_2\bar{x}^2 + \bar{x}^4 = 1332$$

Calculation of β_1 and β_2

$$\beta_1 = \frac{\mu_3^2}{\mu_2^3} = 0.492377 \qquad \beta_2 = \frac{\mu_4}{\mu_2^2} = 1.573398$$

Example 11. *The first four moments of a distribution about the value '4' of the variable are -1.5, 17, -30 and 108. Find the moments about mean, about origin ; β_1 and β_2. Also, find the moments about the point $x = 2$.*

(A.K.T.U. 2017)

Sol. We have $A = 4, \mu'_1 = -1.5, \mu'_2 = 17, \mu'_3 = -30, \mu'_4 = 108$

Moments about mean

$$\mu_1 = 0$$
$$\mu_2 = \mu'_2 - \mu_1'^2 = 14.75$$
$$\mu_3 = \mu'_3 - 3\mu'_2\,\mu'_1 + 2\mu_1'^3 = 39.75$$
$$\mu_4 = \mu'_4 - 4\mu'_3\,\mu'_1 + 6\mu'_2\mu_1'^2 - 3\mu_1'^4 = 142.3125$$

Also, $\bar{x} = \mu'_1 + A = -1.5 + 4 = 2.5$

Moments about origin

$$v_1 = \bar{x} = 2.5$$
$$v_2 = \mu_2 + \bar{x}^2 = 14.75 + (2.5)^2 = 21$$
$$v_3 = \mu_3 + 3\mu_2\bar{x} + \bar{x}^3 = 166$$
$$v_4 = \mu_4 + 4\mu_3\bar{x} + 6\mu_2\bar{x}^2 + \bar{x}^4 = 1132$$

Calculation of β_1 and β_2

$$\beta_1 = \frac{\mu_3^2}{\mu_2^3} = 0.492377 \qquad \beta_2 = \frac{\mu_4}{\mu_2^2} = 0.654122$$

Moments about the point $x = 2$

$$\mu'_1 = \bar{x} - A = 2.5 - 2 = 0.5$$
$$\mu'_2 = \mu_2 + \mu_1'^2 = 14.75 + (.5)^2 = 15$$
$$\mu'_3 = \mu_3 + 3\mu'_2\,\mu'_1 - 2\mu_1'^3 = 39.75 + 3\,(15)\,(.5) - 2\,(.5)^3 = 62$$
$$\mu'_4 = \mu_4 + 4\mu'_3\mu'_1 - 6\mu'_2\mu_1'^2 + 3\mu_1'^4 = 244$$

TEST YOUR KNOWLEDGE

1. (*i*) Calculate the first four moments about mean, for the following individual series:

   5, 5, 5, 5, 5, 5.

   (*ii*) Find the first four moments about the mean of the following series:

   1, 3, 7, 9, 10.

   (*iii*) Calculate $\mu_1, \mu_2, \mu_3, \mu_4$ for the series : 4, 7, 10, 13, 16, 19, 22.

2. (*i*) Find the first four moments for the following frequency distribution:

x	1	2	3	4	5	6	7	8	9
f	1	2	3	4	5	4	3	2	1

   (*ii*) Calculate the first four moments of the following distribution about the mean and hence find β_1 and β_2.

x	0	1	2	3	4	5	6	7
f	1	8	28	56	70	56	28	8

   (*iii*) The number of flowers on Sunflower plants are given below:

No. of flowers	3	6	12	16	25
No. of plants	1	2	3	4	5

   Calculate the first four moments about mean. [*M.T.U. (B. Pharma) 2011*]

3. (*i*) Find the first four moments about mean for the following frequency distribution :

Marks	0–10	10–20	20–30	30–40	40–50
No. of students	5	10	40	20	25

(*ii*) Calculate the first four moments about the mean for the following:

Classes	5–15	15–25	25–35	35–45	45–55
f	14	22	36	18	10

(*iii*) Calculate the first four moments about the mean for the following data:

Class-interval	0–10	10–20	20–30	30–40	40–50
f	10	20	40	20	10

(*M.T.U. 2014*)

4. Calculate the first four moments about $x = 15$ and hence find the moments about the mean of the following distribution :

x	10	11	12	13	14	15	16	17	18	19	20	21
f	9	36	75	105	116	107	88	66	45	30	18	5

5. (*i*) The first three moments of a distribution about the value 4 of the variable are 1.5, 17 and – 30. Find the moments about mean.

(*ii*) The first four moments of a distribution about $x = 4$ are 1, 4, 10 and 45. Show that the mean is 5, the variance is 3, μ_3 is 0 and μ_4 is 26.

(*iii*) The first four moments of a distribution about the value 5 of the variable are 2, 20, 40 and 50. Calculate mean, variance, μ_3 and μ_4. (*G.B.T.U. 2011*)

6. If the first four moments of a distribution about the value 5 are equal to – 4, 22, – 117 and 560. Determine the corresponding moments :

(*i*) about the mean, and (*ii*) about zero.

7. Compute first four moments of the data 3, 5, 7, 9 about the mean. Also, compute the first four moments about the point 4.

8. In a frequency distribution, the mean is 1.5, variance is 0.64, β_2 is 2.5 and γ_1 is 0.3. Find μ_3 and μ_4 and also the first four moments about the origin.

9. The first four moments of a distribution about the value '0' are – 0.20, 1.76, – 2.36 and 10.88. Find the moments about the mean. (*U.P.T.U. 2009*)

Answers

1. (*i*) $\mu_1 = 0, \mu_2 = 0, \mu_3 = 0, \mu_4 = 0$ (*ii*) $\mu_1 = 0, \mu_2 = 12, \mu_3 = -12, \mu_4 = 208.8$
(*iii*) $\mu_1 = 0, \mu_2 = 36, \mu_3 = 0, \mu_4 = 2268$

2. (*i*) $\mu_1 = 0, \mu_2 = 4, \mu_3 = 0, \mu_4 = 37.6$ (*ii*) $\mu_1 = 0, \mu_2 = 2, \mu_3 = 0, \mu_4 = 11, \beta_1 = 0, \beta_2 = 2.75$
(*iii*) $\mu_1 = 0, \mu_2 = 54.8, \mu_3 = -49.6, \mu_4 = 5475.6$

3. (*i*) $\mu_1 = 0, \mu_2 = 125, \mu_3 = -300, \mu_4 = 37625$ (*ii*) $\mu_1 = 0, \mu_2 = 134.56, \mu_3 = 126.14, \mu_4 = 41840.82$
(*iii*) $\mu_1 = 0, \mu_2 = 120, \mu_3 = 0, \mu_4 = 36000$

4. $\mu_1 = 0, \mu_2 = 5.5, \mu_3 = 4.4, \mu_4 = 77.8$

5. (*i*) $\mu_1 = 0, \mu_2 = 14.75, \mu_3 = -99.75$ (*iii*) 7, 16, – 64, 162

6. (*i*) 0, 6, 19, 32 (*ii*) 1, 7, 38, 145

7. $\mu_1 = 0, \mu_2 = 5, \mu_3 = 0, \mu_4 = 41; \mu_1' = 2, \mu_2' = 9, \mu_3' = 38.25, \mu_4' = 177$

8. $\mu_3 = 0.1536, \mu_4 = 1.024, \nu_1 = 1.5, \nu_2 = 2.89, \nu_3 = 6.4086, \nu_4 = 15.6481$

9. $\mu_1 = 0, \mu_2 = 1.72, \mu_3 = -1.32, \mu_4 = 9.4096$.

3.13 MOMENT GENERATING FUNCTION [*U.P.T.U. 2014 ; G.B.T.U. 2012, M.T.U. 2013*]

For certain theoretical developments, an indirect method for computing moments is used. The method depends on the finding of the moment generating function.

3.13.1 In Case of a Continuous Variable *x*, it is defined as

$$M(t) = \int_a^b e^{tx} f(x)\, dx \qquad \text{...(1)}$$

where integral is a function of parameter t only. The limits a, b can be $-\infty$ and ∞ respectively. It is possible to associate a moment generating function with the distribution only when all the moments of the distribution are finite.

Let us see how M(t) generates moments. For this, let us assume that $f(x)$ is a distribution function for which the integral given by (1) exists.

Then e^{tx} may be expanded in a power series and the integration may be performed term by term. It follows that

$$M(t) = \int_a^b \left(1 + tx + \frac{t^2}{2!}x^2 + \cdots\right) f(x)\, dx$$

$$= \int_a^b f(x)\, dx + t\int_a^b xf(x)\, dx + \cdots$$

$$= \nu_0 + \nu_1 t + \nu_2 \cdot \frac{t^2}{2!} + \cdots \qquad \text{...(2)}$$

Obviously, the coefficient of $\frac{t^r}{r!}$ in (2) is the rth moment about the origin.

Also,
$$\left|\frac{d^r}{dt^r} M(t)\right|_{t=0} = \left|\frac{\nu_r}{r!} r! + \nu_{r+1}\, t + \ldots\right|_{t=0} = \nu_r \qquad \text{...(3)}$$

Thus, ν_r about origin = r^{th} derivative of M(t) with $t = 0$.

Although the moment generating function (m.g.f.) has been defined for the variable x only, the definition can be generalized so that it holds for a variable z where z is a function of x. *e.g.*, if $z = x - m$ (m is mean), the rth moment about z will give r^{th} moment of x about the mean m.

Moment generating function for z will clearly be given as

$$M_z(t) = \int_a^b e^{tz} f(x)\, dx$$

$$M_{x-m}(t) = \int_a^b e^{t(x-m)} f(x)\, dx = e^{-mt} \int_a^b e^{tx} f(x) dx = e^{-mt} M_x(t).$$

3.13.2 In Case of Discrete Distribution of the Variable x

We know that, for a variable x,

$$\nu_r = \Sigma x^r . \text{P}$$

where P is the probability that the variable takes on the value x.

If z is any function of x, we get r^{th} moment for z by the relation

$$\nu_r = \Sigma z^r \text{P}$$

and the moment generating function is given by

$$\text{M}_z(t) = \Sigma e^{tz} \text{P} \qquad ...(1)$$

To verify that this function generates moments, we will expand e^{tz} and then sum term by term,

$$\therefore \qquad \text{M}_z(t) = \sum \left(1 + tz + \frac{t^2}{2!} z^2 + \cdots\right) \text{P} = \Sigma\text{P} + t\Sigma z\text{P} + \frac{t^2}{2!} \Sigma z^2 . \text{P} + \cdots$$

$$= \nu_0 + t\nu_1 + \frac{t^2}{2!} \nu_2 + \cdots$$

In this case, we can also show that $\nu_r = \left| \dfrac{d^r}{dt^r} \text{M}_z(t) \right|_{t=0}$

M(t) is clearly the expected value of e^{tx} and hence can be written as E(e^{tx}) which gives the moment generating function in case of discrete as well as continuous variables.

Expectation of any function $\phi(x)$ is given by

$$\text{E}\{\phi(x)\} = \sum_i \phi(x_i) f(x_i) \qquad | \text{ for discrete distribution}$$

or,

$$\text{E}\{\phi(x)\} = \int_{-\infty}^{\infty} \phi(x) f(x)\, dx \qquad | \text{ for continuous distribution}$$

Eqn. (1) can also be rewritten as

$$\text{M}_{x-a}(t) = \text{E}\, [e^{t(x-a)}] = \sum_i e^{t(x_i - a)} \text{P}_i = e^{-at} \sum_i e^{tx_i} \text{P}_i = e^{-at} \text{M}_0(t)$$

Therefore the moment generating function about the point 'a' is equal to e^{-at} times the moment generating function about the origin.

Note. m.g.f. is not always defined since E{| e^{tx} |} does not always exist for all values of t.

e.g., if $f(x) = \dfrac{6}{\pi^2 x^2}$, $x = 1, 2, 3, \ldots$ then m.g.f. does not exist.

m.g.f. always exists for $t = 0$ since $\text{M}_{x=0}(0) = 1$.

3.13.3 Properties of Moment Generating Function *(M.T.U. 2013)*

(1) *The moment generating function of the sum of two independent chance variables is the product of their respective moment generating functions.*

Symbolically, $M_{x+y}(t) = M_x(t) \times M_y(t)$ *provided that x and y are independent random variables.*

Proof. Let x and y be two independent random variables so that $x + y$ is also a random variable.

The m.g.f. of the sum $x + y$ w.r.t. origin is

$$M_{x+y}(t) = E\{e^{t(x+y)}\} = E(e^{tx} . e^{ty}) = E(e^{tx}) . E(e^{ty})$$

Since x and y are independent variables and so are e^{tx} and e^{ty}.

$$\therefore \quad M_{x+y}(t) = M_x(t) . M_y(t)$$

Hence the theorem.

(2) *Effect of change of origin and scale on m.g.f.*

$$\boxed{M_u(t) = e^{-at/h} M_x(t/h)} \qquad \text{where } u = \frac{x-a}{h}.$$

Proof. Let u be a new random variable given by $u = \frac{x-a}{h}$ so that $x = a + hu$

then by definition, the effect of linear transformation on m.g.f. is governed by

$$M_x(t) = E(e^{tx}) = E[e^{t(a+hu)}] = E(e^{at} \cdot e^{thu})$$

$$= e^{at} E(e^{thu}) = e^{at} M_u(th)$$

Also,
$$M_u(t) = E(e^{tu})$$

$$= E\left[e^{t\left(\frac{x-a}{h}\right)}\right] = e^{-\frac{at}{h}} M_x\left(\frac{t}{h}\right)$$

(3) $M_{cx}(t) = M_x(ct)$, *c being a constant.*

Proof. By definition,

$$\text{LHS} = M_{cx}(t) = E(e^{tcx}) = M_x(ct) = \text{RHS}.$$

Hence the result.

ILLUSTRATIVE EXAMPLES

Example 1. *Find the moment generating function of the exponential distribution*

$$f(x) = \frac{1}{c} e^{-x/c} ; 0 \le x \le \infty, c > 0 \qquad \text{(M.T.U. 2014)}$$

Hence find its mean and standard deviation.

Sol. Moment generating function about the origin is given by

$$M_x(t) = \int_0^\infty e^{tx} \cdot \frac{1}{c} e^{-x/c} dx$$

$$= \frac{1}{c}\int_0^\infty e^{\left(t-\frac{1}{c}\right)x} dx = \frac{1}{c}\left[\frac{e^{\left(t-\frac{1}{c}\right)x}}{\left(t-\frac{1}{c}\right)}\right]_0^\infty$$

$$= (1 - ct)^{-1} = 1 + ct + c^2t^2 + c^3t^3 + \ldots$$

$$\therefore \quad \nu_1 = \left[\frac{d}{dt} M_x(t)\right]_{t=0} = (c + 2c^2t + 3c^3t^2 + \ldots)_{t=0} = c$$

and $$\nu_2 = \left[\frac{d^2}{dt^2} M_x(t)\right]_{t=0} = 2c^2$$

Now, mean $\bar{x} = \nu_1 = c$

Variance $\mu_2 = \nu_2 - \bar{x}^2 = \nu_2 - \nu_1^2 = 2c^2 - c^2 = c^2$

$\therefore$ Standard deviation $= \sqrt{\mu_2} = c.$

Example 2. *Obtain the moment generating function of the random variable x having probability distribution*

$$f(x) = \begin{cases} x, & \text{for } 0 < x < 1 \\ 2 - x, & \text{for } 1 \le x < 2 \\ 0, & \text{elsewhere} \end{cases}$$

[M.T.U. 2012; G.B.T.U. (C.O.) 2011; G.B.T.U. 2013]

Also determine mean ν_1, ν_2 *and variance* μ_2.

Sol. $M_x(t) = E(e^{tx})$

$$= \int_0^1 x \,.\, e^{tx}\, dx + \int_1^2 (2 - x)\, e^{tx}\, dx + \int_2^\infty 0 \,.\, e^{tx}\, dx$$

$$= \left(\frac{xe^{tx}}{t} - \frac{e^{tx}}{t^2}\right)_0^1 + \left(\frac{2e^{tx}}{t} - \frac{xe^{tx}}{t} + \frac{e^{tx}}{t^2}\right)_1^2$$

$$= \frac{e^t}{t} - \frac{e^t}{t^2} + \frac{1}{t^2} + \left[\left(\frac{2e^{2t}}{t} - \frac{2e^{2t}}{t} + \frac{e^{2t}}{t^2}\right) - \left(\frac{2e^t}{t} - \frac{e^t}{t} + \frac{e^t}{t^2}\right)\right] = \frac{e^{2t} - 2e^t + 1}{t^2}$$

$$= \left(\frac{e^t - 1}{t}\right)^2 = \frac{\left(t + \frac{t^2}{2!} + \frac{t^3}{3!} + \ldots\right)^2}{t^2} = 1 + t + t^2 + \ldots$$

$$\text{Mean} = \nu_1 = \left[\frac{d}{dt} M_x(t)\right]_{t=0} = 1$$

Similarly, $\nu_2 = 2$, $\mu_2 = \nu_2 - \bar{x}^2 = \nu_2 - \nu_1^2 = 2 - (1)^2 = 1 =$ Variance.

Example 3. *Find the moment generating function of the random variable whose moments are* $\nu_r = (r + 1)\,!\, 2^r$.

Sol. $$M_x(t) = E(e^{tx}) = \sum_{x=0}^{\infty} e^{tx}\, P(X = x)$$

$$= \sum_{r=0}^{\infty} \frac{t^r}{r!} \nu_r = \sum_{r=0}^{\infty} \frac{t^r}{r!} (r+1)!\,.\,2^r = \sum_{r=0}^{\infty} (r+1)\,(2t)^r$$

$$= 1 + 2.2t + 3\,.\,(2t)^2 + \ldots. = (1 - 2t)^{-2}.$$

Example 4. *Find the moment generating function of the probability distribution function*

$$f(z) = e^{-z}\,(1 + e^{-z})^{-2}, \quad -\infty < z < \infty.$$

Sol. $M_z(t) = E(e^{tz})$

$$= \int_{-\infty}^{\infty} e^{tz} \,.\, e^{-z} (1 + e^{-z})^{-2}\, dz$$

$$= \int_1^\infty u^2 (u-1)^{-t}\, du \quad \text{where } 1 + e^{-z} = u \;\Rightarrow\; -e^{-z}\, dz = du$$

$$= \int_0^1 v^{-t} (1-v)^t\, dv \quad \text{where } v = 1 - \frac{1}{u} \;\Rightarrow\; dv = \frac{1}{u^2}\, du$$

$$= \beta(1-t, 1+t)\; ; \; 1 - t > 0$$

$$= \pi t \operatorname{cosec} \pi t,\; t < 1.$$

Example 5. *Find the moment generating function of the negative exponential function* $f(x) = \lambda e^{-\lambda x}\; ; \; x, \lambda > 0$.

Sol. $$M_x(t) = \lambda \int_0^\infty e^{tx} . e^{-\lambda x}\, dx = \lambda \int_0^\infty e^{(t-\lambda)x}\, dx = \lambda \int_0^\infty e^{-(\lambda - t)x}\, dx$$

$$= \frac{\lambda}{\lambda - t} = \left(1 - \frac{t}{\lambda}\right)^{-1} = \sum_{r=0}^{\infty} \left(\frac{t}{\lambda}\right)^r \; ; \lambda > t$$

Example 6. *Find the moment generating function of the discrete binomial distribution given by*

$$P(x) = {}^nc_x\, p^x q^{n-x} \quad (\textit{where } q = 1 - p)$$

Also find the first and second moments about the mean.

Sol. Moment generating function about the origin is given by

$$M_x(t) = \Sigma\, e^{tx} . {}^nc_x . p^x q^{n-x}$$

$$= \Sigma\, {}^nc_x (pe^t)^x q^{n-x} = (q + pe^t)^n$$

$$\nu_1 = \left[\frac{d}{dt} M_x(t)\right]_{t=0} = [n(q + pe^t)^{n-1} . pe^t\,]_{t=0} = np \qquad |\text{Since } q + p = 1$$

$$\nu_2 = \left[\frac{d^2}{dt^2} M_x(t)\right]_{t=0}$$

$$= [np\{e^t . (n-1)(q + pe^t)^{n-2} pe^t + (q + pe^t)^{n-1} . e^t\}]_{t=0}$$

$$= [np\,(q + pe^t)^{n-2} . e^t \{(n-1)pe^t + (q + pe^t)\}]_{t=0}$$

$$= [np(q + pe^t)^{n-2} . e^t(q + npe^t)]_{t=0}$$

$$= np(q + np) \qquad |\because\; q + p = 1$$

$$= npq + n^2 p^2$$

Hence first and second moments about the mean are given by

$$\mu_1 = 0$$

Since $\bar{x} = \nu_1 = np$

$\therefore$ $\mu_2 = \nu_2 - \bar{x}^2 = \nu_2 - \nu_1^2 = npq + n^2p^2 - n^2p^2 = npq$

Hence, mean $= np$, S.D. $= \sqrt{\mu_2} = \sqrt{npq}$.

Example 7. *Find the moment generating function of the discrete Poisson distribution given by* $P(x) = e^{-\lambda} \cdot \dfrac{\lambda^x}{x\,!}$. *Also find the first and second moments about the mean.*

(M.T.U. 2013)

Sol. Moment generating function about the origin is given by

$$M_x(t) = \Sigma e^{tx} . e^{-\lambda} . \frac{\lambda^x}{x\,!} = e^{-\lambda}\, \Sigma \frac{(\lambda e^t)^x}{x\,!} = e^{-\lambda} . e^{\lambda e^t} = e^{\lambda(e^t - 1)}$$

$$\nu_1 = \left[\frac{d}{dt} M_x(t)\right]_{t=0} = [e^{\lambda(e^t-1)}\, \lambda e^t]_{t=0} = \lambda$$

$$\nu_2 = \left[\frac{d^2}{dt^2} M_x(t)\right]_{t=0} = [\lambda\, \{e^t \,.\, e^{\lambda(e^t-1)} \,.\, \lambda e^t + e^{\lambda(e^t-1)} e^t\}]_{t=0}$$

$$= [\lambda e^{\lambda(e^t-1)}\, e^t\, (\lambda\, e^t + 1)]_{t=0} = \lambda(\lambda + 1)$$

Hence, the first and second moments about the mean are given by

$$\mu_1 = 0$$

Since $\quad \nu_1 = \bar{x} = \lambda$

$\therefore \quad \mu_2 = \nu_2 - \bar{x}^2 = \nu_2 - \nu_1{}^2 = \lambda\,(\lambda + 1) - \lambda^2 = \lambda$

Example 8. *Find the moment generating function of the continuous normal distribution given by* $f(x) = \dfrac{1}{\sigma\sqrt{2\pi}}\, e^{-\frac{1}{2\sigma^2}(x-\mu)^2}$; $-\infty < x < \infty$.

Sol. Moment generating function about the origin is defined as

$$M_x(t) = E\,(e^{tx}) = \int_{-\infty}^{\infty} e^{tx} \cdot \frac{1}{\sigma\sqrt{2\pi}}\, e^{-\frac{1}{2}\left(\frac{x-\mu}{\sigma}\right)^2} dx$$

$$= \frac{e^{\mu t}}{\sqrt{2\pi}} \int_{-\infty}^{\infty} e^{-\frac{1}{2}z^2} \,.\, e^{-t\sigma z}\, dz \qquad \text{where } z = \frac{x-\mu}{\sigma}$$

$$= \frac{1}{\sqrt{2\pi}}\, e^{\left(\mu t + \frac{1}{2}t^2\sigma^2\right)} \int_{-\infty}^{\infty} e^{-\frac{1}{2}(z-t\sigma)^2}\, dz$$

$$= e^{\mu t + \frac{1}{2}t^2\sigma^2} \,.\, 1 = e^{\mu t + \frac{1}{2}t^2\sigma^2}. \qquad \left|\; \because \int_0^{\infty} e^{-z^2}\, dz = \frac{\sqrt{\pi}}{2}\right.$$

Example 9. *The random variable X assuming only non-negative values has a Gamma probability distribution if its probability distribution is given by*

$$f(x) = \begin{cases} \dfrac{\alpha^{\beta}}{\Gamma\beta} x^{\beta-1}\, e^{-\alpha x} & ; x > 0, \alpha > 0, \beta > 1 \\ 0, & \text{elsewhere} \end{cases}$$

Find the moment generating function of Gamma probability distribution.

Sol. $\quad M_x(t) = E(e^{tx})$

$$= \int_0^{\infty} e^{tx} \cdot \frac{\alpha^{\beta}}{\Gamma\beta} \cdot x^{\beta-1} e^{-\alpha x}\, dx = \frac{\alpha^{\beta}}{\Gamma\beta} \int_0^{\infty} x^{\beta-1}\, e^{-x(\alpha-t)}\, dx$$

$$= \frac{\alpha^{\beta}}{(\alpha-t)^{\beta}\,\Gamma\beta} \int_0^{\infty} y^{\beta-1}\, e^{-\beta}\, dy \qquad |\text{ where } y = x(\alpha - t) \text{ so that } dy = (\alpha - t)\, dx$$

$$= \frac{1}{\left(1-\frac{t}{\alpha}\right)^{\beta}} \cdot \frac{1}{\Gamma\beta}\, \Gamma\beta = \left(1-\frac{t}{\alpha}\right)^{-\beta} ; \;|\; t \;|< \alpha.$$

Example 10. *Let the random variable X assume the value 'n' with the probability law $p(X = n) = pq^{n-1}$, $n = 1, 2, 3, \ldots$. Find the moment generating function and hence mean and variance.* (G.B.T.U. 2010)

Sol. The given distribution is a discrete distribution.

$$M_n(t) = \Sigma e^{tn} pq^{n-1} = \frac{p}{q} \Sigma (e^t q)^n = \frac{p}{q}(1 - e^t q)^{-1} = \frac{p}{q(1 - qe^t)}$$

which is the moment generating function.

$$\nu_1 = \left[\frac{d}{dt} M_n(t)\right]_{t=0} = \frac{p}{q}\left[\frac{qe^t}{(1-qe^t)^2}\right]_{t=0} = \frac{p}{(1-q)^2} = \frac{p}{p^2} = \frac{1}{p}$$

$$\nu_2 = \left[\frac{d^2}{dt^2} M_n(t)\right]_{t=0} = p\left[\frac{d}{dt}\left\{\frac{e^t}{(1-qe^t)^2}\right\}\right]_{t=0}$$

$$= p\left[\frac{(1-qe^t)^2 . e^t - e^t . 2(1-qe^t)(-qe^t)}{(1-qe^t)^4}\right]_{t=0}$$

$$= p\left[\frac{(1-q)^2 + 2q(1-q)}{(1-q)^4}\right] = \frac{1}{p} + \frac{2q}{p^2}$$

$$\text{Mean} = \bar{x} = \nu_1 = \frac{1}{p}$$

$$\text{Variance} = \mu_2 = \nu_2 - \bar{x}^2 = \frac{1}{p} + \frac{2q}{p^2} - \frac{1}{p^2} = \frac{q}{p^2}$$

TEST YOUR KNOWLEDGE

1. Define moment generating function. Find the moment generating function of a random variable X whose probability function is given by:

   $P(X = x) = p(1-p)^x,\ x = 0, 1, 2, \ldots, \infty$ *(G.B.T.U. 2012)*
2. Define moment generating function and two properties of moment generating function with proof. *(M.T.U. 2013)*
3. The probability density function of the random variable X is $f(x) = \frac{1}{2\theta} \exp\left(-\frac{|x-\theta|}{\theta}\right),\ -\infty < x < \infty.$

   Find moment generating function of X. Hence find the mean E(X) and variance V(X). *(M.T.U. 2013)*

   $\left[\textbf{Hint: } M_x(t) = \frac{1}{2\theta}\int_{-\infty}^{\theta} \exp\left(-\frac{\theta - x}{\theta}\right) e^{tx}\, dx + \frac{1}{2\theta}\int_{\theta}^{\infty} \exp\left(-\frac{x-\theta}{\theta}\right) e^{tx}\, dx\right]$
4. Show that the moment generating function of random variable X having the p.d.f.

   $$f(x) = \begin{Bmatrix} \frac{1}{3}, & -1 < x < 2 \\ 0, & \text{elsewhere} \end{Bmatrix} \text{ is } M_X(t) = \begin{cases} \frac{e^{2t} - e^{-t}}{3t}, & t \neq 0 \\ 1, & t = 0 \end{cases}$$
5. Find the moment generating function for triangular distribution defined by

   $$f(x) = \begin{Bmatrix} x, & 0 \le x \le 1 \\ 2 - x, & 1 \le x \le 2 \end{Bmatrix}$$ *(M.T.U. 2013)*

6. If $P(X = x) = \dfrac{1}{2^x}$, $x = 1, 2, 3, \ldots$, find the moment generating function of x. Hence obtain the variance. *(U.P.T.U. 2014)*

Answers

1. $\dfrac{p}{1 - e^t(1-p)}$

3. $\dfrac{e^{\theta t}}{1-\theta^2 t^2}$ or $1 + \theta t + \dfrac{3\theta^2 t^2}{2!} + \ldots$; $E(X) = \theta$, $V(x) = 2\theta^2$

5. $M_x(t) = 1 + t + \dfrac{1}{18}t^2 + \ldots$

6. $\dfrac{e^t}{2-e^t}$; $\dfrac{1}{2}$.

3.14 SKEWNESS

For a symmetrical distribution, the frequencies are symmetrically distributed about the mean *i.e.,* variates equidistant from the mean have equal frequencies. Also, the mean, mode and median coincide and median lies half-way between the two quartiles.

Thus, $M = M_0 = M_d$ and $Q_3 - M = M - Q_1$.

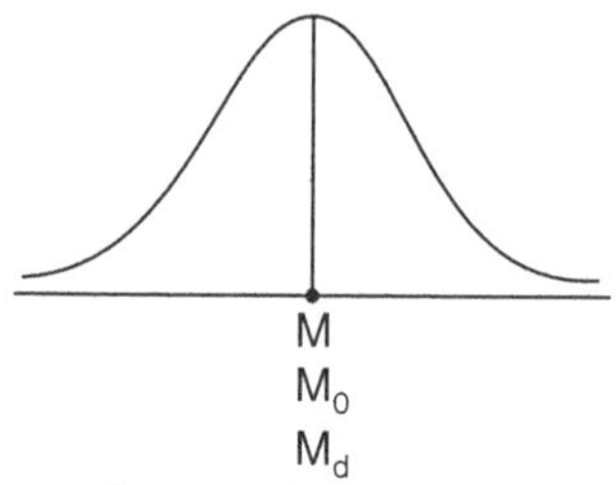

Symmetrical distribution

3.15 MEANING OF SKEWNESS *(U.P.T.U. 2015)*

If the curve of the distribution is not symmetrical, it may admit of tail on either side of the distribution. **Skewness means lack of symmetry or lopsidedness in a frequency distribution.**

The object of measuring skewness is to estimate the extent to which a distribution is distorted from a perfectly symmetrical distribution. Skewness indicates whether the curve is turned more to one side than to other *i.e.,* whether the curve has a longer tail on one side. Skewness can be positive as well as negative.

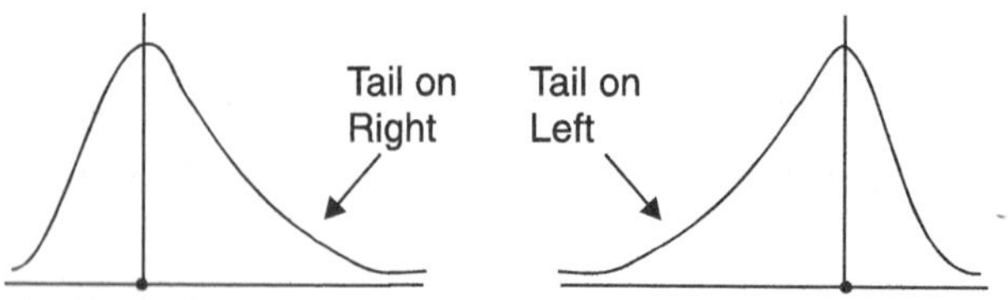

Positively Skewed Distribution   Negatively Skewed Distribution

Skewness is positive if the longer tail of the distribution lies towards the right and negative if it lies towards the left.

3.16 TESTS OF SKEWNESS

1. If A.M. = Mode = Median, then there is no skewness in the distribution. In other words, the curve of the frequency distribution would be symmetrical, bell-shaped.

2. If A.M. is less than (greater than), the value of mode, the tail would be on left (right) side, *i.e.,* the distribution is negatively (positively) skewed.

3. If sum of frequencies of values less than mode is equal to the sum of frequencies of values greater than mode, then there would be no skewness.

4. If quartiles are equidistant from median, then there would be no skewness.

3.17 METHODS OF MEASURING SKEWNESS

Relative measures of skewness are called the **coefficient of skewness**. They are independent of the units of measurement and as such, they are pure numbers.

Following are the methods of measuring skewness:

1. Karl Pearson's Method
2. Bowley's Method
3. Kelly's Method
4. Method of Moments.

Here, we will discuss Karl Pearson's method and the method of moments only.

3.17.1 Karl Pearson's Method

This method is based on the fact that in a symmetrical distribution, the value of A.M. is equal to that of mode. As we have already noted that the distribution is positively skewed if A.M. > Mode and negatively skewed if A.M. < Mode. The Karl Pearson's coefficient of skewness is given by:

$$\textbf{Karl Pearson's coefficient of skewness} = \frac{\textbf{A.M.} - \textbf{Mode}}{\textbf{S.D.}}$$

We have already studied the methods of calculating A.M., Mode and S.D. of frequency distributions. If mode is ill-defined in some frequency distribution, then the value of empirical mode is used in the formula.

$$\text{Empirical mode} = 3 \text{ Median} - 2 \text{ A.M.}$$

$$\therefore \quad \text{Coeff. of skewness} = \frac{\text{A.M.} - \text{Mode}}{\text{S.D.}}$$

$$= \frac{\text{A.M.} - (3 \text{ Median} - 2 \text{ A.M.})}{\text{S.D.}} = \frac{3\text{A.M.} - 3 \text{ Median}}{\text{S.D.}}$$

$$\therefore \quad \textbf{Karl Pearson's coefficient of skewness} = \frac{\textbf{3(A.M.} - \textbf{Median)}}{\textbf{S.D.}}$$

The coefficient of skewness as calculated by using this method gives magnitude as well as direction of skewness, present in the distribution. Practically, its value lies between – 1 and 1. For a symmetrical distribution, its value comes out to be zero.

The Karl Pearson's coefficient of skewness is generally denoted by 'SK_P'.

(*i*) If $SK_P = 0$ $\quad \Leftrightarrow \quad \dfrac{\text{Mean} - \text{Mode}}{\text{S.D.}} = 0$

$\Leftrightarrow$ Mean = Mode

$\Leftrightarrow$ Distribution is symmetrical.

Thus a distribution is a symmetrical distribution iff $SK_P = 0$.

(*ii*) If $SK_P > 0$ $\Leftrightarrow \frac{\text{Mean} - \text{Mode}}{\text{S.D.}} > 0$

$\Leftrightarrow$ Mean – Mode > 0

$\Leftrightarrow$ Mean > Mode

$\Leftrightarrow$ Distribution is positively skewed.

Thus a distribution is a positively skewed distribution iff $SK_P > 0$.

(*iii*) If $SK_P < 0$ $\Leftrightarrow \frac{\text{Mean} - \text{Mode}}{\text{S.D.}} < 0$

$\Leftrightarrow$ Mean – Mode < 0

$\Leftrightarrow$ Mean < Mode

$\Leftrightarrow$ Distribution is negatively skewed.

Thus a distribution is a negatively skewed distribution iff $SK_P < 0$.

ILLUSTRATIVE EXAMPLES

Example 1. *Karl Pearson's coefficient of skewness of a distribution is 0.32, its standard deviation is 6.5 and mean is 29.6. Find the mode of the distribution.*

Sol. We have $SK_P = 0.32$, S.D. = 6.5, $\bar{x} = 29.6$.

Now $$SK_P = \frac{\bar{x} - \text{Mode}}{\text{S.D.}}$$

$\therefore$ $$0.32 = \frac{29.6 - \text{Mode}}{6.5}$$

$\Rightarrow$ $$29.6 - \text{Mode} = 0.32 \times 6.5 = 2.08$$

$\Rightarrow$ $$\text{Mode} = 29.6 - 2.08 = 27.52.$$

Example 2. *For a moderately skewed data, the arithmetic mean is 100, the variance is 35 and Karl Pearson's coefficient of skewness is 0.2. Find its mode and median.*

Sol. We have $\bar{x} = 100$, Variance = 35, $SK_P = 0.2$.

Now $$SK_P = \frac{\bar{x} - \text{Mode}}{\sigma}$$

$\therefore$ $$0.2 = \frac{100 - \text{Mode}}{\sqrt{35}} \qquad (\because \text{ S.D.} = \sqrt{\text{Variance}})$$

$\Rightarrow$ $$100 - \text{Mode} = 0.2 \times 5.92 = 1.184$$

$\Rightarrow$ $$\text{Mode} = 100 - 1.184 = 98.816.$$

Also, $$\text{Mode} = 3 \text{ Median} - 2\bar{x} \Rightarrow 98.816 = 3 \text{ Median} - 2(100).$$

$\therefore$ $$3 \text{ Median} = 98.816 + 200 = 298.816$$

$\therefore$ $$\text{Median} = \frac{298.816}{3} = 99.61.$$

Example 3. *In a certain distribution, the following results were obtained:*

A.M. = 45, Median = 48, Coefficient of skewness = – 0.4. The person who gave you this data, failed to give the value of S.D. You are required to estimate it with the help of available data.

Sol. We have

coeff. of skewness = – 0.4, A.M. = 45, median = 48.

Now, coeff. of skewness $= \frac{3(\bar{x} - \text{Median})}{\text{S.D.}}$

$$\Rightarrow \quad -\frac{4}{10} = \frac{3(45-48)}{\text{S.D.}} = \frac{-9}{\text{S.D.}}$$

$$\Rightarrow \quad 4 \text{ S.D.} = 90$$

$$\text{S.D.} = \frac{90}{4} = 22.5.$$

Example 4. *The sum of 20 observations is 300 and sum of their squares is 5000. The median is 15. Find the Karl Pearson's coefficient of skewness.*

Sol. Let 'x' be the variable under consideration.

We have $n = 20$, $\Sigma x = 300$, $\Sigma x^2 = 5000$, median $= 15$.

Now,
$$\bar{x} = \frac{\Sigma x}{n} = \frac{300}{20} = 15$$

$$\text{S.D.} = \sqrt{\frac{\Sigma x^2}{n} - \bar{x}^2} = \sqrt{\frac{5000}{20} - (15)^2} = \sqrt{250-225} = \sqrt{25} = 5$$

Now, Karl Pearson's coeff. of skewness

$$= \frac{3(\bar{x} - \text{Median})}{\text{S.D.}} = \frac{3(15-15)}{5} = \frac{0}{5} = 0.$$

Example 5. *Find the coefficient of skewness by Karl Pearson's method for the following data:*

Value	*6*	*12*	*18*	*24*	*30*	*36*	*42*
Frequency	*4*	*7*	*9*	*18*	*15*	*10*	*3*

Sol. **Calculation of $\bar{x}$, S.D.**

Value x	f	$d = x - A$, $A = 24$	$u = d/h$, $h = 6$	fu	fu^2
6	4	– 18	– 3	– 12	36
12	7	– 12	– 2	– 14	28
18	9	– 6	– 1	– 9	9
24	18	0	0	0	0
30	15	6	1	15	15
36	10	12	2	20	40
42	3	18	3	9	27
	N = 66			$\Sigma fu = 9$	$\Sigma fu^2 = 155$

$$\textbf{A.M. } \bar{x} = A + \left(\frac{\Sigma fu}{N}\right)h = 24 + \left(\frac{9}{66}\right)6 = 24.82$$

$$\textbf{S.D.} = \sqrt{\frac{\Sigma fu^2}{N} - \left(\frac{\Sigma fu}{N}\right)^2} \times h = 6 \times \sqrt{\frac{155}{66} - \left(\frac{9}{66}\right)^2} = 8.94$$

Mode. **Grouping Table**

x	I f	II	III	IV	V	VI
6	4					
12	7	11				
18	9		16	20		
24	**18**	**27**			**34**	
30	15		**33**	**43**		**42**
36	10	25			28	
42	3		13			

Analysis Table

Column	*24*	*18*	*30*	*36*	*12*
I	1				
II	1	1			
III	1		1		
IV	1		1	1	
V	1	1			
VI	1	1	1		1
Total	6	3	3	1	1

∴ Mode = 24

∴ $$SK_P = \frac{\bar{x} - \text{Mode}}{\text{S.D.}} = \frac{24.82 - 24}{8.94} = \frac{0.82}{8.94} = 0.092.$$

Example 6. *Calculate Karl Pearson's coefficient of skewness for the following data:*

Income (in ₹)	*500—600*	*600—700*	*700—800*	*800—900*	*900—1000*	*1000—1100*
No. of employees	*8*	*12*	*4*	*2*	*1*	*1*

Sol. **Calculation of $\bar{x}$, Mode, S.D.**

Income (in ₹)	*No. of employees* f	*Midpoints of classes* x	$d = x - A$ $A = 750$	$u = d/h$ $h = 100$	fu	fu^2
500—600	8	550	– 200	– 2	– 16	32
600—700	12	650	– 100	– 1	– 12	12
700—800	4	750	0	0	0	0
800—900	2	850	100	1	2	2
900—1000	1	950	200	2	2	4
1000—1100	1	1050	300	3	3	9
	N = 28				$\Sigma fu = -21$	$\Sigma fu^2 = 59$

A.M. $\bar{x} = A + \left(\frac{\Sigma fu}{N}\right)h = 750 + \left(-\frac{21}{28}\right)(100) = 750 - 75 = ₹\,675$

Mode. By inspection, modal class is 600–700

$\therefore$ $$\text{Mode} = l + \left(\frac{\Delta_1}{\Delta_1 + \Delta_2}\right)h$$

Here $l = 600, \Delta_1 = 12 - 8 = 4, \Delta_2 = 12 - 4 = 8, h = 100$

$\therefore$ $$\text{Mode} = 600 + \left(\frac{4}{4+8}\right)(100) = 600 + 33.33 = ₹\,633.33$$

$$\textbf{S.D.} = \left[\sqrt{\frac{\Sigma fu^2}{N} - \left(\frac{\Sigma fu}{N}\right)^2}\right] \times h = \left[\sqrt{\frac{59}{28} - \left(-\frac{21}{28}\right)^2}\right] \times 100$$

$$= \sqrt{2.1071 - 0.5625} \times 100 = \sqrt{1.5446} \times 100 = 1.2428 \times 100 = ₹\,124.28$$

Now, Karl Pearson's coeff. of skewness $= \frac{\bar{x} - \text{Mode}}{\text{S.D.}} = \frac{675 - 633.33}{124.28} = 0.34.$

TEST YOUR KNOWLEDGE

1. A frequency distribution gives the following results:

Coeff. of variation = 5

Karl Pearson's Coeff. of Skewness = 0.5

S.D. = 2

Find A.M and Mode of the distribution.

2. Find Pearson's coeff. of skewness from the following frequency distribution:

Height (in inches)	60–62	63–65	66–68	69–71	72–74
Frequency	5	18	42	27	8

3. From the following data, calculate the coefficient of skewness based on mean, median and S.D.

Variable	100–110	110–120	120–130	130–140	140–150	150–160	160–170	170–180
Frequency	4	16	36	52	64	40	32	11

4. From the following data, find out the Karl Pearson's coefficient of skewness:

Measurement	10	11	12	13	14	15
Frequency	2	4	10	8	5	1

5. Calculate Karl Pearson's coefficient of skewness for the following frequency distribution:

Marks more than	0	10	20	30	40	50	60	70
No. of students	100	90	75	50	25	15	5	0

6. For the following frequency distribution, calculate the value of Karl Pearson's coeff. of skewness:

Temp. (°C)	– 40 to – 30	– 30 to – 20	– 20 to – 10	– 10 to 0	0 to 10	10 to 20	20 to 30
No. of days	10	28	30	42	65	180	10

7. From the following data, calculate Karl Pearson's coefficient of skewness:

Marks (above)	0	10	20	30	40	50	60	70	80
No. of students	150	140	100	80	80	70	30	14	0

8. Find out the mean wage and coefficient of skewness from the following data:

35 men gets at the rate of ₹ 4.5 per man
40 men gets at the rate of ₹ 5.5 per man
48 men gets at the rate of ₹ 6.5 per man
100 men gets at the rate of ₹ 7.5 per man
125 men gets at the rate of ₹ 8.5 per man
87 men gets at the rate of ₹ 9.5 per man
43 men gets at the rate of ₹ 10.5 per man
22 men gets at the rate of ₹ 11.5 per man.

9. Calculate Karl Pearson's coefficient of skewness:

Wages (in ₹)	70–80	80–90	90–100	100–110	110–120	120–130	130–140	140–150
No. of workers	12	18	35	42	50	45	20	8

10. Find the mean, mode, S.D. and Karl Pearson's coeff. of skewness for the following:

Yrs. under	10	20	30	40	50	60
No. of persons	15	32	51	78	97	109

11. Compute the coeff. of skewness from the following figures : 25, 15, 23, 40, 27, 25, 23, 25, 20.

12. In a discrete series of 20 terms, the sum of the terms is 200, the sum of the squares of the terms is 5000 and the median is 15. Find Karl Pearson's coefficient of skewness.

13. Calculate the coefficient of skewness based on mean, median and standard deviation from the following data:

C.I.	0–10	10–20	20–30	30–40	40–50	50–60	60–70	70–80
f	3	6	11	24	28	16	9	3

[*M.T.U. (MBA) 2012*]

14. Which of the following two series is symmetrical?

Series α: Mean = 22, Median = 24, S.D. = 10
Series β: Mean = 22, Median = 25, S.D. = 12

15. Following table gives the data relating to marks obtained by students who appeared for B. Tech. III Semester examination in Mathematics III at a centre. Calculate Karl Pearson's coefficient of skewness from the said data:

Marks	0 to 10	10 to 20	20 to 30	30 to 40	40 to 50	50 to 60	60 to 70	70 to 80
No. of students	10	40	20	0	10	40	16	14

[**Hint.** Since max. frequency corresponds to two classes very far from each other so mode is ill-defined and $SK_P = \frac{3(\text{Mean} - \text{Median})}{\text{S.D.}}$].

Answers

1. Mean = 40, Mode = 39	**2.** 0.0356	**3.** – 0.0087	**4.** 0.3604
5. – 0.0627	**6.** – 0.6617	**7.** – 0.7539	
8. Mean wage = ₹ 8.07, Coeff. of skewness = – 0.2422			**9.** – 0.3314
10. Mean = 29.95, Mode = 35, S.D. = 15.49, Coeff. of skewness = – 0.32			**11.** – 0.03
12. – 1.22	**13.** – 0.08608	**14.** Series α	**15.** 0.754.

3.17.2 Method of Moments

In this method, second and third central moments of the distribution are used. This measure of skewness is called the **Moment coefficient of skewness** and is given by:

$$\textbf{Moment coefficient of skewness} = \frac{\mu_3}{\sqrt{\mu_2^3}}. \qquad [\textit{G.B.T.U. (C.O.) 2009, 2011}]$$

For a symmetrical distribution, its value would come out to be zero. The coefficient of skewness as calculated by this method gives the magnitude as well as direction of the skewness present in the distribution.

In Statistics, we define $\beta_1 = \frac{\mu_3^2}{\mu_2^3}$.

$\therefore$ Moment coefficient of skewness can also be written as $= \frac{\mu_3}{\sqrt{\mu_2^3}} = \pm\sqrt{\beta_1}$.

The sign with $\sqrt{\beta_1}$ is to be taken as that of μ_3. The moment coefficient of skewness is also denoted by γ_1. The moment coefficient of skewness is generally denoted by 'SK_M'.

ILLUSTRATIVE EXAMPLES

Example 1. *The first three central moments of a distribution are 0, 15, – 31. Find the moment coefficient of skewness.*

Sol. We have $\mu_1 = 0, \mu_2 = 15$ and $\mu_3 = -31$

$$\text{Moment coefficient of skewness} = \frac{\mu_3}{\sqrt{\mu_2{}^3}} = \frac{-31}{\sqrt{(15)^3}} = -\frac{31}{58.09} = -0.53.$$

Example 2. *The first four moments of a distribution about the value 5 of the variable are 2, 20, 40 and 50. Calculate the moment coefficient of skewness.*

Sol. We have $A = 5, \mu_1' = 2, \mu_2' = 20, \mu_3' = 40$ and $\mu_4' = 50$

Now, $\mu_2 = \mu_2' - (\mu_1')^2 = 20 - (2)^2 = 16$

$$\mu_3 = \mu_3' - 3\mu_1'\mu_2' + 2\mu_1'^3 = 40 - 3(2)(20) + 2(2)^3$$
$$= 40 - 120 + 16 = -64$$

$$\text{Moment coefficient of skewness} = \frac{\mu_3}{\sqrt{\mu_2{}^3}} = \frac{-64}{\sqrt{(16)^3}} = \frac{-64}{64} = -1.$$

Example 3. *Calculate the moment coefficient of skewness for the following distribution:*

Classes	*2.5–7.5*	*7.5–12.5*	*12.5–17.5*	*17.5–22.5*	*22.5–27.5*	*27.5–32.5*	*32.5–37.5*
Frequency	*8*	*15*	*20*	*32*	*23*	*17*	*5*

Sol. **Calculation of Moment Coefficient of Skewness**

Classes	f	*Mid-pts.* x	$d = x - A$ $A = 20$	$u = d/h$ $h = 5$	fu	fu^2	fu^3
2.5–7.5	8	5	– 15	– 3	– 24	72	– 216
7.5–12.5	15	10	– 10	– 2	– 30	60	– 120
12.5–17.5	20	15	– 5	– 1	– 20	20	– 20
17.5–22.5	32	20	0	0	0	0	0
22.5–27.5	23	25	5	1	23	23	23
27.5–32.5	17	30	10	2	34	68	136
32.5–37.5	5	35	15	3	15	45	135
	N = 120				$\Sigma fu = -2$	$\Sigma fu^2 = 288$	$\Sigma fu^3 = -62$

Now, $$\mu_1' = \left(\frac{\Sigma fu}{N}\right)h = \left(\frac{-2}{120}\right) 5 = -0.083$$

$$\mu_2' = \left(\frac{\Sigma fu^2}{N}\right)h^2 = \left(\frac{288}{120}\right) 5^2 = 60$$

$$\mu_3' = \left(\frac{\Sigma fu^3}{N}\right)h^3 = \left(\frac{-62}{162}\right) 5^3 = -64.583$$

Now, $$\mu_2 = \mu_2' - \mu_1'^2 = 60 - (-0.083)^2 = 59.993$$
$$\mu_3 = \mu_3' - 3\mu_1'\mu_2' + 2\mu_1'^3 = -64.583 - 3(-0.083)(60) + 2(-0.083)^3$$
$$= -49.644.$$

$$\therefore \quad \text{Moment coefficient of skewness} = \frac{\mu_3}{\sqrt{\mu_2^3}} = \frac{-49.644}{\sqrt{(59.993)^3}} = -0.1068.$$

TEST YOUR KNOWLEDGE

1. The first-three central moments of a distribution are 0, 2.5, 0.7. Find the value of the moment coefficient of skewness. *(U.P.T.U. 2015)*
2. In a certain distribution, the first four moments about the point 4 are – 1.5, 17, – 30 and 308. Calculate the moment coefficient of skewness. *(U.P.T.U. 2014)*
3. The first three moments of a frequency distribution about origin '5' are – 0.55, 4.46 and – 0.43. Find the moment coefficient of skewness.
4. Calculate the moment coefficient of skewness for the following data:

Marks	0–10	10–20	20–30	30–40	40–50	50–60	60–70
No. of students	8	12	20	30	15	10	5

5. Calculate the moment coefficient of skewness from the following data:

x	0	1	2	3	4	5	6	7	8
f	1	8	28	56	70	56	28	8	1

6. For the following frequency distribution, find the first four moments about the mean. Also, find the value of β_1. Is it a symmetrical distribution?

x	2	3	4	5	6
f	1	3	7	3	1

7. Compute the coefficient of skewness from the following data:

x	6	7	8	9	10	11	12
f	3	6	9	13	8	5	4

8. In two frequency distributions, the second moments about mean are 36 and 49 respectively while third moments about mean are 43.2 and 85.75. Compare the skewness in the two frequency distributions. *(G.B.T.U. 2012)*
9. The first three moments about the origin are given by $\nu_1 = \frac{n+1}{2}$, $\nu_2 = \frac{(n+1)(2n+1)}{6}$, and

   $\nu_3 = \frac{n(n+1)^2}{4}$. Examine the skewness of the data.
10. (*i*) Define skewness of a distribution. *(U.P.T.U. 2015)*

    (*ii*) Define the coefficients of skewness. *(M.T.U. 2012, 2013)*

Answers

1. 0.17708 **2.** 0.7017 **3.** 0.7781 **4.** 0.0726

5. 0 **6.** 0, 0.933, 0, 2.533, Yes **7.** 0.0903.

8. γ_1 (for I distribution) = 0.2, γ_2 (for II distribution) = 0.25

Second distribution is more positively skewed than the first.

9. data is symmetrical.

3.18 KURTOSIS [*A.K.T.U. 2015*]

Given two frequency distributions which have the same variability as measured by the standard deviation, they may be relatively more or less flat topped than the normal curve. A frequency curve may be symmetrical but it may not be equally flat topped with the normal curve. The relative flatness of the top is called **kurtosis** and is measured by β_2. Kurtosis refers to the bulginess of the curve of a frequency distribution.

Curves which are neither flat nor sharply peaked are called normal curves or **mesokurtic curves.**

Curves which are flatter than the normal curve are called **platykurtic curves.**

Curves which are more sharply peaked than the normal curve are called **leptokurtic curves.**

3.19 MEASURE OF KURTOSIS [*G.B.T.U. (C.O.) 2011*]

The measure of kurtosis is denoted by β_2 and is defined as

$$\boxed{\beta_2 = \frac{\mu_4}{\mu_2^2}}$$

where μ_2 and μ_4 are respectively the second and fourth moments about mean of the distribution.

If $\boldsymbol{\beta_2 > 3}$, the distribution is **leptokurtic.** If $\boldsymbol{\beta_2 = 3}$, the distribution is ***mesokurtic***. If $\boldsymbol{\beta_2 < 3}$, the distribution is **platykurtic.** The kurtosis of a distribution is also measured by using Greek letter 'γ_2' which is defined as $\gamma_2 = \beta_2 - 3$.

$\therefore \quad \gamma_2 > 0 \Rightarrow \beta_2 - 3 > 0 \Rightarrow \beta_2 > 3 \Rightarrow$ the distribution is **leptokurtic.**

Similarly, if $\gamma_2 = 0$, then $\beta_2 = 3 \Rightarrow$ The distribution is **mesokurtic.**

$\gamma_2 < 0 \Rightarrow \beta_2 < 3 \Rightarrow$ the distribution is **platykurtic.**

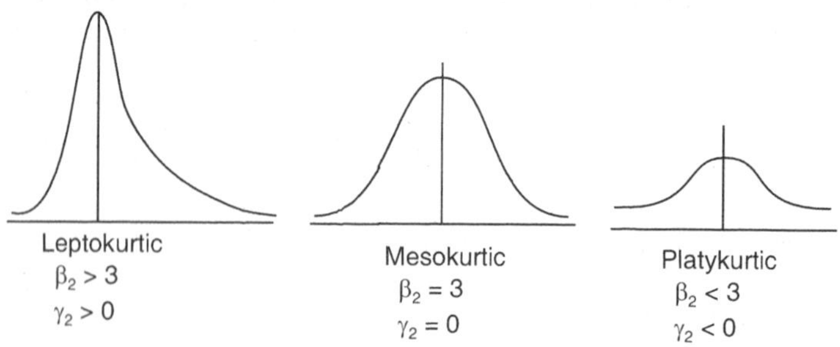

3.19.1 Steps for Computing β_2

I. *If the value of* μ_2 *and* μ_4 *are given, then find* β_2 *by using the formula:* $\beta_2 = \dfrac{\mu_4}{\mu_2^2}$.

II. *If raw moments* μ_1', μ_2', μ_3' *and* μ_4' *are given, then calculate:*

$\mu_2 = \mu_2' - \mu_1'^2$ *and* $\mu_4 = \mu_4' - 4\mu_3'\mu_1' + 6\mu_2'\mu_1'^2 - 3\mu_1'^4$

Now, find $\beta_2 = \dfrac{\mu_4}{\mu_2^2}$.

III. *If moments are not given, then first find μ_2 and μ_4 by using the given data and then use the formula:* $\beta_2 = \dfrac{\mu_4}{\mu_2^2}$.

IV. *The given distribution is leptokurtic, mesokurtic and platykurtic according as $\beta_2 > 3$, $\beta_2 = 3$ and $\beta_2 < 3$ respectively.*

ILLUSTRATIVE EXAMPLES

Example 1. *The first four moments about mean of a frequency distribution are 0, 100, – 7 and 35000. Discuss the kurtosis of the distribution.*

Sol. We have, $\mu_1 = 0, \mu_2 = 100, \mu_3 = -7$ and $\mu_4 = 35000$

Now,
$$\beta_2 = \frac{\mu_4}{\mu_2^2} = \frac{35000}{(100)^2} = 3.5 > 3$$

$\therefore$ The distribution is **leptokurtic.**

Example 2. *The first four moments of a distribution about the value '4' of the variable are – 1.5, 17,– 30 and 108. State whether the distribution is leptokurtic or platykurtic.*

(U.P.T.U. 2014)

Sol. We have, $\mu_1' = -1.5, \mu_2' = 17, \mu_3' = -30, \mu_4' = 108$

Moments about mean:

$$\mu_2 = \mu_2' - \mu_1'^2 = 17 - (-1.5)^2 = 14.75$$
$$\mu_4 = \mu_4' - 4\mu_3'\mu_1' + 6\mu_2'\mu_1'^2 - 3\mu_1'^4$$
$$= 108 - 4(-30)(-1.5) + 6(17)(-1.5)^2 - 3(-1.5)^4 = 142.3125$$

Kurtosis:
$$\beta_2 = \frac{\mu_4}{\mu_2^2} = \frac{142.3125}{(14.75)^2} = 0.6541$$

Since $\beta_2 < 3$, the distribution is **platykurtic.**

Example 3. *The first four moments of a distribution about $x = 4$ are 1, 4, 10 and 45. Obtain the various characteristics of the distribution on the basis of the given information. Comment upon the nature of the distribution.* (A.K.TU. 2016, 2018)

Sol. We have $A = 4, \mu_1' = 1, \mu_2' = 4, \mu_3' = 10$ and $\mu_4' = 45$

Moments about mean:

$$\mu_1 = 0 \text{ (always)}$$
$$\mu_2 = \mu_2' - \mu_1'^2 = 4 - (1)^2 = 3$$
$$\mu_3 = \mu_3' - 3\mu_2'\mu_1' + 2\mu_1'^3 = 10 - 3(4)(1) + 2(1)^3 = 0$$
$$\mu_4 = \mu_4' - 4\mu_3'\mu_1' + 6\mu_2'\mu_1'^2 - 3\mu_1'^4$$
$$= 45 - 4(10)(1) + 6(4)(1)^2 - 3(1)^4 = 26$$

Skewness: Moment coefficient of skewness, $\gamma_1 = \dfrac{\mu_3}{\sqrt{\mu_2^3}} = \dfrac{0}{\sqrt{(3)^3}} = 0$.

$\therefore$ The distribution is **symmetrical.**

Kurtosis: $\beta_2 = \dfrac{\mu_4}{\mu_2^2} = \dfrac{26}{(3)^2} = 2.89 < 3$ $\therefore$ The distribution is **platykurtic.**

Example 4. *The standard deviation of a symmetric distribution is 5. What must be the value of the fourth moment about the mean in order that the distribution be*

(i) leptokurtic (ii) mesokurtic (iii) platykurtic?

Sol. We have, $\sigma = 5 \Rightarrow \sigma^2 = 25 \Rightarrow \mu_2 = 25$ | $\because \mu_2 = \sigma^2$

Now, $$\beta_2 = \frac{\mu_4}{\mu_2^2} = \frac{\mu_4}{625}$$

Thus, the distribution will be

(*i*) Leptokurtic if $\beta_2 > 3 \Rightarrow \frac{\mu_4}{625} > 3 \Rightarrow \mu_4 > 1875$

(*ii*) Mesokurtic if $\beta_2 = 3 \Rightarrow \frac{\mu_4}{625} = 3 \Rightarrow \mu_4 = 1875$

(*iii*) Platykurtic if $\beta_2 < 3 \Rightarrow \frac{\mu_4}{625} < 3 \Rightarrow \mu_4 < 1875.$

Example 5. *The first four moments about the working mean 28.5 of a distribution are 0.294, 7.144, 42.409 and 454.98. Calculate the moments about the mean. Also evaluate β_1, β_2 and comment upon the skewness and kurtosis of the distribution.* (A.K.T.U. 2016)

Sol. We have, $\mu_1' = 0.294, \mu_2' = 7.144, \mu_3' = 42.409, \mu_4' = 454.98$

Moments about mean

$$\mu_1 = 0$$
$$\mu_2 = \mu_2' - \mu_1'^2 = 7.144 - (.294)^2 = 7.0576$$
$$\mu_3 = \mu_3' - 3\mu_2'\mu_1' + 2\mu_1'^3$$
$$= 42.409 - 3(7.144)(.294) + 2(.294)^3 = 36.1588$$
$$\mu_4 = \mu_4' - 4\mu_3'\mu_1' + 6\mu_2'\mu_1'^2 - 3\mu_1'^4$$
$$= 454.98 - 4(42.409)(.294) + 6(7.144)(.294)^2 - 3(.294)^4$$
$$= 408.7896$$

Calculation of β_1 and β_2

$$\beta_1 = \frac{\mu_3^2}{\mu_2^3} = 3.7193 \qquad \beta_2 = \frac{\mu_4}{\mu_2^2} = 8.2070$$

Skewness

Since β_1 is positive, $\gamma_1 = 1.9285$ | μ_3 is positive

$\therefore$ The distribution is **positively skewed.**

Kurtosis

Since $\beta_2 = 8.2070 > 3$

$\therefore$ The distribution is **leptokurtic.**

Example 6. *The first four moments of a distribution about the value '0' are – 0.20, 1.76, – 2.36 and 10.88. Find the moments about the mean and measure the kurtosis.*

Sol. We have, $\mu'_1 = -0.20, \mu'_2 = 1.76, \mu'_3 = -2.36, \mu'_4 = 10.88$

Moments about the mean:

$$\mu_1 = 0$$
$$\mu_2 = \mu'_2 - \mu_1'^2 = 1.76 - (-0.20)^2 = 1.72$$
$$\mu_3 = \mu'_3 - 3\mu'_2\mu'_1 + 2\mu_1'^3$$
$$= -2.36 - 3(1.76)(-0.20) + 2(-0.20)^3 = -1.32$$

$$\mu_4 = \mu'_4 - 4\mu'_3\mu'_1 + 6\mu'_2\mu_1'^2 - 3\mu_1'^4$$
$$= 10.88 - 4(-2.36)(-0.20) + 6(1.76)(-0.20)^2 - 3(-0.20)^4$$
$$= 9.4096$$

Kurtosis:

$$\beta_2 = \frac{\mu_4}{\mu_2^2} = 3.180638$$

Since, $\beta_2 > 3$ hence the distribution is **leptokurtic.**

Example 7. *The following table represents the height of a batch of 100 students. Calculate kurtosis.* (A.K.T.U. 2018)

Height (in cm)	*59*	*61*	*63*	*65*	*67*	*69*	*71*	*73*	*75*
No. of students	*0*	*2*	*6*	*20*	*40*	*20*	*8*	*2*	*2*

Sol. To calculate β_2, we will have to first find the values of μ_2 and μ_4.

Moments about 67

$$\mu_1' = \left(\frac{\Sigma fu}{N}\right)h = \left(\frac{12}{100}\right)(2) = 0.24$$

$$\mu_2' = \left(\frac{\Sigma fu^2}{N}\right)h^2 = \left(\frac{164}{100}\right)(4) = 6.56$$

Height (cm) x	*No. of students f*	$u = \frac{x-67}{2}$	fu	fu^2	fu^3	fu^4
59	0	– 4	0	0	0	0
61	2	– 3	– 6	18	– 54	162
63	6	– 2	– 12	24	– 48	96
65	20	– 1	– 20	20	– 20	20
67	40	0	0	0	0	0
69	20	1	20	20	20	20
71	8	2	16	32	64	128
73	2	3	6	18	54	162
75	2	4	8	32	128	512
	$N = \Sigma f = 100$		$\Sigma fu = 12$	$\Sigma fu^2 = 164$	$\Sigma fu^3 = 144$	$\Sigma fu^4 = 1100$

$$\mu_3' = \left(\frac{\Sigma fu^3}{N}\right)h^3 = \frac{144}{100} \times 8 = 11.52$$

$$\mu_4' = \left(\frac{\Sigma fu^4}{N}\right)h^4 = \frac{1100}{100} \times 16 = 176$$

Moments about mean

$$\mu_2 = \mu_2' - \mu_1'^2 = 6.56 - (.24)^2 = 6.5024$$

$$\mu_4 = \mu_4' - 4\mu_3'\mu_1' + 6\mu_2'\mu_1'^2 - 3\mu_1'^4$$

$$= 176 - 4\,(11.52)\,(0.24) + 6\,(6.56)\,(0.24)^2 - 3\,(0.24)^4 = 167.19798$$

Kurtosis

Measure of kurtosis $\beta_2 = \dfrac{\mu_4}{\mu_2^2} = 3.9544 > 3$

Hence, the distribution is **leptokurtic.**

TEST YOUR KNOWLEDGE

1. The first four moments about mean of a frequency distribution are 0, 60, – 50 and 8020 respectively. Discuss the kurtosis of the distribution.
2. The μ_2 and μ_4 for a distribution are found to be 2 and 12 respectively. Discuss the kurtosis of the distribution.
3. (*i*) The first four central moments of a distribution are 0, 2.5, 0.7 and 18.75. Test the kurtosis of the distribution. (*M.T.U. 2013*)

   (*ii*) Define skewness and kurtosis of a distribution. The first four moments of a distribution are 0, 2.5, 0.7 and 18.71. Find the coefficient of skewness and kurtosis. (*U.P.T.U. 2015*)
4. The standard deviation of symmetric distribution is 4. What must be the value of μ_4 so that the distribution may be mesokurtic?
5. (*i*) If the first four moments about the value '5' of the variable are – 4, 22, – 117 and 560, find the value of β_2 and discuss the kurtosis.

   (*ii*) The first four moments of a distribution about the value 5 of the variable are 2, 20, 40 and 50. Calculate the moments about the mean and comment upon the skewness and kurtosis of the distribution. (*G.B.T.U. 2011*)
6. (*i*) Calculate the value of β_2 for the following distribution:

Class	2.5–7.5	7.5–12.5	12.5–17.5	17.5–22.5	22.5–27.5	27.5–32.5	32.5–37.5
Frequency	8	15	20	32	23	17	5

   (*ii*) Compute the value of β_2 for the following distribution. Is the distribution platykurtic?

Class	10–20	20–30	30–40	40–50	50–60	60–70	70–80
Frequency	1	20	69	108	78	22	2

7. (*i*) Calculate μ_1, μ_2, μ_3, μ_4 for the frequency distribution of heights of 100 students given in the following table and hence find coefficient of skewness and kurtosis.

Height (cm) *Class interval*	144.5 – 149.5	149.5 – 154.5	154.5 – 159.5	159.5 – 164.5	164.5 – 169.5	169.5 – 174.5	174.5 – 179.5
Frequency	2	4	13	31	32	15	3

(*G.B.T.U. 2011*)

(*ii*) Find all four central moments and discuss skewness and kurtosis for the frequency distribution given in the following table:

Range of Expenditure (in ₹ 100 per month)	2–4	4–6	6–8	8–10	10–12
No. of families	38	292	389	212	69

[*G.B.T.U. 2013; M.T.U. 2012*]

8. (*i*) Find the measures of skewness and kurtosis on the basis of moments for the following distribution:

x	1	3	5	7	9
f	1	4	6	4	1

[*G.B.T.U. (C.O.) 2011*]

(*ii*) Find the measure of skewness and kurtosis on the basis of moments for the following distribution and draw your conclusion: [*A.K.T.U. 2018*]

Marks	5–15	15–25	25–35	35–45	45–55
No. of Students	1	3	5	7	4

9. Calculate β_1 and β_2 from the following data:

Profit (in lakhs of ₹)	10–20	20–30	30–40	40–50	50–60
No. of companies	18	20	30	22	10

Indicate the nature of frequency curve.

10. Prove that the frequency distribution curve of the following frequency distribution is leptokurtic.

Class	10–15	15–20	20–25	25–30	30–35	35–40	40–45	45–50	50–55
Frequency	1	4	8	19	35	20	7	5	1

11. Calculate the first four moments about the mean of the following distribution:

x	2	2.5	3	3.5	4	4.5	5
f	5	38	65	92	70	40	10

Also find the measures of skewness and kurtosis. (*M.T.U. 2012*)

12. Calculate the first four moments about the mean for the following frequency distribution and hence find the coefficient of skewness and kurtosis and comment upon the nature of the distribution.

Class-interval	5–10	10–15	15–20	20–25	25–30	30–35	35–40
Frequency	6	8	17	21	15	11	2

(*G.B.T.U. 2013*)

13. Define the coefficients of kurtosis. [*M.T.U. 2014; G.B.T.U. (C.O.) 2011*]

14. (*i*) What do you mean by kurtosis? Explain in brief.

(*ii*) Define kurtosis of a distribution.

Answers

1. $\beta_2 = 2.2278$, Platykurtic **2.** $\beta_2 = 3$, Mesokurtic

3. (*i*) $\beta_2 = 3$, Mesokurtic (*ii*) 0.17708, 2.9936 **4.** $\mu_4 = 768$

5. (*i*) $\beta_2 = 0.8889$, Platykurtic

(*ii*) $\mu_1 = 0, \mu_2 = 16, \mu_3 = -64, \mu_4 = 162$

$\gamma_1 = -1, \beta_2 = 0.6328$; Negatively skewed and platykurtic

6. (*i*) $\beta_2 = 2.3216$, Platykurtic (*ii*) $\beta_2 = 2.7240$, Yes

7. (*i*) $\mu_1 = 0, \mu_2 = 36.66, \mu_3 = -85.104, \mu_4 = 4373.3832, \gamma_1 = -0.3834, \beta_2 = 3.2541$

(*ii*) $\mu_1 = 0, \mu_2 = 37267.04, \mu_3 = 1746530.688, \mu_4 = 3567851989$

$y_1 = 0.24275, \beta_2 = 2.5689$, positively skewed and platykurtic.

8. (*i*) $\gamma_1 = 0, \beta_2 = 2.5$

(*ii*) $\mu_1 = 0, \mu_2 = 125, \mu_3 = -600, \mu_4 = 37625, \gamma_1 = -0.4293, \beta_2 = 2.408$, negatively skewed and platykurtic.

9. $\beta_1 = 0.0001, \beta_2 = 2.047$, Platykurtic.

11. $\mu_1 = 0, \mu_2 = 0.45328125, \mu_3 = 0.009890625, \mu_4 = 0.502111743, \gamma_1 = 0.0324, \beta_2 = 2.44379$, positively skewed and platykurtic.

12. $\mu_1 = 0, \mu_2 = 56, \mu_3 = -176.5625, \mu_4 = 7502.9375, \gamma_1 = -0.4213, \beta_2 = 2.3925$; negatively skewed and platykurtic.

3.20 CURVE FITTING

Let there be two variables x and y which give us a set of n pairs of numerical values (x_1, y_1), (x_2, y_2), ..., (x_n, y_n). In order to have an approximate idea about the relationship of these two variables, we plot these n paired points on a graph thus, we get a diagram showing the simultaneous variation in values of both the variables called *scatter or dot diagram.* From scatter diagram, we get only an approximate non-mathematical relation between two variables. Curve fitting means an exact relationship between two variables by algebraic equations, infact this relationship is the equation of the curve. Therefore, curve fitting means to form an equation of the curve from the given data. Curve fitting is considered of immense importance both from the point of view of theoretical and practical statistics.

Theoretically, it is useful in the study of correlation and regression. Practically, it enables us to represent the relationship between two variables by simple algebraic expressions *e.g.*, polynomials, exponential or logarithmic functions.

It is also used to estimate the values of one variable corresponding to the specified values of the other variable.

The constants occurring in the equation of approximate curve can be found by following methods:

(*i*) Graphical method (*ii*) Method of group averages

(*iii*) Method of least squares (*iv*) Method of moments.

Out of the above four methods, we will only discuss and study here *method of least squares.*

3.21 METHOD OF LEAST SQUARES

Method of least squares provides a unique set of values to the constants and hence suggests a curve of best fit to the given data.

Suppose we have m-paired observations $(x_1, y_1), (x_2, y_2), \ldots\ldots, (x_m, y_m)$ of two variables x and y. It is required to fit a polynomial of degree n of the type

$$y = a + bx + cx^2 + \ldots\ldots + kx^n \qquad \ldots(1)$$

of these values. We have to determine the constants $a, b, c, \ldots, k$ such that it represents the curve of best fit of that degree.

In case $m = n$, we get in general a unique set of values satisfying the given system of equations.

But if $m > n$ then, we get m equations by putting different values of x and y in equation (1) and we want to find only the values of n constants. Thus, there may be no such solution to satisfy all m equations.

Therefore, we try to find out those values of $a, b, c, \ldots\ldots, k$ which satisfy all the equations as nearly as possible. We apply the method of least squares in such cases.

Putting $x_1, x_2, \ldots, x_m$ for x in (1), we get

$$\begin{aligned} y_1' &= a + bx_1 + cx_1^2 + \ldots\ldots + kx_1^n \\ y_2' &= a + bx_2 + cx_2^2 + \ldots\ldots + kx_2^n \\ &\vdots \qquad \vdots \\ y_m' &= a + bx_m + cx_m^2 + \ldots\ldots + kx_m^n \end{aligned}$$

where $y_1', y_2', \ldots\ldots, y_m'$ are the expected values of y for $x = x_1, x_2, \ldots\ldots, x_m$ respectively. The values $y_1, y_2, \ldots\ldots, y_m$ are called observed values of y corresponding to $x = x_1, x_2, \ldots\ldots, x_m$ respectively.

The expected values are different from observed values, the difference $y_r - y_r'$ for different values of r are called *residuals.*

Introduce a new quantity U such that

$$U = \Sigma(y_r - y_r')^2 = \Sigma(y_r - a - bx_r - cx_r^2 - \ldots\ldots - kx_r^n)^2$$

The constants $a, b, c, \ldots\ldots, k$ are chosen in such a way that the sum of the squares of residuals is minimum.

Now the condition for U to be maximum or minimum is $\dfrac{\partial U}{\partial a} = 0 = \dfrac{\partial U}{\partial b} = \dfrac{\partial U}{\partial c} = \ldots\ldots = \dfrac{\partial U}{\partial k}$.

On simplifying these relations, we get

$$\begin{aligned} \Sigma y &= ma + b\Sigma x + \ldots\ldots + k\Sigma x^n \\ \Sigma xy &= a\Sigma x + b\Sigma x^2 + \ldots\ldots + k\,\Sigma x^{n+1} \\ \Sigma x^2 y &= a\Sigma x^2 + b\Sigma x^3 + \ldots\ldots + k\,\Sigma x^{n+2} \\ &\vdots \qquad \vdots \\ \Sigma x^n y &= a\Sigma x^n + b\Sigma x^{n+1} + \ldots\ldots + k\,\Sigma x^{2n} \end{aligned}$$

These are known as *Normal equations* and can be solved as simultaneous equations to give the values of the constants $a, b, c, \ldots\ldots, k$. These equations are $(n + 1)$ in number.

If we calculate the second order partial derivatives and these values are put, they give a positive value of the function, so U is minimum.

This method does not help us to choose the degree of the curve to be fitted but helps us in finding the values of the constants when the form of the curve has already been chosen.

3.22 FITTING A STRAIGHT LINE

Let (x_i, y_i), $i = 1, 2, ..., n$ be n sets of observations of related data and

$$y = a + bx \qquad ...(1)$$

be the straight line to be fitted. The residual at $x = x_i$ is

$$E_i = y_i - f(x_i) = y_i - a - bx_i$$

Introduce a new quantity U such that

$$U = \sum_{i=1}^{n} E_i^2 = \sum_{i=1}^{n} (y_i - a - bx_i)^2$$

By the principle of Least squares, U is minimum

$$\therefore \quad \frac{\partial U}{\partial a} = 0 \quad \text{and} \quad \frac{\partial U}{\partial b} = 0$$

$$\therefore \quad 2\sum_{i=1}^{n} (y_i - a - bx_i)(-1) = 0 \quad \text{or} \quad \boxed{\Sigma y = na + b\Sigma x} \qquad ...(2)$$

and

$$2\sum_{i=1}^{n} (y_i - a - bx_i)(-x_i) = 0 \quad \text{or} \quad \boxed{\Sigma xy = a\Sigma x + b\Sigma x^2} \qquad ...(3)$$

Since x_i, y_i are known, equations (2) and (3) result two equations in a and b. Solving these, the best values for a and b can be known and hence equation (1).

Note. In case of change of origin,

if n is odd then, $u = \dfrac{x - \text{(middle term)}}{\text{interval } (h)}$

but if n is even then, $u = \dfrac{x - \text{(mean of two middle terms)}}{\frac{1}{2}\text{(interval)}}$.

ILLUSTRATIVE EXAMPLES

Example 1. *By the method of least squares, find the straight line that best fits the following data:*

x:	1	2	3	4	5
y:	14	27	40	55	68.

Sol. Let the straight line of best fit be $y = a + bx$ ...(1)

Normal equations are $\Sigma y = ma + b\Sigma x$ ...(2)

and $\Sigma xy = a\Sigma x + b\Sigma x^2$ ...(3)

Here $m = 5$

Table is as below:

x	y	xy	x^2
1	14	14	1
2	27	54	4
3	40	120	9
4	55	220	16
5	68	340	25
$\Sigma x = 15$	$\Sigma y = 204$	$\Sigma xy = 748$	$\Sigma x^2 = 55$

Substituting in (2) and (3), we get

$$204 = 5a + 15b$$
$$748 = 15a + 55b$$

Solving, we get $a = 0, b = 13.6$

Hence required straight line is $y = 13.6x$

Example 2. *Fit a straight line to the following data by least square method:*

x:	0	1	2	3	4
y:	1	1.8	3.3	4.5	6.3.

(U.K.T.U. 2011)

Sol. Let the straight line obtained from the given data be $y = a + bx$ then the normal equations are

$$\Sigma y = ma + b\,\Sigma x \quad ...(1)$$
$$\Sigma xy = a\Sigma x + b\Sigma x^2 \quad ...(2)$$

Here, $m = 5$

x	y	xy	x^2
0	1	0	0
1	1.8	1.8	1
2	3.3	6.6	4
3	4.5	13.5	9
4	6.3	25.2	16
$\Sigma x = 10$	$\Sigma y = 16.9$	$\Sigma xy = 47.1$	$\Sigma x^2 = 30$

From (1) and (2), $16.9 = 5a + 10b$

and $47.1 = 10a + 30b$

Solving, we get $a = 0.72, b = 1.33$

$\therefore$ Required line is $y = 0.72 + 1.33\,x$.

Example 3. *Show that the best fitting linear function for the points $(x_1, y_1), (x_2, y_2), \ldots\ldots, (x_n, y_n)$ may be expressed in the form*

$$\begin{vmatrix} x & y & 1 \\ \Sigma x_i & \Sigma y_i & n \\ \Sigma x_i^2 & \Sigma x_i y_i & \Sigma x_i \end{vmatrix} = 0 \qquad (i = 1, 2, \ldots\ldots, n)$$

Show that the line passes through the mean point $(\bar{x}, \bar{y})$.

Sol. Let the best fitting linear function be $y = a + bx$ ...(1)

Then the normal equations are

$$\Sigma y_i = na + b\Sigma x_i \quad ...(2)$$

and

$$\Sigma x_i y_i = a\Sigma x_i + b\Sigma x_i^2 \quad ...(3)$$

Equations (1), (2), (3) may be rewritten as

$$bx - y + a = 0$$
$$b\Sigma x_i - \Sigma y_i + na = 0$$

and

$$b\Sigma x_i^2 - \Sigma x_i y_i + a\Sigma x_i = 0$$

Eliminating a and b between these equations

$$\begin{vmatrix} x & y & 1 \\ \Sigma x_i & \Sigma y_i & n \\ \Sigma x_i^2 & \Sigma x_i y_i & \Sigma x_i \end{vmatrix} = 0 \qquad \ldots(4)$$

which is the required best fitting linear function for the mean point $(\bar{x}, \bar{y})$,

$$\bar{x} = \frac{1}{n}\Sigma x_i, \quad \bar{y} = \frac{1}{n}\Sigma y_i\,.$$

Clearly, the line (4) passes through point $(\bar{x}, \bar{y})$ as two rows of determinant being equal make it zero.

TEST YOUR KNOWLEDGE

1. Fit a straight line to the following data regarding x as the independent variable:

(i)

x	1	2	3	4	5	6
y	1200	900	600	200	110	50

(ii)

x	71	68	73	69	67	65	66	67
y	69	72	70	70	68	67	68	64

(iii)

x	0	5	10	15	20	25
y	12	15	17	22	24	30

2. (i) Find the best values of a and b so that $y = a + bx$ fits the given data:

x	0	1	2	3	4
y	1.0	2.9	4.8	6.7	8.6

(ii) Fit a straight line of the form $y = a_0 + a_1x$ to the data:

x	1	2	3	4	6	8
y	2.4	3.1	3.5	4.2	5.0	6.0

3. Fit a straight line approximate to the data:

x	1	2	3	4
y	3	7	13	21

4. A simply supported beam carries a concentrated load P(*lb*) at its midpoint. Corresponding to various values of P, the maximum deflection Y (*in*) is measured. The data are given below. Find a law of the type $Y = a + bP$

P	100	120	140	160	180	200
Y	0.45	0.55	0.60	0.70	0.80	0.85

5. What straight line best fits the following data in the least square sense?

x	1	2	3	4
y	0	1	1	2

6. The weight of a calf taken at weekly intervals are given below. Fit a straight line using method of least squares and calculate the average rate of growth per week.

Age	1	2	3	4	5	6	7	8	9	10
Weight	52.5	58.7	65	70.2	75.4	81.1	87.2	95.5	102.2	108.4

7. Find the least square line for the data points

(– 1, 10), (0, 9), (1, 7), (2, 5), (3, 4), (4, 3), (5, 0) and (6, –1).

8. Find the least square line $y = a + bx$ for the data:

x_i	– 2	– 1	0	1	2
y_i	1	2	3	3	4

9. If P is the pull required to lift a load W by means of a pulley block, find a linear law of the form $P = mW + c$ connecting P and W, using the data:

P	12	15	21	25
W	50	70	100	120

where P and W are taken in kg-wt.

10. (*i*) Using the method of least squares, fit a straight line to the following data:

x	1	2	3	4	5
y	2	4	6	8	10

(*ii*)

x	0	2	4	5	6
y	5.012	10	15	21	30

11. Find the least square line that fits the following data, assuming that x-values are free from error:

x	1	2	3	4	5	6
y	5.04	8.12	10.64	13.18	16.20	20.04

Answers

1. (*i*) $y = 1361.97 - 243.42\,x$ (*ii*) $y = 39.5454 + 0.4242\,x$ (*iii*) $y = 11.285 + 0.7\,x$

2. (*i*) $y = 1 + 1.9x$ (*ii*) $y = 2.0253 + 0.502\,x$ **3.** $y = -4 + 6x$

4. $Y = 0.004P + 0.048$ **5.** $y = -0.5 + 0.6x$ **6.** $y = 45.74 + 6.16x$, 6.16

7. $y = -1.6071429x + 8.6428571$ **8.** $y = 2.6 + (0.7)\,x$ **9.** $P = 2.2759 + 0.1879\,W$

10. (*i*) $y = 2x$ (*ii*) $y = 3.07734 + 3.86031\,x$ **11.** $y = 2.0253 + 2.908\,x$.

3.23 FITTING OF AN EXPONENTIAL CURVE $y = ae^{bx}$

Taking logarithm on both sides, we get

$$\log_{10} y = \log_{10} a + bx \log_{10} e$$

i.e.,
$$\boxed{Y = A + BX} \quad \ldots(1)$$

where $Y = \log_{10} y$, $A = \log_{10} a$, $B = b \log_{10} e$ and $X = x$

The normal equations for (1) are

$$\Sigma Y = nA + B\Sigma X \quad \text{and} \quad \Sigma XY = A\Sigma X + B\Sigma X^2$$

Solving these, we get A and B.

Then $a = \text{antilog } A$ and $b = \dfrac{B}{\log_{10} e}$.

3.24 FITTING OF THE CURVE $y = ax^b$

Taking logarithm on both sides, we get

$$\log_{10} y = \log_{10} a + b \log_{10} x$$

i.e.,
$$\boxed{Y = A + BX} \quad \ldots(1)$$

where $Y = \log_{10} y$, $A = \log_{10} a$, $B = b$ and $X = \log_{10} x$.

The normal equations to (1) are

$$\Sigma Y = nA + B\Sigma X$$

and
$$\Sigma XY = A\Sigma X + B\Sigma X^2$$

which results A and B on solving and $a = \text{antilog } A$, $b = B$.

3.25 FITTING OF THE CURVE $y = ab^x$

Taking logarithm on both sides, we get

$$\log y = \log a + x \log b$$

$\Rightarrow$
$$\boxed{Y = A + BX} \quad \ldots(1)$$

where $Y = \log y$, $A = \log a$, $B = \log b$, $X = x$.

This is a linear equation in Y and X.

For estimating A and B, normal equations are

$$\Sigma Y = nA + B\ \Sigma X$$

and
$$\Sigma XY = A\ \Sigma X + B\ \Sigma X^2$$

where n is the number of pairs of values of x and y.

Ultimately, $a = \text{antilog } (A)$ and $b = \text{antilog } (B)$.

3.26 FITTING OF THE CURVE $pv^{\gamma} = k$

$$pv^{\gamma} = k \quad \Rightarrow \quad v = k^{1/\gamma} p^{-1/\gamma}$$

Taking logarithm on both sides, we get

$$\log v = \frac{1}{\gamma} \log k - \frac{1}{\gamma} \log p$$

$$\Rightarrow \qquad \boxed{Y = A + BX}$$

where $Y = \log v$, $A = \frac{1}{\gamma} \log k$, $B = -\frac{1}{\gamma}$ and $X = \log p$

γ and k are determined by above equations. Normal equations are obtained as that of the straight line.

3.27 FITTING OF THE CURVE OF TYPE $xy = b + ax$ *(A.K.T.U. 2017)*

$$xy = b + ax \qquad \Rightarrow \quad y = \frac{b}{x} + a$$

$$\Rightarrow \qquad Y = bX + a, \qquad \text{where } X = \frac{1}{x}.$$

Normal equations are $\Sigma Y = na + b\Sigma X$ and $\Sigma XY = a\Sigma X + b\Sigma X^2$.

3.28 FITTING OF THE CURVE $y = ax^2 + \frac{b}{x}$

Let the n points be $(x_1, y_1), (x_2, y_2), \ldots, (x_n, y_n)$

Error of estimate for i^{th} point (x_i, y_i) is

$$E_i = \left(y_i - ax_i^2 - \frac{b}{x_i}\right)$$

By principle of Least squares, the values of a and b are such that

$$U = \sum_{i=1}^{n} E_i^2 = \sum_{i=1}^{n} \left(y_i - ax_i^2 - \frac{b}{x_i}\right)^2 \text{ is minimum.}$$

Normal equations are given by

$$\frac{\partial U}{\partial a} = 0 \quad \Rightarrow \quad \sum_{i=1}^{n} x_i^2 y_i = a \sum_{i=1}^{n} x_i^4 + b \sum_{i=1}^{n} x_i$$

and

$$\frac{\partial U}{\partial b} = 0 \quad \Rightarrow \quad \sum_{i=1}^{n} \frac{y_i}{x_i} = a \sum_{i=1}^{n} x_i + b \sum_{i=1}^{n} \frac{1}{x_i^2}$$

or Dropping the suffix i, normal equations are

$$\boxed{\Sigma x^2 y = a\, \Sigma x^4 + b \Sigma x} \quad \text{and} \quad \boxed{\sum \frac{y}{x} = a\, \Sigma x + b \sum \frac{1}{x^2}.}$$

3.29 FITTING OF THE CURVE $y = ax + bx^2$ (A.K.T.U. 2019)

Error of estimate for i^{th} point (x_i, y_i) is $E_i = (y_i - ax_i - bx_i^2)$

By principle of Least squares, the values of a and b are such that

$$U = \sum_{i=1}^{n} E_i^2 = \sum_{i=1}^{n} (y_i - ax_i - bx_i^2)^2 \text{ is minimum.}$$

Normal equations are given by

$$\frac{\partial U}{\partial a} = 0 \Rightarrow \sum_{i=1}^{n} x_i y_i = a \sum_{i=1}^{n} x_i^2 + b \sum_{i=1}^{n} x_i^3$$

and
$$\frac{\partial U}{\partial b} = 0 \Rightarrow \sum_{i=1}^{n} x_i^2 y_i = a \sum_{i=1}^{n} x_i^3 + b \sum_{i=1}^{n} x_i^4$$

or Dropping the suffix i, normal equations are

$$\boxed{\Sigma xy = a\,\Sigma x^2 + b\Sigma x^3} \quad \text{and} \quad \boxed{\Sigma x^2 y = a\,\Sigma x^3 + b\Sigma x^4.}$$

3.30 FITTING OF THE CURVE $y = ax + \frac{b}{x}$ (A.K.T.U. 2018)

Error of estimate for i^{th} point (x_i, y_i) is

$$E_i = y_i - ax_i - \frac{b}{x_i}$$

By principle of Least squares, the values of a and b are such that

$$U = \sum_{i=1}^{n} E_i^2 = \sum_{i=1}^{n} \left(y_i - ax_i - \frac{b}{x_i}\right)^2 \text{ is minimum.}$$

Normal equations are given by

$$\frac{\partial U}{\partial a} = 0$$

$$\Rightarrow \quad 2\sum_{i=1}^{n} \left(y_i - ax_i - \frac{b}{x_i}\right)(-x_i) = 0$$

$$\Rightarrow \quad \sum_{i=1}^{n} x_i y_i = a \sum_{i=1}^{n} x_i^2 + nb \quad \text{...(1)}$$

and
$$\frac{\partial U}{\partial b} = 0$$

$$\Rightarrow \quad 2\sum_{i=1}^{n} \left(y_i - ax_i - \frac{b}{x_i}\right)\left(-\frac{1}{x_i}\right) = 0$$

$$\Rightarrow \quad \sum_{i=1}^{n} \frac{y_i}{x_i} = na + b \sum_{i=1}^{n} \frac{1}{x_i^2} \quad \text{...(2)}$$

Dropping the suffix i, normal equations are

$$\boxed{\Sigma xy = a\Sigma x^2 + nb} \quad \text{and} \quad \boxed{\sum \frac{y}{x} = na + b\sum \frac{1}{x^2}}$$

where n is the no. of pairs of values of x and y.

3.31 FITTING OF THE CURVE $y = a + \frac{b}{x} + \frac{c}{x^2}$

Normal equations are

$$\Sigma y = ma + b\sum \frac{1}{x} + c\sum \frac{1}{x^2}$$

$$\sum \frac{y}{x} = a\sum \frac{1}{x} + b\sum \frac{1}{x^2} + c\sum \frac{1}{x^3}$$

$$\sum \frac{y}{x^2} = a\sum \frac{1}{x^2} + b\sum \frac{1}{x^3} + c\sum \frac{1}{x^4}$$

where m is number of pairs of values of x and y.

3.32 FITTING OF THE CURVE $y = \frac{c_0}{x} + c_1\sqrt{x}$ *(A.K.T.U. 2016)*

Error of estimate for i^{th} point (x_i, y_i) is

$$E_i = y_i - \frac{c_0}{x_i} - c_1\sqrt{x_i}$$

By principle of Least squares, the values of a and b are such that

$$U = \sum_{i=1}^{n} E_i^2 = \sum_{i=1}^{n} \left(y_i - \frac{c_0}{x_i} - c_i\sqrt{x_i}\right)^2 \text{ is minimum.}$$

Normal equations are given by

$$\frac{\partial U}{\partial c_0} = 0 \quad \text{and} \quad \frac{\partial U}{\partial c_1} = 0$$

Now, $$\frac{\partial U}{\partial c_0} = 0$$

$$\Rightarrow \quad 2\sum_{i=1}^{n} \left(y_i - \frac{c_0}{x_i} - c_1\sqrt{x_i}\right)\left(-\frac{1}{x_i}\right) = 0$$

$$\Rightarrow \quad \sum_{i=1}^{n} \frac{y_i}{x_i} = c_0\sum_{i=1}^{n} \frac{1}{x_i^2} + c_1\sum_{i=1}^{n} \frac{1}{\sqrt{x_i}} \qquad \ldots(1)$$

Also, $$\frac{\partial U}{\partial c_1} = 0$$

$$\Rightarrow \quad 2\sum_{i=1}^{n} \left(y_i - \frac{c_0}{x_i} - c_1\sqrt{x_i}\right)(-\sqrt{x_i}) = 0$$

$$\Rightarrow \qquad \sum_{i=1}^{n} y_i \sqrt{x_i} = c_0 \sum_{i=1}^{n} \frac{1}{\sqrt{x_i}} + c_1 \sum_{i=1}^{n} x_i \qquad ...(2)$$

Dropping suffix i, normal equations (1) and (2) become

$$\sum \frac{y}{x} = c_0 \sum \frac{1}{x^2} + c_1 \sum \frac{1}{\sqrt{x}}$$

and

$$\sum y\sqrt{x} = c_0 \sum \frac{1}{\sqrt{x}} + c_1 \sum x.$$

3.33 FITTING OF THE CURVE $2^x = ax^2 + bx + c$

Normal equations are

$$\Sigma\, 2^x x^2 = a\Sigma x^4 + b\Sigma x^3 + c\Sigma x^2$$

$$\Sigma\, 2^x x = a\Sigma x^3 + b\Sigma x^2 + c\Sigma x$$

and

$$\Sigma\, 2^x = a\Sigma x^2 + b\Sigma x + mc$$

where, m is number of points (x_i, y_i)

3.34 FITTING OF THE CURVE $y = ae^{-3x} + be^{-2x}$

Normal equations are

$$\Sigma ye^{-3x} = a\ \Sigma e^{-6x} + b\ \Sigma e^{-5x}$$

and

$$\Sigma ye^{-2x} = a\ \Sigma e^{-5x} + b\ \Sigma e^{-4x}$$

ILLUSTRATIVE EXAMPLES

Example 1. *Find the curve of best fit of the type $y = ae^{bx}$ to the following data by the method of Least squares:*

x:	1	5	7	9	12
y:	10	15	12	15	21.

Sol. The curve to be fitted is $y = ae^{bx}$

or $\qquad Y = A + BX$, where, $Y = \log_{10} y$, $A = \log_{10} a$, $X = x$ and $B = b \log_{10} e$

$\therefore$ The normal equations are $\Sigma Y = 5A + B\Sigma X$

and $\qquad \Sigma XY = A\Sigma X + B\Sigma X^2$

$X = x$	y	$Y = \log_{10} y$	X^2	XY
1	10	1.0000	1	1
5	15	1.1761	25	5.8805
7	12	1.0792	49	7.5544
9	15	1.1761	81	10.5849
12	21	1.3222	144	15.8664
$\Sigma X = 34$		$\Sigma Y = 5.7536$	$\Sigma X^2 = 300$	$\Sigma XY = 40.8862$

Substituting the above values in the normal equations, we get

$$5.7536 = 5A + 34B$$

and

$$40.8862 = 34A + 300B$$

On solving, $A = 0.9766$; $B = 0.02561$

$$\therefore \quad a = \text{antilog}_{10} A = 9.4754 \; ; \; b = \frac{B}{\log_{10} e} = 0.059$$

Hence the required curve is $y = 9.4754e^{0.059x}$.

Example 2. *Determine the constants a and b by the method of least squares such that* $y = ae^{bx}$ *fits the following data:*

x	2	4	6	8	10
y	4.077	11.084	30.128	81.897	222.62

Sol. $y = ae^{bx}$

Taking log on both sides

$$\log y = \log a + bx \log e$$

or $$Y = A + BX,$$

where, $Y = \log y, \quad A = \log a, \quad B = b \log_{10} e, \quad X = x.$

Normal equations are

$$\Sigma Y = mA + B\Sigma X \qquad ...(1)$$

and $$\Sigma XY = A\Sigma X + B\Sigma X^2. \qquad ...(2)$$

Here, $m = 5$.

Table is as follows:

x	y	X	Y	XY	X^2
2	4.077	2	.61034	1.22068	4
4	11.084	4	1.04469	4.17876	16
6	30.128	6	1.47897	8.87382	36
8	81.897	8	1.91326	15.30608	64
10	222.62	10	2.347564	23.47564	100
		$\Sigma X = 30$	$\Sigma Y = 7.394824$	$\Sigma XY = 53.05498$	$\Sigma X^2 = 220$

Substituting these values in equations (1) and (2), we get

$$7.394824 = 5A + 30B$$

and $$53.05498 = 30A + 220B.$$

Solving, we get $A = 0.1760594$ and $B = 0.2171509$

$\therefore$ $a = \text{antilog}\,(A) = \text{antilog}\,(0.1760594) = 1.49989$

and $$b = \frac{B}{\log_{10} e} = \frac{0.2171509}{0.4342945} = 0.50001$$

Hence the required equation is

$$y = 1.49989\, e^{0.50001x}.$$

Example 3. *Obtain a relation of the form $y = ab^x$ for the following data by the method of least squares:* [G.B.T.U. MCA (SUM) 2010]

x	2	3	4	5	6
y	8.3	15.4	33.1	65.2	127.4

Sol. The curve to be fitted is $y = ab^x$

or $\quad Y = A + Bx,$

where, $\quad A = \log_{10} a,\ B = \log_{10} b \quad \text{and} \quad Y = \log_{10} y.$

$\therefore$ The normal equations are $\Sigma Y = 5A + B\Sigma x$

and $\quad \Sigma xY = A\Sigma x + B\Sigma x^2.$

x	y	$Y = \log_{10} y$	x^2	xY
2	8.3	0.9191	4	1.8382
3	15.4	1.1872	9	3.5616
4	33.1	1.5198	16	6.0792
5	65.2	1.8142	25	9.0710
6	127.4	2.1052	36	12.6312
$\Sigma x = 20$		$\Sigma Y = 7.5455$	$\Sigma x^2 = 90$	$\Sigma xY = 33.1812$

Substituting the above values, we get

$$7.5455 = 5A + 20B \quad \text{and} \quad 33.1812 = 20A + 90B.$$

On solving $\quad A = 0.31$ and $B = 0.3$

$\therefore \quad a = \text{antilog } A = 2.04 \quad \text{and} \quad b = \text{antilog } B = 1.995.$

Hence the required curve is $\quad y = 2.04(1.995)^x.$

Example 4. *Obtain the least squares fit of the form $f(t) = a\, e^{-3t} + be^{-2t}$ for the data:*

t:	0.1	0.2	0.3	0.4
$f(t)$:	0.76	0.58	0.44	0.35

Sol. Normal equations to the curve $f(t) = a\, e^{-3t} + be^{-2t}$ are:

$$\Sigma f(t)\, e^{-3t} = a\, \Sigma\, e^{-6t} + b\, \Sigma\, e^{-5t} \quad \ldots(1)$$

$$\Sigma f(t)\, e^{-2t} = a\, \Sigma\, e^{-5t} + b\, \Sigma\, e^{-4t} \quad \ldots(2)$$

See art. 3.34

Table of values is

t	$f(t)$	e^{-4t}	e^{-5t}	e^{-6t}	$f(t)\, e^{-2t}$	$f(t)\, e^{-3t}$
0.1	0.76	0.6703	0.6065	0.5488	0.6222	0.5630
0.2	0.58	0.4493	0.3679	0.3012	0.3888	0.3183
0.3	0.44	0.3012	0.2231	0.1653	0.2415	0.1789
0.4	0.35	0.2019	0.1353	0.0907	0.1573	0.1054
Total		Σe^{-4t} = 1.6227	Σe^{-5t} = 1.3328	Σe^{-6t} = 1.106	$\Sigma f(t)\, e^{-2t}$ = 1.4098	$\Sigma f(t)\, e^{-3t}$ = 1.1656

Substituting values in (1) and (2), we get

$$1.106\,a + 1.3328\,b = 1.1656$$
$$1.3328\,a + 1.6227\,b = 1.4098$$

On solving, we get $a = 0.6778$, $b = 0.3121$.

Hence the least squares fit is $f(t) = 0.6778\,e^{-3t} + 0.3121e^{-2t}$.

Example 5. *By the method of least squares, find the curve $y = ax + bx^2$ that best fits the following data:* (U.P.T.U. 2014)

x	1	2	3	4	5
y	1.8	5.1	8.9	14.1	19.8

Sol. Normal equations are

$$\Sigma xy = a\Sigma x^2 + b\Sigma x^3 \quad ...(1)$$

and $$\Sigma x^2y = a\Sigma x^3 + b\Sigma x^4 \quad ...(2)$$

Let us form a table as below:

x	y	x^2	x^3	x^4	xy	x^2y
1	1.8	1	1	1	1.8	1.8
2	5.1	4	8	16	10.2	20.4
3	8.9	9	27	81	26.7	80.1
4	14.1	16	64	256	56.4	225.6
5	19.8	25	125	625	99	495
Total		$\Sigma x^2 = 55$	$\Sigma x^3 = 225$	$\Sigma x^4 = 979$	$\Sigma xy = 194.1$	$\Sigma x^2y = 822.9$

Substituting these values in equations (1) and (2), we get

$$194.1 = 55\,a + 225\,b$$

and $$822.9 = 225\,a + 979\,b$$

$$\Rightarrow \quad a = \frac{83.85}{55} \simeq 1.52 \quad \text{and} \quad b = \frac{317.4}{664} \simeq .49$$

Hence required parabolic curve is $y = 1.52\,x + 0.49\,x^2$.

Example 6. *Fit the curve $pv^\gamma = k$ to the following data:*

p (kg/cm^2)	0.5	1	1.5	2	2.5	3
v (litres)	1620	1000	750	620	520	460

Sol. $$pv^\gamma = k$$

$$\Rightarrow \quad v = \left(\frac{k}{p}\right)^{1/\gamma} = k^{1/\gamma} p^{-1/\gamma}$$

Taking log, $$\log v = \frac{1}{\gamma}\log k - \frac{1}{\gamma}\log p$$

which is of the form $$Y = A + BX$$

where $Y = \log v$, $X = \log p$, $A = \dfrac{1}{\gamma}\log k$ and $B = -\dfrac{1}{\gamma}$

p	v	X	Y	XY	X^2
0.5	1620	– 0.30103	3.20952	– 0.96616	0.09062
1	1000	0	3	0	0
1.5	750	0.17609	2.87506	0.50627	0.03101
2	620	0.30103	2.79239	0.84059	0.09062
2.5	520	0.39794	2.716	1.08080	0.15836
3	460	0.47712	2.66276	1.27046	0.22764
Total		$\Sigma X = 1.05115$	$\Sigma Y = 17.25573$	$\Sigma XY = 2.73196$	$\Sigma X^2 = 0.59825$

Here, $m = 6$

Normal equations are

$$17.25573 = 6\text{A} + 1.05115\ \text{B}$$

and
$$2.73196 = 1.05115\ \text{A} + 0.59825\ \text{B}$$

Solving these, we get

$$\text{A} = 2.99911 \quad \text{and} \quad \text{B} = -\,0.70298$$

$$\therefore \qquad \gamma = -\frac{1}{\text{B}} = \frac{1}{0.70298} = 1.42252$$

Again,
$$\log k = \gamma\text{A} = 4.26629$$

$$\therefore \qquad k = \text{antilog}\,(4.26629) = 18462.48$$

Hence, required curve is

$$pv^{1.42252} = 18462.48.$$

Example 7. *The pressure of the gas corresponding to various volumes V is measured, given by the following data:*

V (cm^3):	*50*	*60*	*70*	*90*	*100*
P (kg cm^{-2}):	*64.7*	*51.3*	*40.5*	*25.9*	*78*

Fit the data to the equation $PV^{\gamma} = C$. (A.K.T.U. 2019)

Sol.
$$\text{PV}^{\gamma} = \text{C}$$

$$\Rightarrow \qquad \text{P} = \text{CV}^{-\gamma}$$

Taking log on both sides, we get

$$\log \text{P} = \log \text{C} - \gamma \log \text{V}$$

$$\Rightarrow \qquad \text{Y} = \text{A} + \text{BX}$$

where, $\text{Y} = \log \text{P}, \text{A} = \log \text{C}, \text{B} = -\,\gamma, \text{X} = \log \text{V}$

Normal equations are

$$\Sigma\text{Y} = m\text{A} + \text{B}\Sigma\text{X} \qquad \ldots(1)$$

and
$$\Sigma\text{XY} = \text{A}\Sigma\text{X} + \text{B}\Sigma\text{X}^2 \qquad \ldots(2)$$

Here $m = 5$

The table is as below:

V	P	$X = log\ V$	$Y = log\ P$	XY	X^2
50	64.7	1.69897	1.81090	3.07666	2.88650
60	51.3	1.77815	1.71012	3.04085	3.16182
70	40.5	1.84510	1.60746	2.96592	3.40439
90	25.9	1.95424	1.41330	2.76193	3.81905
100	78.0	2	1.89209	3.78418	4
		$\Sigma X = 9.27646$	$\Sigma Y = 8.43387$	$\Sigma XY = 15.62954$	$\Sigma X^2 = 17.27176$

From Normal equations, we have

$$8.43387 = 5A + 9.27646\ B$$

and
$$15.62954 = 9.27646\ A + 17.27176\ B$$

Solving these, we get

$$A = 2.22476,\ \ B = -\ 0.28997$$

$\therefore$
$$\gamma = -\ B = 0.28997$$

$$C = \text{antilog } (A) = \text{antilog } (2.22476) = 167.78765$$

Hence, the required equation of curve is

$$PV^{0.28997} = 167.78765.$$

Example 8. (*i*) *Given the following experimental values:*

x:	0	1	2	3
y:	2	4	10	15

Fit by the method of Least squares a parabola of the type $y = a + bx^2$.

(*ii*) *Find the Least squares fit of the form* $y = a_0 + a_1x^2$ *to the following data:*

x:	– 1	0	1	2
y:	2	5	3	0

Sol. (*i*) Error of estimate for i^{th} point (x_i, y_i) is $E_i = (y_i - a - bx_i^2)$

By method of Least squares, the values of a, b are chosen such that

$$U = \sum_{i=1}^{4} E_i^2 = \sum_{i=1}^{4} (y_i - a - bx_i^2)^2 \text{ is minimum.}$$

Normal equation are given by

$$\frac{\partial U}{\partial a} = 0 \ \Rightarrow\ \Sigma y = ma + b\Sigma x^2 \qquad ...(1)$$

and
$$\frac{\partial U}{\partial b} = 0 \ \Rightarrow\ \Sigma x^2 y = a\Sigma x^2 + b\Sigma x^4 \qquad ...(2)$$

x	y	x^2	x^2y	x^4
0	2	0	0	0
1	4	1	4	1
2	10	4	40	16
3	15	9	135	81
Total	$\Sigma y = 31$	$\Sigma x^2 = 14$	$\Sigma x^2y = 179$	$\Sigma x^4 = 98$

Here $m = 4$

From (1) and (2), $31 = 4a + 14b$ and $179 = 14a + 98b$

Solving for a and b, we get $a = 2.71,\ b = 1.44$

Hence the required curve is $y = 2.71 + 1.44\, x^2$.

(*ii*) Normal equations are

$$\Sigma y = ma_0 + a_1\, \Sigma x^2 \qquad ...(1)$$

and

$$\Sigma x^2 y = a_0\, \Sigma x^2 + a_1\, \Sigma x^4 \qquad ...(2)$$

The table is as follows:

x	y	x^2	x^2y	x^4
– 1	2	1	2	1
0	5	0	0	0
1	3	1	3	1
2	0	4	0	16
	$\Sigma y = 10$	$\Sigma x^2 = 6$	$\Sigma x^2y = 5$	$\Sigma x^4 = 18$

Here, $m = 4$

From (1) and (2), $10 = 4a_0 + 6a_1$

$5 = 6a_0 + 18a_1$

$\Rightarrow$ $a_0 = 4.1667,\ a_1 = -\ 1.1111$

Hence, the required curve is $y = 4.1667 - 1.1111\, x^2$

Example 9. *Use the method of Least squares to fit the curve:* $y = \dfrac{c_0}{x} + c_1\sqrt{x}$ *to the following table of values:* [A.K.T.U. 2016, 2018]

x:	*0.1*	*0.2*	*0.4*	*0.5*	*1*	*2*
y:	*21*	*11*	*7*	*6*	*5*	*6.*

Sol. As derived in art. 3.32, normal equations to the curve $y = \dfrac{c_0}{x} + c_1\sqrt{x}$ are

$$\sum \frac{y}{x} = c_0 \sum \frac{1}{x^2} + c_1 \sum \frac{1}{\sqrt{x}} \qquad ...(1)$$

and

$$\sum y\sqrt{x} = c_0 \sum \frac{1}{\sqrt{x}} + c_1 \sum x \qquad ...(2)$$

Table is as below:

x	y	y/x	$y\sqrt{x}$	$\frac{1}{\sqrt{x}}$	$\frac{1}{x^2}$
0.1	21	210	6.64078	3.16228	100
0.2	11	55	4.91935	2.23607	25
0.4	7	17.5	4.42719	1.58114	6.25
0.5	6	12	4.24264	1.41421	4
1	5	5	5	1	1
2	6	3	8.48528	0.70711	0.25
$\Sigma x = 4.2$		$\Sigma(y/x) = 302.5$	$\Sigma y\sqrt{x} = 33.71524$	$\sum \frac{1}{\sqrt{x}} = 10.10081$	$\sum \frac{1}{x^2} = 136.5$

From equations (1) and (2), we have

$$302.5 = 136.5\, c_0 + 10.10081\, c_1$$

and

$$33.71524 = 10.10081\, c_0 + 4.2\, c_1$$

Solving these, we get

$$c_0 = 1.97327 \quad \text{and} \quad c_1 = 3.28182$$

Hence the required equation of curve is

$$y = \frac{1.97327}{x} + 3.28182\sqrt{x}.$$

TEST YOUR KNOWLEDGE

1. (*i*) Using the method of least squares, fit the non-linear curve of the form $y = ae^{bx}$ to the following data:

x	0	2	4
y	5.012	10	31.62

(*ii*) The voltage V across a capacitor at time t seconds is given by the following table. Use the principle of least squares to fit a curve of the form $V = ae^{kt}$ to the data:

t	0	2	4	6	8
V	150	63	28	12	5.6

(*iii*) For the data given below, find the equation to the best fitting exponential curve of the form $y = ae^{bx}$.

x	1	2	3	4	5	6
y	1.6	4.5	13.8	40.2	125	300

2. (*i*) State some important curve-fitting procedures.

(*ii*) Derive the least square equations for fitting a curve of the type $y = ax + \frac{b}{x}$ to a set of n points (x_i, y_i) ; $i = 1, 2, \ldots, n$.

3. (*i*) Fit a curve $y = ax^b$ to the following data:

x	1	2	3	4	5	6
y	2.98	4.26	5.21	6.1	6.8	7.5

(*ii*) Fit a least square geometric curve $y = ax^b$ to the following data:

x	1	2	3	4	5
y	0.5	2	4.5	8	12.5

(*iii*) Fit a curve of the form $y = ax^b$ to the data given below:

x	1	2	3	4	5
y	7.1	27.8	62.1	110	161

4. (*i*) Fit a curve of the form $y = ab^x$ in least square sence to the data given below:

x	2	3	4	5	6
y	144	172.8	207.4	248.8	298.5

(*ii*) Fit an exponential curve of the form $y = ab^x$ to the following data:

x	1	2	3	4	5	6	7	8
y	1	1.2	1.8	2.5	3.6	4.7	6.6	9.1

5. The pressure and volume of a gas are related by the equation $pv^a = b$ where a and b are constants. Fit this equation to the following set of data:

p (kg/cm^2)	0.5	1	1.5	2	2.5	3
v (*litres*)	1.62	1	0.75	0.62	0.52	0.46

6. (*i*) Determine the constants of the curve $y = ax + bx^2$ for the following data:

x	0	1	2	3	4
y	2.1	2.4	2.6	2.7	3.4

(*ii*) Using method of least squares, derive the normal equations to fit the curve $y = ax^2 + bx$. Hence fit this curve to the following data: (*G.B.T.U. 2011*)

x	1	2	3	4	5	6	7	8
y	1	1.2	1.8	2.5	3.6	4.7	6.6	9.1

7. Fit a curve of the type $xy = ax + b$ to the following data:

x	1	3	5	7	9	10
y	36	29	28	26	24	15

8. Fit a relation $y = ax + \frac{b}{x}$ which satisfies the following data, using method of least squares:

x	1	2	3	4	5	6	7	8
y	5.4	6.2	8.2	10.3	12.6	14.8	17.2	19.5

(G.B.T.U. 2010)

9. Derive the least square equations for fitting a curve of the type $y = ax^2 + \frac{b}{x}$ to a set of n points.

Hence fit a curve of this type to the data:

x	1	2	3	4
y	– 1.51	0.99	3.88	7.66

10. Derive the least squares approximations of the type $ax^2 + bx + c$ to the function 2^x at the points $x_i = 0, 1, 2, 3, 4$.

11. A person runs the same race track for 5 consecutive days and is timed as follows:

Day (x)	1	2	3	4	5
Time (y)	15.3	15.1	15	14.5	14

Make a least square fit to the above data using a function $a + \frac{b}{x} + \frac{c}{x^2}$.

12. Use the method of least squares to fit the curve $y = c_0 x + \frac{c_1}{\sqrt{x}}$ for the following data:

x	0.2	0.3	0.5	1	2
y	16	14	11	6	3

13. Experiments with a periodic process gave the following data:

$t°$	0	50	100	150	200	250	300	350
y	0.754	1.762	2.041	1.412	0.303	– 0.484	– 0.38	0.520

Estimate the parameters a and b in the model $y = b + a \sin t$ using the least squares approximation.

14. A physicist wants to approximate the following data:

x	0	0.5	1	2
$f(x)$	0	0.57	1.46	5.05

using a function $a\, e^{bx} + c$. He believes that $b \approx 1$. Compute the values of a and c that give the best least squares approximation assuming that $b = 1$.

15. Determine the normal equations if the cubic polynomial $y = a_0 + a_1x + a_2x^2 + a_3x^3$ is fitted to the data (x_i, y_i); $0 \le i \le m$.

16. Estimate y at $x = 5$ by fitting a least squares curve of the form $y = \frac{b}{x(x-a)}$ to the following data:

x	3.6	4.8	6	7.2	8.4	9.6	10.8
y	0.83	0.31	0.17	0.10	0.07	0.05	0.04

$\left(\textbf{Hint:} \text{ Rewrite the equation as } \frac{1}{y} = -\frac{a}{b}x + \frac{1}{b}x^2\right)$

Answers

1. (*i*) $y = 4.642\, e^{0.46x}$ (*ii*) $V = 146.3\, e^{-0.4118t}$ (*iii*) $y = 0.5580\, e^{1.0631\,x}$

3. (*i*) $y = 2.978\, x^{0.5143}$ (*ii*) $y = 0.5012\, x^{1.9977}$ (*iii*) $y = 7.173x^{1.952}$

4. (*i*) $y = 99.86\,(1.2)^x$ (*ii*) $y = 0.6823\,(1.384)^x$ **5.** $pv^{1.42} = 0.99$

6. (*i*) $a = 1.97, b = -0.298$ (*ii*) $y = 0.107798\, x^2 + 0.217125\, x$ **7.** $xy = 16.18x + 40.78$

8. $y = 2.39188\, x + \frac{2.98195}{x}$ **9.** $y = 0.509x^2 - \frac{2.04}{x}$

10. $y = 1.143x^2 - 0.971x + 1.286$ **11.** $y = 13.0065 + \frac{6.7512}{x} - \frac{4.4738}{x^2}$

12. $y = -1.1836\, x + \frac{7.5961}{\sqrt{x}}$ **13.** $a = 1.312810, b = 0.752575$

14. $a = 0.784976,\quad b = -0.733298$

15. $\Sigma y = m\, a_0 + a_1\, \Sigma x + a_2\, \Sigma x^2 + a_3\, \Sigma x^3$

$\Sigma xy = a_0\, \Sigma x + a_1\, \Sigma x^2 + a_2\, \Sigma x^3 + a_3\, \Sigma x^4$

$\Sigma x^2 y = a_0\, \Sigma x^2 + a_1\, \Sigma x^3 + a_2\, \Sigma x^4 + a_3\, \Sigma x^5$

$\Sigma x^3 y = a_0\, \Sigma x^3 + a_1\, \Sigma x^4 + a_2\, \Sigma x^5 + a_3\, \Sigma x^6.$

16. $y = \frac{3.774}{x(x-2)}$; $y(5) = 0.2516$

3.35 CORRELATION

In a bivariate distribution, if the change in one variable affects a change in the other variable, the variables are said to be *correlated.*

If the two variables deviate in the same direction *i.e.,* if the increase (or decrease) in one results in a corresponding increase (or decrease) in the other, correlation is said to be *direct or positive.*

e.g., the correlation between income and expenditure is positive.

If the two variables deviate in opposite direction *i.e.,* if the increase (or decrease) in one results in a corresponding decrease (or increase) in the other, correlation is said to be *inverse or negative.*

e.g., the correlation between volume and the pressure of a perfect gas or the correlation between price and demand is negative.

Correlation is said to be *perfect* if the deviation in one variable is followed by a corresponding **proportional deviation** in the other.

3.36 REASONS RESPONSIBLE FOR THE EXISTENCE OF CORRELATION

1. Due to mere chance. The correlation between variables may be due to mere chance. Consider the data regarding six students selected at random from a college.

Students:	A	B	C	D	E	F
% of marks obtained in: previous exam.	43%	47%	60%	80%	55%	40%
Height (in inches):	60	62	65	70	64	59

Here the variables are moving in the same direction and a high degree of correlation is expected between the variables. We cannot expect this degree of correlation to hold good for any other sample drawn from the concerned population. In this case, the correlation has occurred just due to chance.

2. Due to the effect of some common cause. The correlation between variables may be due to the effect of some common cause. For example, positive correlation between the number of girls seeking admission in colleges A and B of a city may be due to the effect of increasing interest of girls towards higher education.

3. Due to the presence of cause-effect relationship between variables. For example, a high degree correlation between 'temperature' and 'sale of coffee' is due to the fact that people like taking coffee in winter season.

4. Due to the presence of interdependent relationship between the variables. For example, the presence of correlation between amount spent on entertainment of family and total expenditure of family is due to fact that both variables affect each other.

3.37 SCATTER OR DOT DIAGRAMS

It is the simplest method of the diagrammatic representation of bivariate data. Let (x_i, y_i), $i = 1, 2, 3,......, n$ be a bivariate distribution. Let the values of the variables x and y be plotted along the x-axis and y-axis on a suitable scale. Then corresponding to every ordered pair, there corresponds a point or dot in the xy-plane. The diagram of dots so obtained is called a *dot or scatter diagram.*

If the dots are very close to each other and the number of observations is not very large, a fairly good correlation is expected. If the dots are widely scattered, a poor correlation is expected.

3.38 KARL PEARSON'S CO-EFFICIENT OF CORRELATION (OR PRODUCT MOMENT CORRELATION CO-EFFICIENT) [*U.P.T.U. 2015*]

Correlation co-efficient between two variables x and y, usually denoted by $r(x, y)$ or r_{xy} is a numerical measure of linear relationship between them and is defined as

$$r_{xy} = \frac{\Sigma(x_i - \overline{x})(y_i - \overline{y})}{\sqrt{\Sigma(x_i - \overline{x})^2 \, \Sigma(y_i - \overline{y})^2}} = \frac{\frac{1}{n}\Sigma(x_i - \overline{x})(y_i - \overline{y})}{\sqrt{\frac{1}{n}\Sigma(x_i - \overline{x})^2 \cdot \frac{1}{n}\Sigma(y_i - \overline{y})^2}} = \frac{\frac{1}{n}\Sigma(x_i - \overline{x})(y_i - \overline{y})}{\sigma_x \sigma_y}.$$

$$\therefore \quad \boxed{r_{xy} = \frac{\Sigma(x - \overline{x})(y - \overline{y})}{n\sigma_x \sigma_y}}$$

Alternate form of $r(x, y)$:

$$r(x, y) = \frac{n\,\Sigma xy - \Sigma x \Sigma y}{\sqrt{n\,\Sigma x^2 - (\Sigma x)^2}\,\sqrt{n\,\Sigma y^2 - (\Sigma y)^2}}$$

Here n is the no. of pairs of values of x and y.

Note. Correlation co-efficient is independent of change of origin and scale.

Let us define two new variables u and v as

$$u = \frac{x-a}{h}, v = \frac{y-b}{k}$$ where a, b, h, k are constants, then $r_{xy} = r_{uv}$.

Then,
$$r(u, v) = \frac{n\,\Sigma uv - \Sigma u \Sigma v}{\sqrt{n\,\Sigma u^2 - (\Sigma u)^2}\,\sqrt{n\,\Sigma v^2 - (\Sigma v)^2}}.$$

ILLUSTRATIVE EXAMPLES

Example 1. *Find the coefficient of correlation between the values of x and y:*

x	1	3	5	7	8	10
y	8	12	15	17	18	20

Sol. Here, $n = 6$. The table is as follows:

x	y	x^2	y^2	xy
1	8	1	64	8
3	12	9	144	36
5	15	25	225	75
7	17	49	289	119
8	18	64	324	144
10	20	100	400	200
$\Sigma x = 34$	$\Sigma y = 90$	$\Sigma x^2 = 248$	$\Sigma y^2 = 1446$	$\Sigma xy = 582$

Karl Pearson's coefficient of correlation is given by

$$r\,(x, y) = \frac{n\,\Sigma\, xy - \Sigma\, x\, \Sigma\, y}{\sqrt{n\,\Sigma\, x^2 - (\Sigma\, x)^2}\,\sqrt{n\,\Sigma\, y^2 - (\Sigma\, y)^2}}$$

$$= \frac{(6 \times 582) - (34 \times 90)}{\sqrt{(6 \times 248) - (34)^2}\,\sqrt{(6 \times 1446) - (90)^2}} = 0.9879$$

Example 2. *The following data regarding the heights (y) and weights (x) of 100 college students are given:*

$\Sigma x = 15000, \Sigma x^2 = 2272500, \Sigma y = 6800, \Sigma y^2 = 463025$ *and* $\Sigma xy = 1022250.$

Find the correlation coefficient between height and weight.

Sol. Here, $n = 100$

Correlation co-efficient $r(x, y)$ is given by

$$r = \frac{n\Sigma xy - \Sigma x\Sigma y}{\sqrt{n\Sigma x^2 - (\Sigma x)^2}\sqrt{n\Sigma y^2 - (\Sigma y)^2}}$$

$$= \frac{(100 \times 1022250) - (15000 \times 6800)}{\sqrt{(100 \times 2272500) - (15000)^2}\sqrt{(100 \times 463025) - (6800)^2}} = 0.6.$$

Example 3. *Find the co-efficient of correlation for the following table:* (U.K.T.U. 2011)

x:	*10*	*14*	*18*	*22*	*26*	*30*
y:	*18*	*12*	*24*	*6*	*30*	*36*

Sol. Let $u = \frac{x - 22}{4}$, $v = \frac{y - 24}{6}$

x	y	u	v	u^2	v^2	uv
10	18	– 3	– 1	9	1	3
14	12	– 2	– 2	4	4	4
18	24	– 1	0	1	0	0
22	6	0	– 3	0	9	0
26	30	1	1	1	1	1
30	36	2	2	4	4	4
Total		$\Sigma u = -3$	$\Sigma v = -3$	$\Sigma u^2 = 19$	$\Sigma v^2 = 19$	$\Sigma uv = 12$

Here, $n = 6$, $\quad \bar{u} = \frac{1}{n}\Sigma u = \frac{1}{6}(-3) = -\frac{1}{2}$; $\bar{v} = \frac{1}{n}\Sigma v = \frac{1}{6}(-3) = -\frac{1}{2}$

$$r_{uv} = \frac{n\,\Sigma uv - \Sigma u\Sigma v}{\sqrt{n\,\Sigma u^2 - (\Sigma u)^2}\sqrt{n\,\Sigma v^2 - (\Sigma v)^2}}$$

$$= \frac{(6 \times 12) - (-3)(-3)}{\sqrt{(6 \times 19) - (-3)^2}\sqrt{(6 \times 19) - (-3)^2}} = \frac{63}{\sqrt{105}\sqrt{105}} = 0.6$$

Hence, $\quad r_{xy} = r_{uv} = 0.6.$

Example 4. *Ten students got the following percentage of marks in Principles of Economics and Statistics:*

Roll Nos.:	*1*	*2*	*3*	*4*	*5*	*6*	*7*	*8*	*9*	*10*
Marks in Economics:	*78*	*36*	*98*	*25*	*75*	*82*	*90*	*62*	*65*	*39*
Marks in Statistics:	*84*	*51*	*91*	*60*	*68*	*62*	*86*	*58*	*53*	*47*

Calculate the co-efficient of correlation.

Sol. Let the marks in the two subjects be denoted by x and y respectively.

x	y	$u = x - 65$	$v = y - 66$	u^2	v^2	uv
78	84	13	18	169	324	234
36	51	– 29	– 15	841	225	435
98	91	33	25	1089	625	825

25	60	– 40	– 6	1600	36	240
75	68	10	2	100	4	20
82	62	17	– 4	289	16	– 68
90	86	25	20	625	400	500
62	58	– 3	– 8	9	64	24
65	53	0	– 13	0	169	0
39	47	– 26	– 19	676	361	494
Total		$\Sigma u = 0$	$\Sigma v = 0$	$\Sigma u^2 = 5398$	$\Sigma v^2 = 2224$	$\Sigma uv = 2734$

Here, $n = 10, \bar{u} = \frac{1}{n} \Sigma u_i = 0, \bar{v} = \frac{1}{n} \Sigma v_i = 0$

$$r_{uv} = \frac{n\Sigma uv - \Sigma u \Sigma v}{\sqrt{n\,\Sigma u^2 - (\Sigma u)^2}\,\sqrt{n\,\Sigma v^2 - (\Sigma v)^2}}$$

$$= \frac{(10 \times 2734) - (0 \times 0)}{\sqrt{(10 \times 5398) - (0)^2}\,\sqrt{(10 \times 2224) - (0)^2}} = 0.789$$

Hence, $r_{xy} = r_{uv} = 0.789$.

Example 5. *A computer while calculating correlation co-efficient between two variables X and Y from 25 pairs of observations obtained the following results :*

$$n = 25, \quad \Sigma X = 125, \quad \Sigma X^2 = 650,$$
$$\Sigma Y = 100, \quad \Sigma Y^2 = 460, \quad \Sigma XY = 508.$$

It was, however, later discovered at the time of checking that he had copied down two pairs as

X	Y
6	14
8	6

while the correct values were

X	Y
8	12
6	8

Obtain the correct value of correlation co-efficient.

Sol. Corrected $\Sigma X = 125 - 6 - 8 + 8 + 6 = 125$

Corrected $\Sigma Y = 100 - 14 - 6 + 12 + 8 = 100$

Corrected $\Sigma X^2 = 650 - 6^2 - 8^2 + 8^2 + 6^2 = 650$

Corrected $\Sigma Y^2 = 460 - 14^2 - 6^2 + 12^2 + 8^2 = 436$

Corrected $\Sigma XY = 508 - 6 \times 14 - 8 \times 6 + 8 \times 12 + 6 \times 8 = 520$

(Subtract the incorrect values and add the corresponding correct values)

$$\bar{X} = \frac{1}{n} \Sigma X = \frac{1}{25} \times 125 = 5\,; \quad \bar{Y} = \frac{1}{n} \Sigma Y = \frac{1}{25} \times 100 = 4$$

Corrected $$r_{xy} = \frac{n\,\Sigma XY - \Sigma X \Sigma Y}{\sqrt{n\,\Sigma X^2 - (\Sigma X)^2}\,\sqrt{n\,\Sigma Y^2 - (\Sigma Y)^2}}$$

$$= \frac{(25 \times 520) - (125 \times 100)}{\sqrt{(25 \times 650) - (125)^2}\,\sqrt{(25 \times 436) - (100)^2}} = 0.67.$$

Example 6. *If $z = ax + by$ and r is the correlation co-efficient between x and y, show that*

$$\sigma_z^2 = a^2\sigma_x^2 + b^2\sigma_y^2 + 2abr\,\sigma_x\,\sigma_y.$$

Sol. $z = ax + by$

$\Rightarrow \quad \bar{z} = a\bar{x} + b\bar{y}, \quad z_i = ax_i + by_i$

$$z_i - \bar{z} = a(x_i - \bar{x}) + b(y_i - \bar{y})$$

Now, $$\sigma_z^2 = \frac{1}{n}\Sigma(z_i - \bar{z})^2 = \frac{1}{n}\Sigma[a(x_i - \bar{x}) + b(y_i - \bar{y})]^2$$

$$= \frac{1}{n}\Sigma[a^2(x_i - \bar{x})^2 + b^2(y_i - \bar{y})^2 + 2ab(x_i - \bar{x})(y_i - \bar{y})]$$

$$= a^2 . \frac{1}{n}\Sigma(x_i - \bar{x})^2 + b^2 . \frac{1}{n}\Sigma(y_i - \bar{y})^2 + 2ab . \frac{1}{n}\Sigma(x_i - \bar{x})(y_i - \bar{y})$$

$$= a^2\sigma_x^2 + b^2\sigma_y^2 + 2abr\,\sigma_x\sigma_y \qquad \because\ r = \frac{\frac{1}{n}\Sigma(x_i - \bar{x})(y_i - \bar{y})}{\sigma_x\sigma_y}$$

Example 7. *Establish the formula:* $\sigma_{x-y}^2 = \sigma_x^2 + \sigma_y^2 - 2r_{xy}\sigma_x\sigma_y$

where r_{xy} *is the correlation coefficient between x and y. Using the above formula, calculate the correlation coefficient from the following data relating to the marks of 10 candidates in aptitude test (x) and Achievement rating (y).*

	Marks									
Aptitude (x):	*22*	*53*	*46*	*67*	*43*	*35*	*88*	*11*	*95*	*13*
Achievement (y):	*18*	*39*	*31*	*42*	*55*	*64*	*82*	*10*	*96*	*14*

Sol. Let $$z = x - y$$

$$\therefore \quad \bar{z} = \bar{x} - \bar{y}$$

$$\therefore \quad z - \bar{z} = (x - \bar{x}) - (y - \bar{y})$$

or, $$(z - \bar{z})^2 = (x - \bar{x})^2 + (y - \bar{y})^2 - 2(x - \bar{x})(y - \bar{y})$$

Summing up for n terms,

$$\Sigma(z - \bar{z})^2 = \Sigma(x - \bar{x})^2 + \Sigma(y - \bar{y})^2 - 2\Sigma(x - \bar{x})(y - \bar{y})$$

or, $$\frac{\Sigma(z - \bar{z})^2}{n} = \frac{\Sigma(x - \bar{x})^2}{n} + \frac{\Sigma(y - \bar{y})^2}{n} - \frac{2\Sigma(x - \bar{x})(y - \bar{y})}{n}$$

$$\Rightarrow \quad \sigma_z^2 = \sigma_x^2 + \sigma_y^2 - 2r\sigma_x\sigma_y, \quad \text{where } r = \frac{\Sigma(x - \bar{x})(y - \bar{y})}{n\sigma_x\sigma_y}$$

$$\Rightarrow \quad \boxed{\sigma_{x-y}^2 = \sigma_x^2 + \sigma_y^2 - 2r\sigma_x\sigma_y} \qquad ...(1)$$

Now, $n = 10$

$$\bar{x} = \frac{22 + 53 + 46 + 67 + 43 + 35 + 88 + 11 + 95 + 13}{10} = \frac{473}{10} = 47.3$$

$$\bar{y} = \frac{18 + 39 + 31 + 42 + 55 + 64 + 82 + 10 + 96 + 14}{10} = \frac{451}{10} = 45.1$$

Now we form the table as

x	y	$z = x - y$	$x - \bar{x}$	$y - \bar{y}$	$z - \bar{z}$	$(x - \bar{x})^2$	$(y - \bar{y})^2$	$(z - \bar{z})^2$
22	18	4	– 25.3	– 27.1	1.8	640.09	734.41	3.24
53	39	14	5.7	– 6.1	11.8	32.49	37.21	139.24
46	31	15	– 1.3	– 14.1	12.8	1.69	198.81	163.84
67	42	25	19.7	– 3.1	22.8	38.09	9.61	519.84
43	55	– 12	– 4.3	9.9	– 14.2	18.49	98.01	201.64
35	64	– 29	– 12.3	18.9	– 31.2	151.29	357.21	973.44
88	82	6	40.7	36.9	3.8	1656.49	1361.61	14.44
11	10	1	– 36.3	– 35.1	– 1.2	1317.69	1232.01	1.44
95	96	– 1	47.7	50.9	– 3.2	2275.29	2590.81	10.24
13	14	– 1	– 34.3	– 31.1	– 3.2	1176.49	967.21	10.24
						$\Sigma(x - \bar{x})^2 = 7658.1$	$\Sigma(y - \bar{y})^2 = 7586.9$	$\Sigma(z - \bar{z})^2 = 2037.6$

where, $\bar{z} = \dfrac{\Sigma z}{n} = \dfrac{22}{10} = 2.2$

Now, $\sigma_x^2 = \dfrac{\Sigma(x - \bar{x})^2}{n} = \dfrac{7658.1}{10} = 765.81, \quad \sigma_y^2 = \dfrac{\Sigma(y - \bar{y})^2}{n} = \dfrac{7586.9}{10} = 758.69$

$$\sigma_{x-y}^2 = \sigma_z^2 = \frac{\Sigma(z - \bar{z})^2}{n} = \frac{2037.6}{10} = 203.76.$$

Substituting the values in the formula (1),

$$\sigma_{x-y}^2 = \sigma_x^2 + \sigma_y^2 - 2r\,\sigma_x\sigma_y$$

$\Rightarrow$ $203.76 = 765.81 + 758.69 - 2r\,(27.67)(27.54)$

$$r = \frac{1524.5 - 203.76}{1524.06} = 0.866.$$

Example 8. *(i) Calculate coefficient of correlation from the following results:*

$n = 10,\ \Sigma X = 100,\ \Sigma Y = 150,\ \Sigma(X - 10)^2 = 180,\ \Sigma(Y - 15)^2 = 215,\ \Sigma(X - 10)(Y - 15) = 60.$

(ii) Calculate Karl Pearson's coefficient of correlation between X and Y for the following information:

$n = 12,\ \Sigma X = 120,\ \Sigma Y = 130,\ \Sigma(X - 8)^2 = 150,\ \Sigma(Y - 10)^2 = 200$ *and* $\Sigma(X - 8)(Y - 10) = 50$

Sol. (*i*) Mean of first series, $\bar{X} = \dfrac{\Sigma X}{n} = \dfrac{100}{10} = 10$

Mean of second series, $\bar{Y} = \dfrac{\Sigma Y}{n} = \dfrac{150}{10} = 15$

Now, $r = \dfrac{\Sigma(X - \bar{X})(Y - \bar{Y})}{\sqrt{\Sigma(X - \bar{X})^2 \,.\, \Sigma(Y - \bar{Y})^2}}$

$$= \frac{\Sigma(X - 10)(Y - 15)}{\sqrt{\Sigma(X - 10)^2 \,.\, \Sigma(Y - 15)^2}} = \frac{60}{\sqrt{180 \times 215}} = 0.305$$

(*ii*)

$$\Sigma x = \Sigma(X - 8) = \Sigma X - \Sigma 8 = 120 - (8 \times 12) = 24$$

$$\Sigma y = \Sigma(Y - 10) = \Sigma Y - \Sigma 10 = 130 - (10 \times 12) = 10$$

$$\Sigma xy = \Sigma(X - 8)(Y - 10) = 50 \quad \text{(given)}$$

$$\Sigma x^2 = \Sigma(X - 8)^2 = 150$$

$$\Sigma y^2 = \Sigma(Y - 10)^2 = 200$$

Now,
$$r = \frac{n\,\Sigma xy - \Sigma x \Sigma y}{\sqrt{n\,\Sigma x^2 - (\Sigma x)^2}\ \sqrt{n\,\Sigma y^2 - (\Sigma y)^2}} = \frac{(12 \times 50) - (24 \times 10)}{\sqrt{(12 \times 150) - (24)^2}\ \sqrt{(12 \times 200) - (10)^2}}$$

$$= \frac{360}{\sqrt{1224}\ \sqrt{2300}} = 0.2146.$$

3.39 CALCULATION OF CO-EFFICIENT OF CORRELATION FOR A BIVARIATE FREQUENCY DISTRIBUTION

If the bivariate data on x and y is presented on a two way correlation table and f is the frequency of a particular rectangle in the correlation table, then

$$r_{xy} = \frac{\Sigma fxy - \frac{1}{n}\Sigma fx\ \Sigma fy}{\sqrt{\left[\Sigma fx^2 - \frac{1}{n}(\Sigma fx)^2\right]\left[\Sigma fy^2 - \frac{1}{n}(\Sigma fy)^2\right]}}$$

Since change of origin and scale do not affect the co-efficient of correlation.

$\therefore$ $r_{xy} = r_{uv}$ where the new variables u, v are properly chosen.

Example 9. *The following table gives according to age the frequency of marks obtained by 100 students in an intelligence test:*

Age (in years) / *Marks*	*18*	*19*	*20*	*21*	*Total*
10–20	*4*	*2*	*2*		*8*
20–30	*5*	*4*	*6*	*4*	*19*
30–40	*6*	*8*	*10*	*11*	*35*
40–50	*4*	*4*	*6*	*8*	*22*
50–60		*2*	*4*	*4*	*10*
60–70		*2*	*3*	*1*	*6*
Total	*19*	*22*	*31*	*28*	*100*

Calculate the co-efficient of correlation between age and intelligence.

Sol. Let age and intelligence be denoted by x and y respectively.

Mid value	x / y	18	19	20	21	f	u	fu	fu^2	fuv
15	10–20	4	2	2		8	– 3	– 24	72	30
25	20–30	5	4	6	4	19	– 2	– 38	76	20
35	30–40	6	8	10	11	35	– 1	– 35	35	9
45	40–50	4	4	6	8	22	0	0	0	0
55	50–60		2	4	4	10	1	10	10	2
65	60–70		2	3	1	6	2	12	24	– 2
	f	19	22	31	28	100	Total	**– 75**	**217**	**59**
	v	– 2	– 1	0	1	Total				
	fv	– 38	– 22	0	28	**– 32**				
	fv^2	76	22	0	28	**126**				
	fuv	56	16	0	– 13	**59**				

Let us define two new variables u and v as $u = \dfrac{y-45}{10}$, $v = x - 20$

$$r_{xy} = r_{uv} = \frac{\Sigma fuv - \frac{1}{n}\Sigma fu\,\Sigma fv}{\sqrt{\left[\Sigma fu^2 - \frac{1}{n}(\Sigma fu)^2\right]\left[\Sigma fv^2 - \frac{1}{n}(\Sigma fv)^2\right]}}$$

$$= \frac{59 - \frac{1}{100}(-75)(-32)}{\sqrt{\left[217 - \frac{1}{100}(-75)^2\right]\left[126 - \frac{1}{100}(-32)^2\right]}} = \frac{59-24}{\sqrt{\frac{643}{4} \times \frac{2894}{25}}} = 0.25.$$

3.40 RANK CORRELATION

Sometimes we have to deal with problems in which data cannot be quantitatively measured but qualitative assessment is possible.

Let a group of n individuals be arranged in order of merit or proficiency in possession of two characteristics A and B. The ranks in the two characteristics are, in general, different. For example, if A stands for intelligence and B for beauty, it is not necessary that the most intelligent individual may be the most beautiful and *vice versa.* Thus an individual who is ranked at the top for the characteristic A *may be* ranked at the bottom for the characteristic B. Let (x_i, y_i), $i = 1, 2, \ldots\ldots n$ be the ranks of the n individuals in the group for the characteristics A and B respectively. Pearsonian co-efficient of correlation between the ranks x_i's and y_i's is called the *rank correlation co-efficient* between the characteristics A and B for that group of individuals.

Thus rank correlation co-efficient

$$r = \frac{\Sigma(x_i - \bar{x})(y_i - \bar{y})}{\sqrt{\Sigma(x_i - \bar{x})^2\ \Sigma(y_i - \bar{y})^2}} = \frac{\frac{1}{n}\Sigma(x_i - \bar{x})(y_i - \bar{y})}{\sigma_x \sigma_y} \quad ...(1)$$

Now x_i's and y_i's are merely the permutations of n numbers from 1 to n. *Assuming that no two individuals are bracketed or tied in either classification i.e., $(x_i, y_i) \neq (x_j, y_j)$ for $i \neq j$, both x and y take all integral values from 1 to n.*

$$\therefore \quad \bar{x} = \bar{y} = \frac{1}{n}(1 + 2 + 3 + \cdots\cdots + n) = \frac{1}{n} \cdot \frac{n(n+1)}{2} = \frac{n+1}{2}$$

$$\Sigma x_i = 1 + 2 + 3 + \cdots\cdots + n = \frac{n(n+1)}{2} = \Sigma y_i$$

$$\Sigma x_i^2 = 1^2 + 2^2 + \cdots\cdots + n^2 = \frac{n(n+1)(2n+1)}{6} = \Sigma y_i^2$$

If D_i denotes the difference in ranks of the ith individual, then

$$D_i = x_i - y_i = (x_i - \bar{x}) - (y_i - \bar{y}) \qquad [\because\ \bar{x} = \bar{y}]$$

$$\frac{1}{n}\Sigma D_i^2 = \frac{1}{n}\Sigma[(x_i - \bar{x}) - (y_i - \bar{y})]^2$$

$$= \frac{1}{n}\Sigma(x_i - \bar{x})^2 + \frac{1}{n}\Sigma(y_i - \bar{y})^2 - 2 \cdot \frac{1}{n}\Sigma(x_i - \bar{x})(y_i - \bar{y})$$

$$= \sigma_x^2 + \sigma_y^2 - 2r\sigma_x\sigma_y \quad ...(2) \quad | \text{ using } (1)$$

But $\quad \sigma_x^2 = \frac{1}{n}\Sigma x_i^2 - \bar{x}^2 = \frac{1}{n}\Sigma y_i^2 - \bar{y}^2 = \sigma_y^2$

$$\therefore \quad \text{From (2)}, \frac{1}{n}\Sigma D_i^2 = 2\sigma_x^2 - 2r\sigma_x^2 = 2(1-r)\sigma_x^2 = 2(1-r)\left[\frac{1}{n}\Sigma x_i^2 - \bar{x}^2\right]$$

$$= 2(1-r)\left[\frac{1}{n} \cdot \frac{n(n+1)(2n+1)}{6} - \frac{(n+1)^2}{4}\right]$$

$$= (1-r)(n+1)\left[\frac{4n+2-3n-3}{6}\right] = \frac{(1-r)(n^2-1)}{6} \quad \text{or} \quad 1 - r = \frac{6\Sigma D_i^2}{n(n^2-1)}$$

Hence, $\quad \boxed{r = 1 - \left[\frac{6\Sigma D_i^2}{n(n^2-1)}\right]}$

Note. This is called *Spearman's Formula for* Rank Correlation.

$$\Sigma d_i = \Sigma(x_i - y_i) = \Sigma x_i - \Sigma y_i = 0 \text{ always.}$$

This serves as a check on calculations.

Example 10. *Compute the rank correlation coefficient for the following data:*

Person:	*A*	*B*	*C*	*D*	*E*	*F*	*G*	*H*	*I*	*J*
Rank in Maths:	9	10	6	5	7	2	4	8	1	3
Rank in Physics:	1	2	3	4	5	6	7	8	9	10

Sol. Here the ranks are given and $n = 10$.

Person	R_1	R_2	$D = R_1 - R_2$	D^2
A	9	1	8	64
B	10	2	8	64
C	6	3	3	9
D	5	4	1	1
E	7	5	2	4
F	2	6	– 4	16
G	4	7	– 3	9
H	8	8	0	0
I	1	9	– 8	64
J	3	10	– 7	49
				$\Sigma D^2 = 280$

$$\therefore \quad r = 1 - \left\{\frac{6\Sigma D^2}{n(n^2-1)}\right\} = 1 - \left\{\frac{6 \times 280}{10(100-1)}\right\} = 1 - 1.697 = -0.697.$$

Example 11. *The marks secured by recruits in the selection test (X) and in the proficiency test (Y) are given below:*

Serial No.:	*1*	*2*	*3*	*4*	*5*	*6*	*7*	*8*	*9*
X:	*10*	*15*	*12*	*17*	*13*	*16*	*24*	*14*	*22*
Y:	*30*	*42*	*45*	*46*	*33*	*34*	*40*	*35*	*39*

Calculate the rank correlation co-efficient.

Sol. Here the marks are given. Therefore, first of all, write down ranks. In each series, the item with the largest size is ranked 1, next largest 2 and so on. Here $n = 9$.

X	10	15	12	17	13	16	24	14	22	Total
Y	30	42	45	46	33	34	40	35	39	
Ranks in X (x)	9	5	8	3	7	4	1	6	2	
Ranks in Y (y)	9	3	2	1	8	7	4	6	5	
$D = x - y$	0	2	6	2	– 1	– 3	– 3	0	– 3	0
D^2	0	4	36	4	1	9	9	0	9	72

$$\therefore \quad r = 1 - \left\{\frac{6\Sigma D^2}{n(n^2-1)}\right\} = 1 - \left\{\frac{6 \times 72}{9 \times 80}\right\} = 1 - .6 = 0.4$$

Example 12. *Rank correlation co-efficient of marks obtained by 10 students in Mathematics and English was found to be 0.5. It was later discovered that the difference in ranks in two subjects obtained by one of the students was wrongly taken as 3 instead of 7. Find the correct rank correlation co-efficient.*

Sol. Incorrect $r_k = 0.5$, $n = 10$

$$\therefore \quad \text{Incorrect } \Sigma D^2 = \frac{n(n^2-1)(1-r_k)}{6} = \frac{10 \times 99 \times 0.5}{6} = 82.5$$

Now, correct ΣD^2 = Incorrect $\Sigma D^2 - (3)^2 + (7)^2 = 82.5 - 9 + 49 = 122.5$

$$\therefore \quad \text{Correct } r_k = 1 - \left\{\frac{6\Sigma D^2}{n(n^2-1)}\right\} = 1 - \left\{\frac{6 \times 122.5}{10(100-1)}\right\} = 0.2575.$$

Example 13. *Ten competitors in a beauty contest were ranked by three judges in the following orders:*

First Judge:	*1*	*6*	*5*	*10*	*3*	*2*	*4*	*9*	*7*	*8*
Second Judge:	*3*	*5*	*8*	*4*	*7*	*10*	*2*	*1*	*6*	*9*
Third Judge:	*6*	*4*	*9*	*8*	*1*	*2*	*3*	*10*	*5*	*7*

Use the method of rank correlation to determine which pair of judges has the nearest approach to common taste in beauty?

Sol. Let R_1, R_2, R_3 be the ranks given by three judges.

Calculation of rank correlation coefficient

Competitor	R_1	R_2	R_3	$D_{12} = R_1 - R_2$	$D_{13} = R_1 - R_3$	$D_{23} = R_2 - R_3$	D_{12}^2	D_{13}^2	D_{23}^2
A	1	3	6	– 2	– 5	– 3	4	25	9
B	6	5	4	1	2	1	1	4	1
C	5	8	9	– 3	– 4	– 1	9	16	1
D	10	4	8	6	2	– 4	36	4	16
E	3	7	1	– 4	2	6	16	4	36
F	2	10	2	– 8	0	8	64	0	64
G	4	2	3	2	1	– 1	4	1	1
H	9	1	10	8	– 1	– 9	64	1	81
I	7	6	5	1	2	1	1	4	1
J	8	9	7	– 1	1	2	1	1	4
$n = 10$							$\Sigma D_{12}^2 = 200$	$\Sigma D_{13}^2 = 60$	$\Sigma D_{23}^2 = 214$

Rank correlation coefficient between first and second judges,

$$r_{k_{12}} = 1 - \left\{\frac{6\Sigma D_{12}^2}{n(n^2-1)}\right\} = 1 - \left\{\frac{6 \times 200}{10\,(99)}\right\} = -\,0.212$$

Rank correlation coefficient between first and third judges,

$$r_{k_{13}} = 1 - \left\{\frac{6\Sigma D_{13}^2}{n(n^2-1)}\right\} = 1 - \left\{\frac{6 \times 60}{10 \times 99}\right\} = 0.636$$

Rank correlation coefficient between second and third judges,

$$r_{k_{23}} = 1 - \left\{ \frac{6\Sigma D_{23}{}^2}{n(n^2-1)} \right\} = 1 - \left\{ \frac{6 \times 214}{10 \times 99} \right\} = -0.297$$

Correlation between first and second judges is negative *i.e.*, their opinions regarding beauty test are opposite to each other. Similarly, opinions of second and third judges are opposite to each other, but the opinions of first and third judges are of similar type as their correlation is positive. It means that their likings and dislikings are very much common.

3.41 TIED RANKS

If any two or more individuals have same rank or the same value in the series of marks, then the above formula fails and requires an adjustment. In such cases, each individual is given an average rank. This common average rank is the average of the ranks which these individuals would have assumed if they were slightly different from each other. Thus, if two individual are ranked equal at the sixth place, they would have assumed the 6th and 7th ranks if they were ranked slightly different. Their common rank $= \frac{6+7}{2} = 6.5$. If three individuals are ranked equal at fourth place, they would have assumed the 4th, 5th and 6th ranks if they were ranked slightly different. Their common rank $= \frac{4+5+6}{3} = 5$.

Adjustment. Add $\frac{1}{12} m(m^2 - 1)$ to ΣD^2 where m stands for the number of times an item is repeated.

This adjustment factor is to be added for each repeated item.

Thus
$$r = 1 - \frac{6\left\{\Sigma D^2 + \frac{1}{12} m_1(m_2^2 - 1) + \frac{1}{12} m_2(m_2^2 - 1) + \ldots\right\}}{n(n^2-1)}.$$

Example 14. *Obtain the rank correlation co-efficient for the following data:*

X:	*68*	*64*	*75*	*50*	*64*	*80*	*75*	*40*	*55*	*64*
Y:	*62*	*58*	*68*	*45*	*81*	*60*	*68*	*48*	*50*	*70.*

Sol. Here, marks are given, so write down the ranks.

X	68	64	75	50	64	80	75	40	55	64	Total
Y	62	58	68	45	81	60	68	48	50	70	
Ranks in X (x)	4	6	2.5	9	6	1	2.5	10	8	6	
Ranks in Y (y)	5	7	3.5	10	1	6	3.5	9	8	2	
$D = x - y$	−1	−1	−1	−1	5	−5	−1	1	0	4	0
D^2	1	1	1	1	25	25	1	1	0	16	72

In the X-series, the value 75 occurs twice. Had these values been slightly different, they would have been given the ranks 2 and 3. Therefore, the common rank given to them is $\frac{2+3}{2} = 2.5$. The value 64 occurs thrice. Had these values been slightly different, they would have been given the ranks 5, 6, and 7. Therefore the common rank given to them is $\frac{5+6+7}{3} = 6$. Similarly, in the Y-series, the value 68 occurs twice. Had these values been slightly different, they would have been given the ranks 3 and 4? Therefore, the common rank given to them is $\frac{3+4}{2} = 3.5$.

Thus, m has the values 2, 3, 2.

$$\therefore \quad r = 1 - \frac{6\left\{\Sigma D^2 + \frac{1}{12} m_1(m_1{}^2 - 1) + \frac{1}{12} m_2(m_2{}^2 - 1) + \frac{1}{12} m_3(m_3{}^2 - 1)\right\}}{n(n^2 - 1)}$$

$$= 1 - \frac{6\left[72 + \frac{1}{12} . 2(2^2 - 1) + \frac{1}{12} . 3(3^2 - 1) + \frac{1}{12} . 2(2^2 - 1)\right]}{10(10^2 - 1)}$$

$$= 1 - \left\{\frac{6 \times 75}{990}\right\} = \frac{6}{11} = 0.545.$$

TEST YOUR KNOWLEDGE

1. Define Karl Pearson's coefficient of correlation. How would you interpret the sign and magnitude of a correlation coefficient?
2. Calculate the coefficient of correlation for the following data:

(*i*)

Husband's age (in yrs.) x	23	27	28	28	29	30	31	33	35	36
Wife's age (in yrs.) y	18	20	22	27	21	29	27	29	28	29

(*ii*)

Height of father (in inches)	65	66	67	67	68	69	70	72
Height of son (in inches)	67	68	65	68	72	72	69	71

3. Find the correlation coefficient between x and y for the following data:

x	60	34	40	50	45	41	22	43
y	75	32	34	40	45	33	12	30

4. Calculate the coefficient of correlation between the marks obtained by 8 students in Mathematics and Statistics:

Students	A	B	C	D	E	F	G	H
Mathematics	25	30	32	35	37	40	42	45
Statistics	08	10	15	17	20	23	24	25

5. Calculate the coefficient of correlation between the following ages of husband (x) and wife (y) by taking 30 and 28 as assumed mean in case of x and y respectively:

x :	24	27	28	28	29	30	32	33	35	35	40
y :	18	20	22	25	22	28	28	30	27	30	32

6. From the data given below, find the number of items n:

$r_{xy} = 0.5$, $\Sigma XY = 120$, $\Sigma X^2 = 90$, $\sigma_y = 8$ where X and Y are deviations from the arithmetic mean.

(*A.K.T.U. 2016*)

7. The marks of the same 15 students in two subjects A and B are analysed. The two numbers within the brackets denote the ranks of the same student in A and B respectively:

(1, 10) (2, 7) (3, 2) (4, 6) (5, 4) (6, 8) (7, 3) (8, 1) (9, 11) (10, 15) (11, 9) (12, 5) (13, 14) (14, 12) (15, 13)

Use Spearman's formula to find the rank correlation coefficient.

8. Ten students got the following percentage of marks in Chemistry and Physics:

Students:	1	2	3	4	5	6	7	8	9	10
Marks in Chemistry:	78	36	98	25	75	82	90	62	65	39
Marks in Physics:	84	51	91	60	68	62	86	58	63	47

Calculate the rank correlation co-efficient.

9. Find the rank correlation coefficient of marks of A and B from the follwoing data:

Marks A:	15	20	27	13	45	60	20	75
Marks B:	50	30	55	30	25	10	30	70

(*A.K.T.U. 2016*)

10. A firm not sure of the response to its product in ten different colour shades decides to produce them in those colour shades, if the ranking of these colour shades by two typical consumer judges is highly correlated. The two judges rank the ten colours in the following order:

Colour :	*Red*	*Green*	*Blue*	*Yellow*	*White*	*Black*	*Pink*	*Purple*	*Orange*	*Ivory*
Ranking by I Judge :	6	4	3	1	2	7	9	8	10	5
Ranking by II Judge :	4	1	6	7	5	8	10	9	3	2

Is there any agreement between the two judges, to allow the introduction of the product by the firm in the market?

11. Two judges in a music competition rank the 12 entries as follows:

x	1	2	3	4	5	6	7	8	9	10	11	12
y	12	9	6	10	3	5	4	7	8	2	11	1

What degree of agreement is there between the judgement of the two judges?

Answers

2. (*i*) 0.82 (*ii*) 0.603 **3.** 0.9158 **4.** 0.9804
5. 0.8926 **6.** 10 **7.** 0.51 **8.** 0.84
9. 0 **10.** 0.22, no **11.** – 0.454

3.42 REGRESSION ANALYSIS (*U.P.T.U. 2015*)

The term 'regression' was first used by Sir Francis Galton (1822–1911), a British Biometrician in connection with the height of parents and their offsprings. He found that the offspring of tall or short parents tend to regress to the average height. In other words, though tall fathers do tend to have tall sons yet the average height of tall fathers is more than the average height of their sons and the average height of short fathers is less than the average height of their sons.

The term 'regression' stands for some sort of functional relationship between two or more related variables. The only fundamental difference, if any, between problems of curve-fitting and regression is that in regression, any of the variables may be considered as independent or dependent while in curve-fitting, one variable cannot be dependent.

Regression measures the nature and extent of correlation. Regression is the estimation or prediction of unknown values of one variable from known values of another variable.

3.43 CURVE OF REGRESSION AND REGRESSION EQUATION

If two variates x and y are correlated *i.e.*, there exists an association or relationship between them, then the scatter diagram will be more or less concentrated round a curve. This curve is called the *curve of regression* and the relationship is said to be expressed by means of *curvilinear regression*.

The mathematical equation of the regression curve is called regression equation.

3.44 LINEAR REGRESSION

When the points of the scatter diagram concentrate round a straight line, the regression is called linear and this straight line is known as the line of regression.

Regression will be called non-linear if there exists a relationship other than a straight line between the variables under consideration.

3.45 LINES OF REGRESSION (*A.K.T.U. 2016*)

A line of regression is the straight line which gives the best fit in the least square sense to the given frequency.

In case of n pairs (x_i, y_i) ; $i = 1, 2, ..., n$ from a bivariate data, we have no reason or justification to assume y as dependent variable and x as independent variable. Either of the two may be estimated for the given values of the other. Thus if we wish to estimate y for given values of x, we shall have the regression equation of the form $y = a + bx$, called the regression line of y on x. If we wish to estimate x for given values of y, we shall have the regression line of the form $x = \text{A} + \text{B}y$, called the regression line of x on y.

Thus it implies, in general, *we always have two lines of regression.*

If the line of regression is so chosen that the sum of squares of deviation parallel to the axis of y is minimised [*See Fig. (a)*], it is called *the line of regression of y on x* and it gives *the best estimate of y for any given value of x.*

If the line of regression is so chosen that the sum of squares of deviations parallel to the axis of x is minimised [*See Fig. (b)*], it is called *the line of regression of x on y* and it gives *the best estimate of x for any given value of y.*

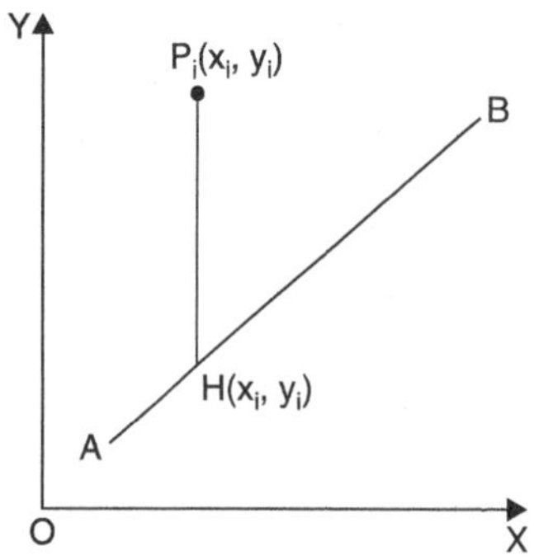

Fig. (*a*)

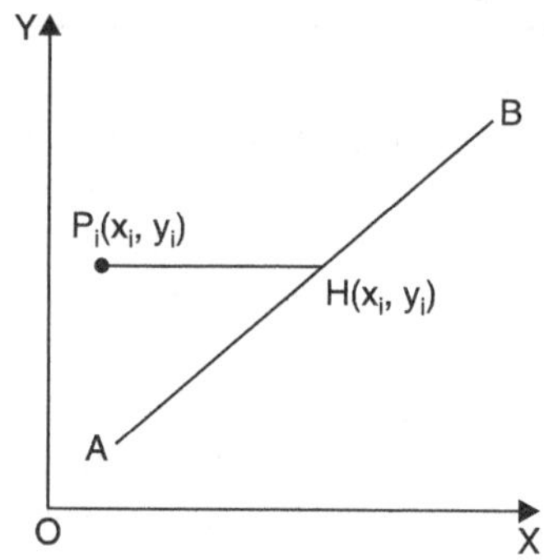

Fig. (*b*)

The independent variable is called *predictor* or regresser or explanator and the dependent variable is called the *predictant* or regressed or explained variable.

3.46 DERIVATION OF LINES OF REGRESSION

3.46.1 Line of Regression of *y* on *x*

To obtain the line of regression of y on x, we shall assume y as dependent variable and x as independent variable. Let $y = a + bx$ be the equation of regression line of y on x.

The residual for i^{th} point is $E_i = y_i - a - bx_i$.

Introduce a new quantity U such that

$$U = \sum_{i=1}^{n} E_i^2 = \sum_{i=1}^{n} (y_i - a - bx_i)^2 \quad \text{...(1)}$$

According to the *principle of Least squares*, the constants a and b are chosen in such a way that the sum of the squares of residuals is minimum.

Now, the condition for U to be maximum or minimum is

$$\frac{\partial U}{\partial a} = 0 \quad \text{and} \quad \frac{\partial U}{\partial b} = 0$$

From (1),

$$\frac{\partial U}{\partial a} = 2\sum_{i=1}^{n} (y_i - a - bx_i)(-1)$$

$$\frac{\partial U}{\partial a} = 0 \text{ gives } 2\sum_{i=1}^{n} (y_i - a - bx_i)(-1) = 0$$

$$\Rightarrow \quad \boxed{\Sigma y = na + b\,\Sigma x} \quad \text{...(2)}$$

Also,

$$\frac{\partial U}{\partial b} = 2\sum_{i=1}^{n} (y_i - a - bx_i)(-x_i)$$

$$\frac{\partial U}{\partial b} = 0 \text{ gives } 2\sum_{i=1}^{n} (y_i - a - bx_i)(-x_i) = 0$$

$$\Rightarrow \quad \boxed{\Sigma xy = a\,\Sigma x + b\,\Sigma x^2} \quad \text{...(3)}$$

Equations (2) and (3) are called *normal equations.*

Solving (2) and (3) for 'a' and 'b', we get

$$b = \frac{\Sigma xy - \frac{1}{n}\Sigma x\,\Sigma y}{\Sigma x^2 - \frac{1}{n}(\Sigma x)^2} = \frac{n\,\Sigma xy - \Sigma x\,\Sigma y}{n\,\Sigma x^2 - (\Sigma x)^2} \quad \text{...(4)}$$

and $$a = \frac{\Sigma y}{n} - b\frac{\Sigma x}{n} = \bar{y} - b\bar{x} \quad \text{...(5)}$$

Eqn. (5) gives $\bar{y} = a + b\bar{x}$

Hence $y = a + bx$ line passes through point $(\bar{x}, \bar{y})$.

Putting $a = \bar{y} - b\bar{x}$ in equation of line $y = a + bx$, we get

$$\boxed{y - \bar{y} = b(x - \bar{x})} \quad \text{...(6)}$$

Equation (6) is called regression line of y on x. 'b' is called the regression coefficient of y on x and is usually denoted by b_{yx}.

Hence eqn. (6) can be rewritten as

$$\boxed{y - \bar{y} = b_{yx}(x - \bar{x})}$$

where $\bar{x}$ and $\bar{y}$ are mean values while

$$b_{yx} = \frac{n\,\Sigma xy - \Sigma x\,\Sigma y}{n\,\Sigma x^2 - (\Sigma x)^2}$$

In equation (3), shifting the origin to $(\bar{x}, \bar{y})$, we get

$$\Sigma(x - \bar{x})(y - \bar{y}) = a\,\Sigma(x - \bar{x}) + b\,\Sigma(x - \bar{x})^2$$

$$\Rightarrow \quad nr\,\sigma_x\sigma_y = a(0) + bn\sigma_x^2$$

$$\Rightarrow \quad b = r\frac{\sigma_y}{\sigma_x}$$

$\because$ $\Sigma(x - \bar{x}) = 0$, $\frac{1}{n}\Sigma(x - \bar{x})^2 = \sigma_x^2$ and $\frac{\Sigma(x - \bar{x})(y - \bar{y})}{n\sigma_x\sigma_y} = r$

Hence, regression coefficient b_{yx} can also be defined as

$$\boxed{b_{yx} = r\frac{\sigma_y}{\sigma_x}}$$

where r is the coefficient of correlation, σ_x and σ_y are the standard deviations of x and y series respectively.

3.46.2 Line of Regression of x on y

Proceeding in the same way as 3.42.1, we can derive the regression line of x on y as

$$\boxed{x - \bar{x} = b_{xy}(y - \bar{y})}$$

where b_{xy} is the regression coefficient of x on y and is given by

$$\boxed{b_{xy} = \frac{n\,\Sigma xy - \Sigma x\,\Sigma y}{n\,\Sigma y^2 - (\Sigma y)^2}}$$

or $$b_{xy} = r\,\frac{\sigma_x}{\sigma_y}$$

where the terms have their usual meanings.

Note. If $r = 0$, the two lines of regression become $y = \bar{y}$ and $x = \bar{x}$ which are two straight lines parallel to x and y axes respectively and passing through their means $\bar{y}$ and $\bar{x}$. They are mutually perpendicular. If $r = \pm 1$, the two lines of regression will coincide.

3.47 USE OF REGRESSION ANALYSIS

(*i*) In the field of Business, this tool of statistical analysis is widely used. Businessmen are interested in predicting future production, consumption, investment, prices, profits and sales etc.

(*ii*) In the field of economic planning and sociological studies, projections of population, birth rates, death rates and other similar variables are of great use.

3.48 COMPARISON OF CORRELATION AND REGRESSION ANALYSIS

[*G.B.T.U. M.B.A. (C.O.) 2011*]

Both the correlation and regression analysis helps us in studying the relationship between two variables yet they differ in their approach and objectives.

(*i*) Correlation studies are meant for studying the covariation of the two variables. They tell us whether the variables under study move in the same direction or in reverse directions. The degree of their covariation is also reflected in the correlation coefficient but the correlation study does not provide the nature of relationship. It does not tell us about the relative movement in the variables and we cannot predict the value of one variable corresponding to the value of other variable. This is possible through regression analysis.

(*ii*) Regression presumes one variable as a cause and the other as its effect. The independent variable is supposed to be affecting the dependent variable and as such we can estimate the values of the dependent variable by projecting the relationship between them. However, correlation between two series is not necessarily a cause-effect relationship.

(*iii*) Coefficient of correlation cannot exceed unity but one of the regression coefficients can have a value higher than unity but the product of two regression coefficients can never exceed unity.

3.49 PROPERTIES OF REGRESSION COEFFICIENTS

Property I. *Correlation coefficient is the geometric mean between the regression coefficients.*

Proof. The coefficients of regression are $\frac{r\sigma_y}{\sigma_x}$ and $\frac{r\sigma_x}{\sigma_y}$.

$$\text{G.M. between them} = \sqrt{\frac{r\sigma_y}{\sigma_x} \times \frac{r\sigma_x}{\sigma_y}} = \sqrt{r^2} = r = \text{coefficient of correlation.}$$

Property II. *If one of the regression coefficients is greater than unity, the other must be less than unity.*

Proof. The two regression coefficients are $b_{yx} = \frac{r\sigma_y}{\sigma_x}$ and $b_{xy} = \frac{r\sigma_x}{\sigma_y}$.

Let $\quad b_{yx} > 1$, then $\dfrac{1}{b_{yx}} < 1 \quad$ ...(1)

Since $\quad b_{yx} \cdot b_{xy} = r^2 \le 1 \quad (\because \ -1 \le r \le 1)$

$\therefore \quad b_{xy} \le \dfrac{1}{b_{yx}} < 1. \quad$ | Using (1)

Similarly, if $b_{xy} > 1$, then $b_{yx} < 1$.

Property III. *Arithmetic mean of regression coefficients is greater than the correlation coefficient.*

Proof. We have to prove that

$$\frac{b_{yx} + b_{xy}}{2} > r$$

or $$r\frac{\sigma_y}{\sigma_x} + r\frac{\sigma_x}{\sigma_y} > 2r$$

or $$\sigma_x^2 + \sigma_y^2 > 2\sigma_x\sigma_y$$

or $$(\sigma_x - \sigma_y)^2 > 0, \quad \text{which is true.}$$

Property IV. *Regression coefficients are independent of the origin but not of scale.*

Proof. Let $\quad u = \dfrac{x-a}{h}, v = \dfrac{y-b}{k}$, where a, b, h and k are constants

$$b_{yx} = \frac{r\sigma_y}{\sigma_x} = r \cdot \frac{k\sigma_v}{h\sigma_u} = \frac{k}{h}\left(\frac{r\sigma_v}{\sigma_u}\right) = \frac{k}{h} b_{vu}$$

Similarly, $\quad b_{xy} = \dfrac{h}{k} b_{uv}.$

Thus, b_{yx} and b_{xy} are both independent of a and b but not of h and k.

Property V. *The correlation coefficient and the two regression coefficients have same sign.*

Proof. Regression coefficient of y on $x = b_{yx} = r\dfrac{\sigma_y}{\sigma_x}$

Regression coefficient of x on $y = b_{xy} = r\dfrac{\sigma_x}{\sigma_y}$

Since σ_x and σ_y are both positive; b_{yx}, b_{xy} and r have same sign.

3.50 ANGLE BETWEEN TWO LINES OF REGRESSION

If θ is the acute angle between the two regression lines in the case of two variables x and y, show that

$$\tan\theta = \frac{1-r^2}{r} \cdot \frac{\sigma_x\sigma_y}{\sigma_x^2 + \sigma_y^2}, \qquad \textit{where } r, \sigma_x, \sigma_y \textit{ have their usual meanings.}$$

Explain the significance of the formula when $r = 0$ and $r = \pm 1$.

[A.K.T.U. 2015, 2017]

Proof. Equations to the lines of regression of y on x and x on y are

$$y - \bar{y} = \frac{r\sigma_y}{\sigma_x}(x - \bar{x}) \qquad \text{and} \qquad x - \bar{x} = \frac{r\sigma_x}{\sigma_y}(y - \bar{y})$$

Their slopes are $m_1 = \dfrac{r\sigma_y}{\sigma_x}$ and $m_2 = \dfrac{\sigma_y}{r\sigma_x}$.

$$\therefore \quad \tan\theta = \pm\frac{m_2 - m_1}{1 + m_2 m_1} = \pm\frac{\dfrac{\sigma_y}{r\sigma_x} - \dfrac{r\sigma_y}{\sigma_x}}{1 + \dfrac{\sigma_y{}^2}{\sigma_x{}^2}}$$

$$= \pm\frac{1 - r^2}{r}\cdot\frac{\sigma_y}{\sigma_x}\cdot\frac{\sigma_x^2}{\sigma_x^2 + \sigma_y^2} = \pm\frac{1 - r^2}{r}\cdot\frac{\sigma_x\sigma_y}{\sigma_x^2 + \sigma_y^2}$$

Since $r^2 \le 1$ and σ_x, σ_y are positive.

∴ +ve sign gives the acute angle between the lines.

Hence
$$\boxed{\tan\theta = \frac{1 - r^2}{r}\cdot\frac{\sigma_x\sigma_y}{\sigma_x^2 + \sigma_y^2}}$$

when $r = 0$, $\theta = \dfrac{\pi}{2}$ ∴ The two lines of regression are perpendicular to each other.

Hence the estimated value of y is the same for all values of x and vice-versa.

When $r = \pm 1$, $\tan\theta = 0$ so that $\theta = 0$ or π

Hence the lines of regression coincide and there is perfect correlation between the two variates x and y.

ILLUSTRATIVE EXAMPLES

Example 1. *If the regression coefficients are 0.8 and 0.2, what would be the value of coefficient of correlation?*

Sol. We know that,

$$r^2 = b_{yx} \,.\, b_{xy} = 0.8 \times 0.2 = 0.16$$

Since r has the same sign as both the regression coefficients b_{yx} and b_{xy}

Hence, $r = \sqrt{0.16} = 0.4$.

Example 2. *Calculate linear regression coefficients from the following:*

x →	1	2	3	4	5	6	7	8
y →	3	7	10	12	14	17	20	24

Sol. Linear regression coefficients are given by

$$b_{yx} = \frac{n\,\Sigma xy - \Sigma x\,\Sigma y}{n\,\Sigma x^2 - (\Sigma x)^2} \quad \text{and} \quad b_{xy} = \frac{n\,\Sigma xy - \Sigma x\,\Sigma y}{n\,\Sigma y^2 - (\Sigma y)^2}$$

Let us prepare the following table:

x	y	x^2	y^2	xy
1	3	1	9	3
2	7	4	49	14
3	10	9	100	30
4	12	16	144	48
5	14	25	196	70
6	17	36	289	102
7	20	49	400	140
8	24	64	576	192
$\Sigma x = 36$	$\Sigma y = 107$	$\Sigma x^2 = 204$	$\Sigma y^2 = 1763$	$\Sigma xy = 599$

Here, $n = 8$

$$\therefore \qquad b_{yx} = \frac{(8 \times 599) - (36 \times 107)}{(8 \times 204) - (36)^2} = \frac{940}{336} = 2.7976$$

and

$$b_{xy} = \frac{(8 \times 599) - (36 \times 107)}{(8 \times 1763) - (107)^2} = \frac{940}{2655} = 0.3540$$

Example 3. *The following table gives age (x) in years of cars and annual maintenance cost (y) in hundred rupees:*

X:	*1*	*3*	*5*	*7*	*9*
Y:	*15*	*18*	*21*	*23*	*22*

Estimate the maintenance cost for a 4 year-old car after finding the regression equation.

Sol.

x	y	xy	x^2
1	15	15	1
3	18	54	9
5	21	105	25
7	23	161	49
9	22	198	81
$\Sigma x = 25$	$\Sigma y = 99$	$\Sigma xy = 533$	$\Sigma x^2 = 165$

Here, $n = 5$

$$\bar{x} = \frac{\Sigma x}{n} = \frac{25}{5} = 5, \ \bar{y} = \frac{\Sigma y}{n} = \frac{99}{5} = 19.8$$

$$\therefore \qquad b_{yx} = \frac{n\,\Sigma xy - \Sigma x\,\Sigma y}{n\,\Sigma x^2 - (\Sigma x)^2} = \frac{(5 \times 533) - (25 \times 99)}{(5 \times 165) - (25)^2} = 0.95$$

Regression line of y on x is given by

$$y - \bar{y} = b_{yx}\,(x - \bar{x})$$

$\Rightarrow \qquad y - 19.8 = 0.95\,(x - 5)$

$\Rightarrow \qquad y = 0.95x + 15.05$

When $x = 4$ years, $\quad y = (0.95 \times 4) + 15.05 = 18.85$ hundred rupees = ₹ 1885.

Example 4. *In a partially destroyed laboratory record of an analysis of a correlation data, the following results only are legible:*

Variance of $x = 9$

Regression equations: $8x - 10y + 66 = 0$, $40x - 18y = 214$.

What were (a) the mean values of x and y (b) the standard deviation of y and the coefficient of correlation between x and y? [A.K.T.U. 2018]

Sol. (*a*) **Since both the lines of regression pass through the point** $(\bar{x}, \bar{y})$ therefore, we have

$$8\bar{x} - 10\bar{y} + 66 = 0 \qquad \ldots(1)$$

$$40\bar{x} - 18\bar{y} - 214 = 0 \qquad \ldots(2)$$

Multiplying (1) by 5, $\quad 40\bar{x} - 50\bar{y} + 330 = 0 \qquad \ldots(3)$

Subtracting (3) from (2), $\quad 32\bar{y} - 544 = 0 \qquad \therefore \quad \bar{y} = 17$

$\therefore$ From (1), $\quad 8\bar{x} - 170 + 66 = 0 \qquad$ or $\quad 8\bar{x} = 104 \quad \therefore \quad \bar{x} = 13$

Hence, $\quad \bar{x} = 13, \quad \bar{y} = 17$

(*b*) Variance of $\quad x = \sigma_x^2 = 9$ (given)

$\therefore \qquad \sigma_x = 3$

The equations of lines of regression can be written as

$$y = 0.8x + 6.6 \quad \text{and} \quad x = 0.45y + 5.35$$

$\therefore$ The regression coefficient of y on x is $\quad \dfrac{r\sigma_y}{\sigma_x} = 0.8 \qquad \ldots(4)$

The regression coefficient of x on y is $\quad \dfrac{r\sigma_x}{\sigma_y} = 0.45 \qquad \ldots(5)$

Multiplying (4) and (5), $r^2 = 0.8 \times 0.45 = 0.36 \qquad \therefore \quad r = 0.6$

(+ve sign with square root is taken because regression coefficients are +ve).

From (4), $\quad \sigma_y = \dfrac{0.8\sigma_x}{r} = \dfrac{0.8 \times 3}{0.6} = 4.$

Example 5. *The regression lines of y on x and x on y are respectively $y = ax + b$, $x = cy + d$. Show that*

$$\frac{\sigma_y}{\sigma_x} = \sqrt{\frac{a}{c}}, \; \bar{x} = \frac{bc + d}{1 - ac} \; \textit{and} \; \bar{y} = \frac{ad + b}{1 - ac}.$$

Sol. The regression line of y on x is

$$y = ax + b \qquad \ldots(1)$$

$\therefore \qquad b_{yx} = a$

The regression line of x on y is

$$x = cy + d \quad \text{...(2)}$$

$\therefore$ $$b_{xy} = c$$

We know that, $$b_{yx} = r\frac{\sigma_y}{\sigma_x} \quad \text{...(3)}$$

and $$b_{xy} = r\frac{\sigma_x}{\sigma_y} \quad \text{...(4)}$$

Dividing eqn. (3) by (4), we get

$$\frac{b_{yx}}{b_{xy}} = \frac{\sigma_y^2}{\sigma_x^2} \Rightarrow \frac{a}{c} = \frac{\sigma_y^2}{\sigma_x^2} \Rightarrow \frac{\sigma_y}{\sigma_x} = \sqrt{\frac{a}{c}}$$

Since both the regression lines pass through the point $(\bar{x}, \bar{y})$ therefore,

$$\bar{y} = a\bar{x} + b \quad \text{and} \quad \bar{x} = c\bar{y} + d$$

$\Rightarrow$ $$a\bar{x} - \bar{y} = -b \quad \text{...(5)}$$

$$\bar{x} - c\bar{y} = d \quad \text{...(6)}$$

Multiplying equation (6) by a and then subtracting from (5), we get

$$(ac - 1)\,\bar{y} = -ad - b \Rightarrow \bar{y} = \frac{ad + b}{1 - ac}$$

Similarly, we get $$\bar{x} = \frac{bc + d}{1 - ac}.$$

Example 6. *For two random variables, x and y with the same mean, the two regression equations are y = ax + b and x = αy + β. Show that* $\frac{b}{\beta} = \frac{1 - a}{1 - \alpha}$. *Find also the common mean.*

(A.K.T.U. 2017)

Sol. Here, $b_{yx} = a,\ b_{xy} = \alpha$

Let the common mean be m, then regression lines are

$$y - m = a\,(x - m)$$

$\Rightarrow$ $$y = ax + m\,(1 - a) \quad \text{...(1)}$$

and $$x - m = \alpha(y - m)$$

$\Rightarrow$ $$x = \alpha y + m\,(1 - \alpha) \quad \text{...(2)}$$

Comparing (1) and (2) with the given equations.

$$b = m\,(1 - a),\ \beta = m\,(1 - \alpha)$$

$\therefore$ $$\frac{b}{\beta} = \frac{1 - a}{1 - \alpha}$$

Since regression lines pass through $(\bar{x}, \bar{y})$

$\therefore$ $\bar{x} = \alpha\bar{y} + \beta$ and $\bar{y} = a\bar{x} + b$ will hold.

$\Rightarrow$ $$m = am + b,\ m = \alpha m + \beta$$

$$\Rightarrow \qquad am + b = \alpha m + \beta$$

$$\Rightarrow \qquad m = \frac{\beta - b}{a - \alpha}.$$

Example 7. (*i*) *Obtain the line of regression of y on x for the data given below:*

x: *1.53* *1.78* *2.60* *2.95* *3.42*

y: *33.50* *36.30* *40.00* *45.80* *53.50.*

(*ii*) *The following data regarding the heights (y) and weights (x) of 100 college students are given:*

$\Sigma x = 15000$, $\Sigma x^2 = 2272500$, $\Sigma y = 6800$, $\Sigma y^2 = 463025$ *and* $\Sigma xy = 1022250$.

Find the equation of regression line of height on weight. *(A.K.T.U. 2019)*

Sol. (*i*) The line of regression of *y* on *x* is given by

$$y - \bar{y} = b_{yx}(x - \bar{x}) \qquad ...(1)$$

where b_{yx} is the coefficient of regression given by

$$b_{yx} = \frac{n\,\Sigma xy - \Sigma x\,\Sigma y}{n\,\Sigma x^2 - (\Sigma x)^2} \qquad ...(2)$$

Now we form the table as,

x	y	x^2	xy
1.53	33.50	2.3409	51.255
1.78	36.30	2.1684	64.614
2.60	40.00	6.76	104
2.95	45.80	8.7025	135.11
3.42	53.50	11.6964	182.97
$\Sigma x = 12.28$	$\Sigma y = 209.1$	$\Sigma x^2 = 32.6682$	$\Sigma xy = 537.949$

Here, $n = 5$

$$b_{yx} = \frac{(5 \times 537.949) - (12.28 \times 209.1)}{(5 \times 32.6682) - (12.28)^2} = 9.726$$

Also, mean $\bar{x} = \frac{\Sigma x}{n} = \frac{12.28}{5} = 2.456$ and $\bar{y} = \frac{\Sigma y}{n} = \frac{209.1}{5} = 41.82$

$\therefore$ From (1), we get

$$y - 41.82 = 9.726(x - 2.456) = 9.726x - 23.887$$

$$y = 17.932 + 9.726x$$

(*ii*) $$\bar{x} = \frac{\Sigma x}{n} = \frac{15000}{100} = 150, \quad \bar{y} = \frac{\Sigma y}{n} = \frac{6800}{100} = 68$$

Regression coefficient of *y* on *x*,

$$b_{yx} = \frac{n\,\Sigma xy - \Sigma x\,\Sigma y}{n\,\Sigma x^2 - (\Sigma x)^2} = \frac{(100 \times 1022250) - (15000 \times 6800)}{(100 \times 2272500) - (15000)^2} = 0.1$$

Regression line of height (y) on weight (x) is given by

$$y - \bar{y} = b_{yx}(x - \bar{x})$$

$$\Rightarrow \quad y - 68 = 0.1(x - 150)$$

$$\Rightarrow \quad y = 0.1x + 53.$$

Example 8. *For 10 observations on price (x) and supply (y), the following data were obtained (in appropriate units):*

$$\Sigma x = 130, \quad \Sigma y = 220, \quad \Sigma x^2 = 2288, \quad \Sigma y^2 = 5506 \text{ and } \Sigma xy = 3467$$

Obtain the two lines of regression and estimate the supply when the price is 16 units.

Sol. Here, $n = 10,\ \bar{x} = \dfrac{\Sigma x}{n} = 13$ and $\bar{y} = \dfrac{\Sigma y}{n} = 22$

Regression coefficient of y on x is

$$b_{yx} = \frac{n\,\Sigma xy - \Sigma x\,\Sigma y}{n\,\Sigma x^2 - (\Sigma x)^2} = \frac{(10 \times 3467) - (130 \times 220)}{(10 \times 2288) - (130)^2} = 1.015$$

∴ Regression line of y on x is

$$y - \bar{y} = b_{yx}(x - \bar{x})$$

$$y - 22 = 1.015(x - 13)$$

$$\Rightarrow \quad y = 1.015x + 8.805 \qquad \text{...(1)}$$

Regression coefficient of x on y is

$$b_{xy} = \frac{n\,\Sigma xy - \Sigma x\,\Sigma y}{n\,\Sigma y^2 - (\Sigma y)^2} = \frac{(10 \times 3467) - (130 \times 220)}{(10 \times 5506) - (220)^2} = 0.9114$$

Regression line of x on y is

$$x - \bar{x} = b_{xy}(y - \bar{y})$$

$$x - 13 = 0.9114(y - 22)$$

$$x = 0.9114y - 7.0508 \qquad \text{...(2)}$$

Since we are to estimate supply (y) when price (x) is given therefore we are to use regression line of y on x here.

When x = 16 units,

$$y = 1.015(16) + 8.805 = 25.045 \text{ units.}$$

Example 9. *The following results were obtained from records of age (x) and systolic blood pressure (y) of a group of 10 men:*

	x	*y*
Mean	*53*	*142*
Variance	*130*	

and $\Sigma(x - \bar{x})(y - \bar{y}) = 1220$

Find the appropriate regression equation and use it to estimate the blood pressure of a man whose age is 45.

Sol. Given

Mean $\bar{x} = 53$ and $\bar{y} = 142$; Variance $\sigma_x^2 = 130$

$n = 10$ and $\Sigma(x - \bar{x})(y - \bar{y}) = 1220$

Since we are to estimate blood pressure (y) of a 45 years old man, we will find regression line of y on x.

Regression coefficient

$$b_{yx} = r\frac{\sigma_y}{\sigma_x} = \frac{\Sigma(x-\bar{x})(y-\bar{y})}{n\,\sigma_x\,\sigma_y}\left(\frac{\sigma_y}{\sigma_x}\right)$$

$$= \frac{\Sigma(x-\bar{x})(y-\bar{y})}{n\,\sigma_x^2} = \frac{1220}{(10)(130)} = 0.93846$$

Regression line of y on x is given by

$$y - \bar{y} = b_{yx}(x - \bar{x})$$

$$\Rightarrow \quad y - 142 = 0.93846\,(x - 53)$$

$$\Rightarrow \quad y = 0.93846\,x + 92.26162$$

When $x = 45$, $\quad y = 134.49$

Hence the required blood pressure = 134.49.

Example 10. *The following results were obtained from marks in Applied Mechanics and Engineering Mathematics in an examination:*

	Applied Mechanics (x)	*Engineering Mathematics (y)*
Mean	*47.5*	*39.5*
Standard Deviation	*16.8*	*10.8*

$r = 0.95.$

Find both the regression equations. Also estimate the value of y for x = 30.

Sol. Here, $\quad \bar{x} = 47.5, \quad \bar{y} = 39.5$

$\sigma_x = 16.8, \quad \sigma_y = 10.8 \quad$ and $\quad r = 0.95.$

Regression coefficients are

$$b_{yx} = r\frac{\sigma_y}{\sigma_x} = 0.95 \times \frac{10.8}{16.8} = 0.6107$$

and

$$b_{xy} = r\frac{\sigma_x}{\sigma_y} = 0.95 \times \frac{16.8}{10.8} = 1.477.$$

Regression line of y on x is

$$y - \bar{y} = b_{yx}(x - \bar{x})$$

$$\Rightarrow \quad y - 39.5 = 0.6107\,(x - 47.5) = 0.6107x - 29.008$$

$$y = 0.6107x + 10.49 \qquad \text{...(1)}$$

Regression line of x on y is

$$x - \bar{x} = b_{xy}(y - \bar{y})$$

$$\Rightarrow \quad x - 47.5 = 1.477\,(y - 39.5)$$

$$x = 1.477y - 10.8415 \qquad \text{...(2)}$$

Putting $x = 30$ in equation (1), we get

$$y = (0.6107)(30) + 10.49 = 28.81.$$

Example 11. *The equations of two regression lines, obtained in a correlation analysis of 60 observations are:*

$$5x = 6y + 24 \text{ and } 1000y = 768x - 3608.$$

What is the correlation coefficient? Show that the ratio of coefficient of variability of x to that of y is $\frac{5}{24}$. *What is the ratio of variances of x and y?* *(A.K.T.U. 2019)*

Sol. Regression line of x on y is

$$5x = 6y + 24$$

$$\Rightarrow \quad x = \frac{6}{5}y + \frac{24}{5}$$

$$\therefore \quad b_{xy} = \frac{6}{5} \quad \text{...(1)}$$

Regression line of y on x is

$$1000y = 768x - 3608$$

$$\Rightarrow \quad y = 0.768x - 3.608$$

$$\therefore \quad b_{yx} = 0.768 \quad \text{...(2)}$$

From (1), $$r\frac{\sigma_x}{\sigma_y} = \frac{6}{5} \quad \text{...(3)}$$

From (2), $$r\frac{\sigma_y}{\sigma_x} = 0.768 \quad \text{...(4)}$$

Multiplying equations (3) and (4), we get

$$r^2 = 0.9216 \quad \Rightarrow \quad r = 0.96 \quad \text{...(5)}$$

Dividing (4) by (3), we get

$$\frac{{\sigma_x}^2}{{\sigma_y}^2} = \frac{6}{5 \times 0.768} = 1.5625.$$

Taking square root, we get

$$\frac{\sigma_x}{\sigma_y} = 1.25 = \frac{5}{4} \quad \text{...(6)}$$

Since the regression lines pass through the point $(\bar{x}, \bar{y})$, we have

$$5\bar{x} = 6\bar{y} + 24$$

$$1000\bar{y} = 768\bar{x} - 3608.$$

Solving the above equations for $\bar{x}$ and $\bar{y}$, we get $\bar{x} = 6$, $\bar{y} = 1$.

Coefficient of variability of $x = \frac{\sigma_x}{\bar{x}}$

Coefficient of variability of $y = \frac{\sigma_y}{\bar{y}}$.

$$\therefore \quad \text{Required ratio} = \frac{\sigma_x}{\bar{x}} \times \frac{\bar{y}}{\sigma_y} = \frac{\bar{y}}{\bar{x}}\left(\frac{\sigma_x}{\sigma_y}\right) = \frac{1}{6} \times \frac{5}{4} = \frac{5}{24}. \quad | \text{ Using (6)}$$

Example 12. *A panel of two judges, A and B, graded seven TV serial performances by awarding marks independently as shown in the following table:*

Performance	*1*	*2*	*3*	*4*	*5*	*6*	*7*
Marks by A	*46*	*42*	*44*	*40*	*43*	*41*	*45*
Marks by B	*40*	*38*	*36*	*35*	*39*	*37*	*41*

The eighth TV performance which judge B could not attend, was awarded 37 marks by judge A. If the judge B had also been present, how many marks would be expected to have been awarded by him to the eighth TV performance?

Use regression analysis to answer this question.

Sol. Let the marks awarded by judge A be denoted by x and the marks awarded by judge B be denoted by y.

Here, $n = 7$; $\bar{x} = \frac{\Sigma x}{n} = \frac{46 + 42 + 44 + 40 + 43 + 41 + 45}{7} = 43$

$$\bar{y} = \frac{\Sigma y}{n} = \frac{40 + 38 + 36 + 35 + 39 + 37 + 41}{7} = 38$$

Let us form the table as

x	y	xy	x^2
46	40	1840	2116
42	38	1596	1764
44	36	1584	1936
40	35	1400	1600
43	39	1677	1849
41	37	1517	1681
45	41	1845	2025
$\Sigma x = 301$	$\Sigma y = 266$	$\Sigma xy = 11459$	$\Sigma x^2 = 12971$

Regression coefficient,

$$b_{yx} = \frac{n\,\Sigma xy - \Sigma x\,\Sigma y}{n\,\Sigma x^2 - (\Sigma x)^2} = \frac{(7 \times 11459) - (301 \times 266)}{(7 \times 12971) - (301)^2} = 0.75$$

Regression line of y on x is given by

$$y - \bar{y} = b_{yx}(x - \bar{x})$$

$$y - 38 = 0.75(x - 43)$$

$$\Rightarrow \quad y = 0.75x + 5.75$$

When $x = 37$, $y = 0.75(37) + 5.75 = 33.5$ marks

Hence, if judge B had also been present, 33.5 marks would be expected to have been awarded to the eighth TV performance.

Example 13. *Two variables x and y have zero means, the same variance* σ^2 *and zero correlation, show that:*

$$u = x \cos \alpha + y \sin \alpha \quad \text{and} \quad v = x \sin \alpha - y \cos \alpha$$

have the same variance σ^2 *and zero correlation.* (U.P.T.U. 2007)

Sol. We are given that

$$r(x, y) = 0 \Rightarrow \text{Cov}(x, y) = 0, \quad \sigma_x^2 = \sigma_y^2 = \sigma^2$$

We have,

$$\begin{aligned} \sigma_u^{\,2} &= \text{V}(x \cos \alpha + y \sin \alpha) \\ &= \cos^2 \alpha \, \text{V}(x) + \sin^2 \alpha \, \text{V}(y) + 2 \sin \alpha \cos \alpha \, \text{Cov}(x, y) \\ &= (\cos^2 \alpha + \sin^2 \alpha) \sigma^2 \qquad | \because \text{Cov}(x, y) = 0 \\ &= \sigma^2 \end{aligned}$$

Similarly, $\sigma_v^2 = \sigma^2$

$$\begin{aligned} \text{Cov}(u, v) &= \text{E}[(u - \bar{u})(v - \bar{v})] \\ &= \text{E}[(x \cos \alpha + y \sin \alpha - \bar{x} \cos \alpha - \bar{y} \sin \alpha) \\ &\qquad (x \sin \alpha - y \cos \alpha - \bar{x} \sin \alpha + \bar{y} \cos \alpha)] \\ &= \text{E}[(x \cos \alpha + y \sin \alpha)(x \sin \alpha - y \cos \alpha)] \qquad | \because \bar{x} = 0 = \bar{y} \\ &= [\text{E}(x^2) - \text{E}(y^2)] \sin \alpha \cos \alpha + \text{E}(xy)(\sin^2 \alpha - \cos^2 \alpha) \\ &= 0 \qquad | \because \sigma_x^{\,2} = \sigma_y^{\,2} = \sigma^2 \text{ and } \text{E}(xy) = 0 \end{aligned}$$

$$\therefore \qquad r = \frac{\text{Cov}(u, v)}{\sigma_u \sigma_v} = 0.$$

TEST YOUR KNOWLEDGE

1. (*i*) Discuss regression and its importance. Given the following data:

x:	1	5	3	2	1	1	7	3
y:	6	1	0	0	1	2	1	5

Find a regression line of x on y.

(*ii*) Find the coefficient of correlation (γ) and obtain the equation to the lines of regression for the following data:

x:	6	2	10	4	8
y:	9	11	5	8	7

(*A.K.T.U. 2017*)

(*iii*) Find the two lines of regression and coefficient of correlation for the data given below:

$n = 18, \Sigma x = 12, \Sigma y = 18, \Sigma x^2 = 60, \Sigma y^2 = 96, \Sigma xy = 48$

(*iv*) From the data given, find the equation of lines of regression of x on y and y on x. Also, calculate the correlation co-efficient.

x:	2	4	6	8	10
y:	5	7	9	8	11

(*U.P.T.U. 2011*)

2. (*i*) Can $\text{Y} = 5 + 2.8\,\text{X}$ and $\text{X} = 3 - 0.5\,\text{Y}$ be the estimated regression equations of Y on X and X on Y respectively? Explain your answer with suitable theoretical arguments.

(*ii*) Find the co-efficient of correlation when the two regression equations are

$$\text{X} = -0.2\,\text{Y} + 4.2, \quad \text{Y} = -0.8\,\text{X} + 8.4$$

3. (*i*) If F is the pull required to lift a load W by means of a pulley block, fit a linear law of the form $\text{F} = m\text{W} + c$ connecting F and W, using the data

W:	50	70	100	120
F:	12	15	21	25

where F and W are in kg wt. Compute F when W = 150 kg wt.

(*ii*) A simply supported beam carries a concentrated load P (kg) at its midpoint. The following table gives maximum deflection y (cm) corresponding to various values of P:

P:	100	120	140	160	180	200
y:	0.45	0.55	0.60	0.70	0.80	0.85

Find a law of the form $y = a + b\text{P}$. Also find the value of maximum deflection when $\text{P} = 150$ kg.

4. (*i*) Find both the lines of regression of following data:

x:	5.60	5.65	5.70	5.81	5.85
y:	5.80	5.70	5.80	5.79	6.01

(*ii*) Obtain regression line of x on y for the given data:

x:	1	2	3	4	5	6
y:	5.0	8.1	10.6	13.1	16.2	20.0

(*iii*) Given that:

x:	1	3	5	7	8	10
y:	8	12	15	17	18	20

Find the equations of both lines of regression.

(*iv*) In a study between the amount of rainfall and the quantity of air pollution removed the following data were collected:

Daily rainfall: (*in 0.01 cm*)	4.3	4.5	5.9	5.6	6.1	5.2	3.8	2.1
Pollution removed: (mg/m^3)	12.6	12.1	11.6	11.8	11.4	11.8	13.2	14.1

Find the regression line of y on x.

5. (*i*) The two regression equations of the variables x and y are $x = 19.13 - 0.87\,y$ and $y = 11.64 - 0.50x$. Find (*a*) mean of x's (*b*) mean of y's and (*c*) correlation coefficient between x and y.

(*ii*) Two random variables have the regression lines with equations $3x + 2y = 26$ and $6x + y = 31$. Find the mean values and the correlation coefficient between x and y.

[*G.B.T.U. (MBA) 2011*]

(*iii*) In a partially destroyed laboratory data, only the equations giving the two lines of regression of y on x and x on y are available and are respectively

$$7x - 16y + 9 = 0, \quad 5y - 4x - 3 = 0.$$

Calculate the coefficient of correlation, $\bar{x}$ and $\bar{y}$.

(*iv*) The regression equations calculated from a given set of observations for two random variables are

$$x = -0.4y + 6.4 \quad \text{and} \quad y = -0.6x + 4.6$$

Calculate (*i*) $\bar{x}$ (*ii*) $\bar{y}$ (*iii*) r. (*A.K.T.U. 2016*)

(*v*) Two lines of regression are given by

$$x + 2y - 5 = 0 \quad \text{and} \quad 2x + 3y - 8 = 0 \quad \text{and} \quad \sigma_x^2 = 12,$$

Calculate:

(*a*) the mean values of x and y (*b*) variance of y

(*c*) the coefficient of correlation between x and y. [*G.B.T.U. (C.O.) 2011*]

6. An analyst for a company was studying travelling expenses (y) in ₹ and duration (x) of these trips for 102 sales trip. He has found relation between x and y linear and data as follows:

$$\Sigma x = 510, \quad \Sigma y = 7140, \quad \Sigma x^2 = 4150, \quad \Sigma xy = 54900, \quad \Sigma y^2 = 740200$$

Calculate (*i*) Two regression lines

(*ii*) A given trip has to take 7 days. How much money should be allowed so that they will not run short of money?

7. Assuming that we conduct an experiment with 8 fields planted with corn, four fields having no nitrogen fertiliser and four fields having 80 kg of nitrogen fertilizer. The resulting corn yields are shown in table in bushels per acre :

Field:	1	2	3	4	5	6	7	8
Nitrogen (kg) x:	0	0	0	0	80	80	80	80
Corn yield y: (acre)	120	360	60	180	1280	1120	1120	760

(*a*) Compute a linear regression equation of y on x.

(*b*) Predict corn yield for a field treated with 60 kg of fertilizer.

8. If the coefficient of correlation between two variables x and y is 0.5 and the acute angle between their lines of regression is $\tan^{-1}\left(\frac{3}{5}\right)$, show that $\sigma_x = \frac{1}{2}\sigma_y$.

9. Given N = 50, Mean of y = 44, Variance of x is $\frac{9}{16}$ of the variance of y.

Regression equation of x on y is $3y - 5x = -180$

Find (*i*) Mean of x (*ii*) Coeff. of correlation between x and y.

10. The means of a bivariate frequency distribution are at (3, 4) and $r = 0.4$. The line of regression of y on x is parallel to the line $y = x$. Find the two lines of regression and estimate value of x when $y = 1$.

11. The following results were obtained in the analysis of data on yield of dry bark in ounces (y) and age in years (x) of 200 cinchona plants:

	x	y
Average:	9.2	16.5
Standard deviation:	2.1	4.2

Correlation coefficient = 0.84

Construct the two lines of regression and estimate the yield of dry bark of a plant of age 8 years.

12. A panel of judges A and B graded 7 debators and independently awarded the following marks:

Debator:	1	2	3	4	5	6	7
Marks by A:	40	34	28	30	44	38	31
Marks by B:	32	39	26	30	38	34	28

An eighth debator was awarded 36 marks by judge A while judge B was not present. If judge B were also present, how many marks would you expect him to award to the eighth debator assuming that the same degree of relationship exists in their judgement?

Answers

1. (*i*) $72x = -20y + 247$ (*ii*) $r = -0.9192$; $y = -0.65x + 11.9$, $x = -1.3y + 16.4$

(*iii*) $y = 0.6923x + 0.53846$; $x = 0.4615y + 0.2051$

(*iv*) $x = 1.3y - 4.4$, $y = 0.65x + 4.1$; $r = 0.9192$

2. (*i*) No (*ii*) $r = -0.4$

3. (*i*) F = 0.18793W + 2.27595; F = 30.4654 kg wt.

(*ii*) $y = 0.04765 + 0.004071$ P; $y = 0.6583$ cm

4. (*i*) Regression line of y on x: $y = 0.74306\, x + 1.56821$

Regression line of x on y: $x = 0.63602\, y + 2.0204$

(*ii*) $x = 0.34195\, y - 0.660355$ (*iii*) $y = 1.3012\, x + 7.6265$; $x = 0.75y - 5.5833$.

(*iv*) $y = -\,0.6842x + 15.5324$

5. (*i*) (*a*) 15.935 (*b*) 3.67 (*c*) – 0.659

(*ii*) $\overline{x} = 4$, $\overline{y} = 7$, $r = -\,0.5$ (*iii*) $r = 0.7395$, $\overline{x} = -\,0.1034$, $\overline{y} = 0.5172$

(*iv*) $\overline{x} = 6$, $\overline{y} = 1$, $r = -\,0.48989$

(*v*) (*a*) $\overline{x} = 1$, $\overline{y} = 2$ (*b*) 4 (*c*) $-\dfrac{\sqrt{3}}{2}$

6. (*i*) $y = 12x + 10$, $x = 0.07986y - 0.59068$ (*ii*) ₹ 94

7. (*a*) $y = 11.125x + 180$ (*b*) 847.5 acre

9. (*i*) 62.4 (*ii*) 0.8

10. $y = x + 1$; $x = 0.16y + 2.36$; $x = 2.52$

11. $y = 1.68x + 1.044$, $x = 0.42y + 2.27$; $y = 14.484$ **12.** 33 marks.

3.51 POLYNOMIAL FIT: NON-LINEAR REGRESSION

Let $$y = a + bx + cx^2 \qquad ...(1)$$
be a second degree parabolic curve of regression of y on x to be fitted for the data (x_i, y_i), $i = 1, 2, \ldots\ldots, n$.

Residual at $x = x_i$ is

$$E_i = y_i - f(x_i) = y_i - a - bx_i - cx_i^2$$

Now, let $$U = \sum_{i=1}^{n} E_i^2 = \sum_{i=1}^{n} (y_i - a - bx_i - cx_i^2)^2$$

By principle of Least squares, U should be minimum for the best values of a, b and c.

For this, $$\frac{\partial U}{\partial a} = 0, \ \frac{\partial U}{\partial b} = 0 \text{ and } \frac{\partial U}{\partial c} = 0$$

$$\frac{\partial U}{\partial a} = 0 \quad \Rightarrow \quad 2\sum_{i=1}^{n} (y_i - a - bx_i - cx_i^2)(-1) = 0$$

$$\Rightarrow \quad \boxed{\Sigma y = na + b\Sigma x + c\Sigma x^2} \qquad ...(1)$$

$$\frac{\partial U}{\partial b} = 0 \quad \Rightarrow \quad 2\sum_{i=1}^{n} (y_i - a - bx_i - cx_i^2)(-x_i) = 0$$

$$\Rightarrow \quad \boxed{\Sigma xy = a\Sigma x + b\Sigma x^2 + c\Sigma x^3} \qquad ...(2)$$

$$\frac{\partial U}{\partial c} = 0 \quad \Rightarrow \quad 2\sum_{i=1}^{n} (y_i - a - bx_i - cx_i^2)(-x_i^2) = 0$$

$$\Rightarrow \boxed{\Sigma x^2 y = a\Sigma x^2 + b\Sigma x^3 + c\Sigma x^4} \quad ...(3)$$

Equations (1), (2) and (3) are the normal equations for fitting a second degree parabolic curve of regression of y on x. Here n is the no. of pairs of values of x and y.

ILLUSTRATIVE EXAMPLES

Example 1. *Fit a second degree parabola to the following data:*

(*a*)

x	0.0	1.0	2.0
y	1.0	6.0	17.0

(*b*)

x	1.0	2.0	3.0	4.0
y	6.0	11.0	18.0	27

(*c*)

x	0.0	1.0	2.0	3.0	4.0
y	1.0	4.0	10.0	17.0	30.0

Sol. The equation of second degree parabola is given by

$$y = a + bx + cx^2 \quad ...(1)$$

Normal equations are

$$\Sigma y = ma + b\Sigma x + c\Sigma x^2 \quad ...(2)$$

$$\Sigma xy = a\Sigma x + b\Sigma x^2 + c\Sigma x^3 \quad ...(3)$$

and

$$\Sigma x^2 y = a\Sigma x^2 + b\Sigma x^3 + c\Sigma x^4 \quad ...(4)$$

(*a*) Here, $m = 3$. Table is as follows:

x	y	x^2	x^3	x^4	xy	x^2y
0	1	0	0	0	0	0
1	6	1	1	1	6	6
2	17	4	8	16	34	68
Total 3	24	5	9	17	40	74

Substituting in eqns. (2), (3) and (4), we get

$$24 = 3a + 3b + 5c \quad ...(5)$$

$$40 = 3a + 5b + 9c \quad ...(6)$$

$$74 = 5a + 9b + 17c \quad ...(7)$$

Solving eqns. (5), (6) and (7), we get $a = 1, b = 2, c = 3$

Hence the required second degree parabola is $y = 1 + 2x + 3x^2$

(*b*) Here, $m = 4$. Table is as follows:

x	y	x^2	x^3	x^4	xy	x^2y
1	6	1	1	1	6	6
2	11	4	8	16	22	44
3	18	9	27	81	54	162
4	27	16	64	256	108	432
$\Sigma x = 10$	$\Sigma y = 62$	$\Sigma x^2 = 30$	$\Sigma x^3 = 100$	$\Sigma x^4 = 354$	$\Sigma xy = 190$	$\Sigma x^2y = 644$

Substituting values in eqns. (2), (3) and (4), we get

$$62 = 4a + 10b + 30c \quad ...(8)$$

$$190 = 10a + 30b + 100c \quad ...(9)$$

$$644 = 30a + 100b + 354c \quad ...(10)$$

Solving equations (8), (9) and (10), we get $a = 3, b = 2, c = 1$

Hence the required second degree parabola is $y = 3 + 2x + x^2$

(*c*) Here, $m = 5$. Table is as follows:

x	y	x^2	x^3	x^4	xy	x^2y
0.0	1.0	0	0	0	0	0
1.0	4.0	1	1	1	4	4
2.0	10.0	4	8	16	20	40
3.0	17.0	9	27	81	51	153
4.0	30.0	16	64	256	120	480
$\Sigma x = 10$	$\Sigma y = 62$	$\Sigma x^2 = 30$	$\Sigma x^3 = 100$	$\Sigma x^4 = 354$	$\Sigma xy = 195$	$\Sigma x^2y = 677$

Substituting values in eqns. (2), (3) and (4), we get

$$62 = 5a + 10b + 30c \quad ...(11)$$

$$195 = 10a + 30b + 100c \quad ...(12)$$

$$677 = 30a + 100b + 354c \quad ...(13)$$

Solving eqns. (11), (12) and (13), we get $a = 1.2, b = 1.1$ and $c = 1.5$

Hence the required second degree parabola is $y = 1.2 + 1.1x + 1.5x^2$.

Example 2. *Fit a second degree parabola to the following data taking y as dependent variable:*

x	1	2	3	4	5	6	7	8	9
y	2	6	7	8	10	11	11	10	9

Sol. Normal equations to fit a second degree parabola of the form $y = a + bx + cx^2$ are

$$\left.\begin{aligned} \Sigma y &= ma + b\Sigma x + c\Sigma x^2 \\ \Sigma xy &= a\Sigma x + b\Sigma x^2 + c\Sigma x^3 \\ \Sigma x^2 y &= a\Sigma x^2 + b\Sigma x^3 + c\Sigma x^4 \end{aligned}\right\} \quad ...(1)$$

and

Here, $m = 9$

x	y	x^2	x^3	x^4	xy	x^2y
1	2	1	1	1	2	2
2	6	4	8	16	12	24
3	7	9	27	81	21	63
4	8	16	64	256	32	128
5	10	25	125	625	50	250
6	11	36	216	1296	66	396
7	11	49	343	2401	77	539
8	10	64	512	4096	80	640
9	9	81	729	6561	81	729
$\Sigma x = 45$	$\Sigma y = 74$	$\Sigma x^2 = 285$	$\Sigma x^3 = 2025$	$\Sigma x^4 = 15333$	$\Sigma xy = 421$	$\Sigma x^2y = 2771$

Putting in (1), we get

$$74 = 9a + 45b + 285c$$

$$421 = 45a + 285b + 2025c$$

$$2771 = 285a + 2025b + 15333c$$

Solving the above equations, we get $a = -1, \quad b = 3.55, \quad c = -0.27$

Hence the required equation of second degree parabola is $y = -1 + 3.55x - 0.27x^2$.

Example 3. *Employ the method of least squares to fit a parabola $y = a + bx + cx^2$ in the data: (x, y): $(-1, 2)$, $(0, 0)$, $(0, 1)$, $(1, 2)$*

Sol. Normal equations to the parabola $y = a + bx + cx^2$ are

$$\Sigma y = ma + b\Sigma x + c\Sigma x^2 \quad ...(1)$$

$$\Sigma xy = a\Sigma x + b\Sigma x^2 + c\Sigma x^3 \quad ...(2)$$

and

$$\Sigma x^2y = a\Sigma x^2 + b\Sigma x^3 + c\Sigma x^4 \quad ...(3)$$

Here, $m = 4$. The table is as follows:

x	y	x^2	x^3	x^4	xy	x^2y
– 1	2	1	– 1	1	– 2	2
0	0	0	0	0	0	0
0	1	0	0	0	0	0
1	2	1	1	1	2	2
$\Sigma x = 0$	$\Sigma y = 5$	$\Sigma x^2 = 2$	$\Sigma x^3 = 0$	$\Sigma x^4 = 2$	$\Sigma xy = 0$	$\Sigma x^2y = 4$

Substituting these values in equations (1), (2) and (3); we get

$$5 = 4a + 2c \quad ...(4)$$

$$0 = 2b \quad ...(5)$$

and

$$4 = 2a + 2c \quad ...(6)$$

Solving (4), (5) and (6), we get $a = 0.5, b = 0$ and $c = 1.5$

Hence the required second degree parabola is $y = 0.5 + 1.5x^2$

Example 4. *Fit a second degree parabola to the following data by Least Squares method:*

x	*1*	*2*	*3*	*4*	*5*
y	*1090*	*1220*	*1390*	*1625*	*1915*

[U.K.T.U. 2010]

Sol. Here, $m = 5$ (odd)

Let $u = x - 3, \quad v = y - 1220$

x	y	u	v	u^2	u^2v	uv	u^3	u^4
1	1090	– 2	– 130	4	– 520	260	– 8	16
2	1220	– 1	0	1	0	0	– 1	1
3	1390	0	170	0	0	0	0	0
4	1625	1	405	1	405	405	1	1
5	1915	2	695	4	2780	1390	8	16
Total		$\Sigma u = 0$	$\Sigma v = 1140$	$\Sigma u^2 = 10$	$\Sigma u^2v = 2665$	$\Sigma uv = 2055$	$\Sigma u^3 = 0$	$\Sigma u^4 = 34$

Putting these values in normal equations, we get

$$1140 = 5a' + 10c', \quad 2055 = 10b', \quad 2655 = 10a' + 34c'$$

$\Rightarrow$ $a' = 173, \quad b' = 205.5, \quad c' = 27.5$

$\therefore$ $v = 173 + 205.5u + 27.5u^2$ ...(1)

Put $u = x - 3$ and $v = y - 1220$

From (1), $y - 1220 = 173 + 205.5\,(x - 3) + 27.5\,(x - 3)^2$

$\Rightarrow$ $y = 27.5x^2 + 40.5x + 1024.$

3.52 MULTIPLE LINEAR REGRESSION

Now we proceed to discuss the case where the dependent variable is a function of two or more linear or non-linear independent variables. Consider such a linear function as

$$y = a + bx + cz \qquad ...(1)$$

The sum of the squares of residual is

$$U = \sum_{i=1}^{n} (y_i - a - bx_i - cz_i)^2 \qquad ...(2)$$

Differentiating U partially w.r.t. a, b, c; we get

$$\frac{\partial U}{\partial a} = 0 \Rightarrow 2\sum_{i=1}^{n} (y_i - a - bx_i - cz_i)(-1) = 0$$

$$\frac{\partial U}{\partial b} = 0 \Rightarrow 2\sum_{i=1}^{n} (y_i - a - bx_i - cz_i)(-x_i) = 0$$

and $$\frac{\partial U}{\partial c} = 0 \Rightarrow 2\sum_{i=1}^{n} (y_i - a - bx_i - cz_i)(-z_i) = 0$$

which on simplification and omitting the suffix i, yields.

$$\Sigma y = ma + b\Sigma x + c\Sigma z$$
$$\Sigma xy = a\Sigma x + b\Sigma x^2 + c\Sigma xz$$
$$\Sigma yz = a\Sigma z + b\Sigma xz + c\Sigma z^2$$

Solving the above three equations, we get values of a, b and c. Consequently, we get the linear function $y = a + bx + cz$ called **regression plane.**

Example 5. *Obtain a regression plane by using multiple linear regression to fit the data given below:*

x	1	2	3	4
z	0	1	2	3
y	12	18	24	30

Sol. Let $y = a + bx + cz$ be the required regression plane where a, b, c are the constants to be determined by following equations:

$$\Sigma y = ma + b\Sigma x + c\Sigma z$$

$$\Sigma yx = a\Sigma x + b\Sigma x^2 + c\Sigma zx$$

and $$\Sigma yz = a\Sigma z + b\Sigma zx + c\Sigma z^2$$

Here, $m = 4$

x	z	y	x^2	z^2	yx	zx	yz
1	0	12	1	0	12	0	0
2	1	18	4	1	36	2	18
3	2	24	9	4	72	6	48
4	3	30	16	9	120	12	90
$\Sigma x = 10$	$\Sigma z = 6$	$\Sigma y = 84$	$\Sigma x^2 = 30$	$\Sigma z^2 = 14$	$\Sigma yx = 240$	$\Sigma zx = 20$	$\Sigma yz = 156$

Substitution yields, $84 = 4a + 10b + 6c$

$$240 = 10a + 30b + 20c$$

and $$156 = 6a + 20b + 14c$$

Solving, we get $a = 10, b = 2, c = 4$

Hence the required regression plane is $y = 10 + 2x + 4z$.

Example 6. *Find the multiple linear regression of X_1 on X_2 and X_3 from the data relating to three variables:*

X_1	4	6	7	9	13	15
X_2	15	12	8	6	4	3
X_3	30	24	20	14	10	4

Sol. Let $X_1 = a + bX_2 + cX_3$ be the required regression plane where a, b, c are the constants, determined by the following normal equations

$$\Sigma X_1 = ma + b\Sigma X_2 + c\Sigma X_3$$

$$\Sigma X_1X_2 = a\Sigma X_2 + b\Sigma X_2^2 + c\Sigma X_2X_3$$

$$\Sigma X_1X_3 = a\Sigma X_3 + b\Sigma X_2X_3 + c\Sigma X_3^2$$

Here, $m = 6$

X_1	X_2	X_3	X_1X_2	X_2^2	X_2X_3	X_1X_3	X_3^2
4	15	30	60	225	450	120	900
6	12	24	72	144	288	144	576
7	8	20	56	64	160	140	400
9	6	14	54	36	84	126	196
13	4	10	52	16	40	130	100
15	3	4	45	9	12	60	16
Total 54	48	102	339	494	1034	720	2188

Substituting the values, we get

$$54 = 6a + 48b + 102c$$

$$339 = 48a + 102b + 1034c$$

$$720 = 102a + 1034b + 2188c$$

On solving, we get, $a = 16.413, \quad b = -0.00536, \quad c = -0.4335$

Hence, $\quad X_1 = 16.413 - 0.00536X_2 - 0.4335X_3$

TEST YOUR KNOWLEDGE

1. Fit a parabola of the form $y = a + bx + cx^2$ to the data:

x	1	2	3	4
y	1.7	1.8	2.3	3.2

by the method of least squares.

2. Find the best values of a_0, a_1, a_2 so that the parabola $y = a_0 + a_1x + a_2x^2$ fits the data:

x	1	1.5	2	2.5	3	3.5	4
y	1.1	1.2	1.5	2.6	2.8	3.3	4.1

3. Using method of least squares, derive the normal equations to fit a parabola $y = a + bx + cx^2$ from the following data:

x:	2	3	4	5	6
y:	14	17	20	24	29

[*A.K.T.U. 2017*]

4. Fit a second degree parabola to the following data:

(*i*)

x	1	2	3	4	5
y	25	28	33	39	46

[*U.P.T.U. (C.O.) 2011*]

(*ii*)

x	1	2	3	4	5	6	7	8	9	10
y	124	129	140	159	228	289	315	302	263	210

(iii)

x	0	1	2	3	4
y	1	5	10	22	38

(iv)

x	1	2	3	4	5	6	7	8	9
y	3	7	8	9	11	12	13	14	15

(v)

x	1	2	3	4	5	6	7	8	9
y	2	6	7	8	10	11	8	13	5

(U.P.T.U. 2015)

5. The profit of a certain company in X^{th} year of its life are given by:

x	1	2	3	4	5
y	1250	1400	1650	1950	2300

Taking $u = x - 3$ and $v = \dfrac{y - 1650}{50}$, show that the parabola of second degree of v on u is $v + 0.086 = 5.3\,u + 0.643u^2$ and deduce that the parabola of second degree of y on x is

$$y = 1144 + 72x + 32.15x^2$$

6. (*i*) The corresponding values of x and y are given below:

x	87	84	79	64	47	37
y	292	283	270	235	197	181

Fit a parabola of the form $y = ax^2 + bx + c$. Also, find the value of y for $x = 80$ correct upto third place of decimal.

(*ii*) Determine the constants a, b and c by the method of least squares such that $y = ax^2 + bx + c$ fits the following data:

x	2	4	6	8	10
y	4.01	11.08	30.12	81.89	222.62

7. The velocity V of a liquid is known to vary with temperature T, according to a quadratic law $V = a + bT + cT^2$. Find the best values of a, b and c for the following table:

T	1	2	3	4	5	6	7
V	2.31	2.01	1.80	1.66	1.55	1.47	1.41

[G.B.T.U. (MCA) 2010]

8. The following table gives the results of the measurements of train resistances, V is the velocity in miles per hour, R is the resistance in pounds per ton:

V	20	40	60	80	100	120
R	5.5	9.1	14.9	22.8	33.3	46

If R is related to V by the relation $R = a + bV + cV^2$; find a, b and c by using the method of least squares.

9. Find the multiple linear regression of X_1 on X_2 and X_3 from the data relating to three variables:

X_1	7	12	17	20
X_2	4	7	9	12
X_3	1	2	5	8

(A.K.T.U. 2017)

10. Using least square method, fit a second degree polynomial from the following data:

x:	0	1	2	3	4	5	6	7	8
y:	12.0	10.5	10	8	7	8	7.5	8.5	9

Also estimate y at $x = 6.5$. *(A.K.T.U. 2016)*

Answers

1. $y = 2 - 0.5\,x + 0.2\,x^2$ **2.** $y = 0.45714 + 0.39286\,x + 0.12857\,x^2$

3. $y = 11 + 0.8428x + 0.3571x^2$

4. (*i*) $y = 22.8 + 1.44x + 0.64x^2$ (*ii*) $y = 18.866 + 66.1576\,x - 4.3333\,x^2$

(*iii*) $y = 1.43 + 0.24x + 2.21x^2$ (*iv*) $y = 1.5238 + 2.38398\,x - 0.10173\,x^2$

(*v*) $y = -1.619 + 4.031\,x - 0.339\,x^2$.

6. (*i*) $y = 0.010626822\,x^2 + 0.908257322\,x + 132.2040143$; 272.876

(*ii*) $a = 5.358035714,\ b = -38.89492857,\ c = 67.56$

7. $V = 2.5928 - 0.3258\,T + 0.02274\,T^2$ **8.** $R = 4.35 + 0.00241\,V + 0.0028705\,V^2$

9. $X_1 = 0.6441 + 1.661X_2 + 0.0169\,X_3$ **10.** $y = 12.1848 - 1.8465x + 0.1829x^2$, 7.91

ASSIGNMENT-III

1. Write the formula of measure of kurtosis β_2. *(A.K.T.U. 2019)*

2. The first three central moments of a distribution are 0, 15, – 31. Find the moment coefficient of skewness. *(A.K.T.U. 2019)*

3. Find the normal equations of a curve $y = ax + bx^2$. *(A.K.T.U. 2019)*

4. The fourth moment about the mean of a frequency distribution is 24. What must be the value of standard deviation in order that the distribution be platykurtic? *(A.K.T.U. 2012)*

5. Define the coefficients of skewness and kurtosis. *(A.K.T.U. 2014)*

6. Define skewness, coefficient of skewness, kurtosis and coefficient of kurtosis. *(A.K.T.U. 2013)*

7. Write a short note on skewness.

8. What is meant by skewness? How is it measured? *(A.K.T.U. 2017)*

9. What do you understand by measures of kurtosis? Disciss in brief. *(A.K.T.U. 2017)*

10. The first four central moments of a distribution are 0, 2.5, 0.7 and 18.75. Comment on the kurtosis of the distribution. *(A.K.T.U 2013)*

11. Define regression lines. *(A.K.T.U 2016)*

12. Discuss, in brief, the types of correlation. *(A.K.T.U 2017)*

13. It covariance between x and y variables is 10 and the variance of x and y are respectively 16 and 9, find the coefficient of correlation. *(A.K.T.U 2016)*

14. Write the normal equations to fit a curve $y = ax^2 + b$ by least square method. *(A.K.T.U 2016, 2018)*

15. Write the normal equation of $y = a + \dfrac{b}{x}$. *(A.K.T.U 2017)*

16. Write the normal equation for the curve $y = \dfrac{a}{x} + bx$. *(A.K.T.U 2018)*

Answers

1. $\beta_2 = \dfrac{\mu_4}{\mu_2^2}$

2. – 0.2847

3. $\left.\begin{aligned} \Sigma\, xy &= a\, \Sigma\, x^2 + b\, \Sigma\, x^3 \\ \Sigma\, x^2 y &= a\, \Sigma\, x^3 + b\, \Sigma\, x^4 \end{aligned}\right\}$

4. $\sigma > 1.6817$

10. $\beta_2 = 3$, Mesokurtic

13. 0.83

14. $\left.\begin{aligned} \Sigma\, x^2 y &= a\, \Sigma\, x^4 + b\, \Sigma\, x^2 \\ \Sigma\, y &= a\, \Sigma\, x^2 + nb \end{aligned}\right\}$

15. $\left.\begin{aligned} \Sigma\, y &= na + b\, \Sigma\, \frac{1}{x} \\ \Sigma\, \frac{y}{x} &= a\, \Sigma\, \frac{1}{x} + b\, \Sigma\, \frac{1}{x^2} \end{aligned}\right\}$

16. $\left.\begin{aligned} \Sigma\, \frac{y}{x} &= a\, \Sigma\, \frac{1}{x^2} + nb \\ \Sigma\, xy &= na + b\, \Sigma\, x^2 \end{aligned}\right\}$

MODULE 4

Statistical Techniques – II

4.1 THEORY OF PROBABILITY

Here we define and explain certain terms which are used frequently.

(*a*) Trial and event. Let an experiment be repeated under essentially the same conditions and let it result in any one of the several possible outcomes. Then, the experiment is called a *trial* and the possible outcomes are known as *events* or *cases*.

For example: (*i*) Tossing of a coin is a trial and the turning up of head or tail is an event.

(*ii*) Throwing a die is a trial and getting 1 or 2 or 3 or 4 or 5 or 6 is an event.

(*b*) Exhaustive events. The total number of all possible outcomes in any trial is known as *exhaustive events* or *exhaustive cases*.

For example: (*i*) In tossing a coin, there are two exhaustive cases, head and tail.

(*ii*) In throwing a dice, there are 6 exhaustive cases, for any one of the six faces may turn up.

(*iii*) In throwing two dice, the exhaustive cases are $6 \times 6 = 6^2$ for any of the 6 numbers from 1 to 6 on one die can be associated with any of the 6 numbers on the other die.

In general, in throwing n dice, the exhaustive cases are 6^n.

(*c*) Favourable events or cases. The cases which entail the happening of an event are said to be *favourable* to the event. It is the total number of possible outcomes in which the specified event happens.

For example: (*i*) In throwing a die, the number of cases favourable to the appearance of a multiple of 3 are two *viz.* 3 and 6 while the number of cases favourable to the appearance of an even number are three, *viz.*, 2, 4 and 6.

(*ii*) In a throw of two dice, the number of cases favourable to getting a sum 6 is 5, *viz.*, (1, 5); (5, 1); (2, 4); (4, 2); (3, 3).

(*d*) Mutually exclusive events. Events are said to be *mutually exclusive* or *incompatible* if the happening of any one of them precludes (*i.e.*, rules out) the happening of all others, *i.e.*, if no two or more than two of them can happen simultaneously in the same trial.

For example: (*i*) In tossing a coin, the events head and tail are mutually exclusive, since if the outcome is head, the possibility of getting tail in the same trial is ruled out.

(*ii*) In throwing a die, all the six faces numbered, 1, 2, 3, 4, 5, 6 are mutually exclusive since any outcome rules out the possibility of getting any other.

(*e*) Equally likely events. Events are said to be *equally likely* if there is no reason to expect any one in preference to any other.

For example: (*i*) When a card is drawn from a well shuffled pack, any card may appear in the draw so that the 52 different cases are equally likely.

(*ii*) In throwing a die, all the six faces are equally likely to come.

(*f*) Independent and dependent events. Two or more events are said to be *independent* if the happening or non-happening of any one does not depend (or is not affected) by the happening or non-happening of any other. Otherwise they are said to be *dependent.*

For example: If a card is drawn from a pack of well shuffled cards and replaced before drawing the second card, the result of the second draw is independent of the first draw. However, if the first card drawn is not replaced, then, the second draw is dependent on the first draw.

4.2 MATHEMATICAL (OR CLASSICAL) DEFINITION OF PROBABILITY

If a trial results in n *exhaustive, mutually exclusive* and *equally likely cases* and m of them are favourable to the happening of an event E, then the probability of happening of E is given by

$$p \text{ or } P(E) = \frac{\text{Favourable number of cases}}{\text{Exhaustive number of cases}} = \frac{m}{n}.$$

Note 1. Since the number of cases favourable to happening of E is m and the exhaustive number of cases in n, therefore, the number of cases unfavourable to happening of E are $n - m$.

Note 2. The probability that the event E will not happen is given by

$$q \text{ or } P(\overline{E}) = \frac{\text{Unfavourable number of cases}}{\text{Exhaustive number of cases}} = \frac{n-m}{n} = 1 - \frac{m}{n} = 1 - p$$

$$p + q = 1 \quad i.e., \quad P(E) + P(\overline{E}) = 1$$

Obviously, p and q are non-negative and cannot exceed unity, *i.e.*, $0 \le p \le 1, 0 \le q \le 1$.

Note 3. If P(E) = 1, E is called a *certain event i.e.*, the chance of its happening is cent per cent. If P(E) = 0, then E is an *impossible event.*

Note 4. If n cases are favourable to E and m cases are favourable to $\overline{E}$ (*i.e.*, unfavourable to E), then exhaustive number of cases = $n + m$.

$$P(E) = \frac{n}{n+m} \quad \text{and} \quad P(\overline{E}) = \frac{m}{n+m}$$

We say that "odds in *favour of E*" are $n : m$ and "odds *against* E" are $m : n$.

4.3. STATISTICAL (OR EMPIRICAL) DEFINITION OF PROBABILITY

If in n trials, an event E happens m times, then the probability of happening of E is given by

$$p = P(E) = \underset{n \to \infty}{\text{Lt}} \frac{m}{n}.$$

ILLUSTRATIVE EXAMPLES

Example 1. *A bag contains 7 white, 6 red and 5 black balls. Two balls are drawn at random. Find the probability that they will both be white.*

Sol. Total number of balls = 7 + 6 + 5 = 18.

Out of 18 balls, 2 can be drawn in $^{18}C_2$ ways.

$$\therefore \quad \text{Exhaustive number of cases} = {}^{18}C_2 = \frac{18 \times 17}{2 \times 1} = 153$$

Out of 7 white balls, 2 can be drawn in $^7C_2 = \frac{7 \times 6}{2 \times 1} = 21$ ways.

$\therefore$ Favourable number of cases = 21

$$\text{Probability} = \frac{21}{153} = \frac{7}{51}.$$

Example 2. *Four cards are drawn from a pack of cards. Find the probability that (i) all are diamonds, (ii) there is one card of each suit, and (iii) there are two spades and two hearts.*

Sol. 4 cards can be drawn from a pack of 52 cards in $^{52}C_4$ ways.

$$\therefore \quad \text{Exhaustive number of cases} = {}^{52}C_4 = \frac{52 \times 51 \times 50 \times 49}{4 \times 3 \times 2 \times 1} = 270725.$$

(*i*) There are 13 diamonds in the pack and 4 can be drawn out of them in $^{13}C_4$ ways.

$$\therefore \quad \text{Favourable number of cases} = {}^{13}C_4 = \frac{13 \times 12 \times 11 \times 10}{4 \times 3 \times 2 \times 1} = 715.$$

$$\text{Required probability} = \frac{715}{270725} = \frac{143}{54145} = \frac{11}{4165}.$$

(*ii*) There are 4 suits, each containing 13 cards.

$$\therefore \quad \text{Favourable number of cases} = {}^{13}C_1 \times {}^{13}C_1 \times {}^{13}C_1 \times {}^{13}C_1 = 13 \times 13 \times 13 \times 13.$$

$$\text{Required probability} = \frac{13 \times 13 \times 13 \times 13}{270725} = \frac{2197}{20825}.$$

(*iii*) 2 spades out of 13 can be drawn in $^{13}C_2$ ways.

2 hearts out of 13 can be drawn in $^{13}C_2$ ways.

$$\therefore \quad \text{Favourable number of cases} = {}^{13}C_2 \times {}^{13}C_2 = 78 \times 78$$

$$\text{Required probability} = \frac{78 \times 78}{270725} = \frac{468}{20825}.$$

Example 3. *A bag contains 50 tickets numbered 1, 2, 3, ..., 50, of which five are drawn at random and arranged in ascending order of magnitude ($x_1 < x_2 < x_3 < x_4 < x_5$). What is the probability that $x_3 = 30$?*

Sol. Exhaustive number of cases $^{50}C_5$.

If $x_3 = 30$, then the two tickets with numbers x_1 and x_2 must come out of 29 tickets numbered 1 to 29 and this can be done in $^{29}C_2$ ways. The other two tickets with numbers x_4 and x_5 must come out of the 20 tickets number 31 to 50 and this can be done in $^{20}C_2$ ways.

$$\therefore \quad \text{Favourable number of cases} = {}^{29}C_2 \times {}^{20}C_2.$$

$$\text{Required probability} = \frac{{}^{29}C_2 \times {}^{20}C_2}{{}^{50}C_5} = \frac{551}{15134}.$$

4.4 RANDOM EXPERIMENT

Occurrences which can be repeated a number of times, essentially under the same conditions, and whose result cannot be predicted before hand are known as **random experiments.** For example, rolling of a die, tossing a coin, taking out balls from an urn.

Sample Space. Out of the several possible outcomes of a random experiment, one and only one can take place in a trial. The set of all these possible outcomes is called the **sample space** for the particular experiment and is denoted by S.

For example, if a coin is tossed, the possible outcomes are H (Head) and T (Tail).

Thus $S = \{H, T\}$.

Sample Point. The elements of S, the sample space, are called **sample points.**

For example, if a coin is tossed and H and T denote 'Head' and 'Tail' respectively, then $S = \{H, T\}$. The two sample points are H and T.

Finite Sample Space. If the number of sample points in a sample space is finite, we call it a **finite sample space.** Here, we shall deal with finite sample spaces only.

Event. Every subset of S, the sample space, is called an event.

Since $S \subset S$, S itself is an event ; called a **certain event.**

Also, $\phi \subset S$, the null set is also an event, called an **impossible event.**

If $e \in S$, then e is called an **elementary event.** Every elementary event contains only one sample point.

4.5 AXIOMS

(*i*) With each event E (*i.e.*, a sample point) is associated a real number between 0 and 1, called the probability of that event and is denoted by P(E). Thus $0 \le P(E) \le 1$.

(*ii*) The sum of the probabilities of all simple (elementary) events constituting the sample space is 1. Thus $P(S) = 1$.

(*iii*) The probability of a compound event (*i.e.*, an event made up of two or more sample events) is the sum of the probabilities of the simple events comprising the compound event.

Thus, if there are *n equally likely* possible outcomes of a random experiment, then the sample space S contains *n* sample points and the probability associated with each sample point is $\frac{1}{n}$. [By Axiom (*ii*)]

Now, if an event E consists of *m* sample points, then the probability of E is

$$P(E) = \frac{1}{n} + \frac{1}{n} + \ldots\ldots m \text{ times} = \frac{m}{n} = \frac{\text{Number of sample points in E}}{\text{Number of sample points in S}}.$$

This closely agrees with the classical definition of probability.

4.6 PROBABILITY OF THE IMPOSSIBLE EVENT IS ZERO, *i.e.*, P(ϕ) = 0

Proof. Impossible event contains no sample point. As such, the sample space S and the impossible event ϕ are *mutually exclusive.*

$$\Rightarrow \quad S \cup \phi = S \quad \Rightarrow \quad P(S \cup \phi) = P(S)$$

$$\Rightarrow \quad P(S) + P(\phi) = P(S) \quad \Rightarrow \quad P(\phi) = 0.$$

4.7. PROBABILITY OF THE COMPLEMENTARY EVENT $\overline{A}$ OF A IS GIVEN BY $P(\overline{A}) = 1 - P(A)$

Proof. $\overline{A}$ and A are disjoint events. Also $A \cup \overline{A} = S$

$$\therefore \quad P(A \cup \overline{A}) = P(S)$$

$$\Rightarrow \quad P(A) + P(\overline{A}) = 1. \text{ Hence } P(\overline{A}) = 1 - P(A).$$

4.8 FOR ANY TWO EVENTS A AND B, P($\bar{A} \cap B$) = P(B) – P(A ∩ B)

Proof. $\bar{A} \cap B = \{p : p \in B \text{ and } p \notin A\}$

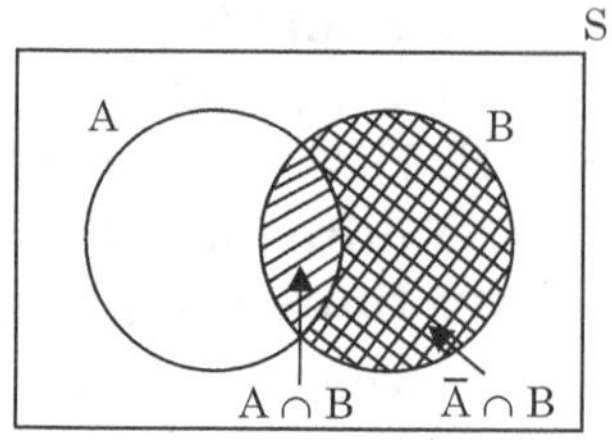

Now $\bar{A} \cap B$ and $A \cap B$ are disjoint sets and

$(\bar{A} \cap B) \cup (A \cap B) = B$

$\Rightarrow \quad P[(\bar{A} \cap B) \cup (A \cap B)] = P(B)$

$\Rightarrow \quad P(\bar{A} \cap B) + P(A \cap B) = P(B)$

$\Rightarrow \quad P(\bar{A} \cap B) = P(B) - P(A \cap B).$

Note. Similarly, it can be proved that $P(A \cap \bar{B}) = P(A) - P(A \cap B)$.

4.9 IF B ⊂ A, THEN (i) P(A ∩ $\bar{B}$) = P(A) – P(B) (ii) P(B) ≤ P(A)

Proof. When $B \subset A$, B and $A \cap \bar{B}$ are disjoint and their union is A.

$\Rightarrow \quad B \cup (A \cap \bar{B}) = A$

$\Rightarrow \quad P[B \cup (A \cap \bar{B})] = P(A)$

$\Rightarrow \quad P(B) + P(A \cap \bar{B}) = P(A)$

$\Rightarrow \quad P(A \cap \bar{B}) = P(A) - P(B) \quad ...(1)$

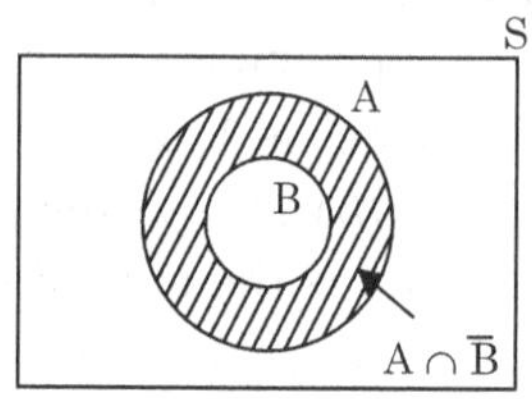

Now, if E is any event,

then $0 \le P(E) \le 1$, *i.e.*, $P(E) \ge 0$

$\therefore \quad P(A \cap \bar{B}) \ge 0 \Rightarrow P(A) - P(B) \ge 0$ [Using (1)]

$\Rightarrow \quad P(B) \le P(A).$

4.10 P(A ∩ B) ≤ P(A) AND P(A ∩ B) ≤ P(B)

Proof. By 3.9 $B \subset A \Rightarrow P(B) \le P(A)$

Since $(A \cap B) \subset A$ and $(A \cap B) \subset B$

$\therefore \quad P(A \cap B) \le P(A)$ and $P(A \cap B) \le P(B)$.

4.11 ADDITION THEOREM OF PROBABILITIES (OR THEOREM OF TOTAL PROBABILITY)

Statement. *If A and B are any two events, then*

$$P(A \cup B) = P(A) + P(B) - P(A \cap B)$$

i.e., $\quad P(A \text{ or } B) = P(A) + P(B) - P(A \text{ and } B).$

Proof. A and $\bar{A} \cap B$ are *disjoint sets* and their union is $A \cup B$.

$\Rightarrow \quad A \cup B = A \cup (\bar{A} \cap B)$

$\Rightarrow \quad P(A \cup B) = P[A \cup (\bar{A} \cap B)] = P(A) + P(\bar{A} \cap B)$

$= P(A) + [P(\bar{A} \cap B) + P(A \cap B) - P(A \cap B)]$

$= P(A) + P[(\bar{A} \cap B) \cup (A \cap B)] - P[(A \cap B)]$

[$\because$ $\bar{A} \cap B$ and $A \cap B$ are disjoint]

$= P(A) + P(B) - P(A \cap B)$ [$\because$ $(\bar{A} \cap B) \cup (A \cap B) = B$]

$P(A \cup B) = P(A) + P(B) - P(A \cap B).$

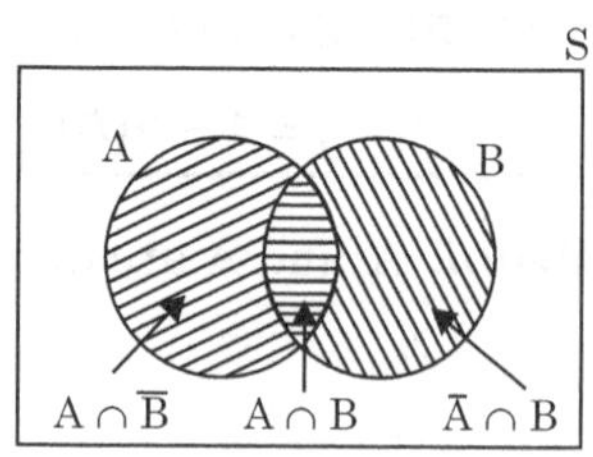

Note 1. If A and B are two mutually disjoint events, then $A \cap B = \phi$, so that $P(A \cap B) = P(\phi) = 0$.

$\therefore \quad P(A \cup B) = P(A) + P(B).$

Note 2. $P(A \cup B)$ is also written as $P(A + B)$. Thus, for mutually disjoint events A and B,

$$P(A + B) = P(A) + P(B).$$

$P(A \cap B)$ is also written as $P(AB)$.

4.12 IF A, B AND C ARE ANY THREE EVENTS, THEN

$$\mathbf{P(A \cup B \cup C) = P(A) + P(B) + P(C) - P(A \cap B) - P(B \cap C) - P(C \cap A) + P(A \cap B \cap C)}$$

Or

$$\mathbf{P(A + B + C) = P(A) + P(B) + P(C) - P(AB) - P(BC) - P(CA) + P(ABC)}$$

Proof. Using the above Art. 4.11 for two events, we have

$$\begin{aligned} P(A \cup B \cup C) &= P[(A \cup B) \cup C] \\ &= P(A \cup B) + P(C) - P[(A \cup B) \cap C] \\ &= [P(A) + P(B) - P(A \cap B)] + P(C) - P[(A \cap C) \cup (B \cap C)] \quad \text{[By distributive law]} \\ &= P(A) + P(B) + P(C) - P(A \cap B) - [P(A \cap C)] \\ &\qquad + P(B \cap C) - P\{(A \cap C) \cap (B \cap C)\} \quad \text{[By 3.11]} \\ &= P(A) + P(B) + P(C) - P(A \cap B) - P(A \cap C) - P(B \cap C) + P(A \cap B \cap C) \\ &\qquad [\because \ (A \cap C) \cap (B \cap C) = A \cap B \cap C] \\ &= P(A) + P(B) + P(C) - P(A \cap B) - P(B \cap C) - P(C \cap A) + P(A \cap B \cap C) \\ &\qquad [\because \ A \cap C = C \cap A] \end{aligned}$$

or $\quad P(A + B + C) = P(A) + P(B) + P(C) - P(AB) - P(BC) - P(CA) + P(ABC).$

4.13 IF $A_1, A_2, ..., A_n$ ARE n MUTUALLY EXCLUSIVE EVENTS, THEN THE PROBABILITY OF THE HAPPENING OF ONE OF THEM IS

$$\mathbf{P(A_1 \cup A_2 \cup \ldots\ldots \cup A_n) = P(A_1 + A_2 + \ldots + A_n) = P(A_1) + P(A_2) + \ldots + P(A_n)}$$

Let N be the total number of mutually exclusive, exhaustive and equally likely cases of which m_1 are favourable to A_1, m_2 are favourable to A_2 and so on.

$$\left.\begin{aligned} &\text{Probability of occurrence of event } A_1 = P(A_1) = \frac{m_1}{N} \\ &\text{Probability of occurrence of event } A_2 = P(A_2) = \frac{m_2}{N} \\ &\ldots\ldots\ldots\ldots\ldots\ldots\ldots\ldots\ldots\ldots\ldots\ldots\ldots \\ &\text{Probability of occurrence of event } A_n = P(A_n) = \frac{m_n}{N} \end{aligned}\right\} \quad \ldots(1)$$

The events being mutually exclusive and equally likely, the number of cases favourable to the event

$$A_1 \text{ or } A_2 \text{ or } \ldots \text{ or } A_n \text{ is } m_1 + m_2 + \ldots + m_n.$$

$\therefore$ Probability of occurrence of one of the events $A_1, A_2, \ldots\ldots, A_n$ is $P(A_1 + A_2 + \ldots + A_n)$

$$\begin{aligned} &= \frac{m_1 + m_2 + \ldots + m_n}{N} = \frac{m_1}{N} + \frac{m_2}{N} + \ldots\ldots + \frac{m_n}{N} \\ &= P(A_1) + P(A_2) + \ldots\ldots + P(A_n) \qquad | \text{ Using (1)} \end{aligned}$$

Example 4. *In a given race, the odds in favour of four horses A, B, C, D are 1 : 3, 1 : 4, 1 : 5, 1 : 6 respectively. Assuming that a dead heat is impossible; find the chance that one of them wins the race.*

Sol. Let p_1, p_2, p_3, p_4 be the probabilities of winning of the horses A, B, C, D respectively.

Since a *dead heat* (in which all the four horses cover *same distance in same time*) is not possible, the events are mutually exclusive.

Odds in favour of A are 1 : 3 $\therefore \quad p_1 = \frac{1}{1+3} = \frac{1}{4}$

Similarly, $\quad p_2 = \frac{1}{5}, p_3 = \frac{1}{6}, p_4 = \frac{1}{7}.$

If p is the chance that one of them wins, then

$$p = p_1 + p_2 + p_3 + p_4 = \frac{1}{4} + \frac{1}{5} + \frac{1}{6} + \frac{1}{7} = \frac{319}{420}.$$

A and B are *not* mutually exclusive.

AB = the event of drawing the ace of spades

$$P(A) = \frac{13}{52}, P(B) = \frac{4}{52}, P(AB) = \frac{1}{52}$$

$$\therefore \quad P(A + B) = P(A) + P(B) - P(AB) = \frac{13}{52} + \frac{4}{52} - \frac{1}{52} = \frac{16}{52} = \frac{4}{13}.$$

4.14 CONDITIONAL PROBABILITY

The probability of the happening of an event E_1 when another event E_2 is known to have already happened is called *Conditional Probability* and is denoted by $P(E_1/E_2)$.

Mutually Independent Events. An event E_1 is said to be independent of an event E_2 if $\quad P(E_1/E_2) = P(E_1)$.

i.e., if the probability of happening of E_1 is independent of the happening of E_2.

4.15 MULTIPLICATIVE LAW OF PROBABILITY (OR THEOREM OF COMPOUND PROBABILITY)

The probability of simultaneous occurrence of two events is equal to the probability of one of the events multiplied by the conditional probability of the other, *i.e.*, **for two events A and B,**

$$\mathbf{P(A \cap B) = P(A) \times P(B/A)}$$

where P(B/A) represents the conditional probability of occurrence of B when the event A has already happened.

Proof. Suppose a trial results in n exhaustive, mutually exclusive and equally likely outcomes, m of them being favourable to the happening of the event A.

$$\therefore \quad \text{Probability of happening of the event A} = P(A) = \frac{m}{n} \qquad ...(1)$$

Out of *m* outcomes favourable to the happening of A, let m_1 be favourable to the happening of the event B.

$$\therefore \quad \text{Conditional probability of B, given that A has happened} = P(B/A) = \frac{m_1}{m} \qquad ...(2)$$

Now out of n exhaustive, mutually exclusive and equally likely outcomes, m_1 are favourable to the happening of '*A and B*'.

$\therefore$ Probability of simultaneous occurrence of A and B

$$= P(A \cap B) = \frac{m_1}{n} = \frac{m_1}{m} \times \frac{m}{n} = \frac{m}{n} \times \frac{m_1}{m}$$

$$= P(A) \times P(B/A) \qquad \text{[Using (1) and (2)]}$$

Hence $\quad P(A \cap B) = P(A) \times P(B/A).$

Note. $P(A \cap B)$ is also written as $P(AB)$.

Thus $\quad P(AB) = P(A) \times P(B/A).$

Cor. 1. Interchanging A and B

$$P(BA) = P(B) \times P(A/B)$$

or $\quad P(AB) = P(B) \times P(A/B) \qquad [\because \ B \cap A = A \cap B]$

Cor. 2. If A and B are independent events, then $P(B/A) = P(B)$

$\therefore \quad P(AB) = P(A) \times P(B).$

Generalisation. If $A_1, A_2, ..., A_n$ are n independent events, then

$$P(A_1 A_2 ... A_n) = P(A_1) \times P(A_2) \times ...... \times P(A_n).$$

Cor. 3. If p is the chance that an event will happen in one trial then the chance that it will happen in a succession of r trials is

$$p.p \text{ ...... } p \text{ . } (r \text{ times}) = p^r.$$

Cor. 4. If $p_1, p_2, ..., p_n$ are the probabilities that certain events happen, then the probabilities of their non-happening are $1 - p_1, 1 - p_2, ..., 1 - p_n$ and, therefore, the probability of all of these failing is

$$(1 - p_1)(1 - p_2)......(1 - p_n).$$

Hence the chance in which at least one of these events much happen is

$$1 - (1 - p_1)(1 - p_2) \text{ ...... } (1 - p_n).$$

Example 5. *A problem in mechanics is given to three students A, B, C whose chances of solving it are* $\frac{1}{2}, \frac{1}{3}, \frac{1}{4}$ *respectively. What is the probability that the problem will be solved?*

Sol. The probabilities of A, B, C solving the problem are $\frac{1}{2}, \frac{1}{3}, \frac{1}{4}$.

The probabilities of A, B, C not solving the problem are $1 - \frac{1}{2}, 1 - \frac{1}{3}, 1 - \frac{1}{4}$ *i.e.,* $\frac{1}{2}, \frac{2}{3}, \frac{3}{4}$.

$\therefore$ The probability that the problem is not solved by any of them $= \frac{1}{2} \times \frac{2}{3} \times \frac{3}{4} = \frac{1}{4}$.

Hence the probability that the problem is solved by at least one of them $= 1 - \frac{1}{4} = \frac{3}{4}$.

Example 6. *The odds that a book will be favourably reviewed by three independent critics are 5 to 2, 4 to 3 and 3 to 4 respectively. What is the probability that, of the three reviews, a majority will be favourable?*

Sol. Let the three critics be A, B, C. The probabilities p_1, p_2, p_3 of the book being favourably reviewed by A, B, C are $\frac{5}{7}, \frac{4}{7}, \frac{3}{7}$ respectively.

$\therefore$ The probabilities that the book is unfavourably reviewed by A, B, C are

$$1 - \frac{5}{7} = \frac{2}{7},\ 1 - \frac{4}{7} = \frac{3}{7},\ 1 - \frac{3}{7} = \frac{4}{7}.$$

A majority will be favourable if the reviews of at least two are favourable.

(*i*) If A, B, C all review favourably, the probability is

$$\frac{5}{7} \times \frac{4}{7} \times \frac{3}{7} = \frac{60}{343} \qquad | \ p_1 p_2 p_3$$

(*ii*) If A, B review favourably and C reviews unfavourably, the probability is

$$\frac{5}{7} \times \frac{4}{7} \times \frac{4}{7} = \frac{80}{343} \qquad | \ p_1 p_2 (1 - p_3)$$

(*iii*) If A, C review favourably and B reviews unfavourably, probability is

$$\frac{5}{7} \times \frac{3}{7} \times \frac{3}{7} = \frac{45}{343} \qquad | \ p_1 (1 - p_2) p_3$$

(*iv*) If B, C review favourably and A reviews unfavourably, the probability is

$$\frac{2}{7} \times \frac{4}{7} \times \frac{3}{7} = \frac{24}{343} \qquad | \ (1 - p_1) p_2 p_3$$

Hence the probability that a majority will be favourable is

$$\frac{60}{343} + \frac{80}{343} + \frac{45}{343} + \frac{24}{343} = \frac{209}{343}.$$

Example 7. *A can hit a target 4 times in 5 shots; B 3 times in 4 shots; C twice in 3 shots. They fire a volley. What is the probability that at least two shots hit?*

Sol. Probability of A's hitting the target $= \frac{4}{5}$

Probability of B's hitting the target $= \frac{3}{4}$

Probability of C's hitting the target $= \frac{2}{3}$.

For at least two hits, we may have

(*i*) A, B, C all hit the target, the probability for which is $\frac{4}{5} \times \frac{3}{4} \times \frac{2}{3} = \frac{24}{60}$.

(*ii*) A, B hit the target and C misses it, the probability for which is

$$\frac{4}{5} \times \frac{3}{4} \times \left(1 - \frac{2}{3}\right) = \frac{4}{5} \times \frac{3}{4} \times \frac{1}{3} = \frac{12}{60}.$$

(*iii*) A, C hit the target and B misses it, the probability for which is

$$\frac{4}{5} \times \left(1 - \frac{3}{4}\right) \times \frac{2}{3} = \frac{4}{5} \times \frac{1}{4} \times \frac{2}{3} = \frac{8}{60}.$$

(*iv*) B, C hit the target and A misses it, the probability for which is

$$\left(1-\frac{4}{5}\right)\times\frac{3}{4}\times\frac{2}{3}=\frac{1}{5}\times\frac{3}{4}\times\frac{2}{3}=\frac{6}{60}.$$

Since these are mutually exclusive events,

$$\text{required probability} \quad =\frac{24}{60}+\frac{12}{60}+\frac{8}{60}+\frac{6}{60}=\frac{50}{60}=\frac{5}{6}.$$

Example 8. *A has 2 shares in a lottery in which there are 3 prizes and 5 blanks ; B has 3 shares in a lottery in which there are 4 prizes and 6 blanks. Show that A's chance of success is to B's as 27 : 35.*

Sol. A can draw two tickets (out of 3 + 5 = 8) in $^8C_3 = 28$ ways.

A will get the blanks in $^5C_2 = 10$ ways. ∴ A can win a prize in 28 – 10 = 18 ways

$$\text{Hence A's chance of success} = \frac{18}{28}=\frac{9}{14}$$

B can draw 3 tickets in $^{10}C_3 = 120$ ways ; B will get all blanks in $^6C_3 = 20$ ways.

∴ B can win a prize in 120 – 20 = 100 ways.

$$\text{Hence B's chance of success} \quad =\frac{100}{120}=\frac{5}{6}.$$

$$\therefore \quad \text{A's chance : B's chance} \quad =\frac{9}{14}:\frac{5}{6}=27:35.$$

Example 9. *A and B throw alternately with a single die, A having the first throw. The person who first throws ace is to win. What are their respective chances of winning?*

Sol. The chance of throwing an ace with a single die $=\frac{1}{6}$

The chance of not throwing an ace with a single die $=1-\frac{1}{6}=\frac{5}{6}$.

If A is to win, he should throw an ace in the first or third or fifth, ..., throws.

If B is to win, he should throw an ace in the second or fourth or sixth, ..., throws.

The chances that an ace is thrown in the first, second, third, ..., throws are

$$\frac{1}{6},\frac{5}{6}\cdot\frac{1}{6},\frac{5}{6}\cdot\frac{5}{6}\cdot\frac{1}{6},\frac{5}{6}\cdot\frac{5}{6}\cdot\frac{5}{6}\cdot\frac{1}{6}\ \ldots\ldots \quad \text{or} \quad \frac{1}{6},\frac{5}{6}\cdot\frac{1}{6},\left(\frac{5}{6}\right)^2\cdot\frac{1}{6},\left(\frac{5}{6}\right)^3\cdot\frac{1}{6},\ldots\ldots$$

$$\therefore \quad \text{A's chance} = \frac{1}{6}+\left(\frac{5}{6}\right)^2\cdot\frac{1}{6}+\left(\frac{5}{6}\right)^4\cdot\frac{1}{6}+\ldots\ldots=\frac{\frac{1}{6}}{1-\left(\frac{5}{6}\right)^2}=\frac{6}{11}$$

| Sum of an infinite G.P. $=\frac{a}{1-r}$

$$\text{B's chance} \quad =1-\frac{6}{11}=\frac{5}{11}.$$

Example 10. *Cards are dealt one by one from a well-shuffled pack until an ace appears. Show that the probability that exactly n cards are dealt before the first ace appears is* $\frac{4(51-n)(50-n)(49-n)}{52\cdot 51\cdot 50\cdot 49}$.

Sol. Let A be the event of drawing n non-ace cards and B, the event of drawing an ace in the $(n + 1)^{th}$ draw.

Consider the event A

n cards can be drawn out of 52 cards in $^{52}C_n$ ways.

$\Rightarrow$ Exhaustive cases $= {}^{52}C_n$

n non-ace cards can be drawn out of 52 cards in $^{48}C_n$ ways.

$\Rightarrow$ Favourable cases $= {}^{48}C_n$

$$\therefore \quad P(A) = {}^{48}C_n/{}^{52}C_n = \frac{48!}{(48-n)!\,n!} \times \frac{(52-n)!\,(n)!}{52!}$$

$$= \frac{48!\,.\,(52-n)(51-n)(50-n)(49-n)(48-n)!}{(48-n)!\cdot 52\cdot 51\cdot 50\cdot 49\cdot (48)!} = \frac{(52-n)(51-n)(50-n)(49-n)}{52\cdot 51\cdot 50\cdot 49}.$$

Consider the event B

n cards have already been drawn in the first n draws.

Exhaustive cases $= {}^{52-n}C_1 = 52 - n$; Favourable cases $= {}^{4}C_1 = 4$

$$\therefore \quad P(B/A) = \frac{4}{52-n}$$

Reqd. Probability = P(A) · P(B/A)

$$= \frac{(52-n)(51-n)(50-n)(49-n)}{52\cdot 51\cdot 50\cdot 49} \times \frac{4}{52-n} = \frac{4(51-n)(50-n)(49-n)}{52\cdot 51\cdot 50\cdot 49}.$$

Example 11. *An urn contains 10 white and 3 black balls, while another urn contains 3 white and 5 black balls. Two balls are drawn from the first urn and put into the second urn and then a ball is drawn from the latter. What is the probability that it is a white ball?*

Sol. The two balls drawn from the first urn may be

(*i*) both white (*ii*) both black (*iii*) one white and one black.

Let these events be denoted by A, B, C respectively.

$$P(A) = \frac{^{10}C_2}{^{13}C_2} = \frac{10\times 9}{13\times 12} = \frac{15}{26}; \qquad P(B) = \frac{^{3}C_2}{^{13}C_2} = \frac{3\times 2}{13\times 12} = \frac{1}{26}$$

$$P(C) = \frac{^{10}C_1 \times {}^{3}C_1}{^{13}C_2} = \frac{10\times 3}{\frac{13\times 12}{2\times 1}} = \frac{10}{26}$$

When two balls are transferred from first urn to second urn, the second urn will contain.

(*i*) 5 white and 5 black balls (*ii*) 3 white and 7 black balls

(*iii*) 4 white and 6 black balls.

Let W denote the event of drawing a white ball from the second urn in the three cases (*i*), (*ii*) and (*iii*).

$$\text{Now,} \qquad P(W/A) = \frac{5}{10}, \quad P(W/B) = \frac{3}{10}, \quad P(W/C) = \frac{4}{10}$$

$$\therefore \quad \text{Reqd. probability} = P(A)\,.\,P(W/A) + P(B)\,.\,P(W/B) + P(C)\,.\,P(W/C)$$

$$= \frac{15}{26}\cdot\frac{5}{10} + \frac{1}{26}\cdot\frac{3}{10} + \frac{10}{26}\cdot\frac{4}{10} = \frac{75+3+40}{260} = \frac{118}{260} = \frac{59}{130}.$$

TEST YOUR KNOWLEDGE

1. In a class of 10 students, 4 are boys and the rest are girls. Find the probability that a student selected will be a girl.
2. What is the chance that a (*i*) non-leap year (*ii*) leap year should have fifty three Sundays?
3. A card is drawn from an ordinary pack and a gambler bets that it is a spade or an ace. What are the odds against his winning the bet?
4. An integer is chosen at random from the first two hundred positive integers. What is the probability that the integer chosen is divisible by 6 or 8?
5. Six cards are drawn at random from a pack of 52 cards. What is the probability that 3 will be red and 3 black?
6. From a set of raffle tickets numbered 1 to 100, three are drawn at random. What is the probability that all are odd numbered?
7. (*a*) If from a lottery of 30 tickets, marked, 1, 2, 3, ..., 30, four tickets be drawn, what is the chance that those marked 1 and 2 are among them?

   (*b*) An urn contains 5 red and 10 black balls. Eight of them are placed in another urn. What is the chance that the latter then contains 2 red and 6 black balls?
8. A party of n persons sit at a round table. Find the odds against two specified individuals sitting next to each other.
9. A five-figured number is formed by the digits 0, 1, 2, 3, 4 (without repetition). Find the probability that the number formed is divisible by 4.
10. Three newspapers A, B, C are published in a city and a survey of readers indicates the following:

    20% read A, 16% read B, 14% read C,

    8% read both A and B, 5% read both A and C,

    4% read both B and C, 2% read all the three.

    For a person chosen at random, find the probability that he reads none of the papers.
11. A problem in Statistics is given to five students. Their chances of solving it are $\frac{1}{2}, \frac{1}{3}, \frac{1}{4}, \frac{1}{4}$ and $\frac{1}{5}$. What is the probability that the problem will be solved?
12. A can hit a target 5 times in 6 shots, B 4 times in 5 shots and C 3 times in 4 shots. They fire a volley. What is the probability that two shots, at least hit the target?
13. Three groups of children contain respectively 3 girls and 1 boy; 2 girls and 2 boys; 1 girl and 3 boys. One child is selected at random from each group. Show that the chance that the three selected consist of 1 girl and 2 boys is $\frac{13}{32}$.
14. Four persons are chosen at random from a group containing 3 men, 2 women and 4 children. Show that the chance that exactly two of them will be children is $\frac{5}{21}$.
15. A bag contains 10 balls, two of which are red, three blue and five black. Three balls are drawn at random from the bag. What is the probability that

    (*i*) the three balls are of different colours, (*ii*) two balls are of the same colour,

    (*iii*) the balls are all of the same colour.
16. It is 8 : 5 against a person who is 40 years old living till he is 70 and 4 : 3 against a person now 50 living till he is 80. Find the probability that one at least of these persons will be alive 30 years hence.
17. Find the chance of throwing 5 or 6 at least once in four throws of a die.
18. A has 3 shares in a lottery where there are 3 prizes and 6 blanks. B has one share in another, where there is just one prize and two blanks. Show that A has a better chance of winning a prize than B in the ratio 16 : 7.

19. A, B and C, in order, toss a coin. The first one to throw a head wins. If A starts, find their respective chances of winning.

20. A speaks truth in 60% cases and B in 70% cases. In what percentages of cases are they likely to contradict each other in stating the same fact?

21. A and B throw alternately with a pair of ordinary dice. A wins if he throws 6 before B throws 7 and B wins if he throws 7 before A throws 6. If A begins, find their respective chances of winning.

(Huyghen's Problem)

22. (*a*) Two cards are randomly drawn from a deck of 52 cards and thrown away. What is the probability of drawing an ace in a single draw from the remaining 50 cards ?

(*b*) A box A contains 2 white and 4 black balls. Another box B contains 5 white and 7 black balls. A ball is transferred from the box A to the box B, then a ball is drawn from the box B. Find the probability that it is white.

23. Of the cigarette smoking population, 70% are men and 30% women, 10% of these men and 20% of these women smoke 'WILLS'. What is the probability that a person seen smoking a 'WILLS' will be a man?

24. A committee consists of 9 students two of which are from 1st year, three from 2nd year and four from 3rd year. Three students are to be removed at random. What is the chance that

(*i*) the three students belong to different classes,

(*ii*) two belong to the same class and third to the different class, and

(*iii*) the three belong to the same class?

25. Five men in a company of twenty are graduates. If 3 men and picked out of 20 at random, what is the probability that

(*i*) they are all graduates? (*ii*) at least one is graduate?

26. If A, B, C are events such that

$$P(A) = 0.3, \quad P(B) = 0.4, \quad P(C) = 0.8,$$
$$P(A \cap B) = 0.08, \quad P(A \cap C) = 0.28, \quad P(A \cap B \cap C) = 0.09$$

If $P(A \cup B \cup C) \geq 0.75$, then show that $0.23 \leq P(B \cap C) \leq 0.48$.

27. For two events A and B, let $P(A) = 0.4$, $P(B) = p$ and $P(A \cup B) = 0.6$

(*i*) Find p so that A and B are independent events.

(*ii*) For what value of p are A and B are mutually exclusive?

28. A husband and wife appear in an interview for two vacancies in the same post. The probability of husband's selection is $\frac{1}{7}$ and that of wife's selection is $\frac{1}{5}$. What is the probability that

(*i*) both of them will be selected, (*ii*) only one of them will be selected, and

(*iii*) none of them will be selected?

29. Two dice are tossed once. Find the probability of getting an even number on the first die or a total of 8.

30. A drawer contains 50 bolts and 150 nuts. Half of the bolts and half of the nuts are rusted. If one item is chosen at random, what is the probability that it is rusted or is a bolt?

31. A purse contains 2 silver and 4 copper coins. A second purse contains 4 silver and 3 copper coins. If a coin is pulled out at random from one of the two purses, what is the probability that it is a silver coin?

32. A man wants to marry a girl having qualities: white complexion—the probability of getting such a girl is one in twenty; handsome dowry—the probability of getting this is one in fifty; westernised manners and etiquettes—the probability here is one in hundred. Find out the probability of his getting married to such a girl when the possession of these three attributes is independent.

33. A class consists of 80 students, 25 of them are girls and 55 boys, 10 of them are rich and the remaining poor, 20 of them are fair complexioned. What is the probability of selecting a fair complexioned rich girl?

34. Of the students attending a lecture, 50% could not see what was written on the board and 40% could not hear what the lecturer was saying. Most unfortunate 30% fell into both of these categories. What is the probability that a student picked at random was able to see and hear satisfactorily?

35. The probabilities of A, B, C solving a problem are $\frac{1}{3}, \frac{2}{7}$ and $\frac{3}{8}$ respectively. If all the three try to solve the problem simultaneously, find the probability that exactly one of them will solve it.

36. A student takes his examination in four subjects $\alpha, \beta, \gamma, \delta$. He estimates his chance of passing in α as $\frac{4}{5}$, in β as $\frac{3}{4}$, in γ as $\frac{5}{6}$ and in δ as $\frac{2}{3}$. To qualify he must pass in α and at least two other subjects. What is the probability that he qualifies?

37. A candidate is called for an interview by three companies. For the first company, there are 12 candidates and for the second, there are 15 candidates and for the third, there are 10 candidates. What are the chances of his getting at least one of the companies? *(G.B.T.U. 2012)*

Answers

1. $\frac{3}{5}$	**2.** (*i*) $\frac{1}{7}$ (*ii*) $\frac{2}{7}$	**3.** 9 : 4	**4.** $\frac{1}{4}$
5. $\frac{13000}{39151}$	**6.** $\frac{4}{33}$	**7.** (*a*) $\frac{2}{145}$, (*b*) $\frac{140}{429}$	**8.** $(n-3) : 2$
9. $\frac{5}{16}$	**10.** $\frac{13}{20}$	**11.** $\frac{17}{20}$	**12.** $\frac{107}{120}$
15. (*i*) $\frac{1}{4}$ (*ii*) $\frac{79}{120}$ (*iii*) $\frac{11}{120}$		**16.** $\frac{59}{91}$	**17.** $\frac{65}{81}$
19. $\frac{4}{7}, \frac{2}{7}, \frac{1}{7}$	**20.** 46%	**21.** $\frac{30}{61}, \frac{31}{61}$	**22.** (*a*) $\frac{1}{13}$, (*b*) $\frac{16}{39}$
23. $\frac{7}{13}$	**24.** (*i*) $\frac{2}{7}$ (*ii*) $\frac{55}{84}$ (*iii*) $\frac{5}{84}$	**25.** (*i*) $\frac{1}{114}$ (*ii*) $\frac{137}{228}$	**27.** (*i*) $\frac{1}{3}$ (*ii*) 0.2
28. (*i*) $\frac{1}{35}$ (*ii*) $\frac{2}{7}$ (*iii*) $\frac{24}{35}$	**29.** $\frac{5}{9}$	**30.** $\frac{5}{8}$	**31.** $\frac{19}{42}$
32. 0.000001	**33.** $\frac{5}{512}$	**34.** $\frac{2}{5}$	**35.** $\frac{25}{56}$
36. $\frac{61}{90}$.	**37.** 0.23		

4.16 BAYE'S THEOREM *(G.B.T.U. 2011)*

If $E_1, E_2, \ldots\ldots, E_n$ are mutually exclusive and exhaustive events with $P(E_i) \neq 0$, $(i = 1, 2, \ldots\ldots, n)$ of a random experiment then for any arbitrary event A of the sample space of the above experiment with $P(A) > 0$, we have

$$P(E_i/A) = \frac{P(E_i)P(A/E_i)}{\sum_{i=1}^{n} P(E_i)P(A/E_i)}.$$

Proof. Let S be the sample space of the random experiment.

The events $E_1, E_2, ..., E_n$ being exhaustive

$$S = E_1 \cup E_2 \cup ...... \cup E_n$$

$$\therefore \quad A = A \cap S \qquad [\because \; A \subset S]$$

$$= A \cap (E_1 \cup E_2 \cup ...... \cup E_n)$$

$$= (A \cap E_1) \cup (A \cap E_2) \cup ...... \cup (A \cap E_n) \qquad \text{[Distributive Law]}$$

$$\Rightarrow \quad P(A) = P(A \cap E_1) + P(A \cap E_2) + ... + P(A \cap E_n)$$

$$= P(E_1)P(A/E_1) + P(E_2)P(A/E_2) + ... + P(E_n)P(A/E_n)$$

$$= \sum_{i=1}^{n} P(E_i)P(A/E_i) \qquad ...(1)$$

Now, $\quad P(A \cap E_i) = P(A)P(E_i/A)$

$$\Rightarrow \quad P(E_i/A) = \frac{P(A \cap E_i)}{P(A)} = \frac{P(E_i)P(A/E_i)}{\sum_{i=1}^{n} P(E_i)P(A/E_i)} \qquad \text{[Using (1)]}$$

Note. The significance of Baye's Theorem may be understood in the following manner:

$P(E_i)$ is the probability of occurrence of E_i. The experiment is performed and we are told that the event A has occurred. With this information, the probability $P(E_i)$ is changed to $P(E_i/A)$. Baye's Theorem enables us to evaluate $P(E_i/A)$ if all the $P(E_i)$ and the conditional probabilities $P(A/E_i)$ are known.

IllUSTRATIVE EXAMPLES

Example 1. *A bag X contains 2 white and 3 red balls and a bag Y contains 4 white and 5 red balls. One ball is drawn at random from one of the bags and is found to be red. Find the probability that it was drawn from bag Y.*

Sol. Let $\quad E_1$: the ball is drawn from bag X; E_2: the ball is drawn from bag Y

and $\quad$ A: the ball is red.

We have to find $P(E_2/A)$. By Baye's Theorem,

$$P(E_2/A) = \frac{P(E_2)P(A/E_2)}{P(E_1)P(A/E_1) + P(E_2)P(A/E_2)} \qquad ...(1)$$

Since the two bags are equally likely to be selected, $P(E_1) = P(E_2) = \frac{1}{2}$

Also $\quad P(A/E_1) = P(\text{a red ball is drawn from bag X}) = \frac{3}{5}$

$$P(A/E_2) = P(\text{a red ball is drawn from bag Y}) = \frac{5}{9}$$

$\therefore$ From (1), we have $P(E_2/A) = \dfrac{\frac{1}{2} \times \frac{5}{9}}{\frac{1}{2} \times \frac{3}{5} + \frac{1}{2} \times \frac{5}{9}} = \dfrac{25}{52}$.

Example 2. *In a bolt factory, machines A, B and C manufacture respectively 25%, 35% and 40% of the total. Of their output 5, 4 and 2 per cent are defective bolts. A bolt is drawn at random from the product and is found to be defective. What is the probability that it was manufactured by machine B?*

Sol. Let E_1, E_2 and E_3 denote the events that a bolt selected at random is manufacture by the machines A, B and C respectively and let H denote the event of its being defective. Then

$$P(E_1) = 0.25,\ P(E_2) = 0.35,\ P(E_3) = 0.40$$

The probability of drawing a defective bolt manufactured by machine A is

$$P(H/E_1) = 0.05$$

Similarly, $\quad P(H/E_2) = 0.04$ and $P(H/E_3) = 0.02$

By Baye's Theorem, we have

$$P(E_2/H) = \frac{P(E_2)P(H/E_2)}{P(E_1)P(H/E_1) + P(E_2)P(H/E_2) + P(E_3)P(H/E_3)}$$

$$= \frac{0.35 \times 0.04}{0.25 \times 0.05 + 0.35 \times 0.04 + 0.40 \times 0.02} = \frac{0.0140}{0.0345} = 0.41.$$

Example 3. *The contents of urns I, II and III are as follows:*

1 white, 2 black and 3 red balls,

2 white, 1 black and 1 red balls, and

4 white, 5 black and 3 red balls.

One urn is chosen at random and two balls drawn. They happen to be white and red. What is the probability that they come from urns I, II or III?

Sol. Let E_1 : urn I is chosen ; E_2 : urn II is chosen ; E_3 : urn III is chosen

and $\quad$ A : the two balls are white and red.

We have to find $P(E_1/A)$, $P(E_2/A)$ and $P(E_3/A)$.

Now $\quad P(E_1) = P(E_2) = P(E_3) = \frac{1}{3}$

$$P(A/E_1) = P\text{ (a white and a red ball are drawn from urn I)} = \frac{{}^1C_1 \times {}^3C_1}{{}^6C_2} = \frac{1}{5}$$

$$P(A/E_2) = \frac{{}^2C_1 \times {}^1C_1}{{}^4C_2} = \frac{1}{3}\ ;\ P(A/E_3) = \frac{{}^4C_1 \times {}^3C_1}{{}^{12}C_2} = \frac{2}{11}$$

By Baye's Theorem, we have

$$P(E_1/A) = \frac{P(E_1)P(A/E_1)}{P(E_1)P(A/E_1) + P(E_2)P(A/E_2) + P(E_3)P(A/E_3)} = \frac{\frac{1}{3} \times \frac{1}{5}}{\frac{1}{3} \times \frac{1}{5} + \frac{1}{3} \times \frac{1}{3} + \frac{1}{3} \times \frac{2}{11}} = \frac{33}{118}$$

Similarly, $\quad P(E_2/A) = \frac{55}{118},\ P(E_3/A) = \frac{15}{59}.$

ASSIGNMENT

1. Two urns contain 4 white, 6 blue and 4 white, 5 blue balls respectively. One of the urns is selected at random and a ball is drawn from it. If the ball drawn is white, find the probability that it is drawn from the

   (*i*) first urn (*ii*) second urn.

2. Three urns contain 6 red, 4 black; 4 red, 6 black and 5 red, 5 black balls respectively. One of the urns is selected at random and a ball is drawn from it. If the ball drawn is red, find the probability that it is drawn from the first urn.
3. A factory has two machines A and B. Past record shows that machine A produced 60% of the items of output and machine B produced 40% of the items. Further, 2% of the items produced by machine A were defective and 1% produced by machine B were defective. If a defective item is drawn at random, what is the probability that it was produced by machine A?
4. An insurance company insured 2000 scooter drivers, 4000 car drivers and 6000 truck drivers. The probability of accident is 0.01, 0.03 and 0.15 respectively. One of the insured persons meets an accident. What is the probability that he is a scooter driver?
5. A company has two plants to manufacture scooters. Plant I manufactures 70% of scooters and plant II manufactures 30%. At plant I, 80% of the scooters are rated standard quality and at plant II, 90% of the scooters are rated standard quality. A scooter is chosen at random and is found to be of standard quality. What is the chance that it has come from plant II?

Answers

1. (*i*) $\frac{9}{19}$ (*ii*) $\frac{10}{19}$ 2. $\frac{2}{5}$ 3. $\frac{3}{4}$

4. $\frac{1}{52}$ 5. $\frac{27}{83}$.

4.17 RANDOM VARIABLE

Let E be a random experiment and S be a sample space associated with it. A function X assigning to every element $s \in S$, a real number, X(s) is called a random variable.

A random variable is any numerical quantity whose value will be determined by the outcome of a random experiment and which have a specific range and a definable probability associated with each value.

In other words, if the numerical values assumed by a variable are the result of some chance factors, so that a particular value cannot be exactly predicted in advance, the variable is then called a *random variable*. A random variable is also called '*chance variable*' or '*stochastic variable*'. Random variables are denoted by capital letters, usually, from the last part of the alphabet, for instance, X, Y, Z, etc.

4.17.1 Definition

A real valued function, X, defined on a sample space, S of a random experiment, E, is called a random variable which assigns to each sample point (or to every outcome of the random experiment E) $s \in S$ one and only one real number $X(s) = x$. The domain of the random variable X is the sample space S and its range, A is the non empty set of real numbers such that

1. The set $[X(s) \le x]$ is an event for any real number x.

2. There corresponds a well defined unique probability of the events, "X assumes the value x" and "X assumes any value in the interval".

e.g., If three coins are tossed, then the sample space contains 8 sample points. Let the random variable X denote "the number of heads". Then X is a real valued function over S with a space A which has four elements 0, 1, 2, 3, 4 as explained below:

Sample Point	*X(s)*	*X = no. of heads*	*Probability*
TTT	X(TTT)	0	$\frac{1}{8}$
HTT, THT TTH	X(HTT), X(THT), X(TTH)	1	$\frac{3}{8}$
THH, HTH, HHT	X(THH), X(HTH), X(HHT)	2	$\frac{3}{8}$
HHH	X(HHH)	3	$\frac{1}{8}$

Thus A the space of X is a set of real numbers given by

$$A = \{x : x = X(s), s \in S\}.$$

In this example, we have

$$S = \{TTT, HTT, THT, TTH, THH, HTH, HHT, HHH\}$$

and $$A = \{0, 1, 2, 3\}.$$

If S has elements which are themselves real numbers then $X(s) = s$ and $A = S$.

As in case of rolling a die S = (1, 2, 3, 4, 5, 6} = A.

4.18 TYPES OF RANDOM VARIABLES

There are two types of random variables:

(1) Discrete Random Variable. A *discrete random variable* is one which can assume only isolated values. For example,

(*i*) the number of heads in 4 tosses of a coin is a discrete random variable as it cannot assume values other than 0, 1, 2, 3, 4.

(*ii*) the number of aces in a draw of 2 cards from a well shuffled deck is a random variable as it can take the values 0, 1, 2 only.

(2) Continuous Random Variable. A *continuous random variable* is one which can assume any value within an interval, *i.e.* all values of a continuous scale. For example (*i*) the weights (in kg) of a group of individuals, (*ii*) the heights of a group of individuals.

4.19 PROBABILITY FUNCTION (OR PROBABILITY MASS FUNCTION)

Let $x_1, x_2, x_3, \ldots$ be the values of a discrete random variable X and let $p_1, p_2, p_3, \ldots, p_i > 0, i = 1, 2, 3, \ldots$ be the corresponding probabilities. That is $P(X = x_i) = p(x_i)$ or simply p_i.

A function $f(x)$ or $p(x)$ defined by

$$P(X = x) = p(x) = \begin{cases} p(x_i) \text{ or } p_i \text{ where } x = x_i, i = 1, 2, \ldots \\ 0 \text{ otherwise} \end{cases}$$

is called the probability function of the (discrete) random variable X.

The probability function $p(x)$ yields the probability that the random variable X assumes any particular value x in its range.

The probability function (or often called frequency function) of the random variable X satisfies the following conditions:

(*i*) $p(x_i) \geq 0$, (*ii*) $\Sigma\, p(x_i) = 1$.

Example. *A random variable X takes values 1, 2, 3, ... with probability mass function* $\frac{\lambda^r}{r!}$, *r = 1, 2, 3, ... ∞. Find the value of λ.*

Sol. $\sum_{r=1}^{\infty} \frac{\lambda^r}{r!}$ should be 1. Thus,

$$\sum_{r=1}^{\infty} \frac{\lambda^r}{r!} = 1$$

$$\Rightarrow \quad \frac{\lambda}{1!} + \frac{\lambda^2}{2!} + \cdots = 1$$

$$\Rightarrow \quad e^{\lambda} - 1 = 1$$

$$\Rightarrow \quad e^{\lambda} = 2$$

$$\lambda = \log_e 2.$$

4.20 PROBABILITY DISTRIBUTION OF A DISCRETE RANDOM VARIABLE

A table or a formula listing all possible values that a random variable can take on together with the respective probabilities, is called a probability distribution of the random variable.

In other words, the set of ordered pairs $[x_i, p(x_i)]$ is called the probability distribution of a discrete random variable X provided $p(x_i) \geq 0$ and $\Sigma\, p(x_i) = 1$.

e.g. (*i*) Let X = The number of points appearing in a throw of a die. Then the probability distribution of X is given by the probability function

$$p(x) = \frac{1}{6}, \; x = 1, 2, 3, 4, 5, 6$$
$$= 0 \text{ otherwise.}$$

This probability distribution may also be expressed as follows:

X:	1	2	3	4	5	6
$p(x) = P(X = x)$:	$\frac{1}{6}$	$\frac{1}{6}$	$\frac{1}{6}$	$\frac{1}{6}$	$\frac{1}{6}$	$\frac{1}{6}$

(*ii*) Suppose a coin is tossed three times. Then the distribution of the number of heads is

X = x:	0	1	2	3
$P(X = x) = p(x)$:	$\frac{1}{8}$	$\frac{3}{8}$	$\frac{3}{8}$	$\frac{1}{8}$

Note. In general, both terms probability function and probability distribution are used interchangeably.

4.21 MEAN AND VARIANCE OF RANDOM VARIABLES

Let

$$X : x_1, \quad x_2, \quad x_3, \ldots\ldots, x_n$$
$$P(X) : p_1, \quad p_2, \quad p_3, \ldots\ldots, p_n$$

be a discrete probability distribution.

We denote the *mean* by μ and define $\mu = \dfrac{\Sigma p_i x_i}{\Sigma p_i} = \Sigma p_i x_i$ $\quad (\because \ \Sigma p_i = 1)$

Other names for the mean are *average or expected value E(X).*

We denote the *variance* by σ^2 and define $\sigma^2 = \Sigma p_i(x_i - \mu)^2$

If μ is not a whole number, then $\sigma^2 = \Sigma p_i x_i^2 - \mu^2$

Standard deviation $\sigma = +\sqrt{\text{Variance}}$.

ILLUSTRATIVE EXAMPLES

Example 1. *Five defective bulbs are accidentally mixed with twenty good ones. It is not possible to just look at a bulb and tell whether or not it is defective. Find the probability distribution of the number of defective bulbs, if four bulbs are drawn at random from this lot.*

Sol. Let X denote the number of defective bulbs in 4. Clearly X can take the values 0, 1, 2, 3 or 4.

Number of defective bulbs = 5

Number of good bulbs = 20

Total number of bulbs = 25

$$P(X = 0) = P\text{ (no defective)} = P\text{ (all 4 good ones)}$$

$$= \frac{^{20}C_4}{^{25}C_4} = \frac{20 \times 19 \times 18 \times 17}{25 \times 24 \times 23 \times 22} = \frac{969}{2530}$$

$$P(X = 1) = P\text{ (one defective and 3 good ones)} = \frac{^5C_1 \times {}^{20}C_3}{^{25}C_4} = \frac{1140}{2530}$$

$$P(X = 2) = P\text{ (2 defectives and 2 good ones)} = \frac{^5C_2 \times {}^{20}C_2}{^{25}C_4} = \frac{380}{2530}$$

$$P(X = 3) = P\text{ (3 defectives and 1 good one)} = \frac{^5C_3 \times {}^{20}C_1}{^{25}C_4} = \frac{40}{2530}$$

$$P(X = 4) = P\text{ (all 4 defectives)} = \frac{^5C_4}{^{25}C_4} = \frac{1}{2530}$$

∴ The probability distribution of the random variable X is

X:	0	1	2	3	4
P(X):	$\frac{969}{2530}$	$\frac{1140}{2530}$	$\frac{380}{2530}$	$\frac{40}{2530}$	$\frac{1}{2530}$.

Example 2. *A die is tossed thrice. A success is 'getting 1 or 6' on a toss. Find the mean and the variance of the number of successes.*

Sol. Let X denote the number of success. Clearly X can take the values 0, 1, 2 or 3.

Probability of success $= \dfrac{2}{6} = \dfrac{1}{3}$; Probability of failure $= 1 - \dfrac{1}{3} = \dfrac{2}{3}$

$$P(X = 0) = P\text{ (no success)} = P\text{ (all 3 failures)} = \frac{2}{3} \times \frac{2}{3} \times \frac{2}{3} = \frac{8}{27}$$

$$P(X = 1) = P\text{ (one success and 2 failures)} = {}^3C_1 \times \frac{1}{3} \times \frac{2}{3} \times \frac{2}{3} = \frac{12}{27}$$

$$P(X = 2) = P \text{ (two successes and one failure)} = {}^3C_2 \times \frac{1}{3} \times \frac{1}{3} \times \frac{2}{3} = \frac{6}{27}$$

$$P(X = 3) = P \text{ (all 3 successes)} = \frac{1}{3} \times \frac{1}{3} \times \frac{1}{3} = \frac{1}{27}$$

∴ The probability distribution of the random variable X is

X:	0	1	2	3
P(X):	$\frac{8}{27}$	$\frac{12}{27}$	$\frac{6}{27}$	$\frac{1}{27}$

To find the mean and variance

x_i	p_i	$p_i x_i$	$p_i x_i^2$
0	$\frac{8}{27}$	0	0
1	$\frac{12}{27}$	$\frac{12}{27}$	$\frac{12}{27}$
2	$\frac{6}{27}$	$\frac{12}{27}$	$\frac{24}{27}$
3	$\frac{1}{27}$	$\frac{3}{27}$	$\frac{9}{27}$
		1	$\frac{5}{3}$

Mean $\mu = \Sigma p_i x_i = 1$

Variance $\sigma^2 = \Sigma p_i x_i^2 - \mu^2 = \frac{5}{3} - 1 = \frac{2}{3}$.

Example 3. *A random variable X has the following probability function:*

Values of X,	*x:*	*0*	*1*	*2*	*3*	*4*	*5*	*6*	*7*
	p(x):	*0*	*k*	*2k*	*2k*	*3k*	k^2	$2k^2$	$7k^2 + k$

(i) Find k, *(ii) Evaluate* $P(X < 6)$, $P(X \geq 6)$, $P(3 < X \leq 6)$

(iii) Find the minimum value of x so that $P(X \leq x) > \frac{1}{2}$.

Sol. (*i*) Since $\sum_{x=0}^{7} p(x) = 1$, we have

$$0 + k + 2k + 2k + 3k + k^2 + 2k^2 + 7k^2 + k = 1$$

$$\Rightarrow \quad 10k^2 + 9k - 1 = 0 \quad \Rightarrow \quad (10k - 1)(k + 1) = 0$$

$$\Rightarrow \quad k = \frac{1}{10} \qquad [\because \; p(x) \geq 0]$$

(*ii*) $P(X < 6) = P(X = 0) + P(X = 1) + \cdots\cdots + P(X = 5)$

$$= 0 + k + 2k + 2k + 3k + k^2 = 8k + k^2 = \frac{8}{10} + \frac{1}{100} = \frac{81}{100}$$

$$P(X \geq 6) = P(X = 6) + P(X = 7) = 2k^2 + 7k^2 + k = \frac{9}{100} + \frac{1}{10} = \frac{19}{100}$$

$$P(3 < X \leq 6) = P(X = 4) + P(X = 5) + P(X = 6) = 3k + k^2 + 2k^2 = \frac{3}{10} + \frac{3}{100} = \frac{33}{100}$$

(*iii*) $P(X \le 1) = k = \frac{1}{10} < \frac{1}{2}$; $P(X \le 2) = k + 2k = \frac{3}{10} < \frac{1}{2}$

$P(X \le 3) = k + 2k + 2k = \frac{5}{10} = \frac{1}{2}$; $P(X \le 4) = k + 2k + 2k + 3k = \frac{8}{10} > \frac{1}{2}$

∴ The maximum value of x so that $P(X \le x) > \frac{1}{2}$ is 4.

TEST YOUR KNOWLEDGE

1. Find the probability distribution of the number of doublets in four throws of a pair of dice.
2. Two bad eggs are mixed accidently with 10 good ones. Find the probability distribution of the number of bad eggs in 3, drawn at random, without replacement, from this lot.
3. A die is tossed twice. Getting a number greater than 4 is considered a success. Find the variance of the probability distribution of the number of successes.
4. Two cards are drawn simultaneously from a well-shuffled deck of 52 cards. Compute the variance for the number of aces.
5. An urn contains 4 white and 3 red balls. Three balls are drawn, with replacement, from this urn. Find μ, σ^2 and σ for the number of red balls drawn.
6. A random variable X has the following probability distribution:

Values of X, x :	0	1	2	3	4	5	6	7	8
$p(x)$:	a	$3a$	$5a$	$7a$	$9a$	$11a$	$13a$	$15a$	$17a$

(*i*) Determine the value of a. (*ii*) Find $P(X < 3)$, $P(X \ge 3)$, $P(2 \le X < 5)$

(*iii*) What is the smallest value of x for which $P(X \le x) > 0.5$?

7. Find the standard deviation for the following discrete distribution:

x :	8	12	16	20	24
$p(x)$:	$\frac{1}{8}$	$\frac{1}{6}$	$\frac{3}{8}$	$\frac{1}{4}$	$\frac{1}{12}$.

Answers

1.

X :	0	1	2	3	4
P(X) :	$\frac{625}{1296}$	$\frac{500}{1296}$	$\frac{150}{1296}$	$\frac{20}{1296}$	$\frac{1}{1296}$

2.

X :	0	1	2
P(X) :	$\frac{12}{22}$	$\frac{9}{22}$	$\frac{1}{22}$

3. $\frac{4}{9}$ **4.** $\frac{400}{2873}$ **5.** $\frac{9}{7}, \frac{36}{49}, \frac{6}{7}$

6. (*i*) $a = \frac{1}{81}$ (*ii*) $\frac{1}{9}, \frac{8}{9}, \frac{7}{27}$ (*iii*) 5 **7.** $2\sqrt{5}$.

4.22 CONTINUOUS DISTRIBUTIONS

So far we have dealt with discrete distributions *i.e.*, the distribution in which the variate takes a finite set of values. But when we deal with variates like temperature, heights and weights, we find that variates can take an infinite set of values in a given interval, say, $a \le x \le b$. Such variates are called continuous variates and their distributions are accordingly known as *continuous distributions*.

$$y = \frac{1}{2} \sin x \qquad (0 \le x \le \pi)$$

is an example of a continuous distribution, as x can assume all values lying between 0 and π.

4.23 PROBABILITY DENSITY FUNCTION

Let the probability of the variate x falling in the infinitesimal interval $(x - \frac{1}{2}dx, x + \frac{1}{2}dx)$ be expressed in the form $f(x)\, dx$, where $f(x)$ is a continuous function of x. Then $f(x)$ is called the *probability density function* or simply *density function*. The continuous curve $y = f(x)$ is called the *probability density curve* or briefly the *probability curve*.

Symbolically it is expressed as

$$P(x - \frac{1}{2}\, dx \le X \le x + \frac{1}{2} dx) = f(x)\, dx.$$

The interval of the variate may be finite or infinite. A function defined only for a finite interval say $f(x) = \phi(x)$ when $a \le x \le b$ can be put in the following form

$$f(x) = 0, \qquad x < a$$
$$f(x) = \phi(x), \qquad a \le x \le b$$
$$f(x) = 0, \qquad x > b.$$

The density function possesses the following two properties:

(1) $f(x) \ge 0$ for every x, as negative probability has no meaning,

(2) $\int_{-\infty}^{\infty} f(x)\, dx = 1$, this corresponds to the fact that the probability of an event that is sure to happen is equal to unity.

For a density function, $f(x)$, the probability that the variate X falls in any interval (p, q) is given by

$$P(p \le X < q) = \int_p^q f(x)\, dx.$$

Note 1. Any positive function of a variate x can be changed to give a probability density function $f(x)$ by multiplying it by a constant which will make the total area under the curve $y = f(x)$ equal to unity. For example we know that

$$\int_0^2 x(2 - x)\, dx = \frac{4}{3},$$

hence if we multiply both sides by $\frac{3}{4}$, we get

$$\int_0^2 \frac{3}{4} x\, (2 - x)\, dx = 1.$$

Hence a probability density function can be formed as given by

$$f(x) = \begin{cases} 0, & x < 0 \\ \dfrac{3x(2-x)}{4}, & 0 \le x \le 2 \\ 0, & x > 2. \end{cases}$$

Note 2. Whenever $f(x)$ is constant throughout its interval, the variable is said to have a rectangular distribution of same probability.

ILLUSTRATIVE EXAMPLES

Example 1. *If the function $f(x)$ is defined by $f(x) = ce^{-x}$, $0 \le x \le \infty$, find the value of c which changes $f(x)$ to a probability density function.*

Sol. In order that $f(x)$ may be density function, we should have

(*a*) $f(x) \ge 0$ for every x (*b*) $\int_{-\infty}^{+\infty} f(x)\, dx = 1$.

Since e^{-x} is always positive for the values of x lying between 0 and ∞, the condition will be satisfied if $c \ge 0$.

The second condition will be satisfied if

$$\int_0^{\infty} ce^{-x}\, dx = 1$$

i.e., if $$\left[-ce^{-x}\right]_0^{\infty} = 1$$

i.e., if $$c = 1.$$

Example 2. *If $f(x)$ has probability density cx^2, $0 < x < 1$, determine c and find the probability that $\frac{1}{3} < x < \frac{1}{2}$ i.e., $P\left(\frac{1}{3} < x < \frac{1}{2}\right)$.*

Sol. $f(x)$ will have a probability density if $\int_0^1 cx^2 dx = 1$

i.e., $$\left[\frac{1}{3}cx^3\right]_0^1 = 1, \text{ i.e., } c = 3.$$

$$P\left(\frac{1}{3} < x < \frac{1}{2}\right) = \int_{\frac{1}{3}}^{\frac{1}{2}} 3x^2 dx = \left[x^3\right]_{\frac{1}{3}}^{\frac{1}{2}} = \left(\frac{1}{8} - \frac{1}{27}\right) = \frac{19}{216}.$$

4.24 VARIOUS MEASURES FOR CONTINUOUS PROBABILITY DISTRIBUTIONS

Let $f_X(x)$ or $f(x)$ be the p.d.f. of a random variable X where X is defined from a to b. Then,

(*i*) **Arithmetic Mean** $= \int_a^b x\, f(x)\, dx$ ...(1)

(*ii*) **Harmonic Mean.** Harmonic mean H is given by: $\frac{1}{H} = \int_a^b \frac{1}{x} . f(x)\, dx$ ...[1(*a*)] (*iii*) **Geometric Mean.** Geometric mean G is given by: $\log G = \int_a^b \log x . f(x)\, dx$ ...[1(*b*)]

(*iv*) v_r (about origin) $= \int_a^b x^r . f(x)\, dx$ ...(2)

μ_r' (about the point $x = A$) $= \int_a^b (x - A)^r . f(x)\, dx$ ...[2(*a*)]

and μ_r (about mean) $= \int_a^b (x - \text{mean})^r . f(x)\, dx$ ...[2(*b*)]

In particular, from (1) and (2), we have

$$\nu_1 \text{ (about origin)} = Mean = \int_a^b x\, f(x)\, dx \quad \text{and} \quad \nu_2 = \int_a^b x^2\, f(x)\, dx$$

Hence $\mu_2 = (\nu_2 - \nu_1^2) = \int_a^b x^2\, f(x)\, dx - \left(\int_a^b x\, f(x)\, dx\right)^2$ ...[2(*c*)]

From (2), on putting $r = 3$ and 4 respectively, we get the values of ν_3 and ν_4 and consequently the moments about mean can be obtained and hence β_1 and β_2 can be computed.

(*v*) **Median.** Median is the point which divides the entire distribution in two equal parts. In case of continuous distribution, median is the point which divides the total area into two equal parts. Thus if M is the median, then

$$\int_a^M f(x)\, dx = \int_M^b f(x)\, dx = \frac{1}{2} \qquad ...(3)$$

Thus solving $\int_a^M f(x)\, dx = \frac{1}{2}$ or $\int_M^b f(x)\, dx = \frac{1}{2}$ ...[3(*a*)]

for M, we get the value of median.

(*vi*) **Mean Deviation.** Mean deviation about the mean μ_1' is given by:

$$\text{M.D.} = \int_a^b |x - \text{mean}|\, f(x)\, dx \qquad ...(4)$$

In general, mean deviation about an average 'A' is given by:

$$\text{M.D. about 'A'} = \int_a^b |x - A|\, f(x)\, dx \qquad ...[4(a)]$$

(*vii*) **Quartiles and Deciles.** Q_1 and Q_3 are given by the equations:

$$\int_a^{Q_1} f(x)\, dx = \frac{1}{4} \text{ and } \int_a^{Q_3} f(x)\, dx = \frac{3}{4} \qquad ...(5)$$

D_i, i^{th} decile is given by: $\int_a^{D_i} f(x)\, dx = \frac{i}{10}$; $i = 1, 2, ..., 9$ ...[5(*a*)]

(*viii*) **Mode.** Mode is the value of x for which $f(x)$ is maximum. Mode is thus the solution of $f'(x) = 0$ and $f''(x) < 0$, provided it lies in $[a, b]$. ...(6)

4.25 CUMULATIVE DISTRIBUTION FUNCTION

If $F(x) = \int_{-\infty}^{x} f(x)\, dx = P\,(X \le x)$, then the function $F(x)$ is the probability that the value of the variate X will be $\le x$.

$$F(b) = P(X \le b).$$

and
$$F(b) - F(a) = \int_a^b f(x)\, dx = P(a \le X \le b).$$

$F(x)$ is called the cumulative distribution function of x or simply the *distribution function.*

The cumulative distribution function has the following properties:

(1) $F'(x) = f(x) \ge 0$, so that $F(x)$ is a non decreasing function. This means that $dF(x) = f(x)\, dx$. This is known as *probability differential of* X.

(2) $F(-\infty) = 0$

(3) $F(\infty) = 1.$

(4) $F(x)$ is a continuous function of x on the right.

Example 3. *The distribution function of a random variable X is given by*

$$F(x) = \begin{cases} 1-(1+x)e^{-x}, & \text{for } x \ge 0 \\ 0, & \text{for } x < 0 \end{cases},$$

find the corresponding density function of random variable X.

Sol. The probability density function $= f(x) = \dfrac{d}{dx}\{F(x)\}$

$$f(x) = -e^{-x} + (1+x)e^{-x}$$

$$= \begin{cases} xe^{-x}, & \text{for } x \ge 0 \\ 0, & \text{for } x < 0. \end{cases}$$

Example 4. *A random variable x has the density function*

$$f(x) = k.\frac{1}{1+x^2},\ -\infty < x < \infty.$$

Determine k and the distribution function.

Sol. It will be a density function if

$$\int_{-\infty}^{\infty} k.\frac{1}{1+x^2} = 1 \text{ i.e., } k.\,\pi = 1 \text{ giving } k = 1/\pi$$

$$F(x) = \int \frac{1}{\pi}.\frac{1}{1+x^2}\,dx = \frac{1}{\pi}\tan^{-1}x + C$$

But $F(-\infty)$ should be zero for distribution function

$$\therefore \qquad \frac{1}{\pi}\left(-\frac{\pi}{2}\right) + C = 0 \text{ giving } C = \frac{1}{2}.$$

$$\therefore \qquad F(x) = \frac{1}{\pi}\tan^{-1}x + \frac{1}{2} \text{ for } -\infty < x < \infty.$$

Example 5. *The diameter, say X, of an electric cable, is assumed to be a continuous random variable with p.d.f.:* $f(x) = 6x(1-x),\ 0 \le x \le 1$

(*i*) *Check that the above is a p.d.f.,*

(*ii*) *Obtain an expression for the c.d.f. of X.,*

(*iii*) *Compute* $P\left(X \le \frac{1}{2} \,\middle|\, \frac{1}{3} \le X \le \frac{2}{3}\right)$, *and*

(*iv*) *Determine the number k such that* $P(X < k) = P(X > k)$.

Sol. (*i*) Since $\int_0^1 f(x)\,dx = \int_0^1 6x(1-x)\,dx = 6\left|\frac{x^2}{2} - \frac{x^3}{3}\right|_0^1 = 1$, $f(x)$ is a p.d.f.

$$\text{(ii) } F(x) = \begin{cases} 0, & \text{if } x \le 0 \\ \int_0^x 6t(1-t)dt = (3x^2 - 2x^3), & 0 < x \le 1 \\ 1, & \text{if } x > 1 \end{cases}$$

$$\text{(iii) } P\left(X \le \frac{1}{2} \,\middle|\, \frac{1}{3} \le X \le \frac{2}{3}\right) = \frac{P\left(\frac{1}{3} \le X \le \frac{1}{2}\right)}{P\left(\frac{1}{3} \le X \le \frac{2}{3}\right)} = \frac{\int_{1/3}^{1/2} 6x(1-x)\,dx}{\int_{1/3}^{2/3} 6x(1-x)\,dx} = \frac{11/54}{13/27} = \frac{11}{26}$$

(*iv*) We have $P(X < k) = P(X > k)$

$$\Rightarrow \quad \int_0^k 6x(1-x)dx = \int_k^1 6x(1-x)\,dx$$

or

$$3k^2 - 2k^3 = 3(1-k^2) - 2(1-k^3)$$

$$\Rightarrow \quad 4k^3 - 6k^2 + 1 = 0 \quad \Rightarrow \quad k = \frac{1}{2}, \frac{1 \pm \sqrt{3}}{2}.$$

The only admissible value of k in the given range is $\frac{1}{2}$. Hence the value of k is $\frac{1}{2}$.

Example 6. *Let X be a continuous random variable with p.d.f. given by:*

$$f(x) = \begin{cases} kx, & 0 \le x < 1 \\ k, & 1 \le x < 2 \\ -kx + 3k, & 2 \le x < 3 \\ 0, & \text{elsewhere} \end{cases}$$

(*i*) *Determine the constant k,* (*ii*) *Determine F(x), the c.d.f., and*

(*iii*) *If* x_1, x_2 *and* x_3 *are three independent observations from X, what is the probability that exactly one of these three numbers is larger than 1.5?*

Sol. (*i*) Since $f(x)$ in the p.d.f. of X, we have:

$$\int_0^3 f(x)\,dx = \int_0^1 f(x)\,dx + \int_1^2 f(x)\,dx + \int_2^3 f(x)\,dx = 1$$

$$\Rightarrow \qquad \int_0^1 k\,x\,dx + \int_1^2 k\,dx + \int_2^3 (-\,kx + 3k) = 1$$

$$\therefore \qquad \left[k\frac{x^2}{2}\right]_0^1 + [kx]_1^2 + \left[-\,k.\frac{x^2}{2} + 3kx\right]_2^3 = 1 \quad \Rightarrow \quad k = \frac{1}{2}.$$

(*ii*) For any x such that $-\infty < x < 0$; $F(x) = \int_{-\infty}^{x} 0\,.\,dt = 0$

For any x, where $0 \le x < 1$; $F(x) = \int_{-\infty}^{0} 0\,.\,dt + \int_0^x \frac{t}{2}dt = \frac{x^2}{4}$

For x, where $1 \le x < 2$, $F(x) = \int_{-\infty}^{0} 0\,.\,dt + \int_0^1 \frac{t}{2}dt + \int_1^x \frac{1}{2}dt = \frac{2x-1}{4}$

For any x, where $2 \le x < 3$,

$$F(x) = \int_{-\infty}^{0} 0\,.\,dt + \int_0^1 \frac{t}{2}dt + \int_1^2 \frac{1}{2}dt + \int_2^x \left(-\frac{t}{2} + \frac{3}{2}\right)dt.$$

$$= \frac{1}{4} + \left(1 - \frac{1}{2}\right) + \left(-\frac{x^2}{4} + \frac{3x}{2} - 2\right) = -\frac{x^2}{4} + \frac{3x}{2} - \frac{5}{4}$$

For any x, where $3 \le x < \infty$

$$F(x) = \int_{-\infty}^{0} 0.\,dt + \int_0^1 \frac{t}{2}\,dt + \int_1^2 \frac{1}{2}dt + \int_2^3 \left(-\frac{t}{2} + \frac{3}{2}\right)dt + \int_3^x 0.\,dt$$

$$= \frac{1}{4} + \left(1 - \frac{1}{2}\right) + \left(-\frac{9}{4} + \frac{9}{2} + 1 - 3\right) = 1$$

Hence the distribution function $F(x)$ is given by:

$$F(x) = \begin{cases} 0, & \text{for } -\infty \le x < 0 \\ \dfrac{x^2}{4}, & \text{for } 0 \le x < 1 \\ \dfrac{2x-1}{4}, & \text{for } 1 \le x < 2 \\ -\dfrac{x^2}{4} + \dfrac{3x}{2} - \dfrac{5}{4}, & \text{for } 2 \le x < 3 \\ 1, & \text{for } 3 \le x < \infty \end{cases}$$

(*iii*) The probability that X is larger than 1.5 is given by:

$$P(X > 1.5) = 1 - P\,(X < 1.5) = 1 - P(1.5) = \left(1 - \frac{3-1}{4}\right) = \frac{1}{2}.$$

$\therefore$ The probability that X is not larger than 1.5 = $P(X < 1.5) = 1 - \frac{1}{2} = \frac{1}{2}$.

Hence out of three numbers x_1, x_2 and x_3, the probability that exactly one is larger than 1.5 is:

$$3\left(\frac{1}{2}\right)\left(\frac{1}{2}\right)\left(\frac{1}{2}\right) = \frac{3}{8}.$$

Example 7. *A petrol pump is supplied with petrol once a day. If its daily volume of sales (X) in thousands of litres is distributed by: $f(x) = 5\,(1-x)^4$, $0 \le x \le 1$, what must be the capacity of its tank in order that the probability that its supply will be exhausted in a given day shall be 0.01?*

Sol. Let the capacity of the tank (in '000 of litres) be 'a' such that

$$P(X \geq a) = 0.01$$

$$\Rightarrow \quad \int_a^1 f(x)\, dx = 0.01$$

$$\Rightarrow \quad \int_a^1 5(1-x)^4\, dx = 0.01$$

$$\Rightarrow \quad (1-a)^5 = \frac{1}{100}$$

or
$$1 - a = \left(\frac{1}{100}\right)^{1/5}$$

$$\therefore \quad a = 1 - \left(\frac{1}{100}\right)^{1/5} = 1 - 0.3981 = 0.6019$$

Hence the capacity of the tank = 0.6019 × 1,000 litres = 601.9 litres.

Example 8. *A continuous random variable X has a p.d.f. $f(x) = 3x^2$, $0 \leq x \leq 1$. Find a and b such that (i) $P(X \leq a) = P(X > a)$, and (ii) $P(X > b) = 0.05$.*

Sol. (*i*) Since $P(X \leq a) = P(X > a)$, each must be equal to $\frac{1}{2}$, because total probability is always unity.

$$\therefore \quad P(X \leq a) = \frac{1}{2}$$

$$\Rightarrow \quad \int_0^a f(x)\, dx = \frac{1}{2}$$

$$\Rightarrow \quad 3\int_0^a x^2\, dx = \frac{1}{2} \Rightarrow 3\left|\frac{x^3}{3}\right|_0^a = \frac{1}{2} \Rightarrow a = \left(\frac{1}{2}\right)^{\frac{1}{3}}$$

(*ii*)
$$P(X > b) = 0.05 \Rightarrow \int_b^1 f(x)\, dx = 0.05$$

$$\Rightarrow \quad 3\left|\frac{x^3}{3}\right|_b^1 = \frac{1}{20} \Rightarrow 1 - b^3 = \frac{1}{20} \Rightarrow b = \left(\frac{19}{20}\right)^{\frac{1}{3}}.$$

Example 9. (*a*) *A random variable X is distributed at random between the values 0 and 1 so that its probability density function is: $f(x) = kx^2(1-x^3)$, where k is a constant. Find the value of k. Using this value of k, find its mean and variance.*

(*b*) *A variable X is distributed at random between the values 0 and 4 and its probability density function is given by: $f(x) = kx^3(4-x)^2$.*

Find the value of k, the mean and standard deviation of the distribution.

Sol. (*a*) Since $\int_{-\infty}^{\infty} f(x)\, dx = 1$,

$$k\int_0^1 (x^2 - x^5)\,dx = 1 \;\Rightarrow\; k\left|\frac{x^3}{3} - \frac{x^6}{6}\right|_0^1 = 1 \;\Rightarrow\; k = 6.$$

$$\text{Mean} = \mu_1' = \int_{-\infty}^{\infty} x\,f(x)\,dx = 6\int_0^1 (x^3 - x^6)\,dx = 6\left|\frac{x^4}{4} - \frac{x^7}{7}\right|_0^1 = \frac{9}{14}$$

$$\mu_2' = \int_{-\infty}^{\infty} x^2 f(x)\,dx = 6\int_0^1 (x^4 - x^7)\,dx = 6\left|\frac{x^5}{5} - \frac{x^8}{8}\right|_0^1 = \frac{9}{20}$$

$$\text{Variance} = \mu_2 = \mu_2' - \mu_1'^2 = \left\{\frac{9}{20} - \left(\frac{9}{14}\right)^2\right\} = \frac{9}{245}.$$

(*b*) Since $\int_{-\infty}^{\infty} f(x)\,dx = 1,\ k\int_0^4 x^3 (4-x)^2\,dx = 1 \;\Rightarrow\; k = \frac{15}{1024}$

$$\text{Mean} = \mu_1' = \int_{-\infty}^{\infty} xf(x)\,dx = k\int_0^4 x^4 (4 - x^2)\,dx = \frac{16}{7}$$

$$\mu_2' = \int_{-\infty}^{\infty} x^2 f(x)\,dx = k\int_0^4 x^5 (4-x)^2\,dx = \frac{40}{7}$$

$$\text{Variance} = \mu_2 = \mu_2' - \mu_1'^2 = \left\{\frac{40}{7} - \left(\frac{16}{7}\right)^2\right\} = \frac{24}{49}$$

$\therefore$ Standard deviation $= \frac{2\sqrt{6}}{7}.$

Example 10. *The time one has to wait for a bus at a downtown bus stop is observed to be random phenomenon (X) with the following probability density function:*

$$f_X(x) = \begin{cases} 0, & \text{for } x < 0 \\ \frac{1}{9}(x+1), & \text{for } 0 \le x < 1 \\ \frac{4}{9}(x - \frac{1}{2}), & \text{for } 1 \le x < \frac{3}{2} \\ \frac{4}{9}\left(\frac{5}{2} - x\right), & \text{for } \frac{3}{2} \le x < 2 \\ \frac{1}{9}(4-x), & \text{for } 2 \le x < 3 \\ \frac{1}{9}, & \text{for } 3 \le x < 6 \\ 0, & \text{for } x \ge 6 \end{cases}$$

Let the events A and B be defined as follows:

A: One waits between 0 to 2 minutes inclusive.

B: One waits between 0 to 3 minutes inclusive.

Show that (a) $P(B \mid A) = \frac{2}{3}$, *(b)* $P(\overline{A} \cap \overline{B}) = \frac{1}{3}$.

Sol. (*a*) $\quad P(A) = P(X \le 2) = \int_0^2 f(x)\,dx$

$$= \int_0^1 \frac{1}{9}(x+1)\,dx + \int_1^{3/2} \frac{4}{9}\left(x - \frac{1}{2}\right)dx + \int_{3/2}^2 \frac{4}{9}\left(\frac{5}{2} - x\right)dx$$

$$= \frac{1}{2}. \text{ (on simplification).}$$

$$\text{P}(\text{A} \cap \text{B}) = \text{P}[(0 \le \text{X} \le 2) \cap (1 \le \text{X} \le 3)] = \text{P}(1 \le \text{X} \le 2) = \int_1^2 f(x)\,dx$$

$$= \int_1^{3/2} \frac{4}{9}\left(x - \frac{1}{2}\right)dx + \int_{3/2}^2 \frac{4}{9}\left(\frac{5}{2} - x\right)dx$$

$$= \frac{4}{9}\left|\frac{x^2}{2} - \frac{x}{2}\right|_1^{3/2} + \frac{4}{9}\left|\frac{5}{2}x - \frac{x^2}{2}\right|_{3/2}^2 = \frac{1}{3} \text{ (on simplification)}$$

$$\therefore \quad \text{P}(\text{B}\,|\,\text{A}) = \frac{\text{P}(\text{A} \cap \text{B})}{\text{P}(\text{A})} = \frac{1/3}{1/2} = \frac{2}{3}.$$

(*b*) $\overline{\text{A}} \cap \overline{\text{B}}$ means that waiting time is more than 3 minutes.

$$\therefore \quad \text{P}(\overline{\text{A}} \cap \overline{\text{B}}) = \text{P}(\text{X} > 3) = \int_3^\infty f(x)\,dx = \int_3^6 f(x)\,dx + \int_6^\infty f(x)\,dx$$

$$= \int_3^6 \frac{1}{9}\,dx = \frac{1}{9}\,|\,x\,|_3^6 = \frac{1}{3}$$

TEST YOUR KNOWLEDGE

1. Find the constant k so that function $f(x)$ defined as follows be a density function:

$$f(x) = \begin{cases} 1/k, & a \le x \le b \\ 0, & \text{elsewhere} \end{cases}$$

2. Find the value of y_0 so that the function $f(x)$ defined as follows be a density function:

$$f(x) = y_0 e^{-x/\sigma}, \qquad 0 \le x \le \infty.$$

3. If $f(x) = \begin{cases} 2x, & 0 \le x \le 1 \\ 0, & \text{otherwise.} \end{cases}$; find the probability $\text{P}\left(\frac{1}{4} \le x \le \frac{1}{2}\right)$.

4. If $f(x) = \begin{cases} 0, & x < 2 \\ \frac{1}{18}(3 + 2x), & 2 \le x \le 4 \\ 0, & x > 4 \end{cases}$

prove that it is a density function. Find the probability that a variate having this density will fall in the interval $2 \le x \le 3$.

5. Define distribution function. Verify that the following is a distribution function:

$$\text{F}(x) = \begin{cases} 0, & x < -a \\ \frac{1}{2}\left(\frac{x}{a} + 1\right), & -a \le x \le a \\ 1, & x > a. \end{cases}$$

6. What do you understand by probability distribution?

7. Find the mean, variance and the coefficients β_1, β_2 of the distribution: $dF = kx^2 e^{-x}\,dx = 1, 0 < x < \infty$.

8. Show that the symmetrical distribution

$$f(x) = \frac{2a}{\pi}\left(\frac{1}{a^2 + x^2}\right), \quad -a \le x \le a$$

represents a probability density function. Also show that

$$\mu_2 = \frac{a^2}{\pi}(4 - \pi) \quad \text{and} \quad \mu_4 = a^4\left(1 - \frac{8}{3\pi}\right).$$

9. The amount of bread (in hundreds of pounds) x that a certain bakery is able to sell in a day is found to be a numerical valued random phenomenon with a probability function specified by the p.d.f. $f(x)$ is given by

$$f(x) = \begin{cases} kx, & 0 \le x < 5 \\ k(10 - x), & 5 \le x < 10 \\ 0, & \text{Otherwise} \end{cases}$$

Find the value of k such that $f(x)$ is a probability density function.

10. The kms X (in thousands of kms) which car owners get with a certain kind of tyre is a random variable having probability density function:

$$f(x) = \begin{cases} \frac{1}{20} e^{-x/20}, & \text{for } x > 0 \\ 0, & \text{for } x \le 0 \end{cases}$$

Find the probabilities that one of these tyres will last (*i*) atmost 10,000 kms, (*ii*) anywhere from 16,000 to 24,000 kms, and (*iii*) at least 30,000 miles.

[**Hint:** (*i*) $P(X \le 10) = \int_0^{10} f(x)\, dx$, (*ii*) $P(16 \le X \le 24)$, (*iii*) $P(X \ge 30)$]

Answers

1. $b - a$ **2.** $y_0 = \frac{1}{\sigma}$ **3.** $\frac{3}{16}$ **4.** $\frac{4}{9}$

7. $k = \frac{1}{2}$, mean = 3, variance = 3, $\beta_1 = \frac{4}{3}$, $\beta_2 = 5$

9. $\frac{1}{25}$ **10.** (*i*) 0.3935 (*ii*) 0.1481 (*iii*) 0.2231

4.26 JOINT PROBABILITY DISTRIBUTION

Let X and Y be two random variables defined on the same sample space.

Case 1. If X and Y are discrete random variables then a probability function

$$P(X = x_i, Y = y_j) = P(x_i, y_j) \text{ or } p_{ij}$$

that yields the probability that X will assume a particular value x_i while at the same time Y assumes a particular value y_j called a joint probability function of X and Y and has the following properties:

(*i*) $p(x_i, y_j) \ge 0$ (*ii*) $\sum_i \sum_j p(x_i, y_j) = 1.$

Case 2. If X and Y are continuous random variables then the probability density function

$$P\left[x - \frac{dx}{2} \le X \le x + \frac{dx}{2}, y - \frac{dy}{2} \le Y \le y + \frac{dy}{2}\right] = f(x, y)$$

that yields the probability that the point (x, y) lies in the infinitesimal rectangular region of area $dx\, dy$ is called the joint probability density function of X and Y and satisfies (*i*) $f(x, y) \geq 0$ for all (x, y) in the given range.

(*ii*) $\int_{-\infty}^{\infty} \int_{-\infty}^{\infty} f(x, y)\, dx\, dy$ exists and is equal to 1.

(*iii*) $P(a \leq X \leq b, c \leq Y \leq d) = \int_{a}^{b} \int_{c}^{d} f(x, y)\, dx\, dy$

The probability distribution defined in terms of joint probability function or joint probability density function is called a joint probability distribution.

4.27 MARGINAL PROBABILITY DISTRIBUTION

Let X and Y be two discrete random variables defined on the same sample space, such that

$$P(X = x_i) = p_i \text{ and } P(Y = y_j) = p_j';\ i = 1, ..., m, j = 1, ..., n$$

and

$$P(X = x_i, Y = y_j) = p_{ij}$$

then their joint occurrence have $m \times n$ distinguished pairs (x_i, y_j) which can be arranged in a rectangular array with their probabilities.

As the variable X assumes a definite value x_i, it is acccompained by one of the n values $y_1, y_2, ..., y_n$ of Y. *i.e.*, $X = x_i$ can occur in n mutually exclusive ways ; therefore

$$P(X = x_i) = P(X = x_i, Y = y_1) + P(X = x_i, Y = y_2) + \cdots + P(X = x_i, Y = y_n)$$

or $p = p_{i1} + p_{i2} + ... + p_{in} = \sum_{j=1}^{n} p_{ij} = g(x_i)$ say, which is called marginal probability of X for $X = x_i$.

and similarly for fixed y_j

$$p_j' = \sum_{i=1}^{n} p_{ij} = h(y_j) \text{ say,}$$

which is called marginal probability of Y for $Y = y_j$.

i.e., by adding the probabilities p_{ij} ... in individual rows and columns, we obtain the marginal probabilities of X and Y which are identical to the individual probabilities of X and Y respectively.

The set of values of a random variable X together with the marginal probabilities is called the *marginal distribution* of that random variable. In fact, marginal distribution is simply the probability distribution of the random variable.

Let X and Y be continuous random variables.

If (X, Y) has the joint density function f(x, y), then the marginal probability density function of X is

$$g(x) = \int_{-\infty}^{\infty} f(x, y) dy$$

and marginal probability density function of Y is

$$h(y) = \int_{-\infty}^{\infty} f(x, y) dx.$$

The corresponding probability distribution is called *Marginal Probability Distribution.*

4.28 CONDITIONAL PROBABILITY DISTRIBUTION

Suppose $P(X = x_i, Y = y_j) = p_{ij}, i = 1, 2,......, m, j = 1, 2,....., n$ is the joint probability distribution of two discrete random variables, $p_j = P(Y = y_j)$ the probability distribution of Y and $p_i = P(X = x_i)$ the probability distribution of X.

The conditional probability of the event "Y = y_j, given that X = x_i (with $p_i > 0$)" is

$$\frac{P(X = x_i, Y = y_j)}{P(X = x_i)}$$

i.e.,
$$P(Y = y_j \mid X = x_i) = \frac{p_{ij}}{p_i} = h(y_j/x_i), \text{ say.}$$

The function $h(y_j/x_i) = \frac{p_{ij}}{p_i}$ $i = 1, 2, \ldots, m, j = 1, 2, \ldots, n$ *that yields the conditional probabilities of the random variable Y is called the conditional distribution of Y for given X, and* $g(x_i/y_j) = P(X = x_i/Y = y_j) = \frac{p_{ij}}{p_j}$ is the conditional probability distribution of X for given Y.

If (X, Y) has the joint probability density function f(x, y), then the conditional probability density function of X for given Y is defined for any y such that h(y) > 0 and is equal to

$$g(x/y) = \frac{f(x, y)}{h(y)}$$

and the conditional p.d.f. of Y for given X is defined for any x such that g(x) > 0 and is equal to $h(y/x) = \frac{f(x, y)}{g(x)}$.

ILLUSTRUTIVE EXAMPLES

Example 1. *The joint probability distribution of two random variables X and Y is given by :* $P(X = 0, Y = 1) = \frac{1}{3}$, $P(X = 1, Y = -1) = \frac{1}{3}$, *and* $P(X = 1, Y = 1) = \frac{1}{3}$.

Find (i) Marginal distributions of X and Y, and (ii) the conditional probability distribution of X given Y = 1.

Sol. (*i*) $P(X = -1) = \sum_y P(X = -1, Y = y)$

$= P(X = -1, Y = -1) + P(X = -1, Y = 0) + P(X = -1, Y = 1) = 0$

Similarly $P(X = 0) = \frac{1}{3}$ and $P(X = 1) = \frac{2}{3}$

Y \ *X*	*– 1*	*0*	*1*	*Marginal Y*
– 1	0	0	$\frac{1}{3}$	$\frac{1}{3}$
0	0	0	0	0
1	0	$\frac{1}{3}$	$\frac{1}{3}$	$\frac{2}{3}$
Marginal (X)	0	$\frac{1}{3}$	$\frac{2}{3}$	1

Thus

Marginal distribution of X is:

Values of X, x :	-1	0	1
$P(X = x)$:	0	$\frac{1}{3}$	$\frac{2}{3}$

Marginal distribution of Y is:

Values of Y, y :	-1	0	1
$P(Y = y)$:	$\frac{1}{3}$	0	$\frac{2}{3}$

(*ii*) The conditional probability distribution of X given Y is:

$$P(X = x \mid Y = y) = \frac{P(X = x, Y = y)}{P(Y = y)}. \text{ Now}$$

$$P(X = -1 \mid Y = 1) = \frac{P(X = -1, Y = 1)}{P(Y = 1)} = 0,$$

$$P(X = 0 \mid Y = 1) = \frac{P(X = 0, Y = 1)}{P(Y = 1)} = \frac{1/3}{2/3} = \frac{1}{2}$$

$$P(X = 1 \mid Y = 1) = \frac{P(X = 1, Y = 1)}{P(Y = 1)} = \frac{1/3}{2/3} = \frac{1}{2}$$

Thus the conditional distribution of X given Y = 1 is :

Values of X = x	-1	0	1
$P(X = x \mid Y = 1)$	0	$\frac{1}{2}$	$\frac{1}{2}$

Example 2. *For the adjoining bivariate probability distribution of X and Y, find:*
(i) $P(X \le 1, Y = 2)$, *(ii)* $P(X \le 1)$,
(iii) $P(Y \le 3)$, *and* *(iv)* $P(X < 3, Y \le 4)$.

X \ *Y*	*1*	*2*	*3*	*4*	*5*	*6*
0	*0*	*0*	$\frac{1}{32}$	$\frac{2}{32}$	$\frac{2}{32}$	$\frac{3}{32}$
1	$\frac{1}{16}$	$\frac{1}{16}$	$\frac{1}{8}$	$\frac{1}{8}$	$\frac{1}{8}$	$\frac{1}{8}$
2	$\frac{1}{32}$	$\frac{1}{32}$	$\frac{1}{64}$	$\frac{1}{64}$	*0*	$\frac{2}{64}$

Sol. The marginal distributions are given below:

X \ Y	1	2	3	4	5	6	$p_X(x)$
0	0	0	$\frac{1}{32}$	$\frac{2}{32}$	$\frac{2}{32}$	$\frac{3}{32}$	$\frac{8}{32}$
1	$\frac{1}{16}$	$\frac{1}{16}$	$\frac{1}{8}$	$\frac{1}{8}$	$\frac{1}{8}$	$\frac{1}{8}$	$\frac{10}{16}$
2	$\frac{1}{32}$	$\frac{1}{32}$	$\frac{1}{64}$	$\frac{1}{64}$	0	$\frac{2}{64}$	$\frac{8}{64}$
$p_Y(y)$	$\frac{3}{32}$	$\frac{3}{32}$	$\frac{11}{64}$	$\frac{13}{64}$	$\frac{6}{32}$	$\frac{16}{64}$	$\Sigma p(x) = 1$ $\Sigma p(y) = 1$

(*i*) $P(X \le 1, Y = 2) = P(X = 0, Y = 2) + P(X = 1, Y = 2) = 0 + \frac{1}{16} = \frac{1}{16}$

(*ii*) $P(X \le 1) = P(X = 0) + P(X = 1) = \frac{8}{32} + \frac{10}{16} = \frac{7}{8}$

(*iii*) $P(Y \le 3) = P(Y = 1) + P(Y = 2) + P(Y = 3) = \frac{3}{32} + \frac{3}{32} + \frac{11}{64} = \frac{23}{64}$

(*iv*) $P(X < 3, Y \le 4) = P(X = 0, Y \le 4) + P(X = 1, Y \le 4) + P(X = 2, Y \le 4)$

$$= \left(\frac{1}{32} + \frac{2}{32}\right) + \left(\frac{1}{16} + \frac{1}{16} + \frac{1}{8} + \frac{1}{8}\right) + \left(\frac{1}{32} + \frac{1}{32} + \frac{1}{64} + \frac{1}{64}\right) = \frac{9}{16}.$$

Example 3. *For the joint probability distribution of two random variables X and Y given below:*

Y / X	1	2	3	4	Total
1	$\frac{4}{36}$	$\frac{3}{36}$	$\frac{2}{36}$	$\frac{1}{36}$	$\frac{10}{36}$
2	$\frac{1}{36}$	$\frac{3}{36}$	$\frac{3}{36}$	$\frac{2}{36}$	$\frac{9}{36}$
3	$\frac{5}{36}$	$\frac{1}{36}$	$\frac{1}{36}$	$\frac{1}{36}$	$\frac{8}{36}$
4	$\frac{1}{36}$	$\frac{2}{36}$	$\frac{1}{36}$	$\frac{5}{36}$	$\frac{9}{36}$
Total	$\frac{11}{36}$	$\frac{9}{36}$	$\frac{7}{36}$	$\frac{9}{36}$	1

Find (i) the marginal distributions of X and Y, and (ii) conditional distribution of X given the value of Y = 1 and that of Y given the value of X = 2.

Sol. (*i*) The marginal distribution of X is defined as:

$$P(X = x) = \sum_y P(X = x, Y = y)$$

$$\therefore \quad P(X = 1) = \sum_y P(X = 1, Y = y)$$

$$= P(X = 1, Y = 1) + P(X = 1, Y = 2) + P(X = 1, Y = 3) + P(X = 1, Y = 4)$$

$$= \frac{4}{36} + \frac{3}{36} + \frac{2}{36} + \frac{1}{36} = \frac{10}{36}.$$

Similarly, $P(X = 2) = \sum_y P(X = 2, Y = y) = \frac{9}{36}$; $P(X = 3) = \sum_y P(X = 3, Y = y) = \frac{8}{36}$

and $\quad P(X = 4) = \sum_y P(X = 4, Y = y) = \frac{9}{36}.$

Similarly, we can obtain the marginal distribution of Y.

Marginal Distribution of X

Values of X, x	1	2	3	4
P(X = x)	$\frac{10}{36}$	$\frac{9}{36}$	$\frac{8}{36}$	$\frac{9}{36}$

Marginal Distribution of Y

Values of Y, y	1	2	3	4
P(Y = y)	$\frac{11}{36}$	$\frac{9}{36}$	$\frac{7}{36}$	$\frac{9}{36}$

(*ii*) The conditional probability function of X given Y is defined as follows:

$$P(X = x \mid Y = y) = \frac{P(X = x, Y = y)}{P(Y = y)}. \text{ Therefore}$$

$$P(X = 1 \mid Y = 1) = \frac{P(X = 1, Y = 1)}{P(Y = 1)} = \frac{4/36}{11/36} = \frac{4}{11}$$

$$P(X = 2 \mid Y = 1) = \frac{P(X = 2, Y = 1)}{P(Y = 1)} = \frac{1/36}{11/36} = \frac{1}{11}$$

$$P(X = 3 \mid Y = 1) = \frac{P(X = 3, Y = 1)}{P(Y = 1)} = \frac{5/36}{11/36} = \frac{5}{11}$$

$$P(X = 4 \mid Y = 1) = \frac{P(X = 4, Y = 1)}{P(Y = 1)} = \frac{1/36}{11/36} = \frac{1}{11}$$

Hence the conditional distribution of X given Y = 1 is:

x :	1	2	3	4
$P(X = x \mid Y = 1)$:	$\frac{4}{11}$	$\frac{1}{11}$	$\frac{5}{11}$	$\frac{1}{11}$

Similarly, we can obtain the conditional distribution of Y for X = 2 as given below:

y :	1	2	3	4
$P(Y = y \mid X = 2)$:	$\frac{1}{9}$	$\frac{1}{3}$	$\frac{1}{3}$	$\frac{2}{9}$

Example 4. *Two discrete random variables X and Y have the joint probability density function:*

$$p_{XY}(x, y) = \frac{\lambda^x e^{-\lambda} p^y (1-p)^{x-y}}{y!(x-y)!}, \; y = 0, 1, 2, \ldots, x \,;\, x = 0, 1, 2, \ldots,$$

where λ, p are constants with λ > 0 and 0 < p < 1.

Find (i) The marginal probability density functions of X and Y.

(ii) The conditional distribution of Y for a given X and of X for a given Y.

Sol. (*i*) The marginal *p.m.f.* of X is given by:

$$p_X(x) = \sum_{y=0}^{x} p(x, y) = \sum_{y=0}^{x} \frac{\lambda^x e^{-\lambda} p^y (1-p)^{x-y}}{y!(x-y)!} = \frac{\lambda^x e^{-\lambda}}{x!} \sum_{y=0}^{x} \frac{x!\, p^y (1-p)^{x-y}}{y!(x-y)!}$$

$$= \frac{\lambda^x e^{-\lambda}}{x!} \sum_{y=0}^{x} {}^xC_y p^y (1-p)^{x-y} = \frac{\lambda^x e^{-\lambda}}{x!}[p + (1-p)]^x = \frac{\lambda^x e^{-\lambda}}{x!}, \; x = 0, 1, 2, \ldots$$

which is the probability function of a Poisson distribution with parameter λ.

$$p_Y(y) = \sum_{x=y}^{\infty} p(x, y) = \sum_{x=y}^{\infty} \frac{\lambda^x e^{-\lambda} p^y (1-p)^{x-y}}{y!(x-y)!} = \frac{(\lambda p)^y e^{-\lambda}}{y!} \sum_{x=y}^{\infty} \frac{[\lambda(1-p)]^{x-y}}{(x-y)!}$$

$$[\because \; y = 0, 1, 2, \ldots, x \;\Rightarrow\; x \le y \;\Rightarrow\; x \ge y]$$

$$= \frac{(\lambda p)^y e^{-\lambda}}{y!} e^{\lambda(1-p)} = \frac{e^{-\lambda p}(\lambda p)^y}{y!}; \; y = 0, 1, 2, \ldots$$

which is the probability function of a Poisson distribution with parameter λp.

(*ii*) The conditional distribution of Y for given X is:

$$P_{Y\,|\,X}(y\mid x) = \frac{p_{XY}(x, y)}{p_X(x)} = \frac{\lambda^x e^{-\lambda} p^y (1-p)^{x-y} x!}{y!(x-y)!\,\lambda^x e^{-\lambda}} = \frac{x!}{y!(x-y)!} p^y (1-p)^{x-y}$$

$$= {}^xC_y\, p^y (1-p)^{x-y},\ x \ge y \quad i.e.,\quad y = 0, 1, 2, ..., x.$$

The conditional probability distribution of X for given Y is:

$$p_{X\,|\,Y}(x\mid y) = \frac{p_{XY}(x, y)}{p_Y(y)} = \frac{\lambda^x e^{-\lambda} p^y (1-p)^{x-y}}{y!\,(x-y)!} \cdot \frac{y!}{e^{-\lambda p}(\lambda\, p)^y} \qquad [c.f.\ \text{Part } (i)]$$

$$= \frac{e^{-\lambda q}(\lambda\, q)^{x-y}}{(x-y)!};\ q = 1-p,\ x \ge y\ i.e.,\ x = y, y+1, y+2, ...$$

Example 5. *If X and Y are two random variables having joint density function:*

$$f(x, y) = \begin{cases} \dfrac{1}{8}(6-x-y); 0 \le x < 2, 2 \le y < 4 \\ 0, \text{otherwise} \end{cases}$$

Find (i) P (X < 1 ∩ Y < 3), (ii) P (X + Y < 3), and (iii) P (X < 1 | Y < 3).

Sol. We have

(*i*) $P(X < 1 \cap Y < 3) = \int_{-\infty}^{1}\int_{-\infty}^{3} f(x, y)\, dx\, dy = \int_0^1\int_2^3 \frac{1}{8}(6-x-y)\, dx\, dy = \frac{3}{8}$

(*ii*) $P(X + Y < 3) = \int_0^1\int_2^{3-x} \frac{1}{8}(6-x-y)\, dx\, dy = \frac{5}{24}$

(*iii*) $P(X < 1 \mid Y < 3) = \frac{P(X<1 \cap Y<3)}{P(Y<3)} = \frac{3/8}{5/8} = \frac{3}{5}$

$$\left[\text{From part } (i) \text{ and } P(Y < 3) = \int_0^2\int_2^3 \frac{1}{8}(6-x-y)\, dx\, dy = \frac{5}{8}\right]$$

Example 6. *Suppose that two-dimensional continuous random variable (X, Y) has joint p.d.f. given by*

$$f(x, y) = \begin{cases} 6x^2 y, 0 < x < 1, 0 < y < 1 \\ 0, \text{elsewhere} \end{cases}$$

(*i*) *Verify that* $\int_0^1\int_0^1 f(x, y)\, dx\, dy = 1.$

(*ii*) *Find* $P\left(0 < X < \frac{3}{4}, \frac{1}{3} < Y < 2\right)$, *P(X + Y < 1), P(X > Y) and P(X < 1 | Y < 2).*

Sol. (*i*) $\int_0^1\int_0^1 f(x, y)\, dx\, dy = \int_0^1\int_0^1 6x^2 y\, dx dy = \int_0^1 6x^2 \left|\frac{y^2}{2}\right|_0^1 dx = \int_0^1 3x^2\, dx = \left| x^3 \right|_0^1 = 1$

(*ii*) $P\left(0 < X < \frac{3}{4}, \frac{1}{3} < Y < 2\right) = \int_0^{3/4}\int_{1/3}^1 6x^2 y\, dx dy + \int_0^{3/4}\int_1^2 0\,.\,dx dy$

$$= \int_0^{3/4} 6x^2 \left|\frac{y^2}{2}\right|_{1/3}^1 dx = \frac{8}{9}\int_0^{3/4} 3x^2\, dx = \frac{8}{9}\left| x^3 \right|_0^{3/4} = \frac{3}{8}.$$

$$P(X + Y < 1) = \int_0^1 \int_1^{1-x} 6x^2 y \, dxdy = \int_0^1 6x^2 \left| \frac{y^2}{2} \right|_0^{1-x} dx = \int_0^1 3x^2(1-x)^2 dx = \frac{1}{10}$$

$$P(X > Y) = \int_0^1 \int_0^x 6x^2 y \, dxdy = \int_0^1 3x^2 \left| y^2 \right|_0^x dx = \int_0^1 3x^4 dx = \frac{3}{5}.$$

$$P(X < 1 \mid Y < 2) = \frac{P(X < 1 \cap Y < 2)}{P(Y < 2)}$$

where $$P(X < 1 \cap Y < 2) = \int_0^1 \int_0^1 6x^2 y \, dxdy + \int_0^1 \int_1^2 0 \, . \, dxdy = 1$$

and $$P(Y < 2) = \int_0^1 \int_0^2 f(x, y) \, dxdy = \int_0^1 \int_0^1 6x^2 y \, dxdy + \int_0^1 \int_1^2 0 \, . \, dxdy = 1$$

$\therefore$ $$P(X < 1 \mid Y < 2) = \frac{P(X < 1 \cap Y < 2)}{P(Y < 2)} = 1.$$

Example 7. *The joint probability density function of a two-dimensional random variable (X, Y) is given by:*

$$f(x, y) = \begin{cases} 2; 0 < x < 1, 0 < y < x; \\ 0, \text{elsewhere} \end{cases}$$

(i) Find the marginal density functions of X and Y.

(ii) Find the conditional density function of Y given X = x and conditional density function of X given Y = y.

(iii) Check for independence of X and Y.

Sol. Evidently $f(x, y) \geq 0$ and $\int_0^1 \int_0^x 2 \, dxdy = 2\int_0^1 x \, dx = 1$.

(*i*) The marginal *p.d.f.*'s of X and Y are given by

$$f_X(x) = \begin{cases} \int_{-\infty}^{\infty} f_{XY}(x, y) \, dy = \int_0^x 2dy = 2x, \; 0 < x < 1 \\ 0, \text{elsewhere} \end{cases}$$

$$f_Y(y) = \begin{cases} \int_{-\infty}^{\infty} f_{XY}(x, y) \, dx = \int_y^1 2dx = 2(1-y), 0 < y < 1 \\ 0, \text{elsewhere} \end{cases}$$

(*ii*) The conditional density function of Y given X, $(0 < x < 1)$ is

$$f_{Y \mid X}(y \mid x) = \frac{f_{XY}(x, y)}{f_X(x)} = \frac{2}{2x} = \frac{1}{x}, \; 0 < y < x.$$

The conditional density function of X given Y, $(0 < y < 1)$ is

$$f_{X \mid Y}(x \mid y) = \frac{f_{XY}(x, y)}{f_Y(y)} = \frac{2}{2(1-y)} = \frac{1}{(1-y)}, \; y < x < 1$$

(*iii*) Since $f_X(x) f_Y(y) = 2(2x)(1-y) \neq f_{XY}(x, y)$, X and Y are not independent.

Example 8. *The joint p.d.f. of two random variables X and Y is given by*

$$f(x, y) = \frac{9(1 + x + y)}{2(1+x)^4 (1+y)^4}; \; 0 \leq x < \infty, \; 0 \leq y < \infty$$

Find the marginal distributions of X and Y, and the conditional distribution of Y for X = x.

Sol. Marginal *p.d.f.* of X is given by:

$$f_X(x) = \int_0^\infty f(x, y)\, dy = \frac{9}{2(1+x)^4} \int_0^\infty \frac{(1+y)+x}{(1+y)^4}\, dy$$

$$= \frac{9}{2(1+x)^4} \int_0^\infty \{(1+y)^{-3} + x(1+y)^{-4}\}\, dy$$

$$= \frac{9}{2(1+x)^4} \left(\left| \frac{-1}{2(1+y)^2} \right|_0^\infty + x \left| \frac{-1}{3(1+y)^3} \right|_0^\infty \right)$$

$$= \frac{9}{2(1+x)^4} \cdot \left(\frac{1}{2} + \frac{x}{3} \right) = \frac{3}{4} \cdot \frac{3+2x}{(1+x)^4};\ 0 < x < \infty$$

Since $f(x, y)$ is symmetric in x and y, the marginal *p.d.f.* of Y is given by:

$$f_Y(y) = \int_0^\infty f(x, y)\, dx = \frac{3}{4} \cdot \frac{3+2y}{(1+y)^4};\ 0 < y < \infty$$

The conditional distribution of Y for X = x is given by:

$$f_{XY}(Y = y \mid X = x) = \frac{f_{XY}(x, y)}{f_X(x)} = \frac{9(1+x+y)}{2(1+x)^4 (1+y)^4} \times \frac{4(1+x)^4}{3(3+2x)} = \frac{6(1+x+y)}{(1+y)^4 (3+2x)};\ 0 < y < \infty.$$

Example 9. *Joint distribution of X and Y is given by:* $f(x, y) = 4xy\, e^{-(x^2 + y^2)};\ x \geq 0,\ y \geq 0.$

Test whether X and Y are independent. For the above joint distribution, find the conditional density of X given Y = y.

Sol. Marginal density of X is given by

$$f_X(x) = \int_0^\infty f_{XY}(x, y)\, dy = \int_0^\infty 4xy\, e^{-(x^2+y^2)}\, dy = 4x\, e^{-x^2} \int_0^\infty y\, e^{-y^2}\, dy$$

$$= 4x\, e^{-x^2} \cdot \int_0^\infty e^{-t} \cdot \frac{dt}{2} = 2x \cdot e^{-x^2} \left| -e^{-t} \right|_0^\infty$$

$$\therefore \quad f_X(x) = 2x\, e^{-x^2};\ x \geq 0$$

Similarly, the marginal *p.d.f.* of Y is given by:

$$f_Y(y) = \int_0^\infty f_{XY}(x, y)\, dx = 2y\, e^{-y^2};\ y \geq 0$$

Since $f_{XY}(x, y) = f_X(x) \cdot f_Y(y)$, X and Y are independently distributed. The conditional distribution of X for given Y is given by :

$$f_{X \mid Y}(X = x \mid Y = y) = \frac{f(x, y)}{f_Y(y)} = 2x\, e^{-x^2};\ x \geq 0.$$

Example 10. *Let X and Y be jointly distributed with p.d.f.:*

$$f_{XY}(x, y) = \begin{cases} \dfrac{1}{4}(1 + xy), |x| < 1, |y| < 1 \\ 0, \text{ otherwise} \end{cases}$$

Show that X and Y are not independent but X^2 and Y^2 are independent.

Sol. $$f_X(x) = \int_{-1}^{1} f(x, y)\,dy = \frac{1}{4}\left| y + \frac{xy^2}{2} \right|_{-1}^{1} = \frac{1}{2}, \; -1 < x < 1;$$

Similarly, $$f_Y(y) = \int_{-1}^{1} f(x, y)\,dx = \frac{1}{2}, \; -1 < y < 1$$

Since $f_{X,\,Y}(x, y) \neq f_X(x)\, f_Y(y)$, X and Y are not independent. However,

$$P(X^2 \le x) = P(|X| \le \sqrt{x}) = \int_{-\sqrt{x}}^{\sqrt{x}} f_X(x)\,dx = \sqrt{x} \qquad ...(1)$$

$$P(X^2 \le x \cap Y^2 \le y) = P(|X| \le \sqrt{x} \cap |Y| \le \sqrt{y})$$

$$= \int_{-\sqrt{x}}^{\sqrt{x}} \left[\int_{-\sqrt{y}}^{\sqrt{y}} f(u, v)\,dv \right] du = \sqrt{x}\,\sqrt{y}$$

$$= P(X^2 \le x) \,.\, P(Y^2 \le y) \qquad \text{[From (1)]}$$

Hence, X^2 and Y^2 are independent.

TEST YOUR KNOWLEDGE

1. Find the mean and variance of p.d.f.

$$f(x) = \begin{cases} \frac{1}{4} e^{-x/4}, & x > 0 \\ 0, & \text{elsewhere} \end{cases}$$

2. Find the mean and variance of exponential distribution $f(x) = \frac{1}{b} e^{-x/b}; x > 0, b > 0$

3. A continuous random variable X has the probability density function:

$$f(x) = A + Bx, \; 0 \le x \le 1.$$

If the mean of the distribution is $\frac{1}{2}$, find A and B.

4. For the probability density function: $f(x) = cx^2(1 - x)$, $0 < x < 1$, find

(*i*) the constant c, and (*ii*) mean.

5. The distribution of a continuous random variable X in range $(-3, 3)$ is given by p.d.f.:

$$f(x) = \begin{cases} \frac{1}{16}(3 + x)^2, & -3 \le x \le -1 \\ \frac{1}{16}(6 - 2x^2), & -1 \le x \le 1 \\ \frac{1}{16}(3 - x)^2, & 1 \le x \le 3 \end{cases}$$

(*i*) Verify that the area under the curve is unity.

(*ii*) Find the mean and variance of the above distribution.

6. The length of time (in minutes) that a certain lady speaks on the telephone is found to be random phenomenon, with a probability function specified by the probability density function $f(x)$ as:

$$f(x) = \begin{cases} A\, e^{-x/5}, & \text{for } x \ge 0 \\ 0, & \text{otherwise} \end{cases}$$

(*a*) Find the value of A that makes $f(x)$ a p.d.f.

(*b*) What is the probability that the number of minutes that she will talk over the phone is: (*i*) more than 10 minutes, (*ii*) less than 5 minutes, and (*iii*) between 5 and 10 minutes.

7. (*a*) Find k so that $f(x, y) = kxy$, $1 \le x \le y \le 2$ will be a probability density function.

(*b*) If $f(x, y) = \begin{Bmatrix} e^{-(x+y)}, & x \ge 0,\ y \ge 0 \\ 0, & \text{elsewhere} \end{Bmatrix}$ is the joint probability density function of random variables X and Y, find

(*i*) P(X < 1), (*ii*) P(X > Y), and (*iii*) P(X + Y < 1).

8. The joint probability density function of the two-dimensional variable (X, Y) is of the form:

$$f(x, y) = \begin{cases} e^{-(x+y)}, & 0 \le y < x < \infty \\ 0, & \text{elsewhere} \end{cases}$$

(*i*) Determine the constant k.
(*ii*) Find the conditional probability density function $f_1(x \mid y)$.
(*iii*) Compute P(Y ≥ 3).

9. Two-dimensional random variable (X, Y) have the joint density function

$$f(x, y) = \begin{cases} 8xy, & 0 < x < y < 1 \\ 0, & \text{otherwise} \end{cases}$$

Find the marginal and conditional distributions.

10. (*a*) The joint probability density function of the two-dimensional random variable (X, Y) is given by:

$$f(x, y) = \begin{cases} \frac{8}{9}xy, & 1 \le x \le y \le 2 \\ 0, & \text{elsewhere} \end{cases}$$

(*i*) Find the marginal density functions of X and Y.
(*ii*) Find the conditional density function of Y given X = x, and conditional density function of X given Y = y.

(*b*) Let $f(x_1, x_2) = \begin{Bmatrix} A(x_1x_2 + e^{x_1}); & 0 < (x_1\ x_2) < 1 \\ 0, & \text{elsewhere} \end{Bmatrix}$ be the joint p.d.f. of (X_1, X_2). Determine A.

11. The joint probability density function of the two dimensional random variable (X, Y) is given by :

$$f(x, y) = \begin{cases} x^3y^3/16, & 0 \le x \le 2, 0 \le y \le 2 \\ 0, & \text{elsewhere} \end{cases}$$

Find the marginal densities of X and Y. Also, find the cumulative distribution functions for X and Y.

12. (*a*) If the joint distribution function of X and Y is given by

$$F_{XY}(x, y) = \begin{Bmatrix} 1 - e^{-x} - e^{-y} + e^{-(x+y)}; & x > 0, y > 0 \\ 0; & \text{elsewhere} \end{Bmatrix}$$

(*i*) Find the marginal densities of X and Y.
(*ii*) Are X and Y independent?
(*iii*) Find P(X ≤ 1 ∩ Y ≤ 1) and P(X + Y ≤ 1).

(*b*) Given $f(x, y) = e^{-(x+y)}$, $0 \le x < \infty$, $0 \le y < \infty$. Are X and Y independent? Also, find P(X > 1).

13. A two-dimensional random variable (X, Y) have a bivariate distribution given by :

$P(X = x, Y = y) = \dfrac{x^2 + y}{32}$, for x = 0, 1, 2, 3 and y = 0, 1. Find the marginal distributions of X and Y.

14. A two-dimensional random variable (X, Y) have a joint probability mass function:

$p(x, y) = \frac{1}{27}(2x + y)$, where x and y can assume only the integer values 0, 1 and 2.

Find the conditional distribution of Y for X = x.

Answers

1. Mean = 4, Variance = 80

2. Mean = b, Variance = b^2

3. A = 1, B = 0

4. (*i*) $c = 12$, (*ii*) Mean = 3/5

5. (*ii*) Mean = 0, Variance = 1

6. (*a*) $A = \frac{1}{5}$ (*b*) (*i*) $\frac{1}{e^2}$ (*ii*) $\frac{e-1}{e}$ (*iii*) $\frac{e-1}{e^2}$

7. (*a*) $k = 8/9$ (*b*) (*i*) $1 - \frac{1}{e}$ (*ii*) $\frac{1}{2}$ (*iii*) $1 - \frac{2}{e}$

8. (*i*) $k = 1$ (*ii*) $f_1(x \mid y) = e^{-x}$ (*iii*) e^{-3}

9. (*i*) $f_X(x) = \begin{cases} 4x(1 - x^2), & 0 < x < 1 \\ 0, & \text{otherwise} \end{cases}$, $f_Y(y) = 4y^3, 0 < y < 1$

(*ii*) $f_{X|Y}(x \mid y) = \frac{2x}{y^2}, 0 < x < y, 0 < y < 1$

$f_{Y|X}(y \mid x) = \frac{2y}{1 - x^2}, x < y < 1, 0 < x < 1$

10. (*a*) (*i*) $f_X(x) = \begin{cases} \frac{4}{9}x(4 - x^2), & 1 \le x \le 2 \\ 0, & \text{otherwise} \end{cases}$

(*ii*) $f_{X|Y}(x \mid y) = \frac{2x}{y^2 - 1}, \quad 1 \le x \le y$

$f_Y(y) = \frac{4}{9}y(y^2 - 1), 1 \le y \le 2$

$f_{Y|X}(y \mid x) = \frac{2y}{4 - x^2}, x \le y \le 2.$

(*b*) $A = \frac{4}{4e - 3}$.

11. $f_X(x) = \frac{x^3}{4}, 0 \le x \le 2$, $f_Y(y) = \frac{y^3}{4}, 0 \le y \le 2$; $F_X(x) = \begin{cases} 0, x < 0 \\ x^4/16, 0 \le x \le 2, \\ 1, x > 2 \end{cases}$ $F_Y(y) = \begin{cases} 0, y < 0 \\ y^4/16, 0 \le y \le 2 \\ 1, y > 2 \end{cases}$

12. (*a*) (*i*) $f_X(x) = e^{-x}, x \ge 0$; $f_Y(y) = e^{-y}, y \ge 0$ (*ii*) Yes

(*iii*) $P(X \le 1 \cap Y \le 1) = \left(1 - \frac{1}{e}\right)^2$, $P(X + Y \le 1) = 1 - \frac{2}{e}$

(*b*) Yes; $\frac{1}{e}$.

13.

Y \ X	0	1	2	3	Marginal distribution of Y, P (Y = y)
0	0	1/32	1/8	9/32	7/16
1	1/32	1/16	5/32	5/16	9/16
Marginal distribution of X, P(X = x)	1/32	3/32	9/32	19/32	1

14. Conditional distribution of Y for X = x.

Y \ X	0	1	2
0	0	1/3	2/3
1	2/9	3/9	4/9
2	4/15	5/15	6/15

4.29 MATHEMATICAL EXPECTATION OR EXPECTED VALUE OF A RANDOM VARIABLE

Once we have constructed the probability distribution for a random variable, we often want to compute the mean or expected value of the random variable. The *expected value* of a discrete random variable is a weighted average of all possible values of the random variable, where the weights are the probabilities associated with the corresponding values.

If x, denotes a discrete random variable which can assume the values $x_1, x_2 ..., x_n$ with respective probabilities $p_1, p_2, ..., p_n$ where $p_1 + p_2 + ... + p_n = 1$, the *mathematical expectation* of x or simply the *expectation of* X, denoted by E(x), is defined as

$$E(x) = p_1x_1 + p_2x_2 + ... + p_nx_n = \sum_{i=1}^{n} p_ix_i = \Sigma px, \Sigma p = 1.$$

Let $\phi(x)$ be a function of the variate x so that it takes the values $\phi(x_1), \phi(x_2), ...$ When x takes the values $x_1, x_2, ...$; and if $p_1, p_2, ...,$ be the respective probabilities, then the expected value of the function $\phi(x)$ is defined as

$$E[\phi(x)] = p_1\phi(x_1) + p_2\phi(x_2) + ... + p_n\phi(x_n), \Sigma p = 1$$

If $\phi(x) = x^r$, then

$$E(x^r) = p_1x_1^r + p_2x_2^r + ... + p_nx_n^r.$$

This is defined as the rth moment of the discrete probability distribution about $x = 0$ and is denoted by $\mu r'$. This $\mu r'$ is the expected value of the rth power of the variate.

Also $$\mu r = E(x - \bar{x})^r = \Sigma p_i(x_i - \bar{x})^r$$

In particular $$\mu_1' = E(x) = p_1x_1 + p_2x_2 + ... = \sum_i p_ix_i$$

Here if p_i is replaced by $\frac{f_i}{N}$ where $\Sigma f_i = N$, then

$$E(x) = \frac{\Sigma fx}{N}, \text{ which is the mean.}$$

$\therefore$ E(x) represents the mean.

Just as in the case of frequency, this moment is called the *mean value* of the variate or of the distributions. More generally it is known as the *expected value* or the *expectation* of the variate x. The relation

$$\mu_2 = E[\{x - E(x)\}^2] = E(x^2) - [E(x)]^2.$$

is called the *variance* of the distribution of x and is denoted by *var* (x). The relation between μ_r and μ_r' are the same as for a frequency distribution. But here we have

$$E[x - E(x)] = 0$$

Thus the expected value of the deviation of variate from its mean is zero.

If x is a continuous random variable with probability density function $f(x)$, then

$$E(x) = \int_{-\infty}^{\infty} xf(x)\,dx = \int_{-\infty}^{\infty} x\,dF(x)$$

is expectation of continuous random variable with certain restrictions.

4.30 LAWS OF EXPECTATION

Theorem 1. *If C is a constant then E(C) = C.*

Proof. Since the constant function C, assigns the value C to each value within its domain, we have

$$P(C = C) = 1$$

$$P(C = D) = 0,\ D \neq C$$

therefore $\quad E(C) = C\,.1 + D \times 0 = C.$

Theorem 2. *If a is a constant, then E(aX) = aE(X).*

Proof. Let X take values $x_1, x_2, \ldots$ with probabilities $p_1, p_2, \ldots$

Then $\quad E(aX) = ax_1p_1 + ax_2p_2 + \ldots \qquad \{\because\ P(aX = ax_i) = P(X = x_i)\}$

$$= a(x_1p_1 + x_2p_2 + \ldots) = aE(X).$$

Theorem 3. *Expectation of the sum of two Random Variables.*

If X and Y are two discrete random variables with finite expectations E(X) and E(Y) respectively, then the expectation of their sum exists and is the sum of their expectations *i.e.*,

$$E(X + Y) = E(X) + E(Y).$$

Corollary 1. *If $X_1, X_2, \ldots, X_n$ are n random variables with finite expectations $E(X_1)$, $E(X_2), \ldots, E(X_n)$, then the expectation of their sum exists and is the sum of their expectations. In symbols*

$$E(X_1 + X_2 + \cdots + X_n) = E(X_1) + E(X_2) + \cdots + E(X_n)$$

or

$$E\left(\sum_{i=1}^{n} X_i\right) = \sum_{i=1}^{n} E(X_i)$$

i.e., the sign E and Σ are interchangeable.

Corollary 2. *If a and b are constants, then E(aX + b) = aE(X) + b.*

Proof. $\quad E(aX + b) = E(aX) + E(b), \qquad \{\because\ E(X + Y) = E(X) + E(Y)\}$

$$= aE(X) + b. \qquad \{\because\ E(aX) = aE(X)\}$$

Theorem 4. *Expectation of the Product of two Independent Random Variables.*

If X and Y are two independent discrete random variables with finite expectations, then the expectation of their product exists and is equal to the product of their expectations. In symbols,

$$E(XY) = E(X)\,E(Y).$$

Corollary. *If $X_1, X_2, \ldots, X_k$ are mutually independent random variables with finite expectations, then the expectation of their product exists, and is equal to the product of their expectations. In symbols*

$$E(X_1, X_2, \ldots, X_k) = E(X_1)\,E(X_2) \ldots E(X_k).$$

Theorem 5. *Expectation of the difference of two random variables.*

If X and Y are two random variables with finite expectations E(X) and E(Y) respectively then the expectation of their difference is equal to the difference of their expectations. In symbols

$$E(X - Y) = E(X) - E(Y).$$

In general

$$E(X_1 \pm X_2 \pm, \ldots, \pm X_k) = E(X_1) \pm E(X_2) \pm \ldots \pm E(X_k).$$

ILLUSTRITIVE EXAMPLES

Example 1. *What is the expected value of the number of points that will be obtained in a single throw with an ordinary die? Find variance also.*

Sol. Here the variate is the number of points showing on a die. It assumes the values 1, 2, 3, 4, 5, 6 with probability $\frac{1}{6}$ in each case.

Hence, $$E(x) = p_1x_1 + p_2x_2 + \ldots + p_6x_6 = \frac{1}{6} \cdot 1 + \frac{1}{6} \cdot 2 + \cdots + \frac{1}{6} \cdot 6 = 3.5$$

Also, $$\text{var}(x) = E(x^2) - [E(x)]^2 = \frac{1}{6}(1^2 + 2^2 + \ldots + 6^2) - \left(\frac{7}{2}\right)^2 = \frac{35}{12}.$$

Example 2. *Thirteen cards are drawn simultaneously from a deck of 52. If aces count 1, face cards 10 and others according to denomination, find the expectation of the total score on the 13 cards.*

Sol. Let x_i be the number corresponding to the ith card, then x_i takes the values 1, 2, 3, 4, 5, 6, 7, 8, 9, 10, 10, 10; 10 each having the probability $\frac{1}{13}$. Hence

$$E(x_i) = \frac{1}{13} \cdot 1 + \frac{1}{13} \cdot 2 + \frac{1}{13} \cdot 3 + \cdots + \frac{1}{13} \cdot 9 + \frac{1}{13} \cdot 10 + \frac{1}{13} \cdot 10 + \frac{1}{13} \cdot 10 + \frac{1}{13} \cdot 10$$

$$E(x_i) = \frac{1}{13}(1 + 2 + 3 + \ldots + 9 + 10 + 10 + 10 + 10) = \frac{85}{13}.$$

Example 3. *In four tosses of a coin, let x be the number of heads. Calculate the expected values of x.*

Sol. There will be all the heads, there is only one way

i.e., $$P(x = 4) = \frac{1}{2} \times \frac{1}{2} \times \frac{1}{2} \times \frac{1}{2} = \frac{1}{16}$$

$$P(x = 3) = \frac{{}^4C_3}{16} = \frac{4}{16} = \frac{1}{4}$$

$$P(x = 2) = \frac{{}^4C_2}{16} = \frac{6}{16} = \frac{3}{8}$$

$$P(x = 1) = \frac{{}^4C_1}{16} = \frac{4}{16} = \frac{1}{4}$$

$$P(x = 0) = \frac{{}^4C_0}{16} = \frac{1}{16}.$$

Hence the probability distribution of x is

x:	0	1	2	3	4
P(x):	$\frac{1}{16}$	$\frac{1}{4}$	$\frac{3}{8}$	$\frac{1}{4}$	$\frac{1}{16}$

$$\therefore \quad E(x) = \sum_{x=0}^{4} x(px) = 0 \cdot \frac{1}{16} + 1 \cdot \frac{1}{4} + 2 \cdot \frac{3}{8} + 3 \cdot \frac{1}{4} + 4 \cdot \frac{1}{16} = 2.$$

Example 4. *Find $E(x)$, $E(x^2)$, $E\{(x - \bar{x})^2\}$ for the following probability distribution:*

x:	8	12	16	20	24
$p(x)$:	$\frac{1}{8}$	$\frac{1}{6}$	$\frac{3}{8}$	$\frac{1}{4}$	$\frac{1}{12}$

$$E(x) = \Sigma x \cdot p(x) = 8 \cdot \frac{1}{8} + 12 \cdot \frac{1}{6} + 16 \cdot \frac{3}{8} + 20 \cdot \frac{1}{4} + 24 \cdot \frac{1}{12} = 16.$$

This represents the *mean* of the distribution.

$$E(x^2) = \Sigma x^2 \cdot p(x) = 8^2 \cdot \frac{1}{8} + 12^2 \cdot \frac{1}{6} + 16^2 \cdot \frac{3}{8} + 20^2 \cdot \frac{1}{4} + 24^2 \cdot \frac{1}{12} = 276.$$

This represents the *second moment* about the origin zero.

$$E\{(x - \bar{x})^2\} = \Sigma(x - \bar{x})^2 \, p(x)$$

$$= (8 - 16)^2 \cdot \frac{1}{8} + (12 - 16)^2 \cdot \frac{1}{6} + (16 - 46)^2 \cdot \frac{3}{8} + (20 - 16)^2 \cdot \frac{1}{4} + (24 - 16)^2 \cdot \frac{1}{12} = 20$$

This represents the *variance of the distribution.*

4.31 INDEPENDENT VARIATES

The variates, x and y, are said to be independent, if the probability that either of them will assume a prescribed value does not depend on the value assumed by the other.

3.31.1 Theorem. *The expectation of the product of two independent variates is equal to the product of their expectations i.e., $E(xy) = E(x)E(y)$.*

4.32 COVARIANCE

If x and y are two variates with the respective expected values (or means) x and y, the covariance between x and y is defined as

$$\text{cov}(x, y) = E[(x - \bar{x})(y - \bar{y})].$$

Thus the expected value of product of the deviations of the two variates from their mean is called their *covariance.*

4.33 THEOREM

The covariance of two independent variances is equal to zero.

If x and y are independent variates, then

$$E(x - \bar{x}) = E(x) - E(\bar{x}) = \bar{x} - \bar{x} = 0,$$

$$E(y - \bar{y}) = E(y) - E(\bar{y}) = \bar{y} - \bar{y} = 0.$$

$$\therefore \quad \text{cov}(x, y) = E(x - \bar{x})E(y - \bar{y}) = 0$$

Note. $\text{cov}(x, y) = E\{(x - \bar{x})(y - \bar{y})\} = E(xy - \bar{x}y - \bar{y}x + \bar{x}\bar{y})$

$$= E(xy) - E(\bar{x}y) - E(\bar{y}x) + E(\bar{x}\bar{y})$$
$$= E(xy) - \bar{x}E(y) - \bar{y}E(x) + \bar{x}\bar{y}$$
$$= E(xy) - \bar{x}\bar{y} - \bar{y}\bar{x} + \bar{x}\bar{y}$$
$$= E(xy) - E(x)E(y).$$

4.34 VARIANCE OF *n* VARIATES

If $x_1, x_2, ..., x_n$ be n variates with variances $\sigma_1^2, \sigma_2^2, ..., \sigma_n^2$ to find the variance of u where

$$u = a_1x_1 + a_2x_2 + ... + a_nx_n \text{ where } a\text{'s are constant.}$$
$$E(u) = E(a_1x_1 + a_2x_2 + ... + a_nx_n)$$
$$= E(a_1x_1) + E(a_2x_2) + ... + E(a_nx_n)$$
$$= a_1E(x_1) + a_2E(x_2) + ... + a_nE(x_n)$$
$$\therefore \quad u - E(u) = a_1[x_1 - E(x_1)] + a_2[x_2 - E(x_2)] + \cdots + a_n[x_n - E(x_n)]$$

Squaring both sides, we get

$$[u - E(u)]^2 = a_1^2[x_1 - E(x_1)]^2 + a_2^2[x_2 - E(x_2)]^2 + ...$$
$$+ a_n^2[x_n - E(x_n)]^2 + 2a_1a_2[x_1 - E(x_1)][x_2 - E(x_2)] + ...$$

Taking expected values of both sides, we get

$$\text{var}(u) = a_1^2 \text{var}(x_1) + a_2^2 \text{var}(x_2) + ... + a_n^2 \text{var}(x_n)$$
$$+ 2a_1a_2 \text{cov}(x_1, x_2) + ... + 2a_{n-1}a_n \text{cov}(x_{n-1}, x_n)$$

Cor. 1. If $a_1 = a_2 = 1, a_3 = a_4 = ... = a_n = 0$, we have

$$\text{var}(x_1 + x_2) = \text{var}(x_1) + \text{var}(x_2) + 2\text{cov}(x_1, x_2).$$

Cor. 2. If $a_1 = 1, a_2 = -1, a_3 = a_4 = ... = a_n = 0$, we have

$$\text{var}(x_1 - x_2) = \text{var}(x_1) + \text{var}(x_2) - 2\text{cov}(x_1, x_2).$$

Cor. 3. If $x_1, x_2, ... x_n$ are independent, we have

$$\text{var}(u) = a_1^2 \text{var}(x_1) + a_2^2 \text{var}(x_2) + \cdots + a_n^2 \text{var}(x_n).$$

ILLUSTRATIVE EXAMPLES

Example 1. *Balls are taken one by one out of an urn containing a white and b black balls until the first white ball is drawn. Show that the expectation of the number of black balls preceding the first white ball is* $\frac{b}{a+1}$.

Sol. Let x_1 denote the number of black balls before the first white ball is drawn and x_i denote the number of black balls following the $(i-1)^{\text{th}}$ white balls and preceding the ith white balls.

Let x_{a+1} denote the number of black balls left after the white balls have been drawn.

Then $\quad x_1 + x_2 + \cdots + x_{a+1} = b,$

Hence $\quad E(x_1 + x_2 + \cdots + x_{a+1}) = E(b),$

or $\quad E(x_1) + E(x_2) + \cdots + E(x_{a+1}) = E(b).$

The probability of every system of numbers $x_1, x_2, ..., x_{a+1}$ is the same and is $\frac{a!b!}{(a+b)!}$

Thus, $E(x_1) = E(x_2) = \ldots = E(x_{a+1}) = K$ say.

Therefore, $(a + 1)K = b$ giving $K = \dfrac{b}{a+1}$.

Example 2. *A box contains a white and b black balls ; c balls are drawn. Show that the expectation of the number of white balls drawn is* $\dfrac{ca}{a+b}$.

Sol. Let a variate x_i be defined as follows

$$x_i = x_1, x_2, x_3, \ldots, x_c$$

and in that case

$$x_i = \begin{cases} 1 \text{ if the } i^{\text{th}} \text{ ball drawn is white} \\ 0 \text{ if the } i^{\text{th}} \text{ ball drawn is black} \end{cases}$$

Then the number of white balls, s, among the c drawn balls is given by

$$s = x_1 + x_2 + \cdots + x_c.$$

Now, $P(x_i = 1) = \dfrac{a}{a+b}$, $P(x_i = 0) = \dfrac{b}{a+b}$

$\therefore$ $$E_i = 1 \,.\, \frac{a}{a+b} + 0 \,.\, \frac{b}{a+b} = \frac{a}{a+b}$$

$\therefore$ $$E(S) = \frac{a}{a+b} + \frac{a}{a+b} + \cdots + \frac{a}{a+b}, \; c \text{ terms} = \frac{ca}{a+b}$$

Example 3. *What is the expectation of the number of failures preceding the first success in an infinite series of independent trials with constant probability p of success in a trial?*

Sol. Let x denote the number of failures preceding the first success we find that x takes the values 0, 1, 2, 3, 4, ... Their respective probabilities are $p, (1-p)p, (1-p)^2p, (1-p)^3 p, \ldots$

Since, $1 - p = q$, these are $p, qp, q^2p, q^3p, \ldots$

$$E(x) = 0 \,.\, p + 1.\, qp + 2q^2p + 3q^3p + \ldots = pq\,(1 + 2q + 3q^2 + \ldots)$$

$$= pq(1-q)^{-2} = pq\,\frac{1}{p^2} = \frac{q}{p} = \frac{1-p}{p} = \frac{1}{p} - 1.$$

Example 4. *Show by an example that the mathematical expectation need not be finite.*

Sol. Let x be the discontinuous random variable having values $0!, 1!, 2!, 3!, \ldots, \infty$ and let the probability law for it be $\dfrac{e^{-1}}{x!}$

$$E(x!) = 0!.\,\frac{e^{-1}}{0!} + 1!.\,\frac{e^{-1}}{1!} + 2!.\,\frac{e^{-1}}{2!} + 3!.\,\frac{e^{-1}}{3!} + \cdots\cdots$$

$$= e^{-1} + e^{-1} + e^{-1} + e^{-1} + \cdots\cdots$$

$$= e^{-1}(1 + 1 + \cdots\cdots \infty)$$

$$= e^{-1}\sum_{0}^{\infty} 1, \text{ which is not finite.}$$

Example 5. *Show that if x takes the values $x_n = (-1)^n 2^n n^{-1}$ for n = 1, 2, ... with probabilities $p_n = 2^{-n}$ then E(x) = – log 2.*

Sol. Here

$$E(x) = \sum_{n=1}^{\infty} 2^{-n} . (-1)^n . 2^n n^{-1}$$

$$= \sum_{n=1}^{\infty} \frac{(-1)^n}{n} = -1 + \frac{1}{2} - \frac{1}{3} + \frac{1}{4} - \ldots$$

$$= -\left(1 - \frac{1}{2} + \frac{1}{3} - \frac{1}{4} + \ldots\right) = -\log 2.$$

Example 6. *What is the mathematical expectation of the sum of points on n dice?*

Sol. Denoting by x_i the number of points on the ith dice, the sum of points on n dice is

$$x_1 + x_2 + \cdots + x_n$$

$$\therefore \quad E(x_1 + x_2 + \cdots + x_n) = E(x_1) + E(x_2) + \cdots + E(x_n)$$

But for every single dice $E(x_i) = \frac{7}{2}$, $i = 1, 2, 3, \ldots, n$

$$\therefore \quad E(x_1 + x_2 + \cdots + x_n) = \frac{7}{2} + \frac{7}{2} + \cdots n \text{ terms} = \frac{7}{2} n.$$

Example 7. *Find the excepted value of the product of points on n dice.*

Sol. Denoting by x_i the number of points on the ith dice the product of points on n dice is

$$x_1, x_2, x_3, \ldots, x_n$$

For every single dice $E(x_i) = \frac{7}{2}$, $i = 1, 2, 3, \ldots, n$

$$\therefore \quad E(x_1 x_2 \ldots x_n) = E(x_1) . E(x_2) \ldots E(x_n) = \frac{7}{2} . \frac{7}{2} . \frac{7}{2} \ldots n \text{ factors} = \left(\frac{7}{2}\right)^n.$$

Example 8. *There is a series of n independent trials, where p_i is the probability for i^{th} trial. Show that the mean number of successes is $\sum_{i=1}^{n} p_i$.*

Sol. Let a variable x_i be associated with i^{th} trial, which has the value 1 in case of success and the value 0 in case of a failure. If m be the number of successes in n trials then

$$m = x_1 + x_2 + \ldots + x_n$$

$$\therefore \quad E(m) = E(x_1 + x_2 + \ldots + x_n) = E(x_1) + E(x_2) + \ldots + E(x_n)$$

Since x_i can take the values 1 or 0 with probabilities p_i or $1 - p_i$, we have

$$E(x_i) = 1, p_i + 0\,(1 - p_i) = p_i$$

$$\therefore \quad E(m) = E(x_1) + E(x_2) + \ldots + E(x_n) = p_1 + p_2 + \ldots + p_n.$$

Example 9. *Find the variance of the number of successes in a series of n independent trials in which the probability of success in the ith trial is p_i.*

Sol. Here

$$m = x_1 + x_2 + \ldots + x_n$$

$$\therefore \quad \text{var}(m) = \text{var}(x_1 + x_2 + \ldots + x_n) = \text{var}(x_1) + \text{var}(x_2) + \ldots + \text{var}(x_n)$$

If there is success, $x_i^2 = 1$, if failure, $x_i^2 = 0$

$$\therefore \quad E(x_i^2) = 1 . p_i + 0 . (1 - p_i) = p_i.$$

Here the variance of the number of successes in the ith trial is given by

$$\begin{aligned}\text{var}(x_i) &= \text{E}[x_i - \text{E}(x_i)]^2 \\ &= \text{E}(x_i)^2 - 2x_i\,\text{E}(x_i) + \{\text{E}(x_i)\}^2 = \text{E}(x_i)^2 - \{\text{E}(x_i)\}^2 \\ &= p_i - p_i^2 = p_i(1-p_i) = p_i\,q_i\end{aligned}$$

$$\therefore \qquad \text{var}(m) = \sum_{i=1}^{n} p_i q_i = p_1q_1 + p_2q_2 + \ldots + p_nq_n.$$

Corollary 1. Show that var $(m) \le \frac{1}{4}$.

Since $pq = p(1-p) = p - p^2 - \frac{1}{4} + \frac{1}{4} = \frac{1}{4} - \left(p - \frac{1}{2}\right)^2 \le \frac{1}{4}$, or $\left(p - \frac{1}{2}\right)^2$ is positive.

Corollary 2. If $p_1 = p_2 = \ldots = p_n = p$, then var $(m) = pq + pq + \ldots + pq = npq$.

Example 10. *From a point on the circumference of a circle of radius a, chord is drawn in a random direction. Show that the expected value of the length of the chord is $4a/\pi$ and that the variance of the length is $2a^2(1 - 8/\pi^2)$. Also show that the chance is $\frac{1}{3}$ and the length of the chord will exceed the length of the side of an equilateral triangle inscribed in the circle.*

Sol. Let C be the centre and A be any point. Let AB be any random chord.

Let $\angle$DAB $= \theta$. Then θ lies from $-\frac{\pi}{2}$ to $\frac{\pi}{2}$.

$$d\text{F}(\theta) = f(\theta)\,d\theta = \frac{d\theta}{\frac{1}{2}\pi - \left(-\frac{1}{2}\pi\right)} = \frac{d\theta}{\pi}, \quad -\frac{\pi}{2} \le \theta \le \frac{\pi}{2}$$

If a be the radius, AB $= 2a \cos\theta$

$$\therefore \qquad \text{E(AB)} = \int_{-\pi/2}^{\pi/2} \text{AB}\,.\,f(\theta) = \frac{2a}{\pi}\int_{-\pi/2}^{\pi/2} \cos\theta\,d\theta = \frac{4a}{\pi}.$$

$$\text{E}[(\text{AB})^2] = \int_{-\pi/2}^{\pi/2} (\text{AB})^2\,f(\theta)d\theta = \frac{4a^2}{\pi}\int_{-\pi/2}^{\pi/2} \cos^2\theta\,d\theta = 2a^2.$$

$$\text{var(AB)} = \text{E}[(\text{AB})^2] - \{\text{E(AB)}\}^2 = 2a^2 - 16a^2/\pi^2 = 2a^2(1 - 8/\pi^3).$$

Now as the inscribed equilateral triangle is $a\sqrt{3}$, therefore

$$\begin{aligned}\text{P}(\text{AB} > a\sqrt{3}) &= \text{P}(2a\cos\theta > a\sqrt{3}) = \text{P}(\cos\theta > \sqrt{3}/2) \\ &= \text{P}\left(|\theta| < \frac{1}{6}\pi\right) = \text{P}\left(-\frac{1}{6}\pi < \theta < \frac{1}{6}\pi\right) \\ &= \int_{-\pi/6}^{\pi/2} f|\theta|\,d\theta = \frac{1}{\pi}\int_{-\pi/6}^{\pi/2} 1\,.\,d\theta = \frac{1}{\pi}\,.\,\frac{\pi}{3} = \frac{1}{3}.\end{aligned}$$

TEST YOUR KNOWLEDGE

If x is a variate and a is a constant, show that:

1. $E(a) = a$.
2. $E(ax) = aE(x)$.
3. $\text{var}(ax + b) = a^2 . \text{var}(x)$.
4. $\text{cov}(ax, by) = ab \text{ cov}(x, y)$.
5. $\text{cov}(x + a, y - b) = \text{cov}(x, y)$.
6. $\text{cov}(x, x) = \text{var}(x)$.
7. Prove that:

$$\text{cov}(ax + by, cx + dy) = ac \text{ var}(x) + bd \text{ var}(y) + (ad + bc) \text{ cov}(x, y).$$

8. If x, y, z are random variables each with expectation 10 and variance 1, 4 and 9 respectively; $r(x, y) = 0, r(y, z) = \frac{1}{4}, r(x, z) = \frac{1}{4}$ obtain

   (*a*) $E(x + y - 2z)$ (*b*) $\text{cov}(x + z, y + z)$

   (*c*) $V(x - 2z)$ (*d*) $\text{cov}(3x, 5z)$.

9. Two unbiased dice are thrown. Find the expected value of the sum of numbers of points drawn.
10. A coin is tossed until a head appears. What is the expectation of the number of tosses?
11. In a lottery m tickets are drawn out of n tickets numbered from 1 to n. What is the expectation and the variance of the sum of the numbers drawn?
12. Three urns contain respectively 3 green and 2 white balls, 5 green and 6 white balls and 2 green and 4 white balls. One ball is drawn from each urn. Find the expected number of white balls drawn out.

$$\left[\textbf{Hint. } 0\left(\frac{3}{5}\cdot\frac{5}{11}\cdot\frac{2}{6}\right) + 1\cdot\left(\frac{2}{5}+\frac{2}{5}\cdot\frac{2}{6}+\frac{3}{5}\cdot\frac{5}{11}\frac{4}{6}+\frac{3}{5}\cdot\frac{6}{11}\cdot\frac{2}{6}\right) + 2\left(\frac{2}{5}\cdot\frac{6}{11}\frac{2}{6}+\frac{2}{5}\cdot\frac{5}{11}\frac{4}{6}+\frac{3}{5}\cdot\frac{6}{11}\cdot\frac{4}{6}\right) + 3\cdot\left(\frac{2}{5}\cdot\frac{6}{11}\cdot\frac{4}{6}\right) = \frac{266}{165}\right]$$

13. Show that the expectation of the number of failures preceding the first success in an indefinite series of independent trials with constant probability p of success, is

$$qp + 2q^2p + 3q^3p + \cdots = \frac{pq}{(1-q)^2} = \frac{q}{p} = \frac{1-p}{p}.$$

14. Define mathematical expectation. Prove that mathematical expectation of the sum of two discrete random variables is equal to the sum of their mathematical expectations.
15. x denotes the profit that a man can make in business. He may earn ₹2,800 with probability $\frac{1}{2}$, he may lose ₹5,000 with probability $\frac{3}{10}$ and he may neither lose nor gain with probability $\frac{1}{5}$. Show that the mathematical expectation is – 100.
16. If x is a random variable for which $E(x) = 10$ and $\text{var}(x) = 25$. Find the positive values of a and b such that $y = ax - b$ has expectation 0 and variance 1.

Answers

9. 7 **10.** 2 **12.** $\frac{266}{165}$

11. $\frac{1}{2} m(n + 1); \frac{1}{12} m(n + 1)(n - 2)$ **16.** $a = \frac{1}{5}, b = 2$.

4.35 THEORETICAL PROBABILITY DISTRIBUTIONS

Generally, frequency distribution are formed from the observed or experimental data. However, frequency distribution of certain populations can be deduced mathematically by fitting theoretical probability distribution under certain assumptions.

Frequency distributions can be classified under two heads:

(*i*) Observed Frequency Distributions.

(*ii*) Theoretical or Expected Frequency Distributions.

Observed frequency distributions are based on actual observation and experimentation. If certain hypothesis is assumed, it is sometimes possible to derive mathematically what the frequency distribution of certain universe should be. Such distributions are called **Theoretical Distributions**.

Theoretical probability distributions are of two types:

(*i*) Discrete probability distribution. Binomial, poisson, geometric, negative binomial, hypergeometric, multinomial, multivariate hypergeometric distributions.

(*ii*) Continuous probability distributions.

Uniform, normal Gamma, exponential, χ^2, Beta, bivariate normal, t, F-distributions.

Here, we will study three important theoretical probability distributions:

1. Binomial Distribution (or Bernoulli's Distribution)
2. Poisson's Distribution
3. Normal Distribution.

4.36 BINOMIAL PROBABILITY DISTRIBUTION *(A.K.T.U. 2018)*

It was discovered by a Swiss Mathematician Jacob James Bernoulli in the year 1700.

This distribution is concerned with trials of a repretitive nature in which only the occurrence or non-occurrence, success or failure, acceptance or rejection, yes or no of a particular event is of interest.

For convenience, we shall call the occurrence of the event 'a success' and its non-occurrence 'a failure'.

Let there be n independent trials in an experiment. Let a random variable X denote the number of successes in these n trials. Let p be the probability of a success and q that of a failure in a single trial so that $p + q = 1$. Let the trials be independent and p be constant for every trial.

Let us find the probability of *r* successes in *n* trials.

r successes can be obtained in n trials in ${}^nC_r$ ways.

$$\therefore \quad P(X = r) = {}^nC_r P(\underbrace{S\,S\,S\ldots\ldots S}_{r \text{ times}} \quad \underbrace{F\,F\,F\ldots\ldots F}_{(n-r) \text{ times}})$$

$$= {}^nC_r \underbrace{P(S)\,P(S)\ldots\ldots P(S)}_{r \text{ factors}} \quad \underbrace{P(F)\,P(F)\ldots\ldots P(F)}_{(n-r) \text{ factors}}$$

$$= {}^nC_r \underbrace{p\,p\,p\ldots\ldots p}_{r \text{ factors}} \quad \underbrace{q\,q\,q\ldots\ldots q}_{(n-r) \text{ factors}}$$

$$= {}^nC_r p^r q^{n-r} \qquad \ldots(1)$$

Hence $\boxed{\mathbf{P(X = r) = {}^nC_r\, p^r q^{n-r}}}$ where $\boldsymbol{p + q = 1}$ and $\boldsymbol{r = 0, 1, 2, ..., n}$.

The distribution (1) is called the *binomial probability distribution* and X is called the *binomial variate.*

Note 1. $P(X = r)$ is usually written as $P(r)$.

Note 2. The successive probabilities $P(r)$ in (1) for $r = 0, 1, 2, ..., n$ are

$$^nC_0 q^n, {}^nC_1 q^{n-1} p, {}^nC_2 q^{n-2} p^2, ..., {}^nC_n p^n$$

which are the successive terms of the binomial expansion of $(q + p)^n$. That is why this distribution is called "binomial" distribution.

Note 3. n and p occurring in the binomial distribution are called the *parameters* of the distribution.

Note 4. In a binomial distribution:

(*i*) n, the number of trials is finite.

(*ii*) each trial has only two possible outcomes usually called success and failure.

(*iii*) all the trials are independent.

(*iv*) p (and hence q) is constant for all the trials.

4.37 RECURRENCE OR RECURSION FORMULA FOR THE BINOMIAL DISTRIBUTION

In a binomial distribution,

$$P(r) = {}^nC_r\, q^{n-r} p^r = \frac{n!}{(n-r)!\,r!}\, q^{n-r} p^r$$

$$P(r+1) = {}^nC_{r+1}\, q^{n-r-1} p^{r+1} = \frac{n!}{(n-r-1)!\,(r+1)!}\, q^{n-r-1} p^{r+1}$$

$$\therefore \quad \frac{P(r+1)}{P(r)} = \frac{(n-r)!}{(n-r-1)!} \times \frac{r!}{(r+1)!} \times \frac{p}{q}$$

$$= \frac{(n-r)\times(n-r-1)!}{(n-r-1)!} \times \frac{r!}{(r+1)\times r!} \times \frac{p}{q} = \left(\frac{n-r}{r+1}\right) \cdot \frac{p}{q}$$

$$\Rightarrow \quad \boxed{P(r+1) = \frac{n-r}{r+1} \cdot \frac{p}{q}\, P(r)}$$

which is the required recurrence formula. Applying this formula successively, we can find P(1), P(2), P(3), ..., if P(0) is known.

4.38 MEAN AND VARIANCE OF THE BINOMIAL DISTRIBUTION [*A.K.T.U. 2018*]

For the binomial distribution, $P(r) = {}^nC_r\, q^{n-r} p^r$

$$\text{Mean } \mu = \sum_{r=0}^{n} r P(r) = \sum_{r=0}^{n} r \,.\, {}^nC_r q^{n-r} p^r$$

$$= 0 + 1 \,.\, {}^nC_1\, q^{n-1} p + 2 \,.\, {}^nC_2\, q^{n-2} p^2 + 3 \,.\, {}^nC_3\, q^{n-3} p^3 + ... + n \,.\, {}^nC_n\, p^n$$

$$= nq^{n-1} p + 2 \,.\, \frac{n(n-1)}{2 \cdot 1}\, q^{n-2} p^2 + 3 \,.\, \frac{n(n-1)(n-2)}{3 \cdot 2 \cdot 1}\, q^{n-3} p^3 + ... + n.p^n$$

$$= nq^{n-1}p + n(n-1)\,q^{n-2}p^2 + \frac{n(n-1)(n-2)}{2.1}\,q^{n-3}p^3 + \ldots\ldots + np^n$$
$$= np[{}^{n-1}C_0q^{n-1} + {}^{n-1}C_1q^{n-2}p + {}^{n-1}C_2q^{n-3}p^2 + \ldots + {}^{n-1}C_{n-1}p^{n-1}]$$
$$= np(q+p)^{n-1} = np \qquad (\because\ p+q=1)$$

Hence mean of the binomial distribution is *np*.

$$\text{Variance } \sigma^2 = \sum_{r=0}^{n} r^2 P(r) - \mu^2 = \sum_{r=0}^{n} [r + r(r-1)]\, P(r) - \mu^2$$

$$= \sum_{r=0}^{n} rP(r) + \sum_{r=0}^{n} r(r-1)\, P(r) - \mu^2 = \mu + \sum_{r=2}^{n} r(r-1)\, {}^nC_r q^{n-r} p^r - \mu^2$$

(since the contribution due to $r = 0$ and $r = 1$ is zero)

$$= \mu + [2\cdot 1 \cdot {}^nC_2q^{n-2}p^2 + 3\cdot 2\cdot {}^nC_3q^{n-3}p^3 + \ldots + n(n-1)\,{}^nC_np^n] - \mu^2$$

$$= \mu + \left[2\cdot 1\cdot \frac{n(n-1)}{2.1}\,q^{n-2}p^2 + 3\cdot 2\cdot \frac{n(n-1)(n-2)}{3.2.1}\,q^{n-3}p^3 + \ldots + n(n-1)p^n\right] - \mu^2$$

$$= \mu + [n(n-1)q^{n-2}p^2 + n(n-1)(n-2)q^{n-3}p^3 + \ldots + n(n-1)p^n] - \mu^2$$
$$= \mu + n(n-1)\,p^2[q^{n-2} + (n-2)q^{n-3}p + \ldots + p^{n-2}] - \mu^2$$
$$= \mu + n\,(n-1)p^2\,[{}^{n-2}C_0q^{n-2} + {}^{n-2}C_1q^{n-3}p + \ldots + {}^{n-2}C_{n-2}p^{n-2}] - \mu^2$$
$$= \mu + n(n-1)p^2(q+p)^{n-2} - \mu^2 = \mu + n(n-1)p^2 - \mu^2 \qquad [\because\ q+p=1]$$
$$= np + n(n-1)p^2 - n^2p^2 = np[1-p] = npq. \qquad [\because\ \mu = np]$$

Hence the variance of the binomial distribution is *npq*.

Standard deviation of the binomial distribution is $\sqrt{npq}$.

4.39 MOMENT GENERATING FUNCTION OF BINOMIAL DISTRIBUTION [A.K.T.U. 2018]

1. About origin

$$M_x(t) = E(e^{tx}) = \sum_{x=0}^{n} e^{tx}\, {}^nC_x\, p^x\, q^{n-x} = \sum_{x=0}^{n} {}^nC_x\, (pe^t)^x\, q^{n-x} = (q + pe^t)^n$$

2. About mean [G.B.T.U. 2012, U.P.T.U. 2015]

$$M_{x-np}(t) = E[e^{t(x-np)}]$$
$$= e^{-npt}\, E(e^{tx}) = e^{-npt}\, M_x(t) = e^{-npt}\,(q + pe^t)^n$$
$$= (qe^{-pt} + pe^{t-pt})^n = (qe^{-pt} + pe^{qt})^n \qquad |\because\ 1-p=q$$

4.40 MOMENTS ABOUT MEAN OF BINOMIAL DISTRIBUTION

$$M_{x-np}(t) = (qe^{-pt} + pe^{qt})^n$$

$$= \left[q\left(1 - pt + \frac{p^2t^2}{2!} - \frac{p^3t^3}{3!} + \ldots\right) + p\left(1 + qt + \frac{q^2t^2}{2!} + \frac{q^3t^3}{3!} + \ldots\right)\right]^n$$

$$= \left[(q+p) + \frac{t^2}{2!}\,pq\,(q+p) + \frac{t^3}{3!}\,pq\,(q^2 - p^2) + \frac{t^4}{4!}\,pq\,(q^3 + p^3) + \ldots\right]^n$$

$$= \left[1 + \left\{\frac{t^2}{2!} \cdot pq + \frac{t^3}{3!} pq (q-p) + \frac{t^4}{4!} qp (1 - 3pq) + \ldots\right\}\right]^n$$

$$= \left[1 + {}^nC_1 \left\{\frac{t^2}{2!} \cdot pq + \frac{t^3}{3!} pq (q-p) + \frac{t^4}{4!} pq (1 - 3pq) + \ldots\right\}\right.$$

$$\left. + {}^nC_2 \left\{\frac{t^2}{2!} \cdot pq + \frac{t^3}{3!} pq (q-p) + \ldots\right\}^2 + \ldots\right]$$

Now,

$$\mu_2 = \text{coefficient of } \frac{t^2}{2!} = npq$$

$$\mu_3 = \text{coefficient of } \frac{t^3}{3!} = npq\,(q-p)$$

$$\mu_4 = \text{coefficient of } \frac{t^4}{4!} = npq\,(1 - 3\,pq) + 3n(n-1)\,p^2q^2$$

$$= 3n^2p^2q^2 + npq\,(1 - 6pq)$$

Hence,

$$\beta_1 = \frac{\mu_3^2}{\mu_2^3} = \frac{(q-p)^2}{npq} = \frac{(1-2p)^2}{npq}$$

$$\therefore \quad \gamma_1 = \frac{1-2p}{\sqrt{npq}}$$

and

$$\beta_2 = \frac{\mu_4}{\mu_2^2} = 3 + \frac{1-6pq}{npq}$$

$$\therefore \quad \gamma_2 = \frac{1-6pq}{npq}$$

Note 1. $\gamma_1 = \dfrac{1-2p}{\sqrt{npq}}$ gives a **measure of skewness** of the binomial distribution. If $p < \frac{1}{2}$, skewness is positive, if $p > \frac{1}{2}$, skewness is negative and if $p = \frac{1}{2}$, it is zero.

$\beta_2 = 3 + \dfrac{1-6pq}{npq}$ gives a **measure of the kurtosis** of the binomial distribution.

Note 2. If n independent trials constitute one experiment and this experiment is repeated N times then the frequency of r successes is $N \,.\, {}^nC_r\, p^r\, q^{n-r}$.

4.41 APPLICATIONS OF BINOMIAL DISTRIBUTION

1. In problems concerning no. of defectives in a sample production line.
2. In estimation of reliability of systems.

3. No. of rounds fired from a gun hitting a target.
4. In Radar detection.

ILLUSTRATIVE EXAMPLES

Example 1. (*i*) *Comment on the following statement:*

For a Binomial distribution, mean is 6 and variance is 9.

(*ii*) *A die is tossed thrice. A success is getting 1 or 6 on a toss. Find the mean and variance of the number of success.*

Sol. (*i*) $$\mu = np = 6 \quad \text{...(1)}$$

$$\sigma^2 = npq = 9 \quad \text{...(2)}$$

Dividing (2) by (1), we get

$$q = \frac{9}{6} = 1.5$$

which is impossible as $0 \le q \le 1$

$\therefore$ The above statement is **False.**

(*ii*) Prob. of getting success (1 or 6) on a toss $= \frac{2}{6} = \frac{1}{3} = p$

$$\therefore \quad q = 1 - \frac{1}{3} = \frac{2}{3}$$

No. of tosses of a die, $n = 3$

(*i*) Mean $= np = 3\left(\frac{1}{3}\right) = 1.$ (*ii*) Variance $= npq = (3)\left(\frac{1}{3}\right)\left(\frac{2}{3}\right) = \frac{2}{3}.$

Example 2. *If 10% of the bolts produced by a machine are defective, determine the probability that out of 10 bolts choosen at random*

(*i*) *1* (*ii*) *None* (*iii*) *at most 2 bolts will be defective.*

Sol. Here, $p(\text{defective}) = \frac{10}{100} = \frac{1}{10}$ (given)

$$\therefore \quad q(\text{non-defective}) = 1 - \frac{1}{10} = \frac{9}{10}$$

Also, $n = 10$, (n is no. of bolts chosen). (given)

The probability of r defective bolts out of n bolts chosen at random is given by

$$P(r) = {}^nC_r p^r q^{n-r} \quad \text{...(1)}$$

(*i*) Here $r = 1$,

$$\therefore \quad P(1) = {}^{10}C_1 \left(\frac{1}{10}\right)^1 \left(\frac{9}{10}\right)^{10-1} \quad \text{| Using (1)}$$

$$= 10\left(\frac{1}{10}\right)\left(\frac{9}{10}\right)^9 = (.9)^9 = 0.3874 \quad \text{...(2)}$$

(*ii*) Here $r = 0$

$$\therefore \quad P(0) = {}^{10}C_0 \left(\frac{1}{10}\right)^0 \left(\frac{9}{10}\right)^{10-0} = \left(\frac{9}{10}\right)^{10} = 0.3486 \quad \text{...(3) | Using (1)}$$

(*iii*) Prob. that at most 2 bolts will be defective $= P(r \le 2) = P(0) + P(1) + P(2)$ ...(4)

$$\text{Now,} \quad P(2) = {}^{10}C_2 \left(\frac{1}{10}\right)^2 \left(\frac{9}{10}\right)^{10-2} \quad \text{| Using (1)}$$

$$= 45\left(\frac{1}{100}\right)(0.43046) = 0.1937$$

$\therefore$ From (4), Required Probability $= P(0) + P(1) + P(2)$

$= 0.3486 + 0.3874 + 0.1937 = 0.9297.$

Example 3. *A binomial variable X satisfies the relation 9P (X = 4) = P(X = 2) when n = 6. Find the value of the parameter p and P(X = 1).*

Sol. We know that

$$P(X = r) = {}^nC_r\, p^r\, q^{n-r} \quad \text{...(1)}$$

$$\therefore \quad P(X = 4) = {}^6C_4\, p^4 q^2 = 15p^4q^2$$

$$\text{and} \quad P(X = 2) = {}^6C_2 p^2 q^4 = 15\, p^2\, q^4 \quad \text{| Since } n = 6$$

The given relation is

$$9\, P(X = 4) = P(X = 2) \quad \Rightarrow \quad 9(15p^4\, q^2) = 15p^2q^4$$

$$\Rightarrow \quad 9p^2 = q^2 = (1 - p)^2 \quad | \because \; p + q = 1$$

$$\Rightarrow \quad 9p^2 = 1 + p^2 - 2p$$

$$\Rightarrow \quad 8p^2 + 2p - 1 = 0 \quad \Rightarrow \quad (4p - 1)(2p + 1) = 0$$

$$\therefore \quad \boxed{p = \frac{1}{4}} \quad | \because \; p \text{ cannot be negative}$$

$$\text{Now,} \quad P(X = 1) = {}^6C_1 \left(\frac{1}{4}\right)^1 \left(\frac{3}{4}\right)^5 = .3559. \quad \left| \because \; q = 1 - p = \frac{3}{4}\right.$$

Example 4. *Fit a binomial distribution to the following frequency data:*

x :	*0*	*1*	*2*	*3*	*4*
f :	*30*	*62*	*46*	*10*	*2*

Sol. The table is as follows:

x	f	fx
0	30	0
1	62	62
2	46	92
3	10	30
4	2	8
	$\Sigma f = 150$	$\Sigma fx = 192$

Mean of observations $= \frac{\Sigma fx}{\Sigma f} = \frac{192}{150} = 1.28$

$\Rightarrow \quad np = 1.28$

$\Rightarrow \quad 4p = 1.28$ (*n* is no. of trials)

$\Rightarrow \quad p = 0.32$

$\therefore \quad q = 1 - p = 1 - 0.32 = 0.68$

Also, $\quad N = 150$ $\quad | \because \quad N = \Sigma f$

Hence the binomial distribution is $= N(q+p)^n = 150\,(0.68 + 0.32)^4$.

Example 5. *A student is given a true-false examination with 8 questions. If he corrects at least 7 questions, he passes the examination. Find the probability that he will pass given that he guesses all questions.*

Sol. Here, n = no. of questions asked = 8

$$p = \frac{1}{2}, q = \frac{1}{2}$$ | Since the question can either be true or false

Probability that he will pass

$$= P(r \geq 7) = P(7) + P(8)$$

$$= {}^8C_7 \left(\frac{1}{2}\right)^7 \left(\frac{1}{2}\right)^{8-7} + {}^8C_8 \left(\frac{1}{2}\right)^8 \left(\frac{1}{2}\right)^{8-8} = 8\left(\frac{1}{2}\right)^7 \left(\frac{1}{2}\right)^1 + 1 . \left(\frac{1}{2}\right)^8$$

$$= \left(\frac{1}{2}\right)^8 (8+1) = \frac{9}{256} = .03516.$$

Example 6. *During war, 1 ship out of 9 was sunk on an average in making a certain voyage. What was the probability that exactly 3 out of a convoy of 6 ships would arrive safely?*

Sol. p, the probability of a ship arriving safely $= 1 - \frac{1}{9} = \frac{8}{9}$; $\quad q = \frac{1}{9}, n = 6$

The probability that exactly 3 ships arrive safely $= P(r = 3) = {}^6C_3 \left(\frac{1}{9}\right)^3 \left(\frac{8}{9}\right)^3 = \frac{10240}{9^6}$.

Example 7. *A policeman fires 6 bullets on a dacoit. The probability that the dacoit will be killed by a bullet is 0.6. What is the probability that dacoit is still alive?*

Sol. Here $\quad n$ = no. of bullets fired = 6, $p = 0.6$, $\quad q = 1 - p = 0.4$

Probability that dacoit is still alive (not killed)

$$= P(r = 0) = {}^nC_0\, p^0\, q^{n-0} = {}^6C_0\,(.6)^0\,(.4)^6 = (.4)^6 = .004096.$$

Example 8. *If the probability of hitting a target is 10% and 10 shots are fired independently. What is the probability that the target will be hit at least once?* (A.K.T.U. 2019)

Sol. Here, $p = \frac{10}{100} = \frac{1}{10}, q = 1 - p = 1 - \frac{1}{10} = \frac{9}{10}, n = 10$

Probability that the target will be hit at least once

$$= P(r \geq 1) = 1 - P(r = 0)$$

$$= 1 - [{}^nC_0\, p^0\, q^n] = 1 - \left[{}^{10}C_0 \left(\frac{1}{10}\right)^0 \left(\frac{9}{10}\right)^{10}\right] = 0.6513.$$

Example 9. *Out of 800 families with 4 children each, how many families would be expected to have (i) 2 boys and 2 girls (ii) at least one boy (iii) no girl (iv) atmost two girls? Assume equal probabilities for boys and girls.* (U.P.T.U. 2014)

Sol. Since probabilities for boys and girls are equal,

$$p = \text{probability of having a boy} = \frac{1}{2};\ q = \text{probability of having a girl} = \frac{1}{2}$$

$$n = 4 \qquad N = 800$$

(*i*) The expected number of families having 2 boys and 2 girls

$$= 800\ {}^4C_2 \left(\frac{1}{2}\right)^2 \left(\frac{1}{2}\right)^2 = 800 \times 6 \times \frac{1}{16} = 300.$$

(*ii*) The expected number of families having *at least one boy*

$$= 800\left[{}^4C_1\left(\frac{1}{2}\right)^3\left(\frac{1}{2}\right) + {}^4C_2\left(\frac{1}{2}\right)^2\left(\frac{1}{2}\right)^2 + {}^4C_3\left(\frac{1}{2}\right)\left(\frac{1}{2}\right)^3 + {}^4C_4\left(\frac{1}{2}\right)^4\right]$$

$$= 800 \times \frac{1}{16}\ [4 + 6 + 4 + 1] = 750$$

(*iii*) The expected number of families having no girl *i.e.*, having 4 boys

$$= 800\ .\ {}^4C_4\left(\frac{1}{2}\right)^4 = 50.$$

(*iv*) The expected number of families having *atmost two girls i.e.*, having *at least 2 boy*

$$= 800\left[{}^4C_2\left(\frac{1}{2}\right)^2\left(\frac{1}{2}\right)^2 + {}^4C_3\left(\frac{1}{2}\right)\left(\frac{1}{2}\right)^3 + {}^4C_4\left(\frac{1}{2}\right)^4\right] = 800 \times \frac{1}{16}\ [6 + 4 + 1] = 550.$$

Example 10. *Six dice are thrown 729 times. How many times do you expect at least three dice to show a five or six?*

Sol. p = the chance of getting 5 or 6 with one die $= \frac{2}{6} = \frac{1}{3}$

$$q = 1 - \frac{1}{3} = \frac{2}{3},\ n = 6,\ N = 729$$

since dice are in sets of 6 and there are 729 sets.

The expected number of times *at least three* dice showing five or six

$$= N\ .\ P(r \geq 3)$$

$$= 729\ [P(3) + P(4) + P(5) + P(6)]$$

$$= 729\left[{}^6C_3\left(\frac{2}{3}\right)^3\left(\frac{1}{3}\right)^3 + {}^6C_4\left(\frac{2}{3}\right)^2\left(\frac{1}{3}\right)^4 + {}^6C_5\left(\frac{2}{3}\right)\left(\frac{1}{3}\right)^5 + {}^6C_6\left(\frac{1}{3}\right)^6\right]$$

$$= \frac{729}{3^6}\ [160 + 60 + 12 + 1] = 233.$$

Example 11. *The probability of a man hitting a target is $\frac{1}{3}$. How many times must he fire so that the probability of his hitting the target at least once is more than 90%?*

Sol. $$p = \frac{1}{3}$$

The probability of not hitting the target in n trials is q^n.

Therefore, to find the smallest n for which the probability of hitting at least once is more than 90%, we have

$$1 - q^n > 0.9$$

$$\Rightarrow \quad 1 - \left(\frac{2}{3}\right)^n > 0.9$$

$$\Rightarrow \quad \left(\frac{2}{3}\right)^n < 0.1$$

The smallest n for which the above inequality holds true is 6 hence he must fire 6 times.

Example 12. *In a bombing action, there is 50% chance that any bomb will strike the target. Two direct hits are needed to destroy the target completely. How many bombs are required to be dropped to give a 99% chance or better of completely destroying the target?*

Sol. We have, $p = \frac{50}{100} = \frac{1}{2}$

Since the probability must be greater than 0.99, if n bombs are dropped, we have

$$^nC_2\left(\frac{1}{2}\right)^n + {^nC_3}\left(\frac{1}{2}\right)^n + {^nC_4}\left(\frac{1}{2}\right)^n + \cdots\cdots + {^nC_n}\left(\frac{1}{2}\right)^n \geq 0.99$$

$$\Rightarrow \quad \left(\frac{1}{2}\right)^n [{^nC_2} + {^nC_3} + {^nC_4} + \cdots\cdots + {^nC_n}] \geq 0.99$$

$$\frac{2^n - n - 1}{2^n} \geq 0.99$$

$$\Rightarrow \quad 1 - \frac{1+n}{2^n} \geq 0.99$$

$$\Rightarrow \quad \frac{1+n}{2^n} \leq 0.01$$

$$\Rightarrow \quad 2^n\ {}^3\ 100n + 100$$

By trial, $n = 11$ satisfies the inequality.

Hence 11 bombs are required to be dropped.

TEST YOUR KNOWLDGE

1. (*i*) Ten coins are tossed simultaneously. Find the probability of getting at least seven heads.

   (*ii*) A die is thrown five times. If getting an odd number is a success, find the probability of getting at least four successes. (*M.T.U. 2012*)

2. (*a*) The probability of any ship of a company being destroyed on a certain voyage is 0.02. The company owns 6 ships for the voyage. What is the probability of :

   (*i*) losing one ship (*ii*) losing atmost two ships (*iii*) losing none?

   (*b*) Assume that on the average one telephone number out of fifteen called between 2 P.M. and 3 P.M. on week-days is busy. What is the probability that if 6 randomly selected telephone numbers are called (*i*) not more than 3 (*ii*) at least 3 of them will be busy?

3. (*i*) The incidence of occupational disease in an industry is such that the workers have a 20% chance of suffering from it. What is the probability that out of six workers chosen at random, four or more will suffer from the disease?

(*ii*) The probability that a man aged 60 will live to be 70 is 0.65. What is the probability that out of 10 men, now 60, at least 7 will live to be 70?

4. (*i*) If the mean of a binomial distribution is 3 and the variance is $\frac{3}{2}$, find the probability of obtaining at least 4 successes.

(*ii*) In a binomial distribution, for $n = 5$ if $P(x = 1) = 0.4096$ and $P(x = 2) = 0.2048$, then find the value of p.

(*iii*) The sum and product of the mean and variance of a binomial distribution are $\frac{25}{3}$ and $\frac{50}{3}$ respectively. Find the distribution.

(*iv*) If the probability of a defective bolt is 0.1, find (*a*) The mean (*b*) The standard deviation for the distribution in a total of 400 bolts.

(*v*) If the moment generating function of a random variable X is $\left(\frac{1}{3}+\frac{2}{3}e^t\right)^5$, find $P(X = 2)$.

5. (*a*) The probability that a bomb dropped from a plane will strike the target is $\frac{1}{5}$. If six bombs are dropped, find the probability that (*i*) exactly two will strike the target, (*ii*) at least two will strike the target.

(*b*) Four persons in a group of 20 are graduates. If 4 persons are selected at random from 20, find the probability that

(*i*) all are graduates (*ii*) at least one is a graduate.

6. A bag contains 5 white, 7 red and 8 black balls. If four balls are drawn, one by one, with replacement, what is the probability that

(*i*) none is white (*ii*) all are white

(*iii*) at least one is white (*iv*) only 2 are white?

7. (*i*) In a hurdle race, a player has to cross 10 hurdles. The probability that he will clear each hurdle is $\frac{5}{6}$. What is the probability that he will knock down fewer than 2 hurdles ?

(*ii*) If on an average one ship in every ten is wrecked, find the probability that out of 5 ships expected to arrive, 4 at least will arrive safely.

8. The prob. that a bulb produced by a factory will fuse after a use of 150 days is 0.05. Find the prob. that out of 5 such bulbs [*M.T.U. (B. Pharma) 2011*]

(*i*) None (*ii*) Atmost one

(*iii*) More than one (*iv*) At least one will fuse after 150 days of use.

9. Five cards are drawn successively with replacement from a well-shuffled deck of 52 cards. What is the probability that

(*i*) all the five cards are spades (*ii*) only three are spades? (*iii*) none is spade?

10. Manish takes a step forward with probability 0.4 and backward with probability 0.6. Find the probability that at the end of 11 steps, he is one step away from the starting point.

[**Hint.** Manish will take either 6 steps forward and 5 backward or 5 steps forward and 6 backward].

11. (*a*) In 800 families with 5 children each, how many families would be expected to have (*i*) 3 boys and 2 girls, (*ii*) 2 boys and 3 girls, (*iii*) no girl (*iv*) at the most two girls. (Assume probabilities for boys and girls to be equal.) (*A.K.T.U. 2017*)

(*b*) Out of 800 families with 5 children each, how many would you expect to have (*i*) 3 boys (*ii*) 5 girls (*iii*) either 2 or 3 boys? Assume equal probabilities for boys and girls.

(*M.T.U. 2012*)

(*c*) Out of 320 families with 5 children each, what percentage would be expected to have (*i*) 2 boys and 3 girls (*ii*) at least one boy? Assume equal probability for boys and girls.

(*G.B.T.U. 2011*)

12. In 100 sets of ten tosses of an unbiased coin, in how many cases do you expect to get

(*i*) 7 heads and 3 tails (*ii*) at least 7 heads?

13. The following data are the number of seeds germinating out of 10 on damp filter for 80 sets of seeds. Fit a binomial distribution to this data:

x:	0	1	2	3	4	5	6	7	8	9	10	Total
f:	6	20	28	12	8	6	0	0	0	0	0	80

[**Hint.** Here $n = 10$, N = 80, Mean $= \frac{\Sigma fx}{\Sigma f} = 2.175 \quad \therefore \quad np = 2.175$ etc.]

14. Fit a binomial distribution for the following data and compare the theoretical frequencies with the actual ones :

x:	0	1	2	3	4	5
f:	2	14	20	34	22	8

15. Fit a binomial distribution to the data given in the following table:

(*i*)

x:	0	1	2	3	4
f:	24	41	28	5	2

(*M.T.U. 2012*)

(*ii*)

x:	0	1	3	4
f:	28	62	10	4

(*U.K.T.U. 2011*)

16. (*i*) Assuming half the population of a town consumes chocolates so that the chance of an individual being consumer is $\frac{1}{2}$ and that 100 investigators each take 10 individuals to see whether they are consumers, how many investigators would you expect to report that 3 people or less were consumers?

(*ii*) Assuming that 20% of the population of a city are literate, so that the chance of an individual being literate is $\frac{1}{5}$ and assuming that 100 investigators each take 10 individuals to see whether they are literate, how many investigators would you expect to report 3 or less were literate?

17. Following results were obtained when 100 batches of seeds were allowed to germinate on damp filter paper in a laboratory : $\beta_1 = \frac{1}{15}, \beta_2 = \frac{89}{30}$. Determine the Binomial distribution. Calculate the expected frequency for $x = 8$ assuming $p > q$.

18. A coffee connoisseur claims that he can distinguish between a cup of instant coffee and a cup of percolator coffee 75% of the time. It is agreed that his claim will be accepted if he correctly identifies at least 5 of the 6 cups. Find his chances of having the claim (*i*) accepted (*ii*) rejected when he does have the ability he claims.

19. A multiple-choice test consists of 8 questions with 3 answers to each question of which only one is correct. A student answers each question by rolling a balanced die and checking the first answer if he gets 1 or 2, the second answer if he gets 3 or 4 and the third answer if he gets 5 or 6. To get a distinction, the student must secure at least 75% correct answers. If there is no negative marking, what is the probability that the student secures a distinction?

20. An irregular six-faced die is thrown and the expectation that in 10 throws, it will give five even numbers is twice the expectation that it will give four even numbers. How many times in 10,000 sets of 10 throws each, would you expect it to give no even number?

Answers

1. (*i*) $\frac{11}{64}$ (*ii*) $\frac{3}{16}$

2. (*a*) (*i*) 0.1085, (*ii*) 0.9997, (*iii*) 0.8858 (*b*) (*i*) 0.9997 (*ii*) 0.005

3. (*i*) $\frac{53}{3125}$ (*ii*) 0.5137

4. (*i*) $\frac{11}{32}$, (*ii*) $\frac{1}{5}$, (*iii*) $\left(\frac{2}{3}+\frac{1}{3}\right)^{15}$

(*iv*) (*a*) 40 (*b*) 6 (*v*) 0.1646

5. (*a*) (*i*) 0.246 (*ii*) 0.345 (*b*) (*i*) 0.0016 (*ii*) 0.5904

6. (*i*) $\frac{81}{256}$, (*ii*) $\frac{1}{256}$, (*iii*) $\frac{175}{256}$, (*iv*) $\frac{27}{128}$ **7.** (*i*) $\frac{5}{2}\left(\frac{5}{6}\right)^9$, (*ii*) 0.91854

8. (*i*) $\left(\frac{19}{20}\right)^5$ (*ii*) $\frac{6}{5}\left(\frac{19}{20}\right)^4$ (*iii*) $1-\frac{6}{5}\left(\frac{19}{20}\right)^4$ (*iv*) $1-\left(\frac{19}{20}\right)^5$

9. (*i*) $\left(\frac{1}{4}\right)^5$ (*ii*) $90\left(\frac{1}{4}\right)^5$ (*iii*) $\left(\frac{3}{4}\right)^5$ **10.** 0.36787

11. (*a*) (*i*) 250 (*ii*) 250 (*iii*) 25 (*iv*) 400

(*b*) (*i*) 250 (*ii*) 25 (*iii*) 500

(*c*) (*i*) 31.25% (*ii*) 96.875%

12. (*i*) 12 nearly (*ii*) 17 nearly **13.** $80\,(0.7825 + 0.2175)^{10}$

14. $100\,(0.432 + 0.568)^5$ **15.** (*i*) $100\,(0.7 + 0.3)^4$, (*ii*) $104\,(0.7404 + 0.2596)^4$

16. (*i*) 17 (*ii*) 88 **17.** $100\left(\frac{1}{4}+\frac{3}{4}\right)^{20}$, 0.075168752

18. (*i*) 0.534 (*ii*) 0.466

19. 0.0197 **20.** 1.

POISSON DISTRIBUTION

4.42 POISSON DISTRIBUTION AS A LIMITING CASE OF BINOMIAL DISTRIBUTION

[*G.B.T.U. 2013 ; M.T.U. 2013, 2014*]

Poisson distribution was discovered by S.D. Poisson in the year 1837.

If the parameters n and p of a binomial distribution are known, we can find the distribution. But in situations where n is very large and p is very small, application of binomial distribution is very labourious. However, if we assume that as $n \to \infty$ and $p \to 0$ such that np always remains finite, say λ, we get the Poisson approximation to the binomial distribution.

Now, for a binomial distribution

$$P(X = r) = {}^nC_r q^{n-r} p^r$$

$$= \frac{n(n-1)(n-2)\ldots(n-r+1)}{r!} \times (1-p)^{n-r} \times p^r$$

$$= \frac{n(n-1)(n-2)\ldots(n-r+1)}{r!} \times \left(1-\frac{\lambda}{n}\right)^{n-r} \times \left(\frac{\lambda}{n}\right)^r \qquad \Big|\ \text{Since } np = \lambda \ \therefore\ p = \frac{\lambda}{n}$$

$$= \frac{\lambda^r}{r!} \times \frac{n(n-1)(n-2)\ldots(n-r+1)}{n^r} \times \frac{\left(1-\frac{\lambda}{n}\right)^n}{\left(1-\frac{\lambda}{n}\right)^r}$$

$$= \frac{\lambda^r}{r!}\left(\frac{n}{n}\right)\left(\frac{n-1}{n}\right)\left(\frac{n-2}{n}\right)\ldots\ldots\left(\frac{n-r+1}{n}\right) \times \frac{\left(1-\frac{\lambda}{n}\right)^n}{\left(1-\frac{\lambda}{n}\right)^r}$$

$$= \frac{\lambda^r}{r!}\left(1-\frac{1}{n}\right)\left(1-\frac{2}{n}\right)\ldots\ldots\left(1-\frac{r-1}{n}\right) \times \frac{\left[\left(1-\frac{\lambda}{n}\right)^{-\frac{n}{\lambda}}\right]^{-\lambda}}{\left(1-\frac{\lambda}{n}\right)^r}$$

As $n \to \infty$, each of the $(r-1)$ factors

$$\left(1-\frac{1}{n}\right), \left(1-\frac{2}{n}\right), \ldots\ldots, \left(1-\frac{r-1}{n}\right) \quad \text{tends to 1. Also } \left(1-\frac{\lambda}{n}\right)^r \text{ tends to 1.}$$

Since $\underset{x \to \pm\infty}{\text{Lt}} \left(1+\frac{1}{x}\right)^x = e$, the Naperian base. $\therefore \left[\left(1-\frac{\lambda}{n}\right)^{-\frac{n}{\lambda}}\right]^{-\lambda} \longrightarrow e^{-\lambda}$ as $n \to \infty$

Hence in the limiting case when $n \to \infty$, we have

$$\boxed{P(X = r) = \frac{e^{-\lambda}.\lambda^r}{r!} \quad (r = 0, 1, 2, 3, \ldots)} \qquad \ldots(1)$$

where λ is a finite number $= np$.

(1) represents a probability distribution which is called the *Poisson probability distribution*.

Note 1. λ is called the parameter of the distribution.

Note 2. The sum of the probabilities $P(r)$ for $r = 0, 1, 2, 3, \ldots\ldots$ is 1, since

$$P(0) + P(1) + P(2) + P(3) + \ldots\ldots = e^{-\lambda} + \frac{\lambda e^{-\lambda}}{1!} + \frac{\lambda^2 e^{-\lambda}}{2!} + \frac{\lambda^3 e^{-\lambda}}{3!} + \ldots\ldots$$

$$= e^{-\lambda}\left(1+\frac{\lambda}{1!}+\frac{\lambda^2}{2!}+\frac{\lambda^3}{3!}+\ldots\ldots\right) = e^{-\lambda}.e^{\lambda} = 1.$$

4.43 RECURRENCE FORMULA FOR THE POISSON DISTRIBUTION

For Poisson distribution, $P(r) = \dfrac{e^{-\lambda}\lambda^r}{r!}$ and $P(r+1) = \dfrac{e^{-\lambda}\lambda^{r+1}}{(r+1)!}$

$$\therefore \qquad \frac{P(r+1)}{P(r)} = \frac{\lambda r!}{(r+1)!} = \frac{\lambda}{r+1}$$

or $$P(r+1) = \frac{\lambda}{r+1} P(r),\ r = 0, 1, 2, 3, \ldots\ldots$$

This is called the *recurrence* or *recursion formula* for the Poisson distribution.

4.44 MEAN AND VARIANCE OF THE POISSON DISTRIBUTION (*A.K.T.U. 2018*)

For the Poisson distribution, $P(r) = \dfrac{e^{-\lambda}\lambda^r}{r!}$

$$\text{Mean } \mu = \sum_{r=0}^{\infty} rP(r) = \sum_{r=0}^{\infty} r \cdot \frac{e^{-\lambda}\lambda^r}{r!}$$

$$= e^{-\lambda} \sum_{r=1}^{\infty} \frac{\lambda^r}{(r-1)!} = e^{-\lambda}\left(\lambda + \frac{\lambda^2}{1!} + \frac{\lambda^3}{2!} + \ldots\ldots\right)$$

$$= \lambda e^{-\lambda}\left(1 + \frac{\lambda}{1!} + \frac{\lambda^2}{2!} + \ldots\ldots\right) = \lambda e^{-\lambda} \cdot e^{\lambda} = \lambda$$

Thus, the mean of the Poisson distribution is equal to the parameter λ.

$$\text{Variance } \sigma^2 = \sum_{r=0}^{\infty} r^2 P(r) - \mu^2 = \sum_{r=0}^{\infty} r^2 \cdot \frac{\lambda^r e^{-\lambda}}{r!} - \lambda^2 = e^{-\lambda} \sum_{r=1}^{\infty} \frac{r^2 \lambda^r}{r!} - \lambda^2$$

$$= e^{-\lambda}\left[\frac{1^2 \cdot \lambda}{1!} + \frac{2^2 \cdot \lambda^2}{2!} + \frac{3^2 \cdot \lambda^3}{3!} + \frac{4^2 \cdot \lambda^4}{4!} + \ldots\ldots\right] - \lambda^2$$

$$= \lambda e^{-\lambda}\left[1 + \frac{2\lambda^2}{1!} + \frac{3\lambda^2}{2!} + \frac{4\lambda^3}{3!} + \ldots\ldots\right] - \lambda^2$$

$$= \lambda e^{-\lambda}\left[1 + \frac{(1+1)\lambda}{1!} + \frac{(1+2)\lambda^2}{2!} + \frac{(1+3)\lambda^3}{3!} + \ldots\ldots\right] - \lambda^2$$

$$= \lambda e^{-\lambda}\left[\left(1 + \frac{\lambda}{1!} + \frac{\lambda^2}{2!} + \frac{\lambda^3}{3!} + \ldots\ldots\right) + \left(\frac{\lambda}{1!} + \frac{2\lambda^2}{2!} + \frac{3\lambda^3}{3!} + \ldots\ldots\right)\right] - \lambda^2$$

$$= \lambda e^{-\lambda}\left[e^{\lambda} + \lambda\left(1 + \frac{\lambda}{1!} + \frac{\lambda^2}{2!} + \ldots\ldots\right)\right] - \lambda^2$$

$$= \lambda e^{-\lambda}[e^{\lambda} + \lambda e^{\lambda}] - \lambda^2 = \lambda e^{-\lambda} \cdot e^{\lambda}(1+\lambda) - \lambda^2 = \lambda(1+\lambda) - \lambda^2 = \lambda.$$

Hence, the variance of the Poisson distribution is also λ.

Thus, the mean and the variance of the Poisson distribution are each equal to the parameter λ.

Note 1. The mean and the variance of the Poisson distribution can also be derived from those of the binomial distribution in the limiting case when $n \to \infty$, $p \to 0$ and $np = \lambda$.

Mean of Binomial distribution is np.

$\therefore$ Mean of Poisson distribution $= \underset{n \to \infty}{\text{Lt}} \; np = \underset{n \to \infty}{\text{Lt}} \; \lambda = \lambda$

Variance of Binomial distribution is $npq = np\,(1 - p)$

$\therefore$ Variance of Poisson distribution $= \underset{n \to \infty}{\text{Lt}} \; np\,(1 - p) = \underset{n \to \infty}{\text{Lt}} \; \lambda \left(1 - \dfrac{\lambda}{n}\right) = \lambda.$

Note 2. For Poisson distribution, $\mu_3 = \lambda$ and $\mu_4 = 3\lambda^2 + \lambda$.

Coefficients of skewness and kurtosis are given by

$$\beta_1 = \frac{1}{\lambda} \text{ and } \gamma_1 = \frac{1}{\sqrt{\lambda}}. \text{ Also, } \beta_2 = 3 + \frac{1}{\lambda} \text{ and } \gamma_2 = \frac{1}{\lambda}$$

Hence Poisson distribution is always a skewed distribution.

Remark. While fitting the Poisson distribution to a given data, we round the figures to the nearest integer but it should be kept in mind that the total of the observed and the expected frequencies should be same.

4.45 MODE OF POISSON DISTRIBUTION

Let $\quad P(x = r) = e^{-\lambda} \dfrac{\lambda^r}{r!}, \; r = 0, 1, 2, \ldots\ldots, \infty$

The value of r which has a greater probability than any other value is the mode of the Poisson distribution. Thus r is mode if

$$P(X = r) \geq P(X = r + 1) \text{ and } P(X = r) \geq P(X = r - 1)$$

$$\Rightarrow \quad \frac{e^{-\lambda} \,.\, \lambda^r}{r!} \geq \frac{e^{-\lambda} \,.\, \lambda^{r+1}}{(r+1)!} \quad \text{and} \quad \frac{e^{-\lambda} \,.\, \lambda^r}{r!} \geq \frac{e^{-\lambda} \,.\, \lambda^{r-1}}{(r-1)!}$$

$$\Rightarrow \quad 1 \geq \frac{\lambda}{r+1} \quad \text{and} \quad \frac{\lambda}{r} \geq 1$$

$$\Rightarrow \quad r \geq \lambda - 1 \quad \text{and} \quad r \leq \lambda \quad i.e., \quad \lambda - 1 \leq r \leq \lambda$$

Case I. If λ is a positive integer, there are two modes $\lambda - 1$ and λ.

Case II. If λ is not a positive integer, there is one mode and is the integral value between $\lambda - 1$ and λ.

4.46 APPLICATIONS OF POISSON DISTRIBUTION

This distribution is applied to problems concerning :

(*i*) Arrival pattern of defective vehicles in a workshop.

(*ii*) Patients in a hospitals.

(*iii*) Telephone calls.

(*iv*) Demand pattern for certain spare parts.

(*v*) Number of fragments from a shell hitting a target.

(*vi*) Emission of radioactive (α) particles.

ILLUSTRATIVE EXAMPLES

Example 1. *If the variance of the Poisson distribution is 2, find the probabilities for $r = 1, 2, 3, 4$ from the recurrence relation of the Poisson distribution. Also, find $P(r \geq 4)$.*

(M.T.U. 2013)

Sol. λ, the parameter of Poisson distribution = Variance = 2

Recurrence relation for the Poisson distribution is

$$P(r+1) = \frac{\lambda}{r+1} P(r) = \frac{2}{r+1} P(r) \qquad \text{...(1)}$$

Now $$P(r) = \frac{e^{-\lambda} \lambda^r}{r!} \Rightarrow P(0) = \frac{e^{-2} (2)^0}{0!} = e^{-2} = 0.1353$$

Putting $r = 0, 1, 2, 3$ in (1), we get

$$P(1) = 2P(0) = 2 \times 0.1353 = 0.2706 \; ; \qquad P(2) = \frac{2}{2} P(1) = 0.2706$$

$$P(3) = \frac{2}{3} P(2) = \frac{2}{3} \times 0.2706 = 0.1804 \; ; \quad P(4) = \frac{2}{4} P(3) = \frac{1}{2} \times 0.1804 = 0.0902.$$

Now, $P(r \geq 4) = 1 - [P(0) + P(1) + P(2) + P(3)]$

$= 1 - [0.1353 + 0.2706 + 0.2706 + 0.1804] = 0.1431.$

Example 2. *Using Poisson distribution, find the probability that the ace of spades will be drawn from a pack of well-shuffled cards at least once in 104 consecutive trials.* (U.P.T.U. 2015)

Sol. $$p = \frac{1}{52}, \; n = 104$$

$\therefore$ $$\lambda = np = 104 \times \frac{1}{52} = 2$$

Prob. (at least once) = $P(r \geq 1) = 1 - P(0)$

$$= 1 - \frac{e^{-\lambda} \cdot \lambda^0}{0!} = 1 - e^{-2} = 1 - 0.135335 \approx 0.8647.$$

Example 3. (*i*) *Fit a Poisson distribution to the following data and calculate theoretical frequencies.*

Deaths:	*0*	*1*	*2*	*3*	*4*
Frequencies:	*122*	*60*	*15*	*2*	*1*

[U.P.T.U. 2014 ; U.K.T.U. 2010]

(*ii*) *The frequency of accidents per shift in a factory is shown in the following table:*

Accident per shift	*Frequency*
0	*192*
1	*100*
2	*24*
3	*3*
4	*1*
Total	320

Calculate the mean number of accidents per shift. Fit a Poisson distribution and calculate theoretical frequencies.

Sol. (*i*) Mean of given distribution $= \dfrac{\Sigma fx}{\Sigma f}$

$$\Rightarrow \qquad \lambda = \frac{60 + 30 + 6 + 4}{200} = 0.5$$

Required Poisson distribution $= \text{N} . \dfrac{e^{-\lambda} . \lambda^r}{r!} = 200 . \dfrac{e^{-.5}\,(.5)^r}{r!} = (121.306)\dfrac{(0.5)^r}{r!}$

r	$N. P(r)$	*Theoretical frequency*
0	$121.306 \times \dfrac{(0.5)^0}{0!} = 121.306$	121
1	$121.306 \times \dfrac{(0.5)^1}{1!} = 60.653$	61
2	$121.306 \times \dfrac{(0.5)^2}{2!} = 15.163$	15
3	$121.306 \times \dfrac{(0.5)^3}{3!} = 2.527$	3
4	$121.306 \times \dfrac{(0.5)^4}{4!} = 0.3159$	0
		Total = 200

(*ii*) Mean number of accidents per shift

$$\lambda = \frac{\Sigma fx}{\Sigma f} = \frac{100 + 48 + 9 + 4}{320} = 0.5031$$

$\therefore$ Required Poisson distribution

$$= \text{N} . \frac{e^{-\lambda} . \lambda^r}{r!} = 320 . \frac{e^{-.5031}\,(.5031)^r}{r!} = \frac{(193.48)(.5031)^r}{r!}$$

r	$N. P(r)$	*Theoretical frequency*
0	193.48	194
1	97.34	97
2	24.38	24
3	4.10	4
4	0.51	1
		Total = 320

Example 4. *(i) Suppose that a book of 600 pages contains 40 printing mistakes. Assume that these errors are randomly distributed throughout the book and r, the number of errors per page has a Poisson distribution. What is the probability that 10 pages selected at random will be free from errors?*

(ii) Wireless sets are manufactured with 25 solders joints each, on the average 1 joint in 500 is defective. How many sets can be expected to be free from defective joints in a consignment of 10000 sets?

Sol. (*i*)
$$p = \frac{40}{600} = \frac{1}{15}, \quad n = 10$$

$$\therefore \quad \lambda = np = 10\left(\frac{1}{15}\right) = \frac{2}{3}$$

$$P(r) = \frac{e^{-\lambda}\,\lambda^r}{r!} = \frac{e^{-2/3}\,(2/3)^r}{r!}$$

$$\therefore \quad P(0) = \frac{e^{-2/3}\,(2/3)^0}{0!} = e^{-2/3} = 0.51.$$

(*ii*)
$$p = \frac{1}{500}, \; n = 25$$

$$\therefore \quad \lambda = np = 25 \times \frac{1}{500} = \frac{1}{20} = 0.05$$

No. of sets in a consignment, N= 10000

Probability of having no defective joint $= P(r = 0) = \dfrac{e^{-.05}\,(0.05)^0}{0!} = 0.9512.$

$\therefore$ The expected no. of sets free from defective joints $= 0.9512 \times 10000 = 9512.$

Example 5. *A manufacturer knows that the condensors he makes contain on an average 1% of defectives. He packes them in boxes of 100. What is the probability that a box picked at random will contain 4 or more faulty condensors?*

Sol.
$$p = 0.01, \; n = 100$$

$$\therefore \quad \lambda = np = 1$$

$$P(r) = \frac{e^{-\lambda}\,\lambda^r}{r!} = \frac{e^{-1}}{r!}$$

$$\begin{aligned} P(4 \text{ or more faulty condensors}) &= P(4) + P(5) + \ldots + P(100) \\ &= 1 - [P(0) + P(1) + P(2) + P(3)] \\ &= 1 - \left[\frac{e^{-1}}{0!} + \frac{e^{-1}}{1!} + \frac{e^{-1}}{2!} + \frac{e^{-1}}{3!}\right] \\ &= 1 - e^{-1}\left[1 + 1 + \frac{1}{2} + \frac{1}{6}\right] = 1 - \frac{8}{3e} = 1 - 0.981 = 0.019. \end{aligned}$$

Example 6. *(i) If the probabilities of a bad reaction from a certain injection is 0.0002, determine the chance that out of 1000 individuals more than two will get a bad reaction.*

(ii) The probability that a man aged 50 years will die within a year is 0.01125. What is the probability that of 12 such men, at least 11 will reach their 51st birthday?

(Given: $e^{-.135} = 0.87366$)

Sol. (*i*) Here, $p = 0.0002, \quad n = 1000$

$\therefore \quad \lambda = np = 1000 \times 0.0002 = 0.2.$

Since the prob. of bad reaction is very small and no. of trials is very high, we use Poisson distribution here.

The prob. that out of 100 individuals, more than 2 will get a bad reaction is

$$= P(r > 2) = 1 - P(r \le 2) = 1 - [P(0) + P(1) + P(2)] \quad ...(1)$$

Now, $$P(0) = \frac{e^{-0.2}(0.2)^0}{0} = 0.8187 \quad (\text{Here } r = 0)$$

$$P(1) = \frac{e^{-0.2}(0.2)^1}{1} = 0.1637 \quad (\text{Here } r = 1)$$

and $$P(2) = \frac{e^{-0.2}(0.2)^2}{2} = 0.0164. \quad (\text{Here } r = 2)$$

$\therefore$ From (1), Reqd. probability $= 1 - [0.8187 + 0.1637 + 0.0164] = 0.0012.$

(*ii*) $p = 0.01125, n = 12$

$\therefore \quad \lambda = np = 12 \times 0.01125 = 0.135$

P (at least 11 survive) = P(atmost 1 dies)

$$= P(0) + P(1) = \frac{e^{-\lambda}.\lambda^0}{0!} + \frac{e^{-\lambda}.\lambda^1}{1!}$$

$$= e^{-.135}(1 + 0.135) = 1.135 \times 0.87366 = 0.9916.$$

Example 7. *A car-hire firm has two cars, which it hires out day by day. The number of demands for a car on each day is distributed as a Poisson distribution with mean 1.5. Calculate the proportion of days on which neither car is used and the proportion of days on which some demand is refused ($e^{-1.5} = 0.2231$).*

Sol. Since the number of demands for a car is distributed as a Poisson distribution with mean $\lambda = 1.5$.

$\therefore$ Proportion of days on which neither car is used

= Probability of there being no demand for the car

$$= \frac{\lambda^0 e^{-\lambda}}{0!} = e^{-1.5} = 0.2231$$

Proportion of days on which some demand is refused

= probability for the number of demands to be more than two

$$= 1 - P(x \le 2) = 1 - \left(e^{-\lambda} + \frac{\lambda e^{-\lambda}}{1!} + \frac{\lambda^2 e^{-\lambda}}{2!}\right)$$

$$= 1 - e^{-1.5}\left(1 + 1.5 + \frac{(1.5)^2}{2}\right) = 0.1912625.$$

Example 8. *Suppose the number of telephone calls on an operator received from 9 : 00 to 9 : 05 follow a Poisson distribution with a mean 3. Find the probability that*

(*i*) *The operator will receive no calls in that time interval tomorrow.*

(*ii*) *In the next three days, the operator will receive a total of 1 call in that time interval.* (*Given: $e^{-3} = 0.04978$*)

Sol. Here, $\lambda = 3$

(*i*) $$P(0) = \frac{e^{-\lambda}.\lambda^0}{0!} = e^{-3} = 0.04978$$

(*ii*) Required probability = P(0) P(0) P(1) + P(0) P(1) P(0) + P(1) P(0) P(0)

$$= 3\left\{\frac{e^{-\lambda}.\lambda^0}{0!}\right\}^2 \frac{e^{-\lambda}.\lambda^1}{1!} = 9(e^{-3})^3 = 0.00111.$$

Example 9. *The no. of arrivals of customers during any day follows Poisson distribution with a mean of 5. What is the probability that the total no. of customers on two days selected at random is less than 2? (Given: $e^{-10} = 4.54 \times 10^{-5}$)*

Sol. $\lambda = 5$

Arrival of Customers

I day	*II day*	*Total*
0	0	0
0	1	1
1	0	1

Required probability = P(0) P(0) + P(0) P(1) + P(1) P(0)

$$= \frac{e^{-5}.5^0}{0!}.\frac{e^{-5}.5^0}{0!} + \frac{e^{-5}.5^0}{0!}.\frac{e^{-5}.5^1}{1!} + \frac{e^{-5}.5^1}{1!}.\frac{e^{-5}.5^0}{0!}$$

$$= e^{-10} + 2\,.\,5\,.\,e^{-10} = 11\,e^{-10} = 11 \times 4.54 \times 10^{-5}$$

$$= 4.994 \times 10^{-4}.$$

Example 10. *An insurance company finds that 0.005% of the population dies from a certain kind of accident each year. What is the probability that the company must pay off no more than 3 of 10,000 insured risks against such incident in a given year?*

Sol. $$p = \frac{0.005}{100} = 0.00005, \quad n = 10000$$

$\therefore$ $$\lambda = np = 10000 \times 0.00005 = 0.5$$

Required Probability = $1 - P(r \le 3) = 1 - [P(0) + P(1) + P(2) + P(3)]$

$$= 1 - \left[\frac{e^{-0.5}(0.5)^0}{0!} + \frac{e^{-0.5}(0.5)^1}{1!} + \frac{e^{-0.5}(0.5)^2}{2!} + \frac{e^{-0.5}(0.5)^3}{3!}\right]$$

$$= 1 - e^{-5}\,[1 + 0.5 + 0.125 + 0.021] = 0.0016.$$

Example 11. *In a certain factory turning out razor blades, there is a small chance of 0.002 for any blade to be defective. The blades are supplied in packets of 10. Calculate the approximate number of packets containing no defective, one defective and two defective blades in a consignment of 10,000 packets. (Given: $e^{-0.02} = 0.9802$)*

Sol. p(defective) = 0.002

$n = 10$ (no. of blades in a packet)

$\therefore$ $\lambda = np = 10 \times 0.002 = 0.02$

No. of packets in the consignment, N = 10,000.

(*i*) Probability of having no defective = $P(0) = \dfrac{e^{-0.02}(0.02)^0}{0!} = 0.9802$ | Here $r = 0$

Approximate no. of packets having zero defective in the consignment = 0.9802×10000
$= 9802$

(*ii*) Probability of having one defective = $P(1) = \dfrac{e^{-0.02}(0.02)^1}{1!} = 0.9802 \times 0.02 = 0.019604$

Approximate no. of packets having one defective in the consignment
$= 0.019604 \times 10000 \approx 196.$

(*iii*) Probability of having two defective blades

$$P(2) = \frac{e^{-0.02}(0.02)^2}{2!} = \frac{(0.980198) \times (0.0004)}{2} = 0.000196.$$

∴ Approximate no. of packet having two defectives in the consignment
$= 0.000196 \times 10000 = 1.96 \approx 2.$

Example 12. (*i*) *Six coins are tossed 6400 times. Using the Poisson distribution, determine the approximate probability of getting six heads x times.*

(*ii*) *A Poisson distribution has a double mode at x = 3 and x = 4. What is the probability that x will have one or the other of these two values?*

Sol. (*i*) Probability of getting one head with one coin $= \dfrac{1}{2}$.

∴ The probability of getting six heads with six coins $= \left(\dfrac{1}{2}\right)^6 = \dfrac{1}{64}$

∴ Average number of six heads with six coins in 6400 throws $= np = 6400 \times \dfrac{1}{64} = 100$

∴ The mean of the Poisson distribution = 100.

Approximate probability of getting six heads x times when the distribution is Poisson

$$= \frac{\lambda^x e^{-\lambda}}{x!} = \frac{(100)^x \cdot e^{-100}}{x!}.$$

(*ii*) Since 2 modes are given when λ is an integer, modes are $\lambda - 1$ and λ.

∴ $\lambda - 1 = 3 \Rightarrow \lambda = 4$

Probability (when $r = 3$) $= \dfrac{e^{-4}(4)^3}{3!}$

Probability (when $r = 4$) $= \dfrac{e^{-4}(4)^4}{4!}$

∴ Required Probability = $P(r = 3 \text{ or } 4) = P(r = 3) + P(r = 4)$

$$= \frac{e^{-4}(4)^3}{3!} + \frac{e^{-4}(4)^4}{4!} = \frac{64}{3}e^{-4} = 0.39073.$$

Example 13. *For a Poisson distribution with mean m, show that*

$$\mu_{r+1} = mr\,\mu_{r-1} + m\,\frac{d\mu_r}{dm} \text{ where, } \mu_r = \sum_{x=0}^{\infty} (x-m)^r \frac{e^{-m}.m^x}{x!}.$$

Sol.
$$\mu_r = \sum_{x=0}^{\infty} (x-m)^r . \frac{e^{-m}.m^x}{x!}$$

$$\frac{d\mu_r}{dm} = \sum_{x=0}^{\infty} \left[\frac{-e^{-m}}{x!}.m^x (x-m)^r + \frac{e^{-m}}{x!}\{xm^{x-1}(x-m)^r - r(x-m)^{r-1}.m^x\}\right]$$

$$\Rightarrow \quad m\frac{d\mu_r}{dm} = \sum_{x=0}^{\infty} \frac{e^{-m}}{x!} m^x (x-m)^{r+1} - rm \sum_{x=0}^{\infty} \frac{e^{-m}}{x!} m^x (x-m)^{r-1} = \mu_{r+1} - mr\,\mu_{r-1}$$

$$\Rightarrow \quad \mu_{r+1} = m\,\frac{d\mu_r}{dm} + mr\,\mu_{r-1}.$$

Example 14. *Show that in a Poisson distribution with unit mean, mean deviation about mean is* $\left(\frac{2}{e}\right)$ *times the standard deviation.* (G.B.T.U. 2012)

Sol. Here, $\lambda = 1$

$$\therefore \quad P(X = x) = \frac{e^{-1}.(1)^x}{x!} = \frac{e^{-1}}{x!} \; ; x = 0, 1, 2, ...$$

Mean deviation about mean 1 is

$$= \sum_{x=0}^{\infty} |x-1|\,p(x) = e^{-1} \sum_{x=0}^{\infty} \frac{|x-1|}{x!} = e^{-1}\left[1 + \frac{1}{2!} + \frac{2}{3!} + \frac{3}{4!} + ...\right] \qquad ...(1)$$

we have, $$\frac{n}{(n+1)!} = \frac{\overline{n+1-1}}{(n+1)!} = \frac{1}{n!} - \frac{1}{(n+1)!}$$

$\therefore$ From (1), Mean deviation about mean $= e^{-1}\left[1 + \left(1 - \frac{1}{2!}\right) + \left(\frac{1}{2!} - \frac{1}{3!}\right) + ...\right]$

$$= e^{-1}(1+1) = \frac{2}{3} \times 1 = \frac{2}{e} \times \text{S.D.} \qquad | \text{ Since, variance = mean = 1}$$

TEST YOUR KNOWLEDGE

1. If X is a Poisson variate such that P(X = 2) = 9P(X = 4) + 90P(X = 6), find the standard deviation.
2. If a random variable has a Poisson distribution such that P(1) = P(2), find
   (*i*) mean of the distribution (*ii*) P(4).
3. Suppose that X has a Poisson distribution. If P(X = 2) = $\frac{2}{3}$ P(X = 1) find, (*i*) P(X = 0) (*ii*) P(X = 3).

4. A certain screw making machine produces on average 2 defective screws out of 100, and packs them in boxes of 500. Find the probability that a box contains 15 defective screws.

5. (*i*) The incidence of occupational disease in an industry is such that the workmen have a 10% chance of suffering from it. What is the probability that in a group of 7, five or more will suffer from it?

(*ii*) The experience shows that 4 industrial accidents occur in a plant on an average per month. Calculate the probabilities of less than 3 accidents in a certain month. Use Poisson distribution. (Given : $e^{-4} = 0.01832$). [*M.T.U. (MBA) 2011*]

6. (*i*) Suppose a book of 585 pages contains 43 typographical errors. If these errors are randomly distributed throughout the book, what is the probability that 10 pages, selected at random, will be free from errors ?

(*ii*) Assume that the probability of an individual coalminer being killed in a mine accident during a year is $\frac{1}{2400}$. Use Poisson's distribution to calculate the probability that in a mine employing 200 miners there will be at least one fatal accident in a year.

7. (*i*) A manufacturer of cotter pins knows that 5% of his product is defective. If he sells cotter pins in boxes of 100 and guarantee that not more than 10 pins will be defective, what is the approximate prob. that a box will fail to meet the guaranteed quality?

(*ii*) An insurance company insures 4000 people against loss of both eyes in a car accident. Based on previous data it was assumed 10 persons out of 1,00,000 will have such type of injury in car accident. What is probability that more than 2 of the insured will collect on their policy in a given year? (*M.T.U. 2013*)

8. Records show that the probability is 0.00002 that a car will have a flat tyre while driving over a certain bridge. Use Poisson distribution to find the probability that among 20,000 cars driven over the bridge, not more than one will have a flat tyre.

9. Between the hours of 2 and 4 P.M., the average no. of phone calls per minute coming into the switch board of a company is 2.5. Find the probability that during a particular minute, there will be no phone call at all. [Given : $e^{-2} = 0.13534$ and $e^{-0.5} = 0.60650$.]

10. (*i*) Fit a Poisson distribution to the following data given the number of yeast cells per square for 400 squares :

No. of cells per sq. :	0	1	2	3	4	5	6	7	8	9	10
No. of squares :	103	143	98	42	8	4	2	0	0	0	0

It is given that $e^{-1.3225} = 0.2665$.

(*ii*) Data was collected over a period of 10 years, showing number of deaths from horse kicks in each of the 200 army corps. The distribution of deaths was as follows:

No. of deaths :	0	1	2	3	4	Total
Frequency :	109	65	22	3	1	200

Fit a Poisson distribution to the data and calculate the theoretical frequencies.

[*M.T.U. (MBA) 2011*]

(*iii*) The following table gives the no. of days in a 50 day period during which automobile accidents occured in a city.

No. of accidents :	0	1	2	3	4
No. of days :	21	18	7	3	1

Fit a Poisson distribution to the data. (*G.B.T.U. 2011*)

11. The number of accidents in a year involving taxi drivers in a city follows a Poisson distribution with mean equal to 3. Out of 1000 taxi drivers, find approximately the number of drivers with

(*i*) no accidents (*ii*) more than 3 accidents in a year.

12. The distribution of the number of road accidents per day in a city is Poisson with mean 4. Find the number of days out of 100 days when there will be

(*i*) no accident (*ii*) at least 2 accidents

(*iii*) atmost 3 accidents (*iv*) between 2 and 5 accidents.

13. It is given that 2% of the electric bulbs manufactured by a company are defective. Using Poisson distribution, find the probability that a sample of 200 bulbs will contain (*i*) no defective bulb (*ii*) two defective bulbs (*iii*) at the most three defective bulbs. (*G.B.T.U. 2011*)

14. A manager accepts the work submitted by his typist only when there is no mistake in the work. The typist has to type on an average 20 letters per day of about 200 words each. Find the chance of her making a mistake if

(*i*) less than 1% of the letters submitted by her are rejected

(*ii*) on 90% days all the letters submitted by her are accepted.

As the probability of making a mistake is small, you may use Poisson distribution. (Take $e = 2.72$)

[**Hint:** (*i*) $e^{-200p} \geq 0.99$ (*ii*) $e^{-4000p} = 0.90$]

15. In a certain factory manufacturing razor blades, there is a small chance of 0.002 for any blade to be defective. The blades are supplied in packets of 10. Use suitable distribution to calculate the approximate number of packets containing no defective, one defective and two defective blades respectively in a consignment of 50,000 packets. (*A.K.T.U. 2018*)

Answers

1. 1 **2.** (*i*) 2 (*ii*) $\frac{2}{3e^2}$ **3.** (*i*) $e^{-(4/3)}$ (*ii*) $\frac{32}{81} e^{-(4/3)}$ **4.** 0.035

5. (*i*) 0.0008 (*ii*) 0.23816`

6. (*i*) 0.4795 (*ii*) 0.08 **7.** (*i*) 0.0136875 (*ii*) 0.007926

8. 0.938 **9.** 0.08208

10. (*i*) Theoretical frequencies are 107, 141, 93, 41, 13, 4, 1, 0, 0, 0, 0.

(*ii*) $108.67 \times \frac{(0.61)^r}{r!}$; 109, 66, 20, 4 and 1 (*iii*) $(50) \frac{e^{-0.9} (0.9)^r}{r!}$; 20, 18, 8, 3 and 1.

11. (*i*) 50 (*ii*) 353

12. (*i*) 2 (*ii*) 91 (*iii*) 43 (*iv*) 39

13. (*i*) 0.018315 (*ii*) 0.146525 (*iii*) 0.43347

14. (*i*) $p \leq 0.0000506$ (*ii*) $p = 0.0000263$

15. 49010, 980, 10

4.47 NORMAL DISTRIBUTION

The normal distribution is a continuous distribution. It can be derived from the binomial distribution in the limiting case when n, the number of trials is very large and p, the probability of a success, is close to $\frac{1}{2}$. The general equation of the normal distribution is given by

$$f(x) = \frac{1}{\sigma\sqrt{2\pi}} e^{-\frac{1}{2}\left(\frac{x-\mu}{\sigma}\right)^2}$$

where the variable x can assume all values from $-\infty$ to $+\infty$. μ and σ, called the parameters of the distribution, are respectively the mean and the standard deviation of the distribution and $-\infty < \mu < \infty$, $\sigma > 0$. x is called the normal variate and $f(x)$ is called probability density function of the normal distribution.

If a variable x has the normal distribution with mean μ and standard deviation σ, we briefly write $x : N(\mu, \sigma^2)$.

The graph of the normal distribution is called the *normal curve*. It is bell-shaped and symmetrical about the mean μ. The two tails of the curve extend to $+\infty$ and $-\infty$ towards the positive and negative directions of the x-axis respectively and gradually approach the x-axis without ever metting it. The curve is unimodal and the mode of the normal distribution coincides with its mean μ. The line $x = \mu$ divides the area under the normal curve above x-axis into two equal parts. Thus, the median of the distribution also coincides with its mean and mode. The area under the normal curve between any two given ordinates $x = x_1$ and $x = x_2$ represents the probability of values falling into the given interval. The total area under the normal curve above the x-axis is 1.

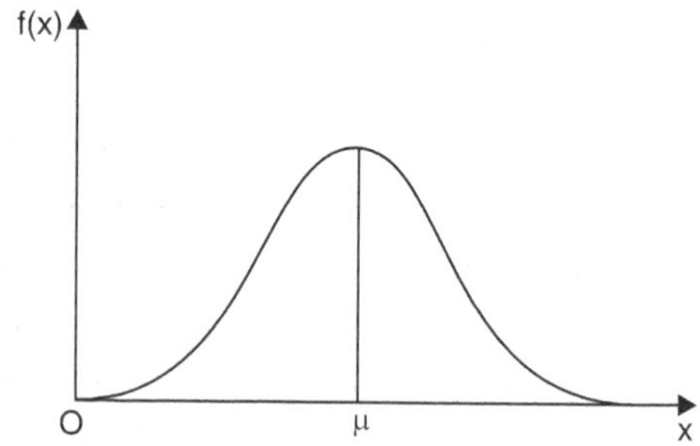

4.48 BASIC PROPERTIES OF THE NORMAL DISTRIBUTION

The probability density function of the normal distribution is given by

$$f(x) = \frac{1}{\sigma\sqrt{2\pi}} e^{-\frac{1}{2}\left(\frac{x-\mu}{\sigma}\right)^2}$$

(*i*) $f(x) \geq 0$ (*ii*) $\int_{-\infty}^{\infty} f(x)\,dx = 1$,

i.e., the total area under the normal curve above the x-axis is 1.

(*iii*) The normal distribution is symmetrical about its mean.

(*iv*) It is a unimodal distribution. The mean, mode and median of this distribution coincide.

4.49 STANDARD FORM OF THE NORMAL DISTRIBUTION

If X is a normal random variable with mean μ and standard deviation σ, then the random variable $Z = \frac{X - \mu}{\sigma}$ has the normal distribution with mean 0 and standard deviation 1. The random variable Z is called the *standardized* (or *standard*) normal random variable.

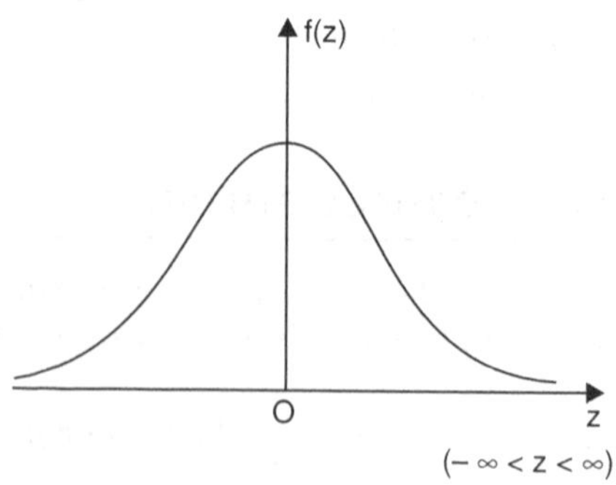

The probability density function for the normal distribution in standard form is given by

$$f(z) = \frac{1}{\sqrt{2\pi}} e^{-\frac{1}{2}z^2}$$

It is free from any parameter. This helps us to compute areas under the normal probability curve by making use of standard tables.

Note 1. If $f(z)$ is the probability density function for the normal distribution, then

$$P(z_1 \le Z \le z_2) = \int_{z_1}^{z_2} f(z)\, dz = F(z_2) - F(z_1), \quad \text{where} \quad F(z) = \int_{-\infty}^{z} f(z)\, dz = P(Z \le z)$$

The function F(z) defined above is called the *distribution function* for the normal distribution.

Note 2. The probabilities $P(z_1 \le Z \le z_2)$, $P(z_1 < Z \le z_2)$, $P(z_1 \le Z < z_2)$ and $P(z_1 < Z < z_2)$ are all regarded to be the same.

Note 3. $F(-z_1) = 1 - F(z_1)$.

4.50 NORMAL DISTRIBUTION AS A LIMITING FORM OF BINOMIAL DISTRIBUTION (when $p = q$)

Let N $(q + p)^n$ be the binomial distribution. If $p = q$ then $p = q = \frac{1}{2}$ (since $p + q = 1$) and consequently the binomial distribution is symmetrical. Let n be an even integer say $2k$, k being an integer. Since $n \to \infty$, the frequencies of r and $r + 1$ successes can be written in following forms:

$$f(r) = N.\ {}^{2k}C_r \left(\frac{1}{2}\right)^{2k}$$

$$f(r+1) = N\ .\ {}^{2k}C_{r+1} \left(\frac{1}{2}\right)^{2k}$$

$$\therefore \qquad \frac{f(r+1)}{f(r)} = \frac{{}^{2k}C_{r+1}}{{}^{2k}C_r} = \frac{2k - r}{r + 1}$$

The frequency of r successes will be greater than the frequency of $(r + 1)$ successes if

$$f(r) > f(r+1)$$

$$\Rightarrow \qquad \frac{f(r+1)}{f(r)} < 1$$

$$\Rightarrow \qquad 2k - r < r + 1$$

$$\Rightarrow \qquad r > k - \frac{1}{2} \qquad \text{...(1)}$$

In a similar way, the frequency of r successes will be greater than the frequencies of $(r - 1)$ successes if $r < k + \frac{1}{2}$ ...(2)

In view of (1) and (2), we observe that if $k - \frac{1}{2} < r < k + \frac{1}{2}$ the frequency corresponding to r successes will be the greatest. Clearly, $r = k$ is the value of the success corresponding to which the frequency is maximum. Suppose it is y_0. Then, we have

$$y_0 = N\ .\ {}^{2k}C_k \left(\frac{1}{2}\right)^{2k} = N\ .\frac{2k\,!}{k\,!\,k\,!}\left(\frac{1}{2}\right)^{2k}$$

Let y_x be the frequency of $k + x$ successes then, we have

$$y_x = N \, . \, {}^{2k}C_{k+x} \left(\frac{1}{2}\right)^{2k} = N \, . \left(\frac{1}{2}\right)^{2k} . \frac{2k\,!}{(k+x)\,!\,(k-x)\,!}$$

Now,
$$\frac{y_x}{y_0} = \frac{k\,!\,k!}{(k+x)\,!\,(k-x)\,!} = \frac{k\,(k-1)\,(k-2)\ldots(k-x+1)}{(k+x)\,(k+x-1)\ldots(k+1)}$$

$$= \frac{\left(1-\frac{1}{k}\right)\left(1-\frac{2}{k}\right)\ldots\left\{1-\frac{x-1}{k}\right\}}{\left(1+\frac{1}{k}\right)\left(1+\frac{2}{k}\right)\ldots\left(1+\frac{x}{k}\right)}$$

Taking log on both sides,

$$\log\frac{y_x}{y_0} = \left[\log\left(1-\frac{1}{k}\right) + \log\left(1-\frac{2}{k}\right) + \cdots + \log\left(1-\frac{x-1}{k}\right)\right]$$

$$-\left[\log\left(1+\frac{1}{k}\right) + \log\left(1+\frac{2}{k}\right) + \cdots + \log\left(1+\frac{x}{k}\right)\right] \quad \ldots(3)$$

Now, writing expression for each term and neglecting higher powers of $\frac{x}{k}$ (very small quantity), we get from (3),

$$\log\frac{y_x}{y_0} = -\frac{1}{k}\{1+2+3+\cdots+(x-1)\} - \frac{1}{k}\{1+2+3+\ldots+(x-1)+x\}$$

$$= -\frac{2}{k}\{1+2+3+\cdots+(x-1)\} - \frac{x}{k}$$

$$= -\frac{2}{k}\frac{(x-1)\,x}{2} - \frac{x}{k} = -\frac{x^2}{k}$$

$\therefore$
$$y_x = y_0\, e^{-x^2/k}$$

$\Rightarrow$
$$\boxed{y_x = y_0\, e^{-x^2/2\sigma^2}} \qquad \Big|\, \because \quad \sigma^2 = npq = \frac{n}{4} = \frac{k}{2}$$

which is **normal distribution.**

4.51 MEAN AND VARIANCE OF NORMAL DISTRIBUTION *(A.K.T.U. 2015, 2018)*

1. The A.M. of a continuous distribution $f(x)$ is given by

$$\text{A.M. } (\overline{x}) = \frac{\int_{-\infty}^{\infty} x\, f(x)\, dx}{\int_{-\infty}^{\infty} f(x)\, dx} \qquad \Big|\text{ By definition}$$

Consider the normal distribution with μ, σ as the parameters then

$$\bar{x} = \int_{-\infty}^{\infty} x.\frac{1}{\sigma\sqrt{2\pi}} e^{-\frac{1}{2}\left(\frac{x-\mu}{\sigma}\right)^2} dx \qquad \text{Since } \int_{-\infty}^{\infty} f(x)\,dx = \text{area under normal curve} = 1$$

Put $\frac{x-\mu}{\sigma} = z$ so that $x = \mu + \sigma z \quad \therefore \quad dx = \sigma dz$

so,
$$\bar{x} = \int_{-\infty}^{\infty} (\mu + \sigma z)\frac{1}{\sigma\sqrt{2\pi}} e^{-\frac{1}{2}z^2} (\sigma\, dz)$$

$$= \mu \int_{-\infty}^{\infty} \frac{1}{\sqrt{2\pi}} e^{-z^2/2} dz + \frac{\sigma}{\sqrt{2\pi}} \int_{-\infty}^{\infty} z\, e^{-\frac{1}{2}z^2} dz$$

$$= \mu + \frac{\sigma}{\sqrt{2\pi}} \int_{-\infty}^{\infty} e^{-z^2/2}\, d\left(\frac{z^2}{2}\right) \qquad \because \int_{-\infty}^{\infty} \frac{1}{\sqrt{2\pi}} e^{-z^2/2} dz = 1$$

$$= \mu + \frac{\sigma}{\sqrt{2\pi}} \left(\frac{e^{-z^2/2}}{-1}\right)_{-\infty}^{\infty}$$

$$\boxed{\bar{x} = \mu}$$

2. By definition,

$$\text{Variance} = \int_{-\infty}^{\infty} (x - \bar{x})^2 f(x)\,dx$$

$$= \int_{-\infty}^{\infty} x^2 f(x)\,dx + \bar{x}^2 \int_{-\infty}^{\infty} f(x)\,dx - 2\bar{x}\int_{-\infty}^{\infty} x f(x)\,dx$$

$$= \int_{-\infty}^{\infty} x^2 f(x)\,dx + \bar{x}^2 - 2\bar{x}\,\bar{x} \qquad \because \int_{-\infty}^{\infty} f(x)\,dx = 1 \text{ and } \int_{-\infty}^{\infty} x f(x)\,dx = \bar{x}$$

$$= \int_{-\infty}^{\infty} x^2 f(x)\,dx - \bar{x}^2 \qquad ...(1)$$

Now, Let $I = \int_{-\infty}^{\infty} x^2 f(x)\,dx = \int_{-\infty}^{\infty} x^2 \frac{1}{\sigma\sqrt{2\pi}} e^{-\frac{1}{2}\left(\frac{x-\bar{x}}{\sigma}\right)^2} dx$

Put $\frac{x-\bar{x}}{\sigma} = z$ so that $x = \bar{x} + \sigma z \quad \therefore \quad dx = \sigma\, dz$

Hence,
$$I = \int_{-\infty}^{\infty} (\bar{x} + \sigma z)^2 \frac{1}{\sigma\sqrt{2\pi}} e^{-z^2/2}\, \sigma\, dz$$

$$= \frac{1}{\sqrt{2\pi}}\left[\sigma^2 \int_{-\infty}^{\infty} z^2 e^{-z^2/2} dz + \bar{x}^2 \int_{-\infty}^{\infty} e^{-z^2/2} dz + 2\sigma\bar{x}\int_{-\infty}^{\infty} z\, e^{-z^2/2} dz\right]$$

$$= \frac{-\sigma^2}{\sqrt{2\pi}} \int_{-\infty}^{\infty} z\, d(e^{-z^2/2}) + \bar{x}^2 \,.\, 1 + 2\sigma\bar{x}\,.\,0$$

$$= -\frac{\sigma^2}{\sqrt{2\pi}}\left(z\,e^{-z^2/2}\right)_{-\infty}^{\infty} + \frac{\sigma^2}{\sqrt{2\pi}}\int_{-\infty}^{\infty} e^{-z^2/2}\,dz + \bar{x}^2$$

$$= 0 + \sigma^2 \,.\, 1 + \bar{x}^2 = \sigma^2 + \bar{x}^2$$

∴ From (1), $\boxed{\text{Variance} = \sigma^2 + \bar{x}^2 - \bar{x}^2 = \sigma^2}$

∴ The standard deviation of the normal distribution is σ.

4.52 AREA UNDER THE NORMAL CURVE

By taking $z = \dfrac{x-\mu}{\sigma}$, standard normal curve is formed. The total area under this curve is 1.

The area under the curve is divided into two equal parts by $z = 0$. The area between the ordinate $z = 0$ and any other ordinate can be noted from the supplied table. It should be noted that mean for the normal distribution is 0.

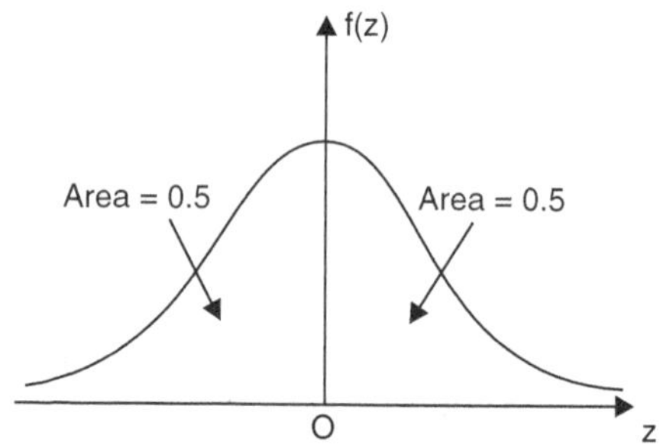

4.53 APPLICATIONS OF NORMAL DISTRIBUTION

De Moivre made the discovery of this distribution in 1733.

This distribution has an important application in the theory of errors made by chance in experimental measurements. Its more applications are in computation of hit probability of a shot and in statistical inference in almost every branch of science.

ILLUSTRATIVE EXAMPLES

Example 1. *A sample of 100 dry battery cells tested to find the length of life produced the following results:*

$$\bar{x} = 12 \text{ hours}, \sigma = 3 \text{ hours}.$$

Assuming the data to be normally distributed, what percentage of battery cells are expected to have life

(i) more than 15 hours *(ii) less than 6 hours* *(iii) between 10 and 14 hours?*

(A.K.T.U. 2018)

Sol. Here x denotes the length of life of dry battery cells.

Also, $z = \dfrac{x-\bar{x}}{\sigma} = \dfrac{x-12}{3}.$

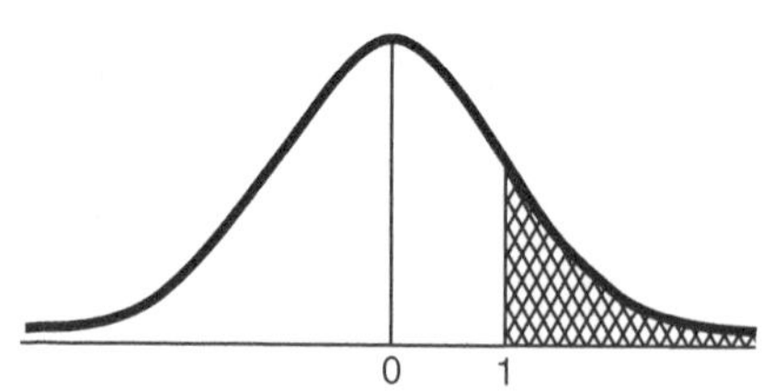

(*i*) When $x = 15, z = 1$

∴ $P(x > 15) = P(z > 1)$

$= P(0 < z < \infty) - P(0 < z < 1)$

$= .5 - 0.3413 = 0.1587 = 15.87\%.$

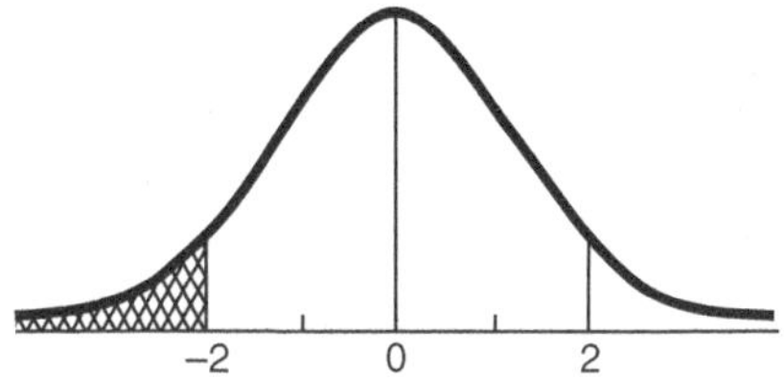

(*ii*) When $x = 6,\ z = -2$

$\therefore \quad P(x < 6) = P(z < -2)$

$= P(z > 2) = P(0 < z < \infty) - P(0 < z < 2)$

$= 0.5 - 0.4772 = 0.0228 = 2.28\%.$

(*iii*) When $x = 10, z = -\frac{2}{3} = -0.67$

When $x = 14, z = \frac{2}{3} = 0.67$

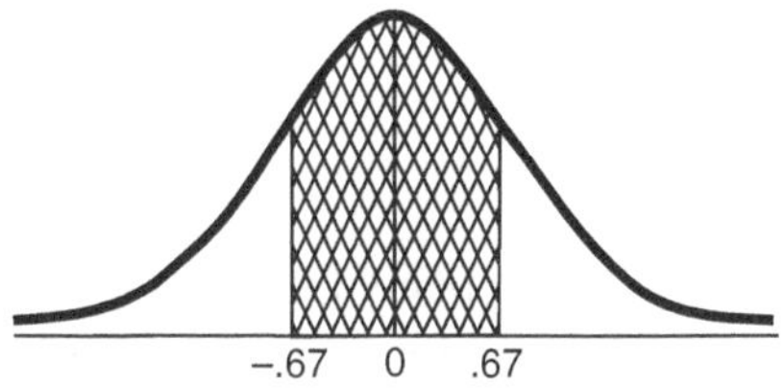

$P(10 < x < 14)$

$= P(-0.67 < z < 0.67)$

$= 2P(0 < z < 0.67) = 2 \times 0.2485$

$= 0.4970 = 49.70\%.$

Example 2. *In a sample of 1000 cases, the mean of a certain test is 14 and S.D. is 2.5. Assuming the distribution to be normal, find*

(*i*) *how many students score between 12 and 15?*

(*ii*) *how many score above 18?*

(*iii*) *how many score below 8?*

(*iv*) *how many score 16?*

Sol. (*i*)
$$z_1 = \frac{x_1 - \mu}{\sigma} = \frac{12 - 14}{2.5} = -0.8$$
$$z_2 = \frac{x_2 - \mu}{\sigma} = \frac{15 - 14}{2.5} = 0.4$$

Area lying between – 0.8 and 0.4

= Area between 0 to 0.8 + Area between 0 to 0.4

$= 0.2881 + 0.1554 = 0.4435$

Reqd. no. of students $= 1000 \times 0.4435 = 444$ (app.)

(*ii*)
$$z = \frac{18 - 14}{2.5} = 1.6$$

Area right to 1.6 = 0.5 – (Area between 0 and 1.6) $= 0.5 - 0.4452 = 0.0548$

Reqd. no. of students $= 1000 \times 0.0548 = 54.8 \approx 55$ (app.)

(*iii*)
$$z = \frac{8 - 14}{2.5} = -2.4$$

Area left to – 2.4 = 0.5 – (Area between 0 and 2.4) $= 0.5 - 0.4918 = 0.0082$

$\therefore$ Reqd. no. of students $= 1000 \times 0.0082 = 8.2 \approx 8$ (app.).

(*iv*)
$$z_1 = \frac{15.5 - 14}{2.5} = 0.6$$
$$z_2 = \frac{16.5 - 14}{2.5} = 1$$

Area between 0.6 and 1 $= 0.3413 - 0.2257 = 0.1156$

$\therefore$ Reqd. no. of students $= 1000 \times 0.1156 = 115.6 \approx 116$ (app.).

Example 3. *Assume mean height of soldiers to be 68.22 inches with a variance of 10.8 inches square. How many soldiers in a regiment of 1,000 would you expect to be over 6 feet tall, given that the area under the standard normal curve between z = 0 and z = 0.35 is 0.1368 and between z = 0 and z = 1.15 is 0.3746.* [G.B.T.U. (C.O.) 2011]

Sol. x = 6 feets = 72 inches

$$\therefore \quad z = \frac{x-\mu}{\sigma} = \frac{72-68.22}{\sqrt{10.8}} = 1.15$$

$$P(x > 72) = P(z > 1.15) = 0.5 - P(0 \le z \le 1.15)$$
$$= 0.5 - 0.3746 = 0.1254$$

$\therefore$ Expected no. of soldiers = 1000 × 0.1254 = 125.4 ≈ 125 (app.).

Example 4. *A large number of measurement is normally distributed with a mean 65.5″ and S.D. of 6.2″. Find the percentage of measurements that fall between 54.8″ and 68.8″.*

Sol. Mean μ = 65.5 inches, S.D. σ = 6.2 inches

x_1 = 54.8 inches, x_2 = 68.8 inches

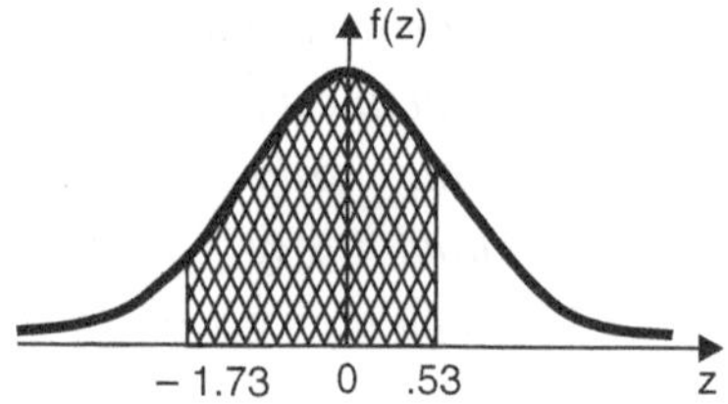

$$\therefore \quad z_1 = \frac{x_1-\mu}{\sigma} = \frac{54.8-65.5}{6.2} = -1.73$$

and $$z_2 = \frac{x_2-\mu}{\sigma} = \frac{68.8-65.5}{6.2} = 0.53$$

Now, $$P(-1.73 \le z \le 0.53) = P(-1.73 \le z \le 0) + P(0 \le z \le 0.53)$$
$$= P(0 \le z \le 1.73) + P(0 \le z \le 0.53)$$
$$= 0.4582 + 0.2019 = 0.6601 \qquad | \text{ By table}$$

$\therefore$ Reqd. percentage of measurements = 66.01%.

Example 5. *A manufacturer knows from experience that the resistance of resistors he produces is normal with mean μ = 100 ohms and standard deviation σ = 2 ohms. What percentage of resistors will have resistance between 98 ohms and 102 ohms?*

Sol. $\mu = 100\ \Omega, \quad \sigma = 2\ \Omega, \quad x_1 = 98\ \Omega, \quad x_2 = 102\ \Omega$

$$\therefore \quad z_1 = \frac{x_1-\mu}{\sigma} = \frac{98-100}{2} = -1$$

and $$z_2 = \frac{x_2-\mu}{\sigma} = \frac{102-100}{2} = 1.$$

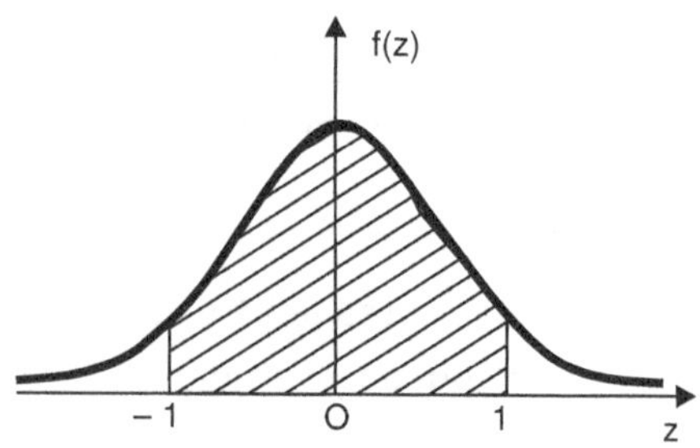

Now, $$P(98 < x < 102) = P(-1 < z < 1)$$
$$= P(-1 \le z \le 0) + P(0 \le z \le 1)$$
$$= P(0 \le z \le 1) + P(0 \le z \le 1)$$
$$= 0.3413 + 0.3413 = 0.6826.$$

$\therefore$ Percentage of resistors having resistance between 98 Ω and 102 Ω = 68.26%.

Example 6. *In a normal distribution, 31% of the items are under 45 and 8% are over 64. Find the mean and standard deviation of the distribution. It is given that if $f(t) = \frac{1}{\sqrt{2\pi}}\int_0^t e^{-\frac{1}{2}x^2}\,dx$ then f(0.5) = 0.19 and f(1.4) = 0.42.* (A.K.T.U. 2019, 2018)

Sol. Let μ and σ be the mean and S.D. respectively.

31% of the items are under 45.

$\Rightarrow$ Area to the left of the ordinate $x = 45$ is 0.31

When $x = 45$, let $z = z_1$

$P(z_1 < z < 0) = 0.5 - 0.31 = 0.19$

From the tables, the value of z corresponding to this area is 0.5

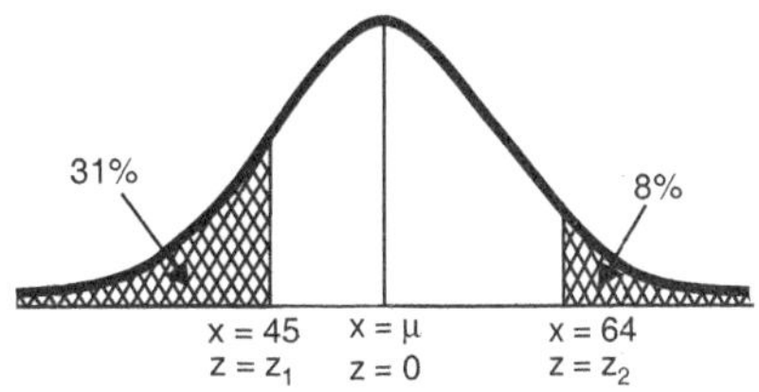

$$\therefore \quad z_1 = -0.5 [z_1 < 0]$$

When $x = 64$, let $z = z_2$

$$P(0 < z < z_2) = 0.5 - 0.08 = 0.42$$

From the tables, the value of z corresponding to this area is 1.4.

$$\therefore \quad z_2 = 1.4$$

Since, $$z = \frac{x - \mu}{\sigma}$$

$$\therefore \quad -0.5 = \frac{45 - \mu}{\sigma} \quad \text{and} \quad 1.4 = \frac{64 - \mu}{\sigma}$$

$$\Rightarrow \quad 45 - \mu = -0.5\sigma \qquad ...(1)$$

and $$64 - \mu = 1.4\sigma \qquad ...(2)$$

Subtracting $$-19 = -1.9\sigma \quad \therefore \quad \sigma = 10$$

From (1), $$45 - \mu = -0.5 \times 10 = -5 \quad \therefore \quad \mu = 50.$$

Example 7. *The life of army shoes is normally distributed with mean 8 months and standard deviation 2 months. If 5000 pairs are insured, how many pairs would be expected to need replacement after 12 months?* $\left[\textit{Given that } P(z \geq 2) = 0.0228 \textit{ and } z = \frac{x - \mu}{\sigma}\right]$.

(A.K.T.U. 2018)

Sol. Mean $(\mu) = 8$, Standard Deviation $(\sigma) = 2$

Number of pairs of shoes = 5000, Total months $(x) = 12$

when $x = 12$, $$z = \frac{x - \mu}{\sigma} = \frac{12 - 8}{2} = 2$$

Area $(z \geq 2) = 0.0228$

Number of pairs whose life is more than 12 months = $5000 \times 0.0228 = 114$

Pair of shoes needing replacement after 12 months = $5000 - 114 = 4886$.

Example 8. *The mean inside diameter of a sample of 200 washers produced by a machine is 0.502 cm and the standard deviation is 0.005 cm. The purpose for which these washers are intended allows a minimum tolerance in the diameter of 0.496 to 0.508 cm, otherwise the washers are considered defective. Determine the percentage of defective washers produced by the machine. Assume the diameters are normally distributed.*

Sol. Given: Mean $\mu = 0.502$ cm, S.D. $\sigma = 0.005$ cm, $x_1 = 0.496$ cm, $x_2 = 0.508$ cm.

Now, $$z_1 = \frac{x_1 - \mu}{\sigma} = \frac{0.496 - 0.502}{0.005} = -1.2$$

$$z_2 = \frac{x_2 - \mu}{\sigma} = \frac{0.508 - 0.502}{0.005} = 1.2$$

Area for non-defective washers

$= P(-1.2 \le z \le 1.2)$

$= P(-1.2 \le z \le 0) + P(0 \le z \le 1.2)$

$= P(0 \le z \le 1.2) + P(0 \le z \le 1.2)$

$= 0.3849 + 0.3849 = 0.7698$

$= 76.98\%$.

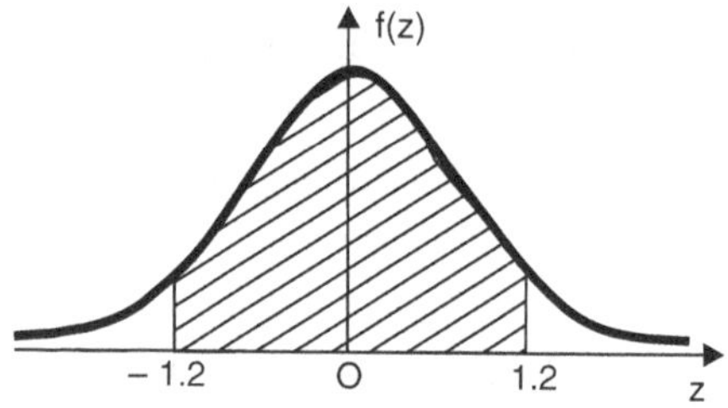

$\therefore$ Percentage of defective washers $= 100 - 76.98$

$= 23.02\%$.

Example 9. *Assuming that the diameters of 1000 brass plugs taken consecutively from a machine, form a normal distribution with mean 0.7515 cm and standard deviation 0.002 cm, how many of the plugs are likely to be rejected if the approved diameter is 0.752 ± 0.004 cm.*

Sol. Tolerance limits of the diameter of non-defective plugs are

$0.752 - 0.004 = 0.748$ cm. and $0.752 + 0.004 = 0.756$ cm.

Standard normal variable, $z = \dfrac{x - \mu}{\sigma}$

If $x_1 = 0.748$, $z_1 = \dfrac{0.748 - 0.7515}{0.002} = -1.75$

If $x_2 = 0.756$, $z_2 = \dfrac{0.756 - 0.7515}{0.02} = 2.25$

Area from $(z_1 = -1.75)$ to $(z_2 = 2.25)$

$= P(-1.75 \le z \le 2.25) = P(-1.75 \le z \le 0) + P(0 \le z \le 2.25)$

$= P(0 \le z \le 1.75) + P(0 \le z \le 2.25) = 0.4599 + 0.4878 = 0.9477$

Number of plugs which are likely to be rejected $= 1000 \times (1 - 0.9477) = 1000 \times .0523 = 52.3$

Hence approximately 52 plugs are likely to be rejected.

Example 10. *If the heights of 300 students are normally distributed with mean 64.5 inches and standard deviation 3.3 inches, find the height below which 99% of the students lie.*

Sol. Mean $\mu = 64.5$ inches, S.D. $\sigma = 3.3$ inches

Area between 0 and $\dfrac{x - 64.5}{3.3} = 0.99 - 0.5 = 0.49$

From the table, for the area 0.49, $z = 2.327$

The corresponding value of x is given by

$$\frac{x - 64.5}{3.3} = 2.327$$

$\Rightarrow$ $x - 64.5 = 7.68$

$\Rightarrow$ $x = 7.68 + 64.5 = 72.18$ inches.

Hence 99% students are of height less than 6 ft. 0.18 inches.

Example 11. *The income of a group of 10,000 persons was found to be normally distributed with mean ₹ 750 p.m. and standard deviation of ₹ 50. Show that, of this group, about 95% had income exceeding ₹ 668 and only 5% had income exceeding ₹ 832. Also find the lowest income among the richest 100.*

Sol. Given: $\mu = 750, \sigma = 50$

Standard normal variable, $z = \dfrac{x - \mu}{\sigma}$

(*i*) If $x_1 = 668$, $z_1 = \dfrac{x_1 - \mu}{\sigma} = \dfrac{668 - 750}{50} = -1.64$

$$\begin{aligned} P(x_1 > 668) &= P(z_1 > -1.64) \\ &= 0.5 + P(-1.64 \le z \le 0) \\ &= 0.5 + P(0 \le z \le 1.64) \\ &= 0.5 + 0.4495 \\ &= 0.9495 \end{aligned}$$

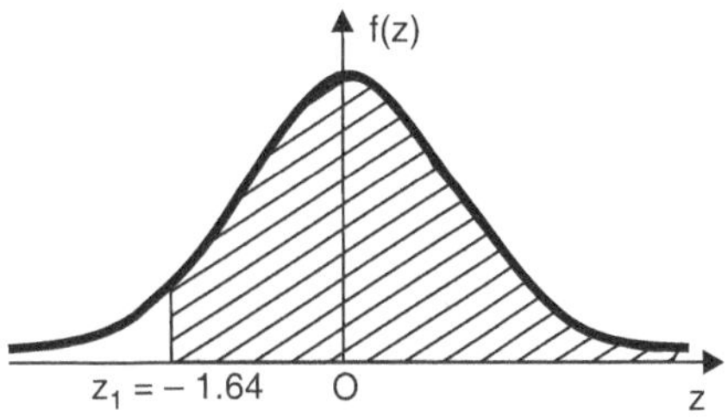

∴ Required percentage of persons having income exceeding ₹ 668 = 94.95% ≈ 95% (approx.)

(*ii*) If $x_2 = 832$, $z_2 = \dfrac{x_2 - \mu}{\sigma} = \dfrac{832 - 750}{50} = 1.64$

$$\begin{aligned} P(x_2 > 832) &= P(z_2 > 1.64) \\ &= 0.5 - P(0 \le z \le 1.64) \\ &= 0.5 - 0.4495 = 0.0505 \end{aligned}$$

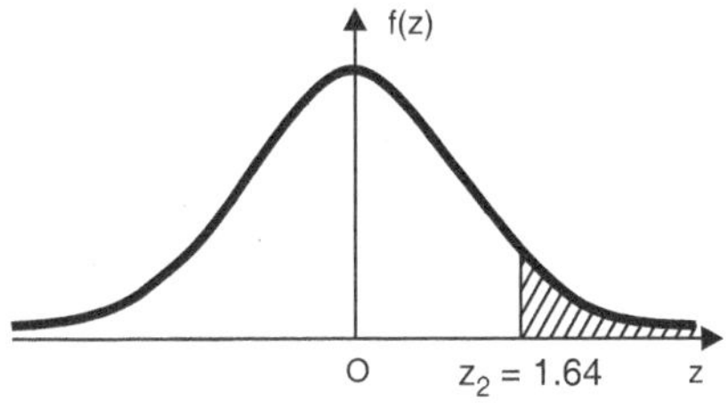

∴ Required percentage of persons having income exceeding ₹ 832 = 5.05% ≈ 5% (approx.)

(*iii*) Let x be the lowest income among the richest 100 persons *i.e.,* 1% of 10,000.

Thus, area between O and $z = 0.49$ (see figure) by Normal distribution table,

$$z = 2.33$$

Thus, $\dfrac{x - \mu}{\sigma} = 2.33$

$\Rightarrow \quad \dfrac{x - 750}{50} = 2.33$

$\Rightarrow \quad x = 866.5$

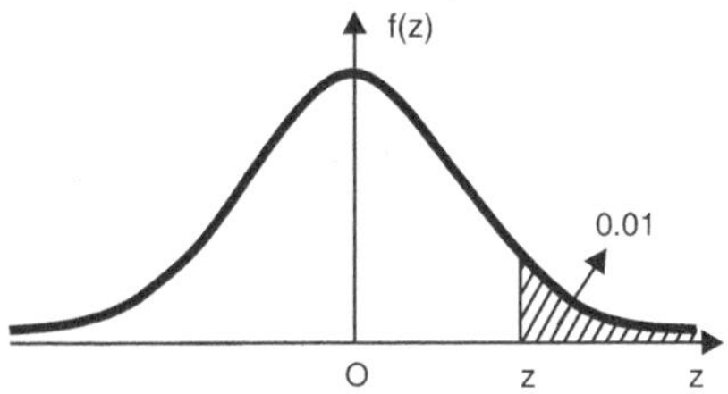

Hence ₹ 866.5 is the minimum income among the richest 100 persons.

Example 12. *255 metal rods were cut roughly 6 inches over size. Finally the lengths of the over size amount, were measured exactly and grouped with 1 inch intervals, there being in all 12 groups* $\frac{1}{2}'' - 1\frac{1}{2}'', 1\frac{1}{2}'' - 2\frac{1}{2}'', \ldots\ldots, 11\frac{1}{2}'' - 12\frac{1}{2}''$.

The frequency distribution for the 255 lengths was as follows:

Length (inches) Central value	*1*	*2*	*3*	*4*	*5*	*6*	*7*	*8*	*9*	*10*	*11*	*12*
Frequency	*2*	*10*	*19*	*25*	*40*	*44*	*41*	*28*	*25*	*15*	*5*	*1*

Fit a normal curve to this data.

Sol. The equation of the normal curve for N observations is

$$y = \frac{N}{\sigma\sqrt{2\pi}} e^{-\frac{1}{2}\left(\frac{x-\mu}{\sigma}\right)^2} \qquad \ldots(1)$$

x	f	$u = x - 6$	fu	fu^2
1	2	– 5	– 10	50
2	10	– 4	– 40	160
3	19	– 3	– 57	171
4	25	– 2	– 50	100
5	40	– 1	– 40	40
6	44	0	0	0
7	41	1	41	41
8	28	2	56	112
9	25	3	75	225
10	15	4	60	240
11	5	5	25	125
12	1	6	6	36
Total	255		66	1300

Mean, $$\mu = a + \frac{\Sigma fu}{\Sigma f} = 6 + \frac{66}{255} = 6.259$$

Variance, $$\sigma^2 = \frac{\Sigma fu^2}{\Sigma f} - \left(\frac{\Sigma fu}{\Sigma f}\right)^2 = \frac{1300}{225} - \left(\frac{66}{255}\right)^2 = 5.031$$

$\therefore$ $$\sigma = 2.243$$

Thus, we have N = 255, Mean, $\mu = 6.259''$, S.D. $\sigma = 2.243''$

Hence the fitted curve is

$$y = \frac{255}{2.243\sqrt{2\pi}} e^{-\frac{1}{2}\left(\frac{x-6.259}{2.243}\right)^2} \quad | \text{ From (1)}$$

$$= \frac{113.68}{\sqrt{2\pi}} e^{-0.099(x-6.259)^2}.$$

Example 13. *Show that the area under the normal curve is unity.*

Sol. Area under the normal curve is given by

$$A = \int_{-\infty}^{\infty} \frac{1}{\sigma\sqrt{2\pi}} e^{-\frac{(x-\mu)^2}{2\sigma^2}} dx$$

Put $\frac{x-\mu}{\sigma} = z$ so $dx = \sigma\, dz$

$\therefore$ $$A = \int_{-\infty}^{\infty} \frac{1}{\sigma\sqrt{2\pi}} e^{-z^2/2} (\sigma dz) = \sqrt{\frac{2}{\pi}} \int_0^{\infty} e^{-z^2/2} dz$$

Now, $$A \cdot A = A^2 = \left(\sqrt{\frac{2}{\pi}} \int_0^{\infty} e^{-x^2/2} dx\right)\left(\sqrt{\frac{2}{\pi}} \int_0^{\infty} e^{-y^2/2} dy\right)$$

$$= \frac{2}{\pi} \int_0^{\infty} \int_0^{\infty} e^{-\left(\frac{x^2+y^2}{2}\right)} dx\, dy \quad | \text{ where } x \text{ and } y \text{ are dummy variables}$$

Put $x = r \cos \theta, y = r \sin \theta$ so that J = r changing to polar coordinates,

$$A^2 = \frac{2}{\pi} \int_0^{\pi/2} \int_0^{\infty} e^{-r^2/2}\, r\, dr\, d\theta = \int_0^{\infty} e^{-r^2/2}\, d\left(\frac{r^2}{2}\right) = 1$$

$\therefore$ A = Area under the normal curve = 1

Example 14. *Prove that for normal distribution, the mean deviation from the mean equals to $\frac{4}{5}$ of the standard deviation approximately.*

Sol. Let μ and σ be the mean and standard deviation of the normal distribution. Then by definition,

Mean deviation from the mean

$$= \int_{-\infty}^{\infty} |x - \mu|\, f(x)\, dx$$

$$= \frac{1}{\sigma\sqrt{2\pi}} \int_{-\infty}^{\infty} |x - \mu|\, e^{-\frac{(x-\mu)^2}{2\sigma^2}}\, dx$$

$$= \frac{1}{\sigma\sqrt{2\pi}} \int_{-\infty}^{\infty} \sigma |z|\, e^{-\frac{1}{2}z^2}\, \sigma\, dz \qquad \left|\ \begin{array}{l} \text{when } \dfrac{x-\mu}{\sigma} = z \\ \Rightarrow\ dx = \sigma\, dz \end{array}\right.$$

$$= \sigma \sqrt{\frac{2}{\pi}} \int_0^{\infty} z\, e^{-z^2/2}\, dz$$

$$= \sigma \sqrt{\frac{2}{\pi}} \left[- e^{-z^2/2}\right]_0^{\infty} = \sqrt{\frac{2}{\pi}}\, \sigma = 0.7979\, \sigma \approx 0.8\sigma \approx \frac{4}{5}\sigma$$

TEST YOUR KNOWLEDGE

1. In a test on 2000 electric bulbs, it was found that the life of a particular make, was normally distributed with an average life of 2040 hours and S.D. of 60 hours, estimate the number of bulbs likely to burn for

   (*i*) more than 2150 hours (*ii*) less than 1950 hours

   (*iii*) more than 1920 hours but less than 2160 hours.

2. An aptitude test for selecting officers in a bank is conducted on 1000 candidates. The average score is 42 and the standard deviation of score is 24. Assuming normal distribution for the scores, find

   (*i*) the number of candidates whose scores exceed 60.

   (*ii*) the number of candidates whose scores lie between 30 and 60.

3. (*i*) In a normal distribution exactly normal, 7% of the items are under 35 and 89% are under 63. What are the mean and standard deviation of the distribution?

   (*ii*) In a normal distribution, 0.0107 of the items lie below 42 and 0.0446 of the items lie above 82. What is the mean and standard deviation of the normal distribution?

4. If Z is a standard normal variable, find the following probabilities:

   (*i*) P(Z < 1.2) (*ii*) P(Z > – 1.2) (*iii*) P(– 1.2 < Z < 1.3).

5. An aptitude test was conducted on 900 employees of the Metro Tyres Limited in which the mean score was found to be 50 units and standard deviation was 20. On the basis of this information, you are required to answer the following questions:

(*i*) What was the number of employees whose mean score was less than 30?

(*ii*) What was the number of employees whose mean score exceeded 70?

(*iii*) What was the number of employees whose mean score were between 30 and 70?

$\frac{x-\mu}{\sigma}$	0.25	0.50	0.70	1.00	1.25	1.50
Area	0.0987	0.1915	0.2734	0.3413	0.3944	0.4332

[*U.P.T.U. (MBA) 2009*]

6. (*a*) Students of a class were given a mechanical aptitude test. Their marks were found to be normally distributed with mean 60 and standard deviation 5. What percent of students scored?

(*i*) more than 60 marks? (*ii*) less than 56 marks? (*iii*) between 45 and 65 marks?

(*b*) 2000 students appeared in an examination. Distribution of marks is assumed to be normal with mean $\mu = 30$ and $\sigma = 6.25$. How many students are expected to get marks?

(*i*) between 20 and 40 (*ii*) less than 35 and (*iii*) above 50.

[*U.P.T.U. (MBA) 2012*]

(*c*) Suppose the weight W of 600 male students are normally distributed with mean $\mu = 70$ kg and standard deviation $\sigma = 5$ kg. Find number of students with weight

(*i*) between 69 and 74 kg (*ii*) more than 76 kg. (*G.B.T.U. 2013*)

7. (*a*) In an intelligence test administered to 1000 students, the average score was 42 and standard deviation 24. Find:

(*i*) the expected number of students scoring more than 50.

(*ii*) the number of students scoring between 30 and 54.

(*iii*) the value of score exceeded by top 100 students. [*G.B.T.U. (MBA) 2010*]

(*b*) The average monthly sales of 5000 firms are normally distributed. Its mean and standard deviation are ₹ 36000 and ₹ 10000 respectively. Find:

(*i*) the no. of firms having sales over ₹ 40000.

(*ii*) the no. of firms having sales between ₹ 30000 and ₹ 40000.

[Given area under normal curve from 0 to z for Z (0.4) = 0.1554 and Z (0.6) = 0.2257]

[*G.B.T.U. (MBA) 2010*]

(*c*) The daily wages of 1000 workers are distributed around a mean of ₹ 140 and with a standard deviation of ₹ 10. Estimate the number of workers whose daily wages will be

(*i*) between ₹ 140 and ₹ 144 (*ii*) less than ₹ 126

(*iii*) more than ₹ 160. (*G.B.T.U. 2012*)

8. (*a*) Records kept by the goods inwards department of a large factory show that the average no. of lorries arriving each week is 248. It is known that the distribution approximates to be normal with a standard deviation of 26.

If this pattern of arrival continues, what percentage of weeks can be expected to have number of arrivals of:

(*i*) less than 229 per week? (*ii*) more than 280 per week?

(*b*) Pipes for tobacco are being packed in fancy plastic boxes. The length of the pipe is normally distributed with $\mu = 5''$ and $\sigma = 0.1''$. The internal length of the boxes is 5.2″. What is the probability that the box would be small for the pipe?

[Given that : $\phi(1.8) = 0.9641$, $\phi(2) = 0.9772$, $\phi(2.5) = 0.9938$]

(*c*) A manufacturer of envelopes knows that the weight of the envelopes is normally distributed with mean 1.9 gm and variance 0.01 square gm. Find how many envelopes weighing

(*i*) 2 gm or more

(*ii*) 2.1 gm or more, can be expected in a given packet of 1000 envelopes? (*U.P.T.U. 2015*)

[Given: if t is the normal variable, then $\phi(0 \le t \le 1) = 0.3413$ and $\phi(0 \le t \le 2) = 0.4772$]

9. (*a*) East-East Airlines has the policy of employing only Indian woman whose height is between 62 inches and 69 inches. If the height of Indian women is approximately normally distributed with a mean of 64 inches and a standard deviation of 3 inches, Out of the 1000 applications received find the number of applicants that would be

(*i*) too tall (*ii*) too short (*iii*) of acceptable height.

(*b*) The mean height of 500 students is 151 cm and the standard deviation is 15 cm. Assuming that the heights are normally distributed, find how many students have heights between 120 and 155 cm?

(*c*) The monthly mess bill of a student who is staying in the hostel follows a normal distribution with a mean of ₹ 2000 and a standard deviation of ₹ 185. What is the probability that in the next month, his bill will go above ₹ 2400? [*U.P.T.U. (MBA) 2009*]

10. In an examination taken by 500 candidates, the average and the standard deviation of marks obtained (normally distributed) are 40% and 10%. Find approximately

(*i*) how many will pass, if 50% is fixed as a minimum?

(*ii*) what should be the minimum if 350 candidates are to pass?

(*iii*) how many have scored marks above 60%?

11. The income distribution of workers in a certain factory was found to be normal with mean ₹ 500 and standard deviation equal to ₹ 50. There were 228 persons getting above ₹ 600. How many workers were there in all?

(Area under the standard normal curve between height at 0 and 2 is 0.4772).

12. In a certain examination, the percentage of passes and distinction were 46 and 9 respectively. Estimate the average marks obtained by the candidates, the minimum pass and distinction marks being 40 and 75 respectively. Assume the distribution of marks to be normal.

$$\left[\textbf{Hint. } \frac{40-\mu}{\sigma} = 0.1 \text{ and } \frac{75-\mu}{\sigma} = 1.34.\right]$$

13. The marks obtained by a no. of students for a certain subject are assumed to be approximately normally distributed with mean value 65 and S.D. of 5. If 3 students are taken at random from this set, what is the probability that exactly 2 of them will have marks over 70?

14. In an examination, it is laid down that a student passes if he secures 30% or more marks. He is placed in the first, second or third division according as he secures 60% or more marks, between 45% and 60% marks and marks between 30% and 45% respectively. He gets distinction incase he secures 80% or more marks. It is noticed from the result that 10% of the students failed in the examination, whereas 5% of them obtained distinction.

Calculate the percentage of students placed in the second division. (Assume normal distribution of marks).

[**Hint.** $P(X < 30) = 0.10$, $P(X \geq 80) = 0.05$]

15. Determine the minimum marks a student must get in order to receive an A grade if the top 10% of the students are awarded A grades in an examination where the mean mark is 72 and standard deviation is 9.

16. (*i*) When the mean of marks was 50% and S.D. 5% then 60% of the students failed in an examination. Determine the 'grace' marks to be awarded in order to show that 70% of the students passed. Assume that the marks are normally distributed.

$$\left[\textbf{Hint. } \frac{x_1 - 0.5}{0.05} = 0.25, \quad \frac{x_2 - 0.5}{0.05} = -0.52\right]$$

(*ii*) In a university examination of a particular year, 60% of the students failed when mean of the marks was 50% and S.D. 5%. University decided to relax the conditions of passing by lowering the pass marks to show its result 70%. Find the minimum marks for a student to pass supposing the marks to be normally distributed and no change in the performance of students takes place.

17. How does a normal distribution differ from a binomial distribution? What are the important properties of normal distribution? *[M.T.U. (MBA) 2012]*

18. If the skulls are classified as A, B and C according as the length-breadth index is under 75, between 75 and 80 or over 80, find approximately (assuming that the distribution is normal) the mean and standard deviation of a series in which A are 58%, B are 38% and C are 4%, being given

that if $f(t) = \frac{1}{\sqrt{2\pi}} \int_0^t e^{-(x^2/2)}\, dx$ then $f(0.20) = 0.08$ and $f(1.75) = 0.46$.

[**Hint:** $P(X < 75) = 0.58$, $P(X > 80) = 0.04$]

19. The following table gives frequencies of occurrence of a variable X between certain limits:

Variable X	*Frequency*
Less than 40	30
40 or more but less than 50	33
50 and more	37

The distribution is exactly normal. Find the distribution and also obtain the fequency between X = 50 and X = 60.

20. The marks X obtained in Mathematics by 1000 students are normally distributed with mean 78% and standard deviation 11%.

Determine:

(*i*) how many students got marks above 90%?

(*ii*) What was the highest marks obtained by the lowest 10% of students?

(*iii*) Within what limits did the middle 90% of the students lie?

Answers

1. (*i*) 67 (*ii*) 134 (*iii*) 1909 **2.** (*i*) 227 (*ii*) 465
3. (*i*) $\bar{x} = 50.3, \sigma = 10.33$ (*ii*) $\mu = 65, \sigma = 10$
4. (*i*) 0.8849 (*ii*) 0.8849 (*iii*) 0.7881
5. (*i*) 143 (*ii*) 143 (*iii*) 614
6. (*a*) (*i*) 50%, (*ii*) 21.2%, (*iii*) 84% (*b*) (*i*) 1781, (*ii*) 1576, (*iii*) 1 (*c*) (*i*) 220, (*ii*) 69
7. (*a*) (*i*) 371, (*ii*) 383, (*iii*) 72.72 (*b*) (*i*) 1723 (*ii*) 1906 (*c*) (*i*) 155, (*ii*) 81, (*iii*) 23
8. (*a*) (*i*) 23% (*ii*) 11% (*b*) 0.0228 (*c*) (*i*) 159 (*ii*) 23
9. (*a*) (*i*) 48 (*ii*) 251 (*iii*) 701 (*b*) 294 (*c*) 0.0154
10. (*i*) 79 (*ii*) 35% (*iii*) 11 **11.** 10,000 **12.** 37.2
13. 0.06357 **14.** 34% **15.** 84 marks
16. (*i*) 3.85 (*ii*) 47.4. **18.** $\mu = 74.35, \sigma = 3.23$ **19.** $\mu = 46.12, \sigma = 11.76$, 25
20. (*i*) 138 (*ii*) 63.92% (*iii*) between 60 and 96.

ASSIGNMENT-IV

(For Section–A)

1. Define a random variable. Define expectation.

2. Two events A and B have probabilities 0.25 and 0.50 respectively. The probability that both events A and B occurs in 0.14. Find the probability that neither A nor B occurs. *(M.T.U. 2013)*

$\left[\textbf{Hint. } P(\overline{A} \cap \overline{B}) = P(\overline{A \cup B}) = 1 - P(A \cup B)\right]$

3. Define probability density function and cumulative distribution function.

4. What is the total probability theorem? *(M.T.U. 2012)*

5. Define marginal and conditional distribution.

6. Find the moment generating function of Poisson distribution. *(M.T.U. 2013)*

7. Find the parameters p and q of the Binomial distribution whose mean is 9 and variance is $\frac{9}{4}$. *(M.T.U. 2012)*

8. If the sum of the mean and variance of a Binomial distribution of 5 trials is $\frac{9}{5}$, find $P(X \geq 1)$. *(M.T.U. 2013)*

9. It has been found that 2% of the tools produced by a certain machine are defective. What is the probability that in a shipment of 400 such tools, 3 or more will be defective? *(M.T.U. 2013)*

10. If $P(X = 0) = P(X = 1) = k$ in a Poisson distribution then what is k?

11. For a Poisson variate X if $P(X = 1) = P(X = 2)$ then find $P(X = 4)$.

12. Find the total area under the curve of *p.d.f.* of a normal curve.

13. If for a Poisson distribution, $P(2) = P(3)$ then what is its probability function?

14. Find the parameters of a binomial distribution with mean = 8 and variance = 4.

15. If X is a normal variate with mean 30 and S.D. 5, find the probabilities that

(*i*) $26 \leq X \leq 40$ (*ii*) $X \geq 45$ and (*iii*) $|X - 30| > 5$

16. Define :

(*i*) Binomial distribution (*ii*) Poisson distribution (*iii*) Normal distribution *(A.K.T.U. 2018)*

17. Find mean, variance and third moment about mean for the binomial distribution. *(A.K.T.U. 2018)*

18. Find mean and variance of poisson distribution. *(A.K.T.U. 2018)*

Answers

2. 0.39 **6.** $M_x(t) = e^{\lambda(e^t - 1)}$ **7.** $q = \frac{1}{4}, p = \frac{3}{4}$ **8.** 0.67232

9. 0.9862 **10.** $\frac{1}{e}$ **11.** $\frac{2}{3e^2}$ **12.** 1

13. $\frac{e^{-3}(3)^x}{x\,!}$ **14.** $n = 16, p = \frac{1}{2}$

15. (*i*) 0.7653 (*ii*) 0.00135 (*iii*) 0.3174. **18.** $\lambda = np$

MODULE 5

Statistical Techniques–III

5.1 POPULATION OR UNIVERSE

An aggregate of objects (animate or inanimate) under study is called **population or universe.** It is thus a collection of individuals or of their attributes (qualities) or of results of operations which can be numerically specified.

A universe containing a finite number of individuals or members is called a **finite inverse.** For example, the universe of the weights of students in a particular class.

A universe with infinite number of members is known as an **infinite universe.** For example, the universe of pressures at various points in the atmosphere.

In some cases, we may be even ignorant whether or not a particular universe is infinite, *e.g.*, the universe of stars.

The universe of concrete objects is an **existent universe.** The collection of all possible ways in which a specified event can happen is called a **hypothetical universe.** The universe of heads and tails obtained by tossing a coin an infinite number of times (provided that it does not wear out) is a hypothetical one.

5.2 SAMPLING

The statistician is often confronted with the problem of discussing universe of which he cannot examine every member *i.e.*, of which complete enumeration is impracticable. For example, if we want to have an idea of the average per capita income of the people of India, enumeration of every earning individual in the country is a very difficult task. Naturally, the question arises : What can be said about a universe of which we can examine only a limited number of members ? This question is the origin of the Theory of Sampling.

A finite subset of a universe is called a **sample**. A sample is thus a small portion of the universe. The number of individuals in a sample is called the **sample size**. The process of selecting a sample from a universe is called **sampling**.

The theory of sampling is a study of relationship existing between a population and samples drawn from the population. The fundamental objective of sampling is to get as much information as possible of the whole universe by examining only a part of it. An attempt is thus made through sampling to give the maximum information about the parent universe with the minimum effort.

Sampling is quite often used in our day-to-day practical life. For example, in a shop we assess the quality of sugar, rice or any other commodity by taking only a handful of it from the bag and then decide whether to purchase it or not. A housewife normally tests the cooked products to find if they are properly cooked and contain the proper quantity of salt or sugar, by taking a spoonful of it.

5.3 SAMPLING METHODOLOGIES

Sampling methodologies are classified under two general categories:

1. Probability sampling and
2. Non-probability sampling

In the former, the researcher knows the exact possibility of selecting each member of the population while in the latter, the chance of being included in the sample is not known. A probability sample tends to be more difficult and costly to conduct. However, probability samples are the only type of samples where the results can be generalized from the sample to the population. In addition, probability samples allow the researcher to calculate the precision of the estimates obtained from the sample and to specify the sampling error.

Non-probability samples, incontrast, donot allow the study's findings to be generalized from the sample to the population. When discussing the results of a non-probability sample, the researchers must limit his/her findings to the persons or elements sampled.

This procedure also does not allow the researcher to calculate sampling statistics that provide information about the precision of the results. The advantage of non-probability sampling is the case in which it can be administered.

Non-probability samples tend to be less complicated and less time consuming than probability samples. If the researcher has no intention of generalizing beyond the sample, one of the non-probability sampling methodologies will provide the desired information.

5.4 NON-PROBABILITY SAMPLING

The three common types of non-probability samples are:

***(i)* Convenience Sampling:** As the name implies, convenience sampling involves choosing respondents at the convenience of the researcher. Examples of convenience sampling include people-in-the street interviews—the sampling of people to which the researcher has easy access, such as a class of students and studies that use people who have volunteered to be questioned as a result of an advertisement or another type of promotion. A drawback to this methodology is the lack of sampling accuracy. Because the probability of inclusion in the sample is unknown for each respondent, none of the reliability or sampling precision statistics can be calculated. Convenience samples, however, are employed by researchers because the time and cost of collecting information can be reduced.

***(ii)* Quota Sampling:** Quota sampling is often confused with stratified and cluster sampling—two probability sampling methodologies. All of these methodologies sample a population that has been subdivided into classes or categories.

The primary differences between the methodologies is that with stratified and cluster sampling, the classes are mutually exclusive and are isolated prior to sampling. Thus, the probability of being selected is known and members of the population selected to be sampled are not arbitrarily disqualified from being included in the results. In quota sampling, the classes cannot be isolated prior to sampling and respondents are categorized into the classes as the survey proceeds. As each class fills or reaches its quota, additional respondents that would have fallen into these classes are rejected or excluded from the results.

An example of a quota sample would be a survey in which the researcher desires to obtain a certain number of respondents from various income categories. Generally, researchers donot know the income of the persons they are sampling until they ask about income.

Therefore, the researcher is unable to subdivide the population from which the sample is drawn into mutually exclusive income categories prior to drawing the sample.

***(iii)* Judgemental Sampling:** In judgemental or purposive sampling, the researcher employs his or her own expert judgement about who to include in the sample frame. Prior knowledge and research skill are used in selecting the respondents or elements to be sampled.

An example of this type of sample would be a study of potential users of a new recreational facility that is limited to those persons who live within two miles of the new facility. Expert judgement based on past experience indicates that most of the use of this type of facility comes from persons living within two miles. However, by limiting the sample to only this group, usage projections may not be reliable if the usage characteristics of the new facility vary from those previously experienced. As with all non-probability sampling methods, the degree and direction of error introduced by the researcher cannot be measured and statistics that measure the precision of the estimates cannot be calculated.

5.5 PROBABILITY SAMPLING

Five methodologies are most commonly used for conducting probability sampling.

***(i)* Simple Random Sampling:** Simple random sampling provides the base from which the other more complex sampling methodologies are derived.

To conduct a simple random sampling, the researcher must first prepare an exhaustive list (sampling frame) of all members of the population of interest. From this list, the sample is drawn so that each person or item has an equal chance of being drawn during each selection round. Samples may be drawn with or without replacement. In practice, however, most simple random sampling for survey research is done without replacement ; that is, a person or item selected for sampling is removed from the population for all subsequent selections. At any draw, the process for a simple random sample without replacement must provide an equal chance of inclusion to any member of the population not already drawn. To draw a simple random sample without introducing researcher bias, computerized sampling programs and random numbers tables are used to impartially select the members of the population to be sampled.

An example of a simple random sample would be a survey of state employees. An exhaustive list of all state employees as of a certain date could be obtained from the Department of Human Resources. If 100 names were selected from this list using a random number table or a computerized sampling program, then a simple random sample would be created. Such a random sampling procedure has the advantage of reducing bias and enables the researcher to estimate sampling errors and the precision of the estimates derived through statistical calculations.

***(ii)* Stratified Random Sampling:** Stratified random sampling involves categorizing the members of the population into mutually exclusive and collectively exhaustive groups. An independent simple random sample is then drawn from each group. Stratified sampling techniques can provide more precise estimates if the population being surveyed is more heterogeneous than the categorized groups, can enable the researcher to determine desired levels of sampling precision for each group, and can provide administrative efficiency.

An example of a stratified sample would be a sample conducted to determine the average income earned by families in India. To obtain more precise estimates of income, the researcher may want to stratify the sample by geographic region and/or stratify the sample by urban, suburban, and rural groupings. If the differences in income among the regions or groupings are greater than the income differences within the regions or groupings, precision of the estimates is improved. In addition, if the research organization has branch offices located in these regions, the administration of the survey can be decentralized and perhaps conducted in a more cost-efficient manner.

***(iii)* Cluster Sampling:** Cluster sampling is similar to stratified sampling because the population to be sampled is subdivided into mutually exclusive groups. However, in cluster sampling, the groups are defined so as to maintain the heterogeneity of the population. It is the researcher's goal to establish clusters that are representative of the population as a whole, although in practice this may be difficult to achieve. After the clusters are established, a simple random sample of the clusters is drawn and the members of the chosen clusters are sampled. If all of the elements (members) of the clusters selected are sampled, then the sampling procedure is defined as **one-stage cluster sampling.** If a random sample of the elements of each selected cluster is drawn, then the sampling procedure is defined as **two-stage cluster sampling**.

Cluster sampling is frequently employed when the researcher is unable to compile a comprehensive list of all the elements in the population of interest. A cluster sample might be used by a researcher attempting to measure the age distribution of persons residing in Mumbai. It would be much more difficult for the researcher to compile a list of every person residing in Mumbai than to compile a list of residential addresses. In this example, each address would represent a cluster of elements (persons) to be sampled. If the elements contained in the clusters are as heterogeneous as the population, then estimates derived from cluster sampling are as precise as those from simple random sampling. However, if the heterogeneity of the clusters is less than that of the population, the estimates will be less precise.

***(iv)* Systematic Sampling:** Systematic sampling, a form of one-stage cluster sampling, is often used in place of simple random sampling. In systematic sampling, the researcher selects every n^{th} member after randomly selecting the first through n^{th} element as the starting point. For example, if the researcher decides to sample every 20th member of the population, a 5 per cent sample, the starting point for the sample is randomly selected from the first 20 members. A systematic sample is a type of cluster sample because each of the first 20 members of the sampling frame defines a cluster that contains 5 percent of the population.

A researcher may choose to conduct a systematic sample instead of a simple random sample for several reasons. Systematic samples tend to be easier to draw and execute. The researcher does not have to jump backward and forward through the sampling frame to draw the members to be sampled. A systematic sample may spread the members selected for measurement more evenly across the entire population than simple random sampling. Therefore, in some cases, systematic sampling may be more representative of the population and more precise.

One of the most attractive aspects of systematic sampling is that this method can allow the researcher to draw a probability sample without complete prior knowledge of the sampling frame. For example, a survey of visitors to the state's publications desk could be conducted by sampling every 10th visitor after randomly selecting the first through 10th visitor as the starting point. By conducting the sample in this manner, it would not be necessary for the researcher to obtain a comprehensive list of visitors prior to drawing the sample.

As with other types of cluster sampling, systematic sampling is as precise as simple random sampling if the members contained in the clusters are as heterogeneous as the population. If this assumption is not valid, then systematic sampling will be less precise than simple random sampling. In conducting systematic sampling, it is also essential that the researcher does not introduce bias into the sample by selecting an inappropriate sampling interval. For instance, when conducting a sample of financial records, or other items that follow a calendar schedule, the researcher would not want to select "7" as the sampling interval because the sample would then be comprised of observations that were all on the same day of the week. Day-of-the-week influences may cause contamination of the sample, giving the researcher biased results.

(*v*) Multi-Stage Sampling: Multi-stage sampling is like cluster sampling, but involves selecting a sample within each chosen cluster, rather than including all units in the cluster. Thus, multi-stage sampling involves selecting a sample in at least two stages. In the first stage, large groups or clusters are selected. These clusters are designed to contain more population units than are required for the final sample.

In the second stage, population units are chosen from selected clusters to derive a final sample. If more than two stages are used, the process of choosing population units within clusters continues until the final sample is achieved.

An example of multi-stage sampling is where, firstly, electoral sub-divisions (clusters) are sampled from a city or state. Secondly, blocks of houses are selected from within the electoral sub-divisions and, thirdly, individual houses are selected from within the selected blocks of houses.

The advantages of multi-stage sampling are convenience, economy and efficiency. Multi-stage sampling does not require a complete list of members in the target population, which greatly reduces sample preparation cost. The list of members is required only for those clusters used in the final stage. The main disadvantage of multi-stage sampling is the same as for cluster sampling: lower accuracy due to higher sampling error.

5.6 PARAMETERS OF STATISTICS

The statistical constants of the population such as mean, the variance etc. are known as the parameters. The statistical concepts of the sample from the members of the sample to estimate the parameters of the population from which the sample has been drawn is known as *statistic*.

Population mean and variance are denoted by μ and σ^2, while those of the samples are given by $\bar{x}$, s^2.

5.7 STANDARD ERROR

The standard deviation of the sampling distribution of a statistic is known as the **standard error (S.E.)**. It plays an important role in the theory of large samples and it forms a basis of the testing of hypothesis. If t is any statistic, for large sample. $z = \dfrac{t - \text{E}(t)}{\text{S.E.}(t)}$ is normally distributed with mean 0 and variance unity.

For large sample, the standard errors of some of the well known statistic are listed below:

n—sample size; σ^2—population variance; s^2—sample variance; p—population proportion; $\text{Q} = 1 - p$; n_1, n_2—are sizes of two independent random samples.

S. No.	Statistic	Standard error
1.	$\bar{x}$	$\sigma/\sqrt{n}$
2.	s	$\sqrt{\sigma^2/2n}$
3.	Difference of two sample means $\bar{x}_1 - \bar{x}_2$	$\sqrt{\dfrac{\sigma_1^2}{n_1} + \dfrac{\sigma_2^2}{n_2}}$
4.	Difference of two sample standard deviation $s_1 - s_2$	$\sqrt{\dfrac{\sigma_1^2}{2n_1} + \dfrac{\sigma_2^2}{2n_2}}$
5.	Difference of two sample proportions $p_1 - p_2$	$\sqrt{\dfrac{P_1Q_1}{n_1} + \dfrac{P_2Q_2}{n_2}}$
6.	Observed sample proportion p	$\sqrt{PQ/n}$

5.8 TEST OF SIGNIFICANCE

An important aspect of the sampling theory is to study the test of significance which will enable us to decide, on the basis of the results of the sample, whether

(i) the deviation between the observed sample statistic and the hypothetical parameter value or

(ii) the deviation between two sample statistics is significant or might be attributed due to chance or the fluctuations of the sampling.

5.9 TESTING OF STATISTICAL HYPOTHESIS

Step 1. Null hypothesis

For applying the tests of significance, we first set up a hypothesis which is a definite statement about the population parameter called Null Hypothesis. It is denoted by H_0.

Null hypothesis is the hypothesis which is tested for possible rejection under the assumption that it is true. First, we set up H_0 in clear terms.

Step 2. Alternative hypothesis

Any hypothesis which is complementary to the null hypothesis (H_0) is called an alternative hypothesis. It is denoted by H_1.

For example, if we want to test the null hypothesis that the population has a specified mean μ_0 then we have

$$H_0 : \mu = \mu_0$$

then the alternative hypothesis will be

(*i*) $H_1 : \mu \neq \mu_0$ (Two tailed alternative hypothesis)

(*ii*) $H_1 : \mu > \mu_0$ (right tailed alternative hypothesis (or) single tailed)

(*iii*) $H_1 : \mu < \mu_0$ (left tailed alternative hypothesis (or) single tailed)

Hence alternative hypothesis helps to know whether the test is two tailed test or one tailed test. Therefore, we set up H_1 for this decision.

Step 3. Level of significance

The probability of the value of the variate falling in the critical region is known as level of significance. A region corresponding to a statistic t in the sample space S which amounts to rejection of the null hypothesis H_0 is called as **critical region** or region of rejection while which amounts to acceptance of H_0 is called **acceptance region.** The probability α that a random value of the statistic t belongs to the critical region is known as the **level of significance.**

$$P(t \in w/H_0) = \alpha$$

i.e., the level of significance is the size of the type I error (refer art. 5.7) or the maximum producer's risk.

We select the appropriate level of significance in advance depending on the reliability of the estimates.

Step 4. Test statistic (or test criterion): We compute the test statistic z under the null hypothesis. For larger samples corresponding to the statistic t, the variable $z = \dfrac{t - E(t)}{S.E.(t)}$ is normally distributed with mean 0 and variance 1. The value of z given above under the null hypothesis is known as **test statistic.**

Step 5. Conclusion: We compare the computed value of z with the critical value z_α at level of significance (α). The critical value of z_α of the test statistic at level of significance α for a two tailed test is given by

$$p(|z| > z_\alpha) = \alpha \qquad \ldots(1)$$

i.e., z_α is the value of z so that the total area of the critical region on both tails is α. Since the normal curve is symmetrical, from equation (1), we get

$$p(z > z_\alpha) + p(z < -z_\alpha) = \alpha; \text{ i.e. } 2p(z > z_\alpha) = \alpha; \text{ i.e., } p(z > z_\alpha) = \alpha/2$$

i.e., the area of each tail is $\alpha/2$.

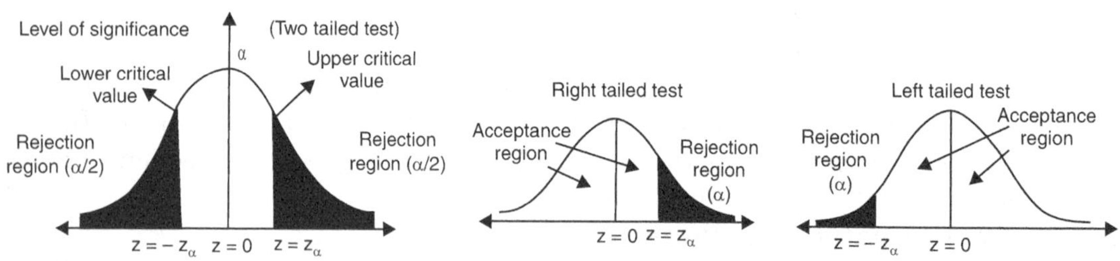

The critical value z_α is that value such that the area to the right of z_α is $\alpha/2$ and the area to the left of $-z_\alpha$ is $\alpha/2$.

In the case of one tailed test,

$p(z > z_\alpha) = \alpha$ if it is right tailed; $p(z < -z_\alpha) = \alpha$ if it is left tailed.

The critical value of z for a single tailed test (right or left) at level of significance α is same as the critical value of z for two tailed test at level of significance 2α.

Using the equation, also using the normal tables, the critical value of z at different levels of significance (α) for both single tailed and two tailed test are calculated and listed below. The equations are

$$p(|z| > z_\alpha) = \alpha; \; p(z > z_\alpha) = \alpha; \; p(z < -z_\alpha) = \alpha$$

	Level of significance		
	1% (0.01)	5% (0.05)	10% (0.1)
Two tailed test	$\lvert z_\alpha \rvert = 2.58$	$\lvert z_\alpha \rvert = 1.96$	$\lvert z_\alpha \rvert = 1.645$
Right tailed test	$z_\alpha = 2.33$	$z_\alpha = 1.645$	$z_\alpha = 1.28$
Left tailed test	$z_\alpha = -2.33$	$z_\alpha = -1.645$	$z_\alpha = -1.28$

If $|z| > z_\alpha$, we reject H_0 and conclude that there is significant difference. If $|z| < z_\alpha$, we accept H_0 and conclude that there is no significant difference.

5.10 ERRORS IN SAMPLING

The main aim of the sampling theory is to draw a valid conclusion about the population parameters on the basis of the sample results. In doing this we may commit the following two types of errors:

Type I Error

When H_0 is true, we may reject it. P(Reject H_0 when it is true) = P(Reject H_0/H_0) = α. α is called the size of the type I error, also referred to as **producer's risk.**

Type II Error: When H_0 is wrong, we may accept it. P(Accept H_0 when it is wrong) = P(Accept H_0/H_1) = β. β is called the size of the type II error, also referred to as **consumer's risk.**

Note. The values of the test statistic which separates the critical region and acceptance region are called the **critical values** or **significant values.** This value is dependent on (*i*) the level of significance used and (*ii*) the alternative hypothesis, whether it is one tailed or two tailed.

5.11 TEST OF SIGNIFICANCE FOR LARGE SAMPLES

If the sample size $n > 30$, the sample is taken as large sample. For such sample we apply distributions, as Binomial, Poisson, which are closely approximated by normal distributions assuming the population as normal.

Under large sample test, the following are the important tests to test the significance:

1. *Testing of significance for single proportion.*
2. *Testing of significance for difference of proportions.*
3. *Testing of significance for single mean.*
4. *Testing of significance for difference of means.*
5. *Testing of significance for difference of standard deviations.*

5.11.1 Testing of Significance for Single Proportion

This test is used to find the significant difference between proportion of the sample and the population. Let X be the number of successes in n independent trials with constant probability P of success for each trial.

$$E(X) = nP;\ V(X) = nPQ;\ Q = 1 - P = \text{Probability of failure.}$$

Let $p = X/n$ called the observed proportion of success.

$$E(p) = E(X/n) = \frac{1}{n}E(X) = \frac{nP}{n} = P$$

$$V(p) = V(X/n) = \frac{1}{n^2}V(X) = \frac{nPQ}{n^2} = PQ/n$$

$$\text{S.E.}(p) = \sqrt{\frac{\text{PQ}}{n}} \ ; \ z = \frac{p - \text{E}(p)}{\text{S.E.}(p)} = \frac{p - \text{P}}{\sqrt{\text{PQ}/n}} \sim \text{N}(0, 1)$$

This z is called test statistic which is used to test the significant difference of sample and population proportion.

Note. 1. The probable limits for the observed proportion of successes are $\text{P} \pm 3\sqrt{\text{PQ}/n}$.

2. If p is not known, the probable limits for the proportion in the population are $p \pm z_\alpha\sqrt{pq/n}$, $q = 1 - p$, where sample proportion, p is taken as an estimate of p and z_α is the significant value of z at level of significance α.

ILLUSTRATIVE EXAMPLES

Example 1. *A coin was tossed 400 times and the head turned up 216 times. Test the hypothesis that the coin is unbiased.*

Sol. Null hypothesis:

H_0: The coin is unbiased *i.e.,* $\text{P} = 0.5$

Alternative hypothesis:

H_1: The coin is biased *i.e.,* $\text{P} \neq 0.5$

Hence we use **two tailed test.**

Here, $n = 400$, X = no. of success = 216

$$\therefore \quad p = \text{proportion of success in the sample} = \frac{\text{X}}{n} = \frac{216}{400} = 0.54$$

P = population proportion = 0.5, $\quad \text{Q} = 1 - \text{P} = 0.5$

Test Statistic:

Under H_0, test statistic $z = \dfrac{p - \text{P}}{\sqrt{\text{PQ}/n}}$

$$|z| = \left| \frac{0.54 - 0.5}{\sqrt{\dfrac{0.5 \times 0.5}{400}}} \right| = 1.6$$

Conclusion:

Since $|z| = 1.6 < 1.96$ *i.e.,* $|z| < z_\alpha$ where z_α is the significant value of z at 5% level of significance.

Hence we accept H_0 and conclude that the coin is unbiased.

Example 2. *A machine is producing bolts of which a certain fraction is defective. A random sample of 400 is taken from a large batch and is found to contain 30 defective bolts. Does this indicate that the proportion of defectives is larger than that claimed by the manufacturer where the manufacturer claims that only 5% of his product are defective. Find 95% confidence limits of the proportion of defective bolts in batch.*

Sol. Null hypothesis H_0: The manufacturer claim is accepted *i.e.,* $P = \frac{5}{100} = 0.05$

$$Q = 1 - P = 1 - 0.05 = 0.95$$

Alternative hypothesis: $P > 0.05$

Hence we use **Right tailed test.**

p = observed proportion of sample $= \frac{30}{400} = 0.075$

Test statistic

Under H_0, the test statistic $z = \frac{p - P}{\sqrt{PQ/n}}$ $\therefore$ $z = \frac{0.075 - 0.05}{\sqrt{\frac{0.05 \times 0.95}{400}}} = 2.2941.$

Conclusion: The tabulated value of z at 5% level of significance for right tailed test is $z_\alpha = 1.645$. Since $|z| = 2.2941 > 1.645$, H_0 is rejected at 5% level of significance. *i.e.,* the proportion of defective is larger than the manufacturer claim.

To find 95% confidence limits of the proportion.

It is given by $P \pm z_\alpha \sqrt{PQ/n}$

i.e., $$0.05 \pm 1.96 \sqrt{\frac{0.05 \times 0.95}{400}} = 0.05 \pm 0.02135 = 0.07136,\ 0.02865$$

Hence 95% confidence limits for the proportion of defective bolts are (0.07136, 0.02865).

TEST YOUR KNOWLEDGE

1. In a hospital 475 female and 525 male babies were born in a week. Do these figures confirm the hypothesis that males and females are born in equal number?
2. In a city a sample of 1000 people were taken and out of them 540 are vegetarian and the rest are non-vegetarian. Can we say that the both habits of eating (vegetarian or non-vegetarian) are equally popular in the city at (*i*) 1% level of significance (*ii*) 5% level of significance?
3. 325 men out of 600 men chosen from a big city were found to be smokers. Does this information support the conclusion that the majority of men in the city are smokers?

Answers

1. H_0 accepted at 5% level **2.** H_0 rejected at 5% level, accepted at 1% level
3. H_0 rejected at 5% level.

5.11.2 Testing of Significance for Difference of Proportions

Consider two samples X_1 and X_2 of sizes n_1 and n_2 respectively taken from two different populations. To test the significance of the difference between the sample proportions p_1 and p_2, the test statistic under the null hypothesis H_0, that there is no significant difference between the two sample proportion, is

$$z = \frac{p_1 - p_2}{\sqrt{PQ\left(\frac{1}{n_1} + \frac{1}{n_2}\right)}}, \quad \text{where} \quad P = \frac{n_1 p_1 + n_2 p_2}{n_1 + n_2} \quad \text{and} \quad Q = 1 - P.$$

ILLUSTRATIVE EXAMPLES

Example 1. *Before an increase in excise duty on tea, 800 people out of a sample of 1000 persons were found to be tea drinkers. After an increase in the duty, 800 persons were known to be tea drinkers in a sample of 1200 people. Do you think that there has been a significant decrease in the consumption of tea after the increase in the excise duty?*

Sol. Here, $n_1 = 800, n_2 = 1200$

$$p_1 = \frac{X_1}{n_1} = \frac{800}{1000} = \frac{4}{5}; \; p_2 = \frac{X_2}{n_2} = \frac{800}{1200} = \frac{2}{3}$$

$$P = \frac{p_1 n_1 + p_2 n_2}{n_1 + n_2} = \frac{X_1 + X_2}{n_1 + n_2} = \frac{800 + 800}{1000 + 1200} = \frac{8}{11}; \; Q = \frac{3}{11}$$

Null hypothesis H_0: $p_1 = p_2$ *i.e.*, there is no significant difference in the consumption of tea before and after increase of excise duty.

Alternative hypothesis H_1 : $p_1 > p_2$

Hence we use **right tailed test.**

Test statistic:

The test statistic is $$z = \frac{p_1 - p_2}{\sqrt{PQ\left(\frac{1}{n_1} + \frac{1}{n_2}\right)}} = \frac{0.8 - 0.6666}{\sqrt{\frac{8}{11} \times \frac{3}{11}\left(\frac{1}{1000} + \frac{1}{1200}\right)}} = 6.842.$$

Conclusion: Since the calculated value of $|z| > 1.645$ and also $|z| > 2.33$, both the significant values of z at 5% and 1% level of significance, hence H_0 is rejected *i.e.*, there is a significant decrease in the consumption of tea due to increase in excise duty.

Example 2. *A machine produced 16 defective articles in a batch of 500. After overhauling it produced 3 defectives in a batch of 100. Has the machine improved?*

Sol. $$p_1 = \frac{16}{500} = 0.032, \quad n_1 = 500$$

$$p_2 = \frac{3}{100} = 0.03, \quad n_2 = 100$$

Null hypothesis:

H_0: The machine has not improved due to overhauling, *i.e.*, $p_1 = p_2$.

Alternative hypothesis:

H_1: $p_1 > p_2$

Hence we use **right tailed test.**

$\therefore$ $$P = \frac{p_1 n_1 + p_2 n_2}{n_1 + n_2} = \frac{19}{600} \cong 0.032$$

Test Statistic:

Under H_0, the test statistic

$$z = \frac{p_1 - p_2}{\sqrt{PQ\left(\frac{1}{n_1} + \frac{1}{n_2}\right)}} = 0.104$$

Conclusion: The calculated value of $|z| < 1.645$ which is the significant value of z at 5% level of significance, H_0 is accepted *i.e.*, the machine has not improved due to overhauling.

TEST YOUR KNOWLEDGE

1. Random sample of 400 men and 600 women were asked whether they would like to have a school near their residence. 200 men and 325 women were in favour of proposal. Test the hypothesis that the proportion of men and women in favour of the proposal are same at 5% level of significance.
2. In a town A, there were 956 births of which 52.5% were males while in towns A and B combined, this proportion in total of 1406 births was 0.496. Is there any significant difference in the proportion of male births in the two towns ?

Answers

1. H_0 : Accepted **2.** H_0 : Rejected.

5.11.3 Testing of Significance for Single Mean

To test whether the difference between sample mean and population mean is significant or not.

Let $X_1, X_2, ..., X_n$ be a random sample of size n from a large population $X_1, X_2, ..., X_N$ of size N with mean μ and variance σ^2.

$\therefore$ the standard error of mean of a random sample of size n from a population with variance σ^2 is $\sigma/\sqrt{n}$.

To test whether the given sample of size n has been drawn from a population with mean μ *i.e.,* to test whether the difference between the sample mean and population mean is significant or not under the null hypothesis that there is no difference between the sample mean and population mean.

The test statistic is $z = \dfrac{\bar{x} - \mu}{\sigma/\sqrt{n}}$, where σ is the standard deviation of the population.

If σ is not known, we use the test statistic $z = \dfrac{\bar{x} - \mu}{s/\sqrt{n}}$, where s is the standard deviation of the sample.

Note. If the level of significance is α and z_α is the critical value $-z_\alpha < | z | = \left| \dfrac{\bar{x} - \mu}{\sigma/\sqrt{n}} \right| < z_\alpha$.

The limits of the population mean μ are given by $\bar{x} - z_\alpha \dfrac{\sigma}{\sqrt{n}} < \mu < \bar{x} + z_\alpha \dfrac{\sigma}{\sqrt{n}}$.

At 5% level of significance, 95% confidence limits are $\bar{x} - 1.96 \dfrac{\sigma}{\sqrt{n}} < \mu < \bar{x} + 1.96 \dfrac{\sigma}{\sqrt{n}}$.

At 1% level of significance, 99% confidence limits are $\bar{x} - 2.58 \dfrac{\sigma}{\sqrt{n}} < \mu < \bar{x} + 2.58 \dfrac{\sigma}{\sqrt{n}}$.

These limits are called **confidence limits** or **fiducial limits.**

ILLUSTRATIVE EXAMPLES

Example 1. *A random sample of 900 members has a mean 3.4 cms. Can it be reasonably regarded as a sample from a large population of mean 3.2 cms and S.D. 2.3 cms?*

Sol. Here $n = 900$, $\bar{x} = 3.4$, $\mu = 3.2$, $\sigma = 2.3$

Null hypothesis:

H_0 : Assume that the sample is drawn from a large population with mean 3.2 and S.D. 2.3

Alternative hypothesis:

$H_1 : \mu \neq 3.2$ (**two tailed test**)

Test statistic:

Under H_0 ; $z = \dfrac{\bar{x} - \mu}{\sigma/\sqrt{n}} = \dfrac{3.4 - 3.2}{2.3/\sqrt{900}} = 0.261.$

Conclusion: As the calculated value of $|z| = 0.261 < 1.96$, the significant value of z at 5% level of significance, H_0 is accepted *i.e.,* the sample is drawn from the population with mean 3.2 and S.D. 2.3.

Example 2. *The mean weight obtained from a random sample of size 100 is 64 gms. The S.D. of the weight distribution of the population is 3 gms. Test the statement that the mean weight of the population is 67 gms at 5% level of significance. Also set up 99% confidence limits of the mean weight of the population.*

Sol. Here $n = 100$, $\mu = 67$, $\bar{x} = 64$, $\sigma = 3$

Null hypothesis:

H_0: There is no significant difference between sample and population mean.

i.e., $\mu = 67$, the sample is drawn from the population with $\mu = 67$.

Alternative hypothesis:

H_1: $\mu \neq 67$ (**Two tailed test**).

Test statistic:

Under H_0, $z = \dfrac{\bar{x} - \mu}{\sigma/\sqrt{n}} = \dfrac{64 - 67}{3/\sqrt{100}} = -10 \quad \therefore \quad |z| = 10.$

Conclusion: Since the calculated value of $|z| > 1.96$, the significant value of z at 5% level of significance, H_0 is rejected *i.e.,* the sample is not drawn from the population with mean 67.

To find 99% confidence limits. It is given by $\bar{x} \pm 2.58\ \sigma/\sqrt{n} = 64 \pm 2.58(3/\sqrt{100}) =$ 64.774, 63.226.

TEST YOUR KNOWLEDGE

1. A sample of 1000 students from a university was taken and their average weight was found to be 112 pounds with a S.D. of 20 pounds. Could the mean weight of students in the population be 120 pounds?
2. A sample of 400 male students is found to have a mean height of 160 cms. Can it be reasonably regarded as a sample from a large population with mean height 162.5 cms and standard deviation 4.5 cms?

3. A random sample of 200 measurements from a large population gave a mean value of 50 and a S.D. of 9. Determine 95% confidence interval for the mean of population.

Answers

1. H_0 is rejected 2. H_0 accepted 3. 48.8 and 51.2.

5.11.4 Test of Significance for Difference of Means of Two Large Samples

Let $\bar{x}_1$ be the mean of a sample of size n_1 from a population with mean μ_1, and variance σ_1^2. Let $\bar{x}_2$ be the mean of an independent sample of size n_2 from another population with mean μ_2 and variance σ_2^2. The test statistic is given by $z = \dfrac{\bar{x}_1 - \bar{x}_2}{\sqrt{\dfrac{\sigma_1^2}{n_1} + \dfrac{\sigma_2^2}{n_2}}}$.

Under the null hypothesis that the samples are drawn from the same population where $\sigma_1 = \sigma_2 = \sigma$ *i.e.*, $\mu_1 = \mu_2$ the test statistic is given by $z = \dfrac{\bar{x}_1 - \bar{x}_2}{\sigma\sqrt{\dfrac{1}{n_1} + \dfrac{1}{n_2}}}$.

Note 1. If σ_1, σ_2 are not known and $\sigma_1 \neq \sigma_2$ the test statistic in this case is $z = \dfrac{\bar{x}_1 - \bar{x}_2}{\sqrt{\dfrac{s_1^2 + s_2^2}{n_1 + n_2}}}$.

Note 2. If σ is not known and $\sigma_1 = \sigma_2$. We use $\sigma^2 = \dfrac{n_1 s_1^2 + n_2 s_2^2}{n_1 + n_2}$ to calculate σ.

ILLUSTRATIVE EXAMPLES

Example 1. *The average income of persons was ₹ 210 with a S.D. of ₹ 10 in sample of 100 people of a city. For another sample of 150 persons, the average income was ₹ 220 with S.D. of ₹ 12. The S.D. of incomes of the people of the city was ₹ 11. Test whether there is any significant difference between the average incomes of the localities.*

Sol. Here $n_1 = 100$, $n_2 = 150$, $\bar{x}_1 = 210$, $\bar{x}_2 = 220$, $s_1 = 10$, $s_2 = 12$.

Null hypothesis: The difference is not significant. *i.e.*, there is no difference between the incomes of the localities. H_0: $\bar{x}_1 = \bar{x}_2$

Alternative hypothesis

$H_1 : \bar{x}_1 \neq \bar{x}_2$ **(two tailed test)**

Test statistic:

Under H_0, $$z = \frac{\bar{x}_1 - \bar{x}_2}{\sqrt{\dfrac{s_1^2}{n_1} + \dfrac{s_2^2}{n_2}}} = \frac{210 - 220}{\sqrt{\dfrac{10^2}{100} + \dfrac{12^2}{150}}} = -7.1428 \quad \therefore \quad |z| = 7.1428.$$

Conclusion: As the calculated value of $|z| > 1.96$, the significant value of z at 5% level of significance, H_0 is rejected *i.e.*, there is significant difference between the average incomes of the localities.

Example 2. *Intelligence tests were given to two groups of boys and girls.*

	Mean	*S.D.*	*Size*
Girls	*75*	*8*	*60*
Boys	*73*	*10*	*100*

Examine if the difference between mean scores is significant.

Sol. Null hypothesis H_0: There is no significant difference between mean scores *i.e.*, $\bar{x}_1 = \bar{x}_2$.

Alternative hypothesis H_1: $\bar{x}_1 \neq \bar{x}_2$ **(two tailed test)**

Test statistic: Under the null hypothesis, $z = \dfrac{\bar{x}_1 - \bar{x}_2}{\sqrt{\dfrac{s_1^2}{n_1} + \dfrac{s_2^2}{n_2}}} = \dfrac{75 - 73}{\sqrt{\dfrac{8^2}{60} + \dfrac{10^2}{100}}} = 1.3912.$

Conclusion: As the calculated value of $|z| < 1.96$, the significant value of z at 5% level of significance, H_0 is accepted *i.e.*, there is no significant difference between mean scores.

TEST YOUR KNOWLEDGE

1. Two random samples of sizes 1000 and 2000 farms gave an average yield of 2000 kg and 2050 kg respectively. The variance of wheat farms in the country may be taken as 100 kg. Examine whether the two samples differ significantly in yield.
2. The means of two large samples of 1000 and 2000 members are 168.75 cms and 170 cms respectively. Can the samples be regarded as drawn from the same population of standard deviation 6.25 cms?
3. In a survey of buying habits, 400 women shoppers are chosen at random in supermarket A. Their average weekly food expenditure is ₹ 250 with a S.D. of ₹ 40. For 500 women shoppers chosen at supermarket B, the average weekly food expenditure is ₹ 220 with a S.D. of ₹ 45. Test at 1% level of significance whether the average food expenditures of the two groups are equal.

Answers

1. Highly significant **2.** Not significant **3.** Highly significant.

5.11.5 Test of Significance for the Difference of Standard Deviations

If s_1 and s_2 are the standard deviations of two independent samples, then under the null hypothesis $H_0 : \sigma_1 = \sigma_2$, *i.e.*, the sample standard deviations don't differ significantly, the statistic

$$z = \frac{s_1 - s_2}{\sqrt{\dfrac{\sigma_1^2}{2n_1} + \dfrac{\sigma_2^2}{2n_2}}}$$, where σ_1 and σ_2 are population standard deviations.

When population standard deviations are not known, then $z = \dfrac{s_1 - s_2}{\sqrt{\dfrac{s_1^2}{2n_1} + \dfrac{s_2^2}{2n_2}}}$.

Example. *Random samples drawn from two countries gave the following data relating to the heights of adult males:*

	Country A	*Country B*
Mean height (in inches)	*67.42*	*67.25*
Standard deviation	*2.58*	*2.50*
Number in samples	*1000*	*1200*

(*i*) *Is the difference between the means significant?*

(*ii*) *Is the difference between the standard deviations significant?*

Sol. Given: $n_1 = 1000$, $n_2 = 1200$, $\bar{x}_1 = 67.42$; $\bar{x}_2 = 67.25$, $s_1 = 2.58$, $s_2 = 2.50$

Since the samples size are large we can take $\sigma_1 = s_1 = 2.58$; $\sigma_2 = s_2 = 2.50$.

(*i*) **Null hypothesis** $H_0 : \mu_1 = \mu_2$ *i.e.*, sample means do not differ significantly.

Alternative hypothesis $H_1 : \mu_1 \neq \mu_2$ **(two tailed test)**

Test statistic:

$$z = \frac{\bar{x}_1 - \bar{x}_2}{\sqrt{\frac{s_1^2}{n_1} + \frac{s_2^2}{n_2}}} = \frac{67.42 - 67.25}{\sqrt{\frac{(2.58)^2}{1000} + \frac{(2.50)^2}{1200}}} = 1.56.$$

Conclusion:

Since $| z | < 1.96$ we accept the null hypothesis at 5% level of significance.

(*ii*) **Null hypothesis:**

H_0: $\sigma_1 = \sigma_2$ *i.e.*, the sample S.D.'s do not differ significantly.

Alternative hypothesis H_1: $\sigma_1 \neq \sigma_2$ **(two tailed test)**

Test statistic:

$$z = \frac{s_1 - s_2}{\sqrt{\frac{\sigma_1^2}{2n_1} + \frac{\sigma_2^2}{2n_2}}} = \frac{s_1 - s_1}{\sqrt{\frac{s_1^2}{2n_1} + \frac{s_2^2}{2n_2}}} = 1.0387.$$

Since $| z | < 1.96$ we accept the null hypothesis at 5% level of significance.

TEST YOUR KNOWLEDGE

1. The mean yield of two sets of plots and their variability are as given. Examine
   (*i*) whether the difference in the mean yield of the two sets of plots is significant.
   (*ii*) whether the difference in the variability in yields is significant.

	Set of 40 plots	*Set of 60 plots*
Mean yield per plot	1258 lb	1243 lb
S.D. per plot	34	28

2. The yield of wheat in a random sample of 1000 farms in a certain area has a S.D. of 192 kg. Another random sample of 1000 farms gives a S.D. of 224 kg. Are the S.D's significantly different?

Answers

1. $z = 2.315$, Difference significant at 5% level ; $z = 1.31$, Difference not significant at 5% level
2. $z = 4.851$. The S.D's are significantly different.

5.12 TEST OF SIGNIFICANCE OF SMALL SAMPLES

When the size of the sample is less than 30, then the sample is called small sample. For such sample it will not be possible for us to assume that the random sampling distribution of a statistic is approximately normal and the values given by the sample data are sufficiently close to the population values and can be used in their place for the calculation of the standard error of the estimate.

5.13 STUDENT'S t-DISTRIBUTION (t-Test)

[G.B.T.U. (MBA) 2011 ; G.B.T.U. (MCA) 2010]

This t-distribution is used when sample size is ≤ 30 and the population standard deviation is unknown.

t-statistic is defined as

$$\boxed{t = \frac{\overline{x} - \mu}{S/\sqrt{n}}} \qquad \text{where,} \qquad S = \sqrt{\frac{\Sigma(x - \overline{x})^2}{n - 1}}$$

$\overline{x}$ is the mean of sample, μ is population mean. S is the standard deviation of population and n is sample size.

If the standard deviation of the sample 's' is given then t-statistic is defined as

$$\boxed{t = \frac{\overline{x} - \mu}{s/\sqrt{n - 1}}}$$

Note. The relation between s and S is $ns^2 = (n - 1)S^2$.

5.13.1 The *t*-Table

The t-table given at the end is the probability integral of t-distribution. The t-distribution has different values for each degrees of freedom and when the degrees of freedom are infinitely large, the t-distribution is equivalent to normal distribution and the probabilities shown in the normal distribution tables are applicable.

5.13.2 Applications of *t*-Distribution

[G.B.T.U. (MBA) 2011]

Some of the applications of t-distribution are given below:

1. *To test if the sample mean* ($\overline{x}$) *differs significantly from the hypothetical value* μ *of the population mean.*
2. *To test the significance between two sample means.*
3. *To test the significance of observed partial and multiple correlation coefficients.*

5.13.3 Critical Value of *t*

The critical value or significant value of t at level of significance α, degrees of freedom γ for two tailed test is given by

$$P[\,|\,t\,| > t_{\gamma}(\alpha)] = \alpha$$

$$P[\,|\,t\,| \leq t_{\gamma}(\alpha)] = 1 - \alpha$$

The significant value of t at level of significance α, for a single tailed test can be got from those of two tailed test by referring to the values at 2α.

5.14 TEST I: *t*-TEST OF SIGNIFICANCE OF THE MEAN OF A RANDOM SAMPLE

To test whether the mean of a sample drawn from a normal population deviates significantly from a stated value when variance of the population is unknown.

H_0: There is no significant difference between the sample mean $\bar{x}$ and the population mean μ, *i.e.*, we use the statistic

$$t = \frac{\bar{x} - \mu}{S/\sqrt{n}} \qquad \text{where} \quad S = \sqrt{\frac{\Sigma(x - \bar{x})^2}{n - 1}}$$

with degree of freedom $n - 1$.

At given level of significance α and degrees of freedom $(n - 1)$, we refer to t-table t_α (two tailed or one tailed). If calculated t value is such that $|t| < t_\alpha$, the null hypothesis is accepted. If $|t| > t_\alpha$, H_0 is rejected.

5.14.1 Fiducial Limits of Population Mean

If t_α is the value of t at level of significance α at $(n - 1)$ degrees of freedom then,

$$\left|\frac{\bar{x} - \mu}{S/\sqrt{n}}\right| < t_\alpha \text{ for acceptance of } H_0.$$

$$\bar{x} - t_\alpha S/\sqrt{n} < \mu < \bar{x} + t_\alpha S/\sqrt{n}$$

95% confidence limits (level of significance 5%) are $\bar{x} \pm t_{0.05} S/\sqrt{n}$.

99% confidence limits (level of significance 1%) are $\bar{x} \pm t_{0.01} S/\sqrt{n}$.

ILLUSTRATIVE EXAMPLES

Example 1. *A random sample of size 16 has 53 as mean. The sum of squares of the deviation' from mean is 135. Can this sample be regarded as taken from the population having 56 as mean? Obtain 95% and 99% confidence limits of the mean of the population.*

Sol. Null hypothesis, H_0: There is no significant difference between the sample mean and hypothetical population mean *i.e.*, $\mu = 56$.

Alternative hypothesis, H_1: $\mu \neq 56$ **(Two tailed test)**

Test statistic Under H_0, test statistic is $t = \dfrac{\bar{x} - \mu}{S/\sqrt{n}}$

Given : $\bar{x} = 53$, $\mu = 56$, $n = 16$, $\Sigma(x - \bar{x})^2 = 135$

$$S = \sqrt{\frac{\Sigma(x - \bar{x})^2}{n - 1}} = \sqrt{\frac{135}{15}} = 3$$

$$t = \frac{\bar{x} - \mu}{S/\sqrt{n}} = \frac{53 - 56}{3/\sqrt{16}} = -4$$

$$|t| = 4$$

$$d.fv. = 16 - 1 = 15.$$

Conclusion: Since $|t| = 4 > t_{0.05} = 2.13$ *i.e.*, the calculated value of t is more than the tabulated value, the null hypothesis is rejected. Hence, the sample mean has not come from a population having 56 as mean.

95% confidence limits of the population mean

$$= \bar{x} \pm \frac{S}{\sqrt{n}} t_{0.05} = 53 \pm \frac{3}{\sqrt{16}} (2.13) = 51.4025, 54.5975$$

99% confidence limits of the population mean

$$= \bar{x} \pm \frac{S}{\sqrt{n}} t_{0.01,} = 53 \pm \frac{3}{\sqrt{16}} (2.95) = 50.7875, 55.2125.$$

Example 2. *The lifetime of electric bulbs for a random sample of 10 from a large consignment gave the following data:*

Item	*1*	*2*	*3*	*4*	*5*	*6*	*7*	*8*	*9*	*10*
Life in '000 hrs.	*4.2*	*4.6*	*3.9*	*4.1*	*5.2*	*3.8*	*3.9*	*4.3*	*4.4*	*5.6*

Can we accept the hypothesis that the average lifetime of bulb is 4000 hrs?

Sol. Null hypothesis H_0 : There is no significant difference in the sample mean and population mean. *i.e.*, $\mu = 4000$ hrs.

Alternative hypothesis: $\mu \neq 4000$ hrs (**Two tailed test**)

Test statistic: Under H_0, the test statistic is $t = \dfrac{\bar{x} - \mu}{S/\sqrt{n}}$

x	4.2	4.6	3.9	4.1	5.2	3.8	3.9	4.3	4.4	5.6
$x - \bar{x}$	– 0.2	0.2	– 0.5	– 0.3	0.8	– 0.6	– 0.5	– 0.1	0	1.2
$(x - \bar{x})^2$	0.04	0.04	0.25	0.09	0.64	0.36	0.25	0.01	0	1.44

$$\bar{x} = \frac{\Sigma x}{n} = \frac{44}{10} = 4.4, \qquad \Sigma(x - \bar{x})^2 = 3.12$$

$$S = \sqrt{\frac{\Sigma(x - \bar{x})^2}{n - 1}} = 0.589$$

$$t = \frac{\bar{x} - \mu}{S/\sqrt{n}} = \frac{4.4 - 4}{\left(\dfrac{0.589}{\sqrt{10}}\right)} = 2.123$$

For $\gamma = 9$, $t_{0.05} = 2.26$.

Conclusion: Since the calculated value of t is less than the tabulated value of t at 5% level of significance.

$\therefore$ The null hypothesis $\mu = 4000$ hrs is accepted *i.e.*, the average lifetime of bulbs could be 4000 hrs.

Example 3. *A sample of 20 items has mean 42 units and S.D. 5 units. Test the hypothesis that it is a random sample from a normal population with mean 45 units.*

Sol. Null hypothesis H_0: There is no significant difference between the sample mean and the population mean. *i.e.*, $\mu = 45$ units

Alternative hypothesis, H_1: $\mu \neq 45$ (**Two tailed test**)

Given: $n = 20$, $\bar{x} = 42$, $s = 5$; $\gamma = 19$ d.f.

Test statistic: Under H_0, the test statistic is

$$t = \frac{\bar{x} - \mu}{s/\sqrt{n-1}} = \frac{42 - 45}{5/\sqrt{19}} = -2.615$$

$\therefore \quad |t| = 2.615$

The tabulated value of t at 5% level for 19 d.f. is $t_{0.05} = 2.09$.

Conclusion: Since the calculated value $|t|$ is greater than the tabulated value of t at 5% level of significance, the null hypothesis H_0 is rejected. *i.e.,* there is significant difference between the sample mean and population mean.

i.e., the sample could not have come from this population.

Example 4. *The 9 items of a sample have the following values:*

45, 47, 50, 52, 48, 47, 49, 53, 51.

Does the mean of these values differ significantly from the assumed mean 47.5?

Sol. Here, $n = 9$, $\mu = 47.5$, $\bar{x} = \frac{\Sigma x}{n} = 49.1$

x	45	47	50	52	48	47	49	53	51
$x - \bar{x}$	– 4.1	– 2.1	0.9	2.9	– 1.1	– 2.1	– 0.1	3.9	1.9
$(x - \bar{x})^2$	16.81	4.41	0.81	8.41	1.21	4.41	0.01	15.21	3.61

$$\Sigma(x - \bar{x})^2 = 54.89,$$

$$S^2 = \frac{\Sigma(x - \bar{x})^2}{n-1} = 6.86$$

$\therefore \quad S = 2.619$

Null hypothesis:

$H_0: \mu = 47.5$

i.e., there is no significant difference between the sample and population means.

Alternative hypothesis:

$H_1: \mu \neq 47.5$

Hence we apply **two-tailed test.**

Test statistic: Under H_0, the test statistic is

$$t = \frac{\bar{x} - \mu}{S/\sqrt{n}} = \frac{49.1 - 47.5}{(2.619/\sqrt{9})} = 1.8327$$

$t_{0.05} = 2.31$ for $\gamma = 8$

Conclusion: Since $|t|_{\text{calculated}} < t_{\text{tabulated}}$ at 5% level of significance, the null hypothesis H_0 is accepted *i.e.,* there is no significant difference between their means.

TEST YOUR KNOWLEDGE

1. Ten individuals are chosen at random from a normal population of students and their marks are found to be 63, 63, 66, 67, 68, 69, 70, 70, 71, 71. In the light of these data, discuss the suggestion that mean mark of the population of students is 66.
2. The following values gives the lengths of 12 samples of Egyptian cotton taken from a consignment: 48, 46, 49, 46, 52, 45, 43, 47, 47, 46, 45, 50. Test if the mean length of the consignment can be taken as 46.
3. A sample of 18 items has a mean 24 units and standard deviation 3 units. Test the hypothesis that it is a random sample from a normal population with mean 27 units.
4. A random sample of 10 boys had the I.Q.'s 70, 120, 110, 101, 88, 83, 95, 98, 107 and 100. Do these data support the assumption of a population mean I.Q. of 160?

5.15 TEST II: t-TEST FOR DIFFERENCE OF MEANS OF TWO SMALL SAMPLES (from a Normal Population)

This test is used to test whether the two samples $x_1, x_2, ..., x_{n_1}, y_1, y_2, ..., y_{n_2}$ of sizes n_1, n_2 have been drawn from two normal populations with mean μ_1 and μ_2 respectively under the assumption that the population variance are equal ($\sigma_1 = \sigma_2 = \sigma$).

H_0: The samples have been drawn from the normal population with means μ_1 and μ_2 *i.e.*, $H_0: \mu_1 \neq \mu_2$.

Let $\bar{x}$, $\bar{y}$ be their means of the two samples.

Under this H_0 the test statistic t is given by $t = \dfrac{(\bar{x} - \bar{y})}{S\sqrt{\dfrac{1}{n_1} + \dfrac{1}{n_2}}}$.

Degree of freedom is $n_1 + n_2 - 2$.

Note 1. If the two sample's standard deviations s_1, s_2 are given then we have $S^2 = \dfrac{n_1 s_1^2 + n_2 s_2^2}{n_1 + n_2 - 2}$.

Note 2. If s_1, s_2 are not given then $S^2 = \dfrac{\Sigma(x_1 - \bar{x}_1)^2 + \Sigma(x_2 - \bar{x}_2)^2}{n_1 + n_2 - 2}$.

ILLUSTRATIVE EXAMPLES

Example 1. *Two samples of sodium vapour bulbs were tested for length of life and the following results were got:*

	Size	*Sample mean*	*Sample S.D.*
Type I	*8*	*1234 hrs*	*36 hrs*
Type II	*7*	*1036 hrs*	*40 hrs*

Is the difference in the means significant to generalise that Type I is superior to Type II regarding length of life?

Sol. Null hypothesis:

$H_0 : \mu_1 = \mu_2$ *i.e.,* two types of bulbs have same lifetime.

Alternative hypothesis:

$H_1 : \mu_1 > \mu_2$ *i.e.,* type I is superior to Type II.

Hence we use **right tailed test.**

$$S^2 = \frac{n_1 s_1^2 + n_2 s_2^2}{n_1 + n_2 - 2} = \frac{8(36)^2 + 7(40)^2}{8 + 7 - 2} = 1659.076$$

$\therefore$ $S = 40.7317$

Test statistic: Under H_0, the test statistic t is given by

$$t = \frac{\bar{x}_1 - \bar{x}_2}{S\sqrt{\frac{1}{n_1} + \frac{1}{n_2}}} = \frac{1234 - 1036}{40.7317\sqrt{\frac{1}{8} + \frac{1}{7}}} = 18.1480$$

$t_{0.05}$ at d.f. $\gamma = n_1 + n_2 - 2 = 13$ is 1.77.

Conclusion: Since calculated $| t | > t_{\text{tabulated}}$ at 5% level of significance, H_0 is rejected.

$\therefore$ Type I is definitely superior to Type II.

Example 2. *Samples of sizes 10 and 14 were taken from two normal populations with S.D. 3.5 and 5.2. The sample means were found to be 20.3 and 18.6. Test whether the means of the two populations are the same at 5% level.*

Sol. We have, $\bar{x}_1 = 20.3$, $\bar{x}_2 = 18.6$, $n_1 = 10$, $n_2 = 14$, $s_1 = 3.5$, $s_2 = 5.2$

$$S^2 = \frac{n_1 s_1^2 + n_2 s_2^2}{n_1 + n_2 - 2} = 22.775$$

$\therefore$ $S = 4.772$

Null hypothesis:

$H_0 : \mu_1 = \mu_2$ *i.e.,* the means of the two populations are the same.

Alternative hypothesis:

$H_1 : \mu_1 \neq \mu_2$

Test statistic: Under H_0, the test statistic is

$$t = \frac{\bar{x}_1 - \bar{x}_2}{S\sqrt{\frac{1}{n_1} + \frac{1}{n_2}}} = \frac{20.3 - 18.6}{4.772\sqrt{\frac{1}{10} + \frac{1}{14}}} = 0.8604$$

The tabulated value of t at 5% level of significance for 22 *d.f.* is $t_{0.05} = 2.0739$

Conclusion:

Since $t = 0.8604 < t_{0.05}$, the null hypothesis H_0 is accepted; *i.e.,* there is no significant difference between their means.

Example 3. *The height of 6 randomly chosen sailors in inches are 63, 65, 68, 69, 71 and 72. Those of 9 randomly chosen soldiers are 61, 62, 65, 66, 69, 70, 71, 72 and 73. Test whether the sailors are on the average taller than soldiers.*

Sol. Let X_1 and X_2 be the two samples denoting the heights of sailors and soldiers.

$n_1 = 6$, $n_2 = 9$

Null hypothesis H_0: $\mu_1 = \mu_2$.

i.e., the mean of both the population are the same.

Alternative hypothesis H_1: $\mu_1 > \mu_2$ (one tailed test)

Calculation of two sample means :

X_1	63	65	68	69	71	72
$X_1 - \overline{X}_1$	– 5	– 3	0	1	3	4
$(X_1 - \overline{X}_1)^2$	25	9	0	1	9	16

$$\overline{X}_1 = \frac{\Sigma X_1}{n_1} = 68; \Sigma(X_1 - \overline{X}_1)^2 = 60$$

X_2	61	62	65	66	69	70	71	72	73
$X_2 - \overline{X}_2$	– 6.66	– 5.66	– 2.66	1.66	1.34	2.34	3.34	4.34	5.34
$(X_2 - \overline{X}_2)^2$	44.36	32.035	7.0756	2.7556	1.7956	5.4756	11.1556	18.8356	28.5156

$$\overline{X}_2 = \frac{\Sigma X_2}{n_2} = 67.66; \Sigma(X_2 - \overline{X}_2)^2 = 152.0002$$

$$S^2 = \frac{1}{n_1 + n_2 - 2}[\Sigma(X_1 - \overline{X}_1)^2 + \Sigma(X_2 - \overline{X}_2)^2] = 16.3077$$

$\therefore$ $\quad S = 4.038$

Test statistic:

Under H_0,
$$t = \frac{\overline{X}_1 - \overline{X}_2}{S\sqrt{\frac{1}{n_1} + \frac{1}{n_2}}} = \frac{68 - 67.666}{4.038\sqrt{\frac{1}{6} + \frac{1}{9}}} = 0.1569$$

The value of t at 5% level of significance for 13 d.f. is 1.77. (d.f. $= n_1 + n_2 - 2$)

Conclusion: Since $t_{calculated} < t_{0.05} = 1.77$, the null hypothesis H_0 is accepted.

i.e., there is no significant difference between their average.

i.e., the sailors are not on the average taller than the soldiers.

TEST YOUR KNOWLEDGE

1. The mean life of 10 electric motors was found to be 1450 hrs with S.D. of 423 hrs. A second sample of 17 motors chosen from a different batch showed a mean life of 1280 hrs with a S.D. of 398 hrs. Is there a significant difference between means of the two samples?
2. The marks obtained by a group of 9 regular course students and another group of 11 part time course students in a test are given below:

Regular :	56	62	63	54	60	51	67	69	58		
Part time :	62	70	71	62	60	56	75	64	72	68	66

Examine whether the marks obtained by regular students and part time students differ significantly at 5% and 1% level of significance.

3. A group of 5 patients treated with the medicine A weigh 42, 39, 48, 60 and 41 kgs. A second group of 7 patients from the same hospital treated with medicine B weigh 38, 42, 56, 64, 68, 69 and 62 kg. Do you agree with the claim that medicine B increases the weight significantly? It is given that the value of t at 10% level of significance for 10 degree of freedom is 1.81.

[*G.B.T.U. (B. Pharm.) 2010*]

4. Two independent samples of sizes 7 and 9 have the following values:

Sample A :	10	12	10	13	14	11	10		
Sample B :	10	13	15	12	10	14	11	12	11

Test whether the difference between the mean is significant.

5. The average number of articles produced by two machines per day are 200 and 250 with standard deviation 20 and 25 respectively on the basis of records of 25 days production. Can you regard both the machines equally efficient at 5% level of significance?

5.16 SNEDECOR'S VARIANCE RATIO TEST OR F-TEST

In testing the significance of the difference of two means of two samples, we assumed that the two samples came from the same population or population with equal variance. The object of the F-test is to discover whether two independent estimates of population variance differ significantly or whether the two samples may be regarded as drawn from the normal populations having the same variance. Hence before applying the t-test for the significance of the difference of two means, we have to test for the equality of population variance by using F-test.

Let n_1 and n_2 be the sizes of two samples with variance s_1^2 and s_2^2. The estimate of the population variance based on these samples are $s_1^2 = \dfrac{n_1 s_1^2}{n_1 - 1}$ and $s_2^2 = \dfrac{n_2 s_2^2}{n_2 - 1}$. The degrees of freedom of these estimates are $\nu_1 = n_1 - 1$, $\nu_2 = n_2 - 1$.

To test whether these estimates s_1^2 and s_2^2 are significantly different or if the samples may be regarded as drawn from the same population or from two populations with same variance σ^2, we set-up the null hypothesis $H_0 : \sigma_1^2 = \sigma_2^2 = \sigma^2$.

i.e., the independent estimates of the common population do not differ significantly.

To carry out the test of significance of the difference of the variances we calculate the test statistic $F = \dfrac{s_1^2}{s_2^2}$, the Numerator is greater than the Denominator. *i.e.,* $s_1^2 > s_2^2$.

Conclusion: If the calculated value of F exceeds $F_{0.05}$ for $(n_1 - 1)$, $(n_2 - 1)$ degrees of freedom given in table we conclude that the ratio is significant at 5% level.

i.e., we conclude that the sample could have come from two normal population with same variance.

The assumptions on which F-test is based are:

1. *The populations for each sample must be normally distributed.*
2. *The samples must be random and independent.*
3. *The ratio of* σ_1^2 to σ_2^2 *should be equal to 1 or greater than 1. That is why we take the larger variance in the Numerator of the ratio.*

5.16.1 Applications: *F-test is used to test*

(*i*) *whether two independent samples have been drawn from the normal populations with the same variance* σ^2.

(*ii*) *Whether the two independent estimates of the population variance are homogeneous or not.*

ILLUSTRATIVE EXAMPLES

Example 1. *Two random samples drawn from 2 normal populations are as follows:*

A	*17*	*27*	*18*	*25*	*27*	*29*	*13*	*17*
B	*16*	*16*	*20*	*27*	*26*	*25*	*21*	

Test whether the samples are drawn from the same normal population.

Sol. To test if two independent samples have been drawn from the same population we have to test (*i*) equality of the means by applying the *t*-test and (*ii*) equality of population variance by applying F-test.

Since the *t*-test assumes that the sample variances are equal, we shall first apply the F-test.

F-test: Null hypothesis H_0: $\sigma_1^2 = \sigma_2^2$ *i.e.*, the population variance do not differ significantly.

Alternative hypothesis H_1: $\sigma_1^2 \neq \sigma_2^2$

Test statistic: $F = \dfrac{s_1^2}{s_2^2}$, (if $s_1^2 > s_2^2$)

Computations for s_1^2 **and** s_2^2

X_1	$X_1 - \overline{X}_1$	$(X_1 - \overline{X}_1)^2$	X_2	$X_2 - \overline{X}_2$	$(X_2 - \overline{X}_2)^2$
17	− 4.625	21.39	16	− 2.714	7.365
27	5.735	28.89	16	− 2.714	7.365
18	− 3.625	13.14	20	1.286	1.653
25	3.375	11.39	27	8.286	68.657
27	5.735	28.89	26	7.286	53.085
29	7.735	54.39	25	6.286	39.513
13	− 8.625	74.39	21	2.286	5.226
17	− 4.625	21.39			

$$\overline{X}_1 = 21.625;\ n_1 = 8;\ \Sigma(X_1 - \overline{X}_1)^2 = 253.87$$

$$\overline{X}_2 = 18.714;\ n_2 = 7;\ \Sigma(X_2 - \overline{X}_2)^2 = 182.859$$

$$s_1^2 = \frac{\Sigma(X_1 - \overline{X}_1)^2}{n_1 - 1} = \frac{253.87}{7} = 36.267;\ s_2^2 = \frac{\Sigma(X_2 - \overline{X}_2)^2}{n_2 - 1} = \frac{182.859}{6} = 30.47$$

$$F = \frac{s_1^2}{s_2^2} = \frac{36.267}{30.47} = 1.190.$$

Conclusion: The table value of F for $v_1 = 7$ and $v_2 = 6$ degrees of freedom at 5% level is 4.21. The calculated value of F is less than the tabulated value of F. ∴ H_0 is accepted. Hence we conclude that the variability in two populations is same.

t-test: Null hypothesis H_0: $\mu_1 = \mu_2$ *i.e.,* the population means are equal.

Alternative hypothesis H_1: $\mu_1 \neq \mu_2$

Test of statistic

$$s^2 = \frac{\Sigma(X_1 - \overline{X}_1)^2 + \Sigma(X_2 - \overline{X}_2)^2}{n_1 + n_2 - 2} = \frac{253.87 + 182.859}{8 + 7 - 2} = 33.594 \quad \therefore \quad s = 5.796$$

$$t = \frac{\overline{X}_1 - \overline{X}_2}{s\sqrt{\frac{1}{n_1} + \frac{1}{n_2}}} = \frac{21.625 - 18.714}{5.796\sqrt{\frac{1}{8} + \frac{1}{7}}} = 0.9704 \sim t(n_1 + n_2 - 2) \text{ d.f.}$$

Conclusion: The tabulated value of t at 5% level of significance for 13 d.f. is 2.16.

The calculated value of t is less than the tabulated value. H_0 is accepted *i.e.,* there is no significant difference between the population mean. *i.e.,* $\mu_1 = \mu_2$. ∴ We conclude that the two samples have been drawn from the same normal population.

Example 2. *Two independent sample of sizes 7 and 6 had the following values:*

Sample A	*28*	*30*	*32*	*33*	*31*	*29*	*34*
Sample B	*29*	*30*	*30*	*24*	*27*	*28*	

Examine whether the samples have been drawn from normal populations having the same variance.

Sol. H_0: The variance are equal. *i.e.,* $\sigma_1^2 = \sigma_2^2$

i.e., the samples have been drawn from normal populations with same variance.

$$H_1: \sigma_1^2 \neq \sigma_2^2$$

Under null hypothesis, the test statistic $F = \frac{s_1^2}{s_2^2} \; (s_1^2 > s_2^2)$

Computations for s_1^2 and s_2^2

X_1	$X_1 - \overline{X}_1$	$(X_1 - \overline{X}_1)^2$	X_2	$X_2 - \overline{X}_2$	$(X_2 - \overline{X}_2)^2$
28	– 3	9	29	1	1
30	– 1	1	30	2	4
32	1	1	30	2	4
33	2	4	24	– 4	16
31	0	0	27	– 1	1
29	– 2	4	28	0	0
34	3	9			
		28			26

$$\overline{X}_1 = 31, \quad n_1 = 7; \quad \Sigma(X_1 - \overline{X}_1)^2 = 28$$

$$\overline{X}_2 = 28, \quad n_2 = 6; \quad \Sigma(X_2 - \overline{X}_2)^2 = 26$$

$$s_1^2 = \frac{\Sigma(X_1 - \overline{X}_1)^2}{n_1 - 1} = \frac{28}{6} = 4.666; \quad s_2^2 = \frac{\Sigma(X_2 - \overline{X}_2)^2}{n_2 - 1} = \frac{26}{5} = 5.2$$

$$F = \frac{s_2^2}{s_1^2} = \frac{5.2}{4.666} = 1.1158. \qquad (\because \; s_2^2 > s_1^2)$$

Conclusion: The tabulated value of F at $\upsilon_1 = 6 - 1$ and $\upsilon_2 = 7 - 1$ d.f. for 5% level of significance is 4.39. Since the tabulated value of F is less than the calculated value, H_0 is accepted *i.e.*, there is no significant difference between the variance. *i.e.*, the samples have been drawn from the normal population with same variance.

Example 3. *The two random samples reveal the following data:*

Sample no.	*Size*	*Mean*	*Variance*
I	*16*	*440*	*40*
II	*25*	*460*	*42*

Test whether the samples come from the same normal population.

Sol. A normal population has two parameters namely the mean μ and the variance σ^2. To test whether the two independent samples have been drawn from the same normal population, we have to test

(*i*) the equality of means (*ii*) the equality of variance.

Since the *t*-test assumes that the sample variance are equal, we first apply F-test.

F-test: Null hypothesis: $\sigma_1^2 = \sigma_2^2$

The population variance do not differ significantly.

Alternative hypothesis: $\sigma_1^2 \neq \sigma_2^2$

Under the null hypothesis the test statistic is given by $F = \frac{s_1^2}{s_2^2}$, $(s_1^2 > s_2^2)$

Given: $n_1 = 16, n_2 = 25; s_1^2 = 40, s_2^2 = 42$

$$\therefore \qquad F = \frac{s_1^2}{s_2^2} = \frac{\dfrac{n_1 s_1^2}{n_1 - 1}}{\dfrac{n_2 s_2^2}{n_2 - 1}} = \frac{16 \times 40}{15} \times \frac{24}{25 \times 42} = 0.9752.$$

Conclusion: The calculated value of F is 0.9752. The tabulated value of F at 16 – 1, 25 – 1 d.f. for 5% level of significance is 2.11.

Since the calculated value is less than that of the tabulated value, H_0 is accepted. *i.e.*, the population variance are equal.

t-test: Null hypothesis H_0 : $\mu_1 = \mu_2$ *i.e.*, the population means are equal.

Alternative hypothesis H_1 : $\mu_1 \neq \mu_2$

Given: $n_1 = 16, n_2 = 25, \overline{X}_1 = 440, \overline{X}_2 = 460$

$$s^2 = \frac{n_1 s_1^2 + n_2 s_2^2}{n_1 + n_2 - 2} = \frac{16 \times 40 + 25 \times 42}{16 + 25 - 2} = 43.333 \quad \therefore \quad s = 6.582$$

$$t = \frac{\overline{X}_1 - \overline{X}_2}{s\sqrt{\frac{1}{n_1} + \frac{1}{n_2}}} = \frac{440 - 460}{6.582\sqrt{\frac{1}{16} + \frac{1}{25}}} = -9.490 \text{ for } (n_1 + n_2 - 2) \text{ d.f.}$$

Conclusion: The calculated value of $|t|$ is 9.490. The tabulated value of t at 39 d.f. for 5% level of significance is 1.96.

Since the calculated value is greater than the tabulated value, H_0 is rejected.

i.e., there is significant difference between means. *i.e.*, $\mu_1 \neq \mu_2$.

Since there is significant difference between means, and no significant difference between variance, we conclude that the samples do not come from the same normal population.

TEST YOUR KNOWLEDGE

1. From the following two sample values, find out whether they have come from the same population:

Sample 1	17	27	18	25	27	29	27	23	17
Sample 2	16	16	20	16	20	17	15	21	

2. The daily wages in Rupees of skilled workers in two cities are as follows:

	Size of sample of workers	*S.D. of wages in the sample*
City A	16	25
City B	13	32

3. The standard deviation calculated from two random samples of sizes 9 and 13 are 2.1 and 1.8 respectively. Can the samples be regarded as drawn from normal populations with the same standard deviation?
4. Two independent samples of size 8 and 9 had the following values of the variables:

Sample I	20	30	23	25	21	22	23	24	
Sample II	30	31	32	34	35	29	28	27	26

Do the estimates of the population variance differ significantly?

Answers

1. rejected **2.** accepted **3.** accepted **4.** accepted.

5.17 CHI-SQUARE (χ^2) TEST [*G.B.T.U. 2010*]

When a coin is tossed 200 times, the theoretical considerations lead us to expect 100 heads and 100 tails. But in practice, these results are rarely achieved. The quantity χ^2 (a Greek letter, pronounced as chi-square) describes the magnitude of discrepancy between theory and observation. If $\chi^2 = 0$, the observed and expected frequencies completely coincide. The greater the discrepancy between the observed and expected frequencies, the greater is the value of χ^2. Thus χ^2 **affords a measure of the correspondence between theory and observation.**

If O_i $(i = 1, 2, \ldots\ldots, n)$ is a set of observed (experimental) frequencies and E_i $(i = 1, 2, \ldots\ldots, n)$ is the corresponding set of expected (theoretical or hypothetical) frequencies, then, χ^2 **is defined as**

$$\chi^2 = \sum_{i=1}^{n} \left[\frac{(O_i - E_i)^2}{E_i} \right]$$

where $\Sigma O_i = \Sigma E_i = N$ (total frequency) and degrees of freedom (d.f.) = $(n - 1)$.

Note. (*i*) If $\chi^2 = 0$, the observed and theoretical frequencies agree exactly.

(*ii*) If $\chi^2 > 0$, they do not agree exactly.

5.18 DEGREES OF FREEDOM

While comparing the calculated value of χ^2 with the tabular value, we have to determine the degrees of freedom.

If we have to choose any four numbers whose sum is 50, we can exercise our independent choice for any three numbers only, the fourth being 50 minus the total of the three numbers selected. Thus, though we were to choose any four numbers, our choice was reduced to three because of one condition imposed. There was only one restraint on our freedom and our degrees of freedom were $4 - 1 = 3$. If two restrictions are imposed, our freedom to choose will be further curtailed and degrees of freedom will be $4 - 2 = 2$.

In general, the number of degrees of freedom is the total number of observations less the number of independent constraints imposed on the observations. Degrees of freedom (d.f.) are usually denoted by ν.

Thus, $\nu = n - k$, where k is the number of independent constraints in a set of data of n observations.

Note. (*i*) For a $p \times q$ contingency table (p columns and q rows), $\nu = (p - 1)(q - 1)$

(*ii*) In the case of a contingency table, the expected frequency of any class

$$= \frac{\text{Total of rows in which it occurs} \times \text{Total of columns in which it occurs}}{\text{Total number of observations}}$$

5.19 APPLICATIONS OF CHI-SQUARE TEST

χ^2 test is one of the simplest and the most general test known. It is applicable to a very large number of problems in practice which can be summed up under the following heads:

(*i*) as a test of goodness of fit.

(*ii*) as a test of independence of attributes.

(*iii*) as a test of homogeneity of independent estimates of the population variance.

(*iv*) as a test of the hypothetical value of the population variance σ^2.

(*v*) as a list to the homogeneity of independent estimates of the population correlation coefficient.

5.20 CONDITIONS FOR APPLYING χ^2 TEST

χ^2 test is an approximate test for large values of n. For the validity of chi-square test of goodness of fit between theory and experiment, the following conditions must be satisfied.

(*a*) The sample observations should be independent.

(*b*) The constraints on the cell frequencies, if any, should be linear *e.g.*, $\Sigma n_i = \Sigma \lambda_i$ or $\Sigma O_i = \Sigma E_i$.

(*c*) N, the total number of frequencies should be reasonably large. It is difficult to say what constitutes largeness, but as an arbitrary figure, we may say that **N should be at least 50,** however, few the cells.

(*d*) No theoretical cell-frequency should be small. Here again, it is difficult to say what constitutes smallness, but 5 should be regarded as the very minimum and **10 is better.** If small theoretical frequencies occur (*i.e.*, < 10), the difficulty is overcome by grouping two or more classes together before calculating (O – E). **It is important to remember that the number of degrees of freedom is determined with the number of classes after regrouping.**

Note 1. If any one of the theoretical frequency is less than 5, then we apply a correction given by F Yates, which is usually known as 'Yates correction for continuity', we add 0.5 to the cell frequency which is less than 5 and adjust the remaining cell frequency suitably so that the marginal total is not changed.

Note 2. It may be noted that the χ^2 test depends only on the set of observed and expected frequencies and on degrees of freedom (d.f.). It does not make any assumption regarding the parent population from which the observations are taken. Since χ^2 does not involve any population parameters, it is termed as a statistic and the test is known as Non-parametric test or Distribution-free test.

5.21 THE χ^2 DISTRIBUTION

For large sample sizes, the sampling distribution of χ^2 can be closely approximated by a continuous curve known as the chi-square distribution. The probability function of χ^2 distribution is given by

$$f(\chi^2) = c(\chi^2)^{(\nu/2-1)}\, e^{-x^2/2}$$

where $e = 2.71828$, ν = number of degrees of freedom; c = a constant depending only on ν.

Symbolically, the degrees of freedom are denoted by the symbol ν or by d.f. and are obtained by the rule $\nu = n - k$, where k refers to the number of independent constraints.

In general, when we fit a binomial distribution the number of degrees of freedom is one less than the number of classes; when we fit a Poisson distribution the degrees of freedom are 2 less than the number of classes, because we use the total frequency and the arithmetic mean to get the parameter of the Poisson distribution. When we fit a normal curve the number of degrees of freedom are 3 less than the number of classes, because in this fitting we use the total frequency, mean and standard deviation.

If the data is given in a series of "n" numbers then degrees of freedom = $n - 1$.

In the case of Binomial distribution d.f. = $n - 1$

In the case of Poisson distribution d.f. = $n - 2$

In the case of Normal distribution d.f. = $n - 3$.

5.22 χ^2 TEST AS A TEST OF GOODNESS OF FIT

χ^2 test enables us to ascertain how well the theoretical distributions such as Binomial, Poisson or Normal etc., fit empirical distributions, *i.e.*, distributions obtained from sample data. If the **calculated value of χ^2 is less than the tabular value** at a specified level (generally 5%) of significance, the **fit is considered to be good** *i.e.*, the divergence between actual and expected frequencies is attributed to fluctuations of simple sampling. If the calculated value of χ^2 is greater than the tabular value, the fit is considered to be poor.

ILLUSTRATIVE EXAMPLES

Example 1. *In experiments on pea breeding, the following frequencies of seeds were obtained:*

Round and yellow	*Wrinkled and yellow*	*Round and green*	*Wrinkled and green*	*Total*
315	*101*	*108*	*32*	*556*

Theory predicts that the frequencies should be in proportions 9 : 3 :: 3 : 1. Examine the correspondence between theory and experiment.

Sol. Null hypothesis

$\mathbf{H_0}$ **:** The experimental result support the theory *i.e.,* there is no significant difference between the observed and theoretical frequency.

Under H_0, The theoretical (expected) frequencies can be calculated as follows:

$$E_1 = \frac{556 \times 9}{16} = 312.75$$

$$E_2 = \frac{556 \times 3}{16} = 104.25$$

$$E_3 = \frac{556 \times 3}{16} = 104.25$$

$$E_4 = \frac{556 \times 1}{16} = 34.75$$

To calculate the value of χ^2:

Observed frequency O_i	315	101	108	32
Expected Frequency E_i	312.75	104.25	104.25	34.75
$\frac{(O_i - E_i)^2}{E_i}$	0.016187	0.101319	0.134892	0.217626

$$\chi^2 = \sum \left[\frac{(O_i - E_i)^2}{E_i}\right] = 0.470024$$

Tabular value of χ^2 at 5% level of significance for $n - 1 = 3$ *d.f.* is 7.815 *i.e.,* $\chi^2_{0.05} = 7.815$.

Conclusion: Since the calculated value of χ^2 is less than that of the tabulated value, hence H_0 is accepted. Therefore, the experimental results support the theory.

Example 2. *The following table gives the number of accidents that took place in an industry during various days of the week. Test if accidents are uniformly distributed over the week.*

Day	*Mon*	*Tue*	*Wed*	*Thu*	*Fri*	*Sat*
No. of accidents	*14*	*18*	*12*	*11*	*15*	*14*

Sol. Null hypothesis H_0 : The accidents are uniformly distributed over the week.

Under this H_0, the expected frequencies of the accidents on each of these days = $\frac{84}{6} = 14$

Observed frequency O_i	14	18	12	11	15	14
Expected frequency E_i	14	14	14	14	14	14
$(O_i - E_i)^2$	0	16	4	9	1	0

$$\chi^2 = \sum \left[\frac{(O_i - E_i)^2}{E_i}\right] = \frac{\Sigma(O_i - E_i)^2}{E_i} = \frac{30}{14} = 2.1428.$$

Tabular value of χ^2 at 5% level for (6 – 1 = 5 d.f.) is 11.09.

Conclusion: Since the calculated value of χ^2 is less than the tabulated value, H_0 is accepted *i.e.,* the accidents are uniformly distributed over the week.

Example 3. *A die is thrown 276 times and the results of these throws are given below:*

No. appeared on the die	*1*	*2*	*3*	*4*	*5*	*6*
Frequency	*40*	*32*	*29*	*59*	*57*	*59*

Test whether the die is biased or not. (A.K.T.U. 2019)

Sol. Null hypothesis H_0: Die is unbiased.

Under this H_0, the expected frequencies for each digit is $\frac{276}{6} = 46.$

To find the value of χ^2

O_i	40	32	29	59	57	59
E_i	46	46	46	46	46	46
$(O_i - E_i)^2$	36	196	289	169	121	169

$$\chi^2 = \sum \left[\frac{(O_i - E_i)^2}{E_i}\right] = \frac{\Sigma(O_i - E_i)^2}{E_i} = \frac{980}{46} = 21.30.$$

Tabulated value of χ^2 at 5% level of significance for (6 – 1 = 5) d.f. is 11.09.

Conclusion: Since the calculated value of $\chi^2 = 21.30 > 11.07$ the tabulated value, H_0 is rejected. *i.e.,* die is not unbiased or die is biased.

Example 4. *Records taken of the number of male and female births in 800 families having four children are as follows:* [U.P.T.U. (MCA) 2009]

No. of male births	*0*	*1*	*2*	*3*	*4*
No. of female births	*4*	*3*	*2*	*1*	*0*
No. of families	*32*	*178*	*290*	*236*	*64*

Test whether the data are consistent with the hypothesis that the Binomial law holds and the chance of male birth is equal to that of female birth, namely p = q = 1/2.

Sol. Null hypothesis H_0: The data are consistent with the hypothesis of equal probability for male and female births. *i.e.*, $p = q = 1/2$.

We use Binomial distribution to calculate theoretical frequency given by:

$$N(r) = N \times P(X = r) = N \times {}^nC_r \, p^r q^{n-r}$$

where N is the total frequency, N(r) is the number of families with r male children.

p and q are probabilities of male and female births respectively, n is the number of children.

$$N(0) = 800 \times {}^4C_0 \left(\frac{1}{2}\right)^4 = 50,$$

$$N(1) = 200, N(2) = 300, \quad N(3) = 200 \text{ and } N(4) = 50$$

Observed frequency O_i	32	178	290	236	64
Expected frequency E_i	50	200	300	200	50
$(O_i - E_i)^2$	324	484	100	1296	196
$\frac{(O_i - E_i)^2}{E_i}$	6.48	2.42	0.333	6.48	3.92

$$\chi^2 = \sum \left[\frac{(O_i - E_i)^2}{E_i}\right] = 19.633.$$

Tabular value of χ^2 at 5% level of significance for $5 - 1 = 4$ d.f. is 9.49.

Conclusion: Since the calculated value of χ^2 is greater than the tabulated value, H_0 is rejected. *i.e.*, the data are not consistent with the hypothesis that the Binomial law holds and that the chance of a male birth is not equal to that of a female birth.

Example 5. *The theory predicts the proportion of beans in the four groups, G_1, G_2, G_3, G_4 should be in the ratio 9 : 3 : 3 : 1. In an experiment with 1600 beans the numbers in the four groups were 882, 313, 287 and 118. Does the experimental result support the theory?*

Sol. Null hypothesis H_0 : The experimental result support the theory. *i.e.*, there is no significant difference between the observed and theoretical frequency.

Under H_0, the theoretical frequency can be calculated as follows:

$$E(G_1) = \frac{1600 \times 9}{16} = 900; \; E(G_2) = \frac{1600 \times 3}{16} = 300;$$

$$E(G_3) = \frac{1600 \times 3}{16} = 300; \; E(G_4) = \frac{1600 \times 1}{16} = 100$$

To calculate the value of χ^2.

Observed frequency O_i	882	313	287	118
Expected frequency E_i	900	300	300	100
$\frac{(O_i - E_i)^2}{E_i}$	0.36	0.5633	0.5633	3.24

$$\chi^2 = \sum\left[\frac{(O_i - E_i)^2}{E_i}\right] = 4.7266.$$

Tablular value of χ^2 at 5% level of significance for 3 d.f. is 7.815.

Conclusion. Since the calculated value of χ^2 is less than that of the tabulated value, hence H_0 is accepted. *i.e.,* the experimental results support the theory.

Example 6. *The following table shows the distribution of digits in numbers chosen at random from a telephone directory:*

Digits	*0*	*1*	*2*	*3*	*4*	*5*	*6*	*7*	*8*	*9*
Frequency	*1026*	*1107*	*997*	*966*	*1075*	*933*	*1107*	*972*	*964*	*853*

Test whether the digits may be taken to occur equally frequently in the directory.

[G.B.T.U. (MCA) 2011]

Sol. Null hypothesis H_0 : The digits taken in the directory occur equally frequently *i.e.,* there is no significant difference between the observed and expected frequency.

Under H_0, the expected frequency $= \frac{10000}{10} = 1000$

Calculation of χ^2

O_i	1026	1107	997	966	1075	933	1107	972	964	853
E_i	1000	1000	1000	1000	1000	1000	1000	1000	1000	1000
$(O_i - E_i)^2$	676	11449	9	1156	5625	4489	11449	784	1296	21609

$$\chi^2 = \sum\left[\frac{(O_i - E_i)^2}{E_i}\right] = \frac{58542}{1000} = 58.542$$

The tabulated value of χ^2 at 5% level of significance for 9 d.f. is 16.919.

Conclusion: Since $\chi^2_{\text{calculated}} > \chi^2_{\text{tabulated}}$, H_0 is rejected *i.e.,* there is significant difference between the observed and theoretical frequencies. Therefore, the digits taken in the directory donot occur equally frequently.

Example 7. *When the first proof of 392 pages of a book of 1200 pages were read, the distribution of printing mistakes were found to be as follows:*

No. of mistakes in a page (x) :	*0*	*1*	*2*	*3*	*4*	*5*	*6*
No. of pages (f) :	*275*	*72*	*30*	*7*	*5*	*2*	*1*

Fit a poisson distribution to the above data and test the goodness of fit.

Sol. Null Hypothesis H_0: Poisson distribution is a good fit to the data.

$$\text{Mean } (\lambda) = \frac{\sum fx}{\sum f} = \frac{189}{392} = 0.4821$$

The frequency of x mistakes per page is given by the poisson law as follows:

$$N(x) = N \cdot P(x)$$

$$= 392\left[\frac{e^{-0.4821}(0.4821)^x}{x!}\right] = \frac{242.05(0.4821)^x}{x!} \quad ; 0 \le x \le 6$$

Under H_0, expected frequencies are,

N(0) = 242.05, N(1) = 116.69, N(2) = 28.13, N(3) = 4.52

N(4) = 0.54, N(5) = 0.052, N(6) = 0.0042

The χ^2-table is as follows:

Mistakes per page (x)	*Observed frequency* (O_i)	*Expected frequency* (E_i) *(correct to one place of decimal)*	$(O_i - E_i)^2$	$\frac{(O_i - E_i)^2}{E_i}$
0	275	242.1	1082.41	4.471
1	72	116.7	1998.09	17.121
2	30	28.1	3.61	0.128
3	7 }	4.5 }		
4	5 } 15	0.5 } 5.1		
5	2 }	0.1 }	98.01	19.217
6	1 }	0 }		
Total	392	392		40.937

$$\chi^2_{cal.} = \sum\left[\frac{(O_i - E_i)^2}{E_i}\right] = 40.937$$

$$\text{d.f.} = 7 - 1 - 1 - 3 = 2$$

One d.f. is lost because of linear constraint $\Sigma O_i = \Sigma E_i$. One d.f. is lost because the parameter λ has been estimated from the given data and is then used for computing the expected frequencies. 3 d.f. are lost because of grouping the last four expected cell frequencies which were less than 5.

Tabulated value of χ^2 for 2 d.f. at 5% level of significance is 5.991.

Conclusion: Since $\chi^2_{cal.} > \chi^2_{tab.}$, the null hypothesis is rejected at 5% level of significance. Hence, we conclude that poisson distribution is not a good fit to the given data.

Example 8. *Fit a Poisson distribution to the following data and best the goodness of fit:*

x :	0	1	2	3	4
f :	109	65	22	3	1

Sol. Null hypothesis, H_0 : Poisson distribution is a good fit to the data.

$$\text{Mean } (\lambda) = \frac{\Sigma fx}{\Sigma f} = \frac{122}{200} = 0.61$$

$$N(x) = N \cdot P(x) = (200) \frac{e^{-0.61}(0.61)^x}{x!} = \frac{(108.67)(0.61)^x}{x!}$$

Under H_0, expected frequencies are

$N(0) = 108.67 \approx 109$, $N(1) = 66.29 \approx 66$, $N(2) = 20.22 \approx 20$

$N(3) = 4.11 \approx 4$, $N(4) = 0.63 \approx 1$

The χ^2-table is as follows:

x	O_i	E_i	$(O_i - E_i)^2$	$\frac{(O_i - E_i)^2}{E_i}$
0	109	109	0	0
1	65	66	1	0.01515
2	22	20	4	0.2
3	3 } 4	4 } 5	1	0.2
4	1	1		
Total	200	200		0.41515

$$\chi^2_{\text{cal.}} = \sum \left[\frac{(O_i - E_i)^2}{E_i}\right] = 0.41515$$

$$\text{d.f.} = 5 - 1 - 1 - 1 = 2$$

Tabulated value of χ^2 for 2 d.f. at 5% level of significance is 5.991.

Conclusion: Since $\chi^2_{\text{cal.}} < \chi^2_{\text{tab.}}$, the null hypothesis H_0 is accepted at 5% level of significance. Hence we conclude that Poisson distribution is a good fit to the given data.

TEST YOUT KNOWLDEDGE

1. A sample analysis of examination results of 500 students, it was found that 220 students have failed, 170 have secured a third class, 90 have secured a second class and the rest, a first class. Do these figures support the general belief that above categories are in the ratio 4 : 3 : 2 : 1 respectively? (The tabular value of χ^2 for d.f. 3 at 5% level of significance is 7.81)

2. What is χ^2–test? *[G.B.T.U. 2010]*

   A die is thrown 90 times with the following results:

Face :	1	2	3	4	5	6	Total
Frequency :	10	12	16	14	18	20	90

   Use χ^2 test to test whether these data are consistent with the hypothesis that die is unbiased.

   Given $\chi^2_{0.05} = 11.07$ for 5 degrees of freedom.

3. (*i*) A survey of 320 families with 5 children shows the following distribution:

No. of boys & girls :	5 boys & 0 girl	4 boys & 1 girl	3 boys & 2 girls	2 boys & 3 girls	1 boy & 4 girls	0 boy & 5 girls	Total
No. of families :	18	56	110	88	40	8	320

Given that values of χ^2 for 5 degrees of freedom are 11.1 and 15.1 at 0.05 and 0.01 significance level respectively, test the hypothesis that male and female births are equally probable.

(*G.B.T.U. 2010*)

(*ii*) A survery of 240 families with 4 children shows the following distribution:

No. of boys :	1	2	3	4	5	6	Total
No. of families :	10	12	16	14	18	20	90

(Given: $\chi^2_{0.05}$ = 9.49 and 11.1 for 4 d.f. and 5 d.f., respectively)

4. A chemical extraction plant processes sea water to collect sodium chloride and magnesium. It is known that sea water contains sodium chloride, magnesium and other elements in the ratio 62 : 4 : 34. A sample of 200 tonnes of sea water has resulted in 130 tonnes of sodium chloride and 6 tonnes of magnesium. Are these data consistent with the known composition of sea water at 5% level of significance? (Given that the tabular value of χ^2 is 5.991 for 2 degree of freedom).

5. The demand for a particular spare part in a factory was found to vary from day-to-day. In a sample study, the following information was obtained:

Days :	Mon	Tue	Wed	Thurs	Fri	Sat
No. of parts demanded :	1124	1125	1110	1120	1126	1115

Test the hypothesis that the number of parts demanded does not depend on the day of the week. [**Given.** The values of chi-square significance at 5, 6, 7 d.f. are respectively 11.07, 12.59, 14.07 at 5% level of significance] (*G.B.T.U. 2011*)

6. The sales in a supermarket during a week are given below. Test the hypothesis that the sales donot depend on the day of the week using a significant level of 0.05.

Days	*Mon*	*Tue*	*Wed*	*Thurs*	*Fri*	*Sat*
Sales (in 1000 ₹)	65	54	60	56	71	84

7. 4 coins were tossed at a time and this operation is repeated 160 times. It is found that 4 heads occur 6 times, 3 heads occur 43 times, 2 heads occur 69 times, one head occur 34 times. Discuss whether the coin may be regarded as unbiased?

8. 200 digits are chosen at random from a set of tables. The frequencies of the digits were:

Digits :	0	1	2	3	4	5	6	7	8	9
Frequency :	18	19	23	21	16	25	22	20	21	15

Use χ^2-test to assess the correctness of the hypothesis that the digits were distributed in equal numbers in the table, given that the value of χ^2 are respectively 16.9, 18.3 and 19.7 for 9, 10 and 11 degrees of freedom at 5% level of significance.

9. A genetical law says that children having one parent of blood group M and the other parent of blood group N will always be one of the three blood groups M, MN, N and that the average no. of children in these groups will be in the ratio 1 : 2 : 1. The report on an experiment states as follows:

"Of 162 children having one M parent and one N parent, 28.4% were found to be of group M, 42% of group MN and the rest of the group N." Do the data in the report conform to the expected genetic ratio 1 : 2 : 1?

10. Every clinical thermometer is classified into one of the four categories A, B, C and D on the basis of inspection and test. From past experience, it is knwon that thermometers produced by a certain manufacturer are distributed among the four categories in the following proportions:

Category :	A	B	C	D
Proportion :	0.87	0.09	0.03	0.01

A new lot of 1336 thermometers is submitted by the manufacturer for inspection and test and the following distribution into four categories results:

Category :	A	B	C	D
No. of thermometers reported :	1188	91	47	10

Does this new lot of thermometers differ from the previous experience with regards to proportion of thermometers in each category?

11. Test for goodness of fit of a poisson distribution at 5% level of significance to the following frequency distribution:

(*i*)

x :	0	1	2	3	4	5	6	7	8
f :	52	151	130	102	45	12	5	1	2

[**Hint.** Group the last three frequencies]

(*ii*)

x :	0	1	2	3	4	5	6	7	8	9	10	11	12	13
f :	3	15	47	76	68	74	46	39	15	9	5	2	0	1

[**Hint.** Group the first two and last four frequencies]

(*iii*)

x :	0	1	2	3	4	5
f :	275	138	75	7	4	1

[**Hint.** Club the last three frequencies]

(*iv*)

x :	0	1	2	3	4
f :	419	352	154	56	19

12. (*i*) Fit a binomial distribution to the data and test for goodness of fit at 5% level of significance.

x :	0	1	2	3	4	5
y :	38	144	342	287	164	25

(*ii*) A random number table of 250 digits showed the following distribution of digits 0, 1, 2, ..., 9.

Digit :	0	1	2	3	4	5	6	7	8	9
Observed Frequency :	17	31	29	18	14	20	35	30	20	36
Expected Frequency :	25	25	25	25	25	25	25	25	25	25

Does the observed distribution differ significantly from expected distributions using a significance level of 0.01? Given that $\chi^2_{0.99}$ for 9 degrees of freedom is 21.7 [*G.B.T.U. MCA (SUM) 2010*]

Answers

1. No **2.** Yes

3. H_0 accepted at 1% level of significance and rejected at 5% level of significance

4. $\chi^2 = 1.025$, Yes **5.** H_0 accepted at 5% level **6.** H_0 accepted

7. Coin is unbiased **8.** H_0 accepted at 5% level **9.** H_0 accepted at 5% level

10. H_0 rejected at 5% level

11. (*i*) H_0 accepted at 5% level, (*ii*) H_0 accepted at 5% level, (*iii*) H_0 rejected at 5% level, not a good fit. (*iv*) H_0 accepted at 5% level

12. (*i*) H_0 accepted at 5% level, provides a good fit. (*ii*) Yes.

5.23 χ^2 TEST AS A TEST OF INDEPENDENCE

With the help of χ^2 test, we can find whether or not, two attributes are associated. We take the null hypothesis that there is no association between the attributes under study, *i.e.*, **we assume that the two attributes are independent. If the calculated value of χ^2 is less than the table value** at a specified level (generally 5%) of significance, the hypothesis holds good, *i.e.*, **the attributes are independent** and do not bear any association. On the other hand, if the calculated value of χ^2 is greater than the table value at a specified level of significance, we say that the results of the experiment do not support the hypothesis. In other words, the attributes are associated. Thus a very useful application of χ^2 test is to investigate the relationship between trials or attributes which can be classified into two or more categories.

The sample data set out into two-way table, called **contingency table.**

Let us consider two attributes A and B divided into r classes $A_1, A_2, A_3, \ldots\ldots, A_r$ and B divided into s classes $B_1, B_2, B_3, \ldots\ldots, B_s$. If (A_i), (B_j) represents the number of persons possessing the attributes A_i, B_j respectively, ($i = 1, 2, \ldots\ldots, r, j = 1, 2, \ldots\ldots, s$) and $(A_i\ B_j)$ represent the number of persons possessing attributes A_i and B_j. Also we have $\sum_{i=1}^{r} A_i = \sum_{j=1}^{s} B_j = N$, where N is the total frequency. The contingency table for $r \times s$ is given as follows:

A / B	A_1	A_2	A_3	$\ldots A_r$	*Total*
B_1	(A_1B_1)	(A_2B_1)	(A_3B_1)	$\ldots\ldots(A_rB_1)$	B_1
B_2	(A_1B_2)	(A_2B_2)	(A_3B_2)	$\ldots\ldots(A_rB_2)$	B_2
B_3	(A_1B_3)	(A_2B_3)	(A_3B_3)	$\ldots\ldots(A_rB_3)$	B_3
......	......	......	......	......	......
......	......	......	......	......	......
B_s	(A_1B_s)	(A_2B_s)	(A_3B_s)	$\ldots\ldots(A_rB_s)$	(B_s)
Total	(A_1)	(A_2)	(A_3)	$\ldots\ldots(A_r)$	N

H_0 : Both the attributes are independent, *i.e.*, A and B are independent under the null hypothesis, we calculate the expected frequency as follows:

$$P(A_i) = \text{Probability that a person possesses the attribute } A_i = \frac{(A_i)}{N} \quad i = 1, 2, \ldots\ldots, r$$

$$P(B_j) = \text{Probability that a person possesses the attribute } B_j = \frac{(B_j)}{N}$$

$$P(A_iB_j) = \text{Probability that a person possesses both attributes } A_i \text{ and } B_j = \frac{(A_iB_j)}{N}$$

If $(A_iB_j)_0$ is the expected number of persons possessing both the attributes A_i and B_j

$$(A_iB_j)_0 = NP(A_iB_j) = NP(A_i)(B_j)$$

$$= N \frac{(A_i)}{N}\frac{(B_j)}{N} = \frac{(A_i)(B_j)}{N} \qquad (\because \text{ A and B are independent})$$

Hence,
$$\boxed{\chi^2 = \sum_{i=1}^{r}\sum_{j=1}^{s}\left[\frac{[(A_iB_j) - (A_iB_j)_0]^2}{(A_iB_j)_0}\right]}$$

which is distributed as a χ^2 variate with $(r-1)(s-1)$ degrees of freedom.

Note 1. For a 2 × 2 contingency table where the frequencies are $\frac{a/b}{c/d}$, χ^2 can be calculated from independent frequencies as $\chi^2 = \frac{(a+b+c+d)(ad-bc)^2}{(a+b)(c+d)(b+d)(a+c)}$.

Note 2. If the contingency table is not 2 × 2, then the above formula for calculating χ^2 can't be used. Hence, we have another formula for calculating the expected frequency $(A_iB_j)_0 = \frac{(A_i)(B_j)}{N}$

i.e., expected frequency in each cell is $= \frac{\text{Product of column total and row total}}{\text{whole total}}$.

Note 3. If $\frac{a|b}{c|d}$ is the 2 × 2 contingency table with two attributes, $Q = \frac{ad-bc}{ad+bc}$ is called the coefficient of association. If the attributes are independent then $\frac{a}{b} = \frac{c}{d}$.

Note 4. Yates's Correction. In a 2 × 2 table, if the frequencies of a cell is small, we make Yates's correction to make χ^2 continuous. Decrease by $\frac{1}{2}$ those cell frequencies which are greater than expected frequencies, and increase by $\frac{1}{2}$ those which are less than expectation. This will not affect the marginal columns. This correction is known as Yates's correction to continuity. After Yates's correction

$$\chi^2 = \frac{N\left(bc - ad - \frac{1}{2}N\right)^2}{(a+c)(b+d)(c+d)(a+b)} \quad \text{when} \quad ad - bc < 0$$

and
$$\chi^2 = \frac{N\left(ad - bc - \frac{1}{2}N\right)^2}{(a+c)(b+d)(c+d)(a+b)} \quad \text{when} \quad ad - bc > 0.$$

ILLUSTRATIVE EXAMPLES

Example 1. *What are the expected frequencies of 2 × 2 contingency tables given below:*

(*i*)

a	*b*
c	*d*

(*ii*)

2	10
6	6

Sol. (i)

Observed frequencies

a	b	$a + b$
c	d	$c + d$
$a + c$	$b + d$	$a + b + c + d = N$

→

Expected frequencies

$\frac{(a+c)(a+b)}{a+b+c+d}$	$\frac{(b+d)(a+b)}{a+b+c+d}$
$\frac{(a+c)(c+d)}{a+b+c+d}$	$\frac{(b+d)(c+d)}{a+b+c+d}$

(ii)

Observed frequencies

2	10	12
6	6	12
8	16	24

→

Expected frequencies

$\frac{8 \times 12}{24} = 4$	$\frac{16 \times 12}{24} = 8$
$\frac{8 \times 12}{24} = 4$	$\frac{16 \times 12}{24} = 8$

Example 2. *From the following table regarding the colour of eyes of father and son, test if the colour of son's eye is associated with that of the father.*

		Eye colour of son	
		Light	*Not light*
Eye colour of father	*Light*	*471*	*51*
	Not light	*148*	*230*

Sol. Null hypothesis H_0: The colour of son's eye is not associated with that of the father, *i.e.*, they are independent.

Under H_0, we calculate

$$\text{the expected frequency in each cell} = \frac{\text{Product of column total and row total}}{\text{whole total}}$$

Expected frequencies are:

Eye colour of son / *Eye colour of father*	*Light*	*Not light*	*Total*
Light	$\frac{619 \times 522}{900} = 359.02$	$\frac{289 \times 522}{900} = 167.62$	522
Not light	$\frac{619 \times 378}{900} = 259.98$	$\frac{289 \times 378}{900} = 121.38$	378
Total	619	289	900

$$\chi^2 = \frac{(471 - 359.02)^2}{359.02} + \frac{(51 - 167.62)^2}{167.62} + \frac{(148 - 259.98)^2}{259.98} + \frac{(230 - 121.38)^2}{121.38} = 261.498.$$

Tabulated value of χ^2 at 5% level for 1 d.f. is 3.841.

Conclusion. Since the calculated value of χ^2 > tabulated value of χ^2, H_0 is rejected. They are dependent *i.e.*, the colour of son's eye is associated with that of the father.

Example 3. *The following table gives the number of good and bad parts produced by each of the three shifts in a factory:*

	Good parts	*Bad parts*	*Total*
Day shift	*960*	*40*	*1000*
Evening shift	*940*	*50*	*990*
Night shift	*950*	*45*	*995*
Total	*2850*	*135*	*2985*

Test whether or not the production of bad parts is independent of the shift on which they were produced.

Sol. Null hypothesis H_0: The production of bad parts is independent of the shift on which they were produced. *i.e.*, the two attributes, production and shifts are independent.

Under H_0,
$$\chi^2 = \sum_{i=1}^{2} \sum_{j=1}^{3} \left[\frac{[(A_iB_j)_0 - (A_iB_j)]^2}{(A_iB_j)_0} \right]$$

Calculation of expected frequencies

Let A and B be the two attributes namely production and shifts. A is divided into two classes A_1, A_2 and B is divided into three classes B_1, B_2, B_3.

$$(A_1B_1)_0 = \frac{(A_1)(B_1)}{N} = \frac{(2850)\times(1000)}{2985} = 954.77$$

$$(A_1B_2)_0 = \frac{(A_1)(B_2)}{N} = \frac{(2850)\times(990)}{2985} = 945.226$$

$$(A_1B_3)_0 = \frac{(A_1)(B_3)}{N} = \frac{(2850)\times(995)}{2985} = 950$$

$$(A_2B_1)_0 = \frac{(A_2)(B_1)}{N} = \frac{(135)\times(1000)}{2985} = 45.27$$

$$(A_2B_2)_0 = \frac{(A_2)(B_2)}{N} = \frac{(135)\times(990)}{2985} = 44.773$$

$$(A_2B_3)_0 = \frac{(A_2)(B_3)}{N} = \frac{(135)\times(995)}{2985} = 45.$$

To calculate the value of χ^2

Class	O_i	E_i	$(O_i - E_i)^2$	$(O_i - E_i)^2/E_i$
(A_1B_1)	960	954.77	27.3529	0.02864
(A_1B_2)	940	945.226	27.3110	0.02889
(A_1B_3)	950	950	0	0
(A_2B_1)	40	45.27	27.7729	0.61349
(A_2B_2)	50	44.773	27.3215	0.61022
(A_2B_3)	45	45	0	0
				1.28126

The tabulated value of χ^2 at 5% level of significance for 2 degrees of freedom $(r-1)(s-1)$ is 5.991.

Conclusion: Since the calculated value of χ^2 is less than the tabulated value, we accept H_0. *i.e.*, the production of bad parts is independent of the shift on which they were produced.

Example 4. *From the following data, find whether hair colour and sex are associated.*

Colour / *Sex*	*Fair*	*Red*	*Medium*	*Dark*	*Black*	*Total*
Boys	*592*	*849*	*504*	*119*	*36*	*2100*
Girls	*544*	*677*	*451*	*97*	*14*	*1783*
Total	*1136*	*1526*	*955*	*216*	*50*	*3883*

Sol. Null hypothesis H_0: The two attributes hair colour and sex are not associated. *i.e.*, they are independent.

Let A and B be the attributes hair colour and sex respectively. A is divides into 5 classes $(r = 5)$. B is divided into 2 classes $(s = 2)$.

$\therefore$ Degrees of freedom $= (r-1)(s-1) = (5-1)(2-1) = 4$

Under H_0, we calculate $\chi^2 = \sum_{i=1}^{5}\sum_{j=1}^{2} \frac{[(A_iB_j)_0 - (A_iB_j)]^2}{(A_iB_j)_0}$

To calculate the expected frequency $(A_i B_j)_0$ as follows:

$$(A_1B_1)_0 = \frac{(A_1)(B_1)}{N} = \frac{1136 \times 2100}{3883} = 614.37$$

$$(A_1B_2)_0 = \frac{(A_1)(B_2)}{N} = \frac{1136 \times 1783}{3883} = 521.629$$

$$(A_2B_1)_0 = \frac{(A_2)(B_1)}{N} = \frac{1526 \times 2100}{3883} = 852.289$$

$$(A_2B_2)_0 = \frac{(A_2)(B_2)}{N} = \frac{1526 \times 1783}{3883} = 700.71$$

$$(A_3B_1)_0 = \frac{(A_3)(B_1)}{N} = \frac{955 \times 2100}{3883} = 516.482$$

$$(A_3B_2)_0 = \frac{(A_3)(B_2)}{N} = \frac{955 \times 1783}{3883} = 483.517$$

$$(A_4B_1)_0 = \frac{(A_4)(B_1)}{N} = \frac{216 \times 2100}{3883} = 116.816$$

$$(A_4B_2)_0 = \frac{(A_4)(B_2)}{N} = \frac{216 \times 1783}{3883} = 99.183$$

$$(A_5B_1)_0 = \frac{(A_5)(B_1)}{N} = \frac{50 \times 2100}{3883} = 27.04$$

$$(A_5B_2)_0 = \frac{(A_5)(B_2)}{N} = \frac{50 \times 1783}{3883} = 22.959$$

Calculation of χ^2

Class	O_i	E_i	$(O_i - E_i)^2$	$\frac{(O_i - E_i)^2}{E_i}$
A_1B_1	592	614.37	500.416	0.8145
A_1B_2	544	521.629	500.462	0.959
A_2B_1	849	852.289	10.8175	0.0127
A_2B_2	677	700.71	562.1641	0.8023
A_3B_1	504	516.482	155.800	0.3016
A_3B_2	451	438.517	155.825	0.3553
A_4B_1	119	116.816	4.7698	0.0408
A_4B_2	97	99.183	4.7654	0.0480
A_5B_1	36	27.04	80.2816	2.9689
A_5B_2	14	22.959	80.2636	3.495
				9.79975

$$\chi^2_{cal.} = 9.799.$$

Tabular value of χ^2 at 5% level of significance for 4 d.f. is 9.488.

Conclusion. Since the calculated value of χ^2 < tabulated value, H_0 is rejected. *i.e.*, the two attributes are not independent. *i.e.*, the hair colour and sex are associated.

Example 5. *Can vaccination be regarded as preventive measure of small pox as evidenced by the following data of 1482 persons exposed to smallpox in a locality. 368 in all were attacked of these 1482 persons and 343 were vaccinated and of these only 35 were attacked.*

Sol. For the given data we form the contingency table. Let the two attributes be vaccination and exposed to small pox. Each attributes is divided into two classes.

Vaccination A / *Disease small-pox B*	*Vaccinated*	*Not*	*Total*
Attacked	35	333	368
Not	308	806	1114
Total	343	1139	1482

Null hypothesis H_0: The two attributes are independent *i.e.,* vaccination can't be regarded as preventive measure of small pox.

Degrees of freedom $\nu = (r-1)(s-1) = (2-1)(2-1) = 1$

Under H_0, $\chi^2 = \sum_{i=1}^{2} \sum_{j=1}^{2} \frac{[(A_iB_j)_0 - (A_iB_j)]^2}{(A_iB_j)_0}$

Calculation of expected frequency

$$(A_1B_1)_0 = \frac{(A_1)(B_1)}{N} = \frac{343 \times 368}{1482} = 85.1713$$

$$(A_1B_2)_0 = \frac{(A_1)(B_2)}{N} = \frac{343 \times 1114}{1482} = 257.828$$

$$(A_2B_1)_0 = \frac{(A_2)(B_1)}{N} = \frac{1139 \times 368}{1482} = 282.828$$

$$(A_2B_2)_0 = \frac{(A_2)(B_2)}{N} = \frac{1139 \times 1114}{1482} = 856.171$$

Calculation of χ^2

Class	O_i	E_i	$(O_i - E_i)^2$	$\frac{(O_i - E_i)^2}{E_i}$
(A_1B_1)	35	85.1713	2517.159	29.554
(A_1B_2)	308	257.828	2517.229	8.1728
(A_2B_1)	333	282.828	2517.2295	7.5592
(A_2B_2)	806	856.171	2517.1292	2.9399
				48.2261

Calculated value of $\chi^2 = 48.2261$.

Tabulated value of χ^2 at 5% level of significance for 1 d.f. is 3.841.

Conclusion: Since the calculated value of χ^2 > tabulated value, H_0 is rejected.

i.e., the two attributes are not independent. *i.e.,* the vaccination can be regarded as preventive measure of small pox.

Example 6. *To test the effectiveness of inoculation against cholera, the following table was obtained:*

	Attacked	*Not attacked*	*Total*
Inoculated	*30*	*160*	*190*
Not inoculated	*140*	*460*	*600*
Total	*170*	*620*	*790*

(The figures represent the number of persons.)

Use χ^2-test to defend or refute the statement that the inoculation prevents attack from cholera. (A.K.T.U. 2009, 2018)

Sol. Null hypothesis H_0: The inoculation does not prevent attack from cholera.

Under H_0, we calculate the expected frequencies as:

	Attacked	*Not attacked*
Inoculated	$\frac{190 \times 170}{790} = 40.886$	$\frac{190 \times 620}{790} = 149.11$
Not inoculated	$\frac{600 \times 170}{790} = 129.11$	$\frac{600 \times 620}{790} = 470.89$

Calculation of χ^2

O_i	30	160	140	460
E_i	40.886	149.11	129.11	470.89
$\frac{(O_i - E_i)^2}{E_i}$	2.898	0.795	0.918	0.252

$$\chi^2_{\text{cal.}} = \sum \frac{(O_i - E_i)^2}{E_i} = 4.863$$

Tabulated value of χ^2 at 5% level of significance for 1 d.f. is 3.841.

Conclusion: Since $\chi^2_{\text{cal.}} > \chi^2_{\text{tab.}}$ at 5% level of significance, null hypothesis H_0 is rejected. Hence we defend the statement that inoculation prevents attack from cholera.

TEST YOUR KNOWLEDGE

1. In a locality 100 persons were randomly selected and asked about their educational achievements. The results are given below:

		Education		
		Middle	*High school*	*College*
Sex	Male	10	15	25
	Female	25	10	15

Based on this information can you say that the education depends on sex.

2. The following data is collected on two characters:

	Smokers	*Non smokers*
Literate	83	57
Illiterate	45	68

Based on this information can you say that there is no relation between habit of smoking and literacy.

3. 500 students at school were graded according to their intelligences and economic conditions of their homes. Examine whether there is any association between economic condition and intelligence, from the following data:

Economic conditions	*Intelligence*	
	Good	*Bad*
Rich	85	75
Poor	165	175

4. In an experiment on the immunisation of goats from anthrox, the following results were obtained. Derive your inferences on the efficiency of the vaccine.

	Died anthrox	*Survived*
Inoculated with vaccine	2	10
Not inoculated	6	6

5. By using χ^2-test, find out whether there is any association between income level and type of schooling:

Income	*Public School*	*Govt. School*
Low	200	400
High	1000	400

(Given for degree of freedom 1, $\chi^2_{0.05} = 3.84$) [*G.B.T.U. (MBA) 2011*]

6. Examine by any suitable method, whether the nature of area is related to voting preference in the election for which the data are tabulated below:

Votes for Area	*A*	*B*	*Total*
Rural	620	480	1100
Urban	380	520	900
Total	1000	1000	2000

7. The groups of 100 people each were taken for testing the use of a vaccine. 15 persons contracted the disease out of the inoculated persons, whicle 25 contracted the diesease in the other group. Test the efficiency of the vaccine using Chi-square test. (The value of χ^2 for one degree of freedom at 5% level of significance is 3.84).

8. By using χ^2-test, find out whether there is any association between income level and type of schooling:

Social Status Health	*Poor*	*Rich*	*Total*
Below Normal	130	20	150
Normal	102	108	210
Above Normal	24	96	120
Total	256	224	480

Answers

1. Yes	**2.** No	**3.** No	**4.** Not effective
5. Yes	**6.** Yes	**7.** Not, associated	**8.** Yes, associated

5.24 ANALYSIS OF VARIANCE (ANOVA)

[*G.B.T.U. (MCA) 2011*]

ANOVA is a statistical technique which can be used to make comparisons among more than two groups. For example, suppose we want to see the effects of three different fertilisers on the yield of a crop. We select 100 identical plots of which 25 are control plots and 75 are experimental plots. Plants are grown in all the 100 plots in the same manner except that no fertiliser is applied in the control plots, while the three fertilisers are applied in the 75 experimental plots. The 75 experimental plots are divided into 3 groups of 25 plots each. Each of these groups is treated with different kinds of fertiliser. When the crop is harvested, the yield of each individual plot is recorded in four groups. We calculate the mean yield of all the 100 plots. Then we apply the technique (analysis of variance) to find out if there is any significant difference between the overall mean yield and the mean yields of the individual groups.

Similarly, suppose we want to test the effectiveness of various doses of a drug on blood pressure. We take a large number of individuals and divide them into several groups depending on the number of doses. For example, if we want to test four different doses, we make five groups (one control group and four experimental groups). Each experimental group is meant for a given dose of the drug.

One way ANOVA is used to see the effects of an independent variable on a given dependent variable. Two way ANOVA is used to see the effects of two independent variables on a given dependent variable.

In ANOVA, we compare the variations existing in the observed values of the dependent variable in the classified groups. We calculate the variance of the total population, variance between the groups and the variance within the groups. Then we examine whether there is any significant difference between the calculated variances. If we find, by ANOVA, that the variance between the groups is not significantly different from the variance within the groups, we conclude that there is no significant difference between the means of the groups, and hence the independent variable (factor) has no effect on the dependent variable.

5.24.1 ANOVA Table

The technique of analysis of variance is referred to as ANOVA. A table showing the source of variation, the sum of squares, degrees of freedom, mean square (variance) and the formula for the F-ratio is known as ANOVA table.

5.24.2 Computation of Test Statistic

The actual analysis of variance is carried out on the basis of ratio between the variances. This ratio forms the test statistic known as *F-statistic, given by*

$$\text{F-statistic} = \frac{\text{Variance between the samples}}{\text{Variance within the samples}}$$

5.24.3 Assumptions

Following are the assumptions for the study of analysis of variance:

(*i*) Each of the samples is a simple random sample.

(*ii*) Population, from which the samples are selected, are normally distributed.

(*iii*) Each of the samples is independent of the other samples.

(*iv*) Each of the population has the same variations and identical means.

(*v*) The effects of various components are additive.

5.25 ONE WAY CLASSIFICATION

(*a*) **Null hypothesis H_0:** $\mu_1 = \mu_2 = ... = \mu_p$, *i.e.*, the means of the populations from which p samples are drawn are equal to one another.

(*b*) **Alternative Hypothesis H_1:** At least two of the means of the populations are unequal, or *all the μ_i's are not equal.*

(*c*) **Computation of test statistic:** To obtain the test statistic known as F-value, we define certain terms as follows:

Sample observation x_{ij} : A sample observation x_{ij} has two subscripts. The subscript i denotes the row or *sample observation*, while the subscript j denotes the column or *population from which the observation came*. Thus x_{ij} is the ith observation from the jth population.

The total observations in the above matrix A is $m \times n$.

The various sums of squares involved in the computation of F-statistics are:

(*i*) **Sum of the Squares of Variations Amongst the Columns (SSC):** It is the sum of the squares of deviation between the column or group means and the grand mean, *i.e.*,

$$SSC = r\,\Sigma(\bar{x}_j - \bar{x})^2 ,$$

where $\bar{x}_j$ = mean of the jth sample, $\bar{x}$ = mean of the sample (column) means, r = number of rows or size of each sample.

Variance amongst columns = $\boxed{MSC = \dfrac{SSC}{c-1}}$, where c is number of columns.

It indicates the degree of explained variance due to sampling variations.

(*ii*) **Sum of the Squares of Variations within Columns (SSE):** It is the sum of the squares of variations between individual items and the column means, *i.e.*,

$$SSE = \sum_i \sum_j (x_{ij} - \bar{x}_j)^2$$

where x_{ij} = i^{th} observation in j^{th} column, $\bar{x}_j$ = mean of jth column

Mean of the square of Column Errors = $\boxed{MSE = \dfrac{SSE}{c(r-1)}}$

where c is number of columns and r is number of rows.

This is called *Unexplained Variance* as it indicates only the chance variation which cannot be explained in terms of variation in population.

(*iii*) **Total Sum of Squares of Variation (SST).**

$$SST = \sum_j \sum_i x_{ij}^2 - C,$$

where C = Correction factor = $\dfrac{T^2}{rc}$, T = grand total of the values in all samples,

r = number of rows, c = number of columns.

It is the sum of squares of observations between the individual values and the grand mean $\bar{x}$.

Also, **SST = SSC + SSE.**

$$\text{Total Variance} = \frac{\text{SST}}{n-1},$$

where $n = r \times c$ = total number of observations in all the samples and $(n - 1)$ is the degree of freedom.

Test Statistic: The test statistic is the **F-value** or **F-statistic** and is given by

$$\textbf{F-statistic or F-value} = \frac{\textbf{Explained variance}}{\textbf{Unexplained variance}} = \frac{\textbf{MSC}}{\textbf{MSE}}.$$

$$\Rightarrow \qquad \mathbf{F} = \frac{\mathbf{MSC}}{\mathbf{MSE}} \text{ with } (c-1) \text{ and } c\,(r-1) \text{ degrees of freedom.}$$

Thus we have

ANOVA Table for one-way classification for equal sample

Source of variation	*Sum of squares*	*Degree of freedom*	*Mean square*	*F*
Between samples (Column Means)	SSC	$c - 1$	$\text{MSC} = \frac{\text{SSC}}{c-1}$	
Within Samples	SSE	$c\,(r-1)$	$\text{MSE} = \frac{\text{SSE}}{c(r-1)}$	$\text{F} = \frac{\text{MSC}}{\text{MSE}}$
Total	SST	$cr - 1$		

(*d*) **Level of significance:** It is denoted by α and the value is taken generally as $\alpha = 0.05$.

(*e*) **Conclusion:** If $\text{F}_{\text{computed}} < \text{F}_{\text{tabular}}$ at α level of significance then the null hypothesis H_0 is accepted and if $\text{F}_{\text{computed}} > \text{F}_{\text{tabular}}$ at α level of significance then the null hypothesis H_0 is rejected and H_1 is accepted.

5.26 PROCEDURE TO PREPARE ANOVA TABLE

The basic structure of ANOVA table is

Source of variation	*Sum of squares*	*Degree of freedom*	*Mean sum of squares*	*Variance ratio or F-statistic*
Between samples				
Within samples				
Total				

(*i*) Find the sum of the values of all the items of all the samples. We call it **Grand Total (G.T.)**

$$\text{G.T.} = \Sigma\text{A} + \Sigma\text{B} + \Sigma\text{C} + \ldots..$$

(*ii*) Calculate the correction factor (C.F.)

$$\text{C.F.} = \frac{(\text{G. T.})^2}{n}$$

where n is the total number of items in all the samples.

(*iii*) Find the total sum of squares. It is obtained as $(\Sigma A^2 + \Sigma B^2 + \Sigma C^2 + ...) - C.F.$

(*iv*) Find the sum of squares between samples.

It is obtained as

$$\left[\frac{(\Sigma A)^2}{n_1} + \frac{(\Sigma B)^2}{n_2} + \frac{(\Sigma C)^2}{n_3} + ...\right] - C.F.$$

(*v*) Calculate the sum of squares within samples.

Sum of squares within samples = (Total sum of squares) – (Sum of squares between samples)

(*vi*) Find total degree of freedom and degree of freedom between samples. Degree of freedom within samples is obtained by subtraction.

(*vii*) Mean sum of squares is obtained by dividing sum of squares by degree of freedom.

(*viii*) Variance ratio is obtained by dividing the corresponding mean sum of squares.

This completes the preparation of ANOVA table.

5.27 CHANGE OF ORIGIN METHOD (Coding Method)

F remains unchanged if all the figures are multiplied or divided by a common factor or if a common factor is added to or subtracted from each figure. This property helps in calculating F-ratio when the figures are large. When such simplifications are done, it is said that data have been coded to simplify calculations.

ILLUSTRATIVE EXAMPLES

Example 1. *It is desired to compare three hospitals with regards to the number of deaths per month. A sample of death records were selected from the records of each hospital and the number of deaths was as given below. From these data, suggest a difference in the no. of the deaths per months among three hospitals:*

Hospitals

A	*B*	*C*
3	6	7
4	3	3
3	3	4
5	4	6
0	4	5

Sol. Null hypothesis H_0: There is no difference in the no. of deaths per months among three hospitals.

Alternate hypothesis H_1: There is a significant difference in the no. of deaths per months among three hospitals.

Level of significance: We use 5% level of significance.

Test statistic: To find the variance ratio, F, we set up an ANOVA table as follows:

Sample totals:

$$\Sigma y_A = 3 + 4 + 3 + 5 + 0 = 15$$

$$\Sigma y_B = 6 + 3 + 3 + 4 + 4 = 20$$

$$\Sigma y_C = 7 + 3 + 4 + 6 + 5 = 25$$

$$\text{Grand total (G.T.)} = \Sigma y_A + \Sigma y_B + \Sigma y_C = 60$$

$$\text{Correction factor (C.F.)} = \frac{(\text{G.T})^2}{n} = \frac{(60)^2}{15} = 240$$

Sum of squares of samples:

$$\Sigma y_A^2 = 3^2 + 4^2 + 3^2 + 5^2 + 0^2 = 59$$

$$\Sigma y_B^2 = 6^2 + 3^2 + 3^2 + 4^2 + 4^2 = 86$$

$$\Sigma y_C^2 = 7^2 + 3^2 + 4^2 + 6^2 + 5^2 = 135$$

$$\text{Total sum of squares} = \Sigma y_A^2 + \Sigma y_B^2 + \Sigma y_C^2 - \text{C.F.}$$

$$= 59 + 86 + 135 - 240 = 40$$

Sum of squares between samples

$$= \frac{(\Sigma y_A)^2}{n_1} + \frac{(\Sigma y_B)^2}{n_2} + \frac{(\Sigma y_C)^2}{n_3} - \text{C.F.}$$

$$= \frac{(15)^2}{5} + \frac{(20)^2}{5} + \frac{(25)^2}{5} - 240 = 10$$

Sum of squares within samples

= Total sum of squares – Sum of squares between samples

= 40 – 10 = 30

Degrees of freedom for total sum of squares $= n - 1 = 15 - 1 = 14$

Degrees of freedom for Hospitals $= k - 1 = 3 - 1 = 2$

Degrees of freedom for error $= n - k = 15 - 3 = 12$

ANOVA TABLE

Source of variation	*Sum of squares*	*Degree of freedom*	*Mean sum of squares*	*Variance ratio or F*
Between samples	10	2	5	$F_{2, 12} = \frac{5}{2.5} = 2$
Within samples	30	12	2.5	
Total	40	14	—	—

The tabular value of F at 5% level of significance with $\nu_1 = 2$, $\nu_2 = 12$ is 3.89

Conclusion: Since $F_{cal.} < F_{tab.}$, the difference is insignificant and we conclude that data donot suggest a difference in the number of deaths per month among the three hospitals.

Example 2. *A manufacturing company purchased three new machines of different makes and wishes to determine whether one of them is faster than the others in producing a certain output. Five hourly production figures are observed at random from each machine and results are given below:*

Observations	A_1	A_2	A_3
1	25	31	24
2	30	39	30
3	36	38	28
4	38	42	25
5	31	35	28

Use ANOVA and determine whether the machines are significantly different in their mean speed. (Given: at 5% level, $F_{2,\,12} = 3.89$*)* **Sol. Null hypothesis H_0:** Machines are not significantly different in their mean speeds *i.e.*, $\mu_1 = \mu_2 = \mu_3$

Alternate hypothesis H_1: Machines are significantly different in their mean speed.

Level of significance: We use 5% level of significance.

Test statistic: To find the variance ratio F, we set up an ANOVA table as follows:

let us shift the origin at 30 *i.e.*, reduce each observation by 30.

Now,

Observations	A_1	A_2	A_3
1	– 5	1	– 6
2	0	9	0
3	6	8	– 2
4	8	12	– 5
5	1	5	– 2

Machine totals:

$$\Sigma A_1 = -5 + 0 + 6 + 8 + 1 = 10$$

$$\Sigma A_2 = 1 + 9 + 8 + 12 + 5 = 35$$

$$\Sigma A_3 = -6 + 0 - 2 - 5 - 2 = -15$$

$$\text{Grand total (G.T.)} = \Sigma A_1 + \Sigma A_2 + \Sigma A_3 = 30$$

$$\text{Correction factor (C.F)} = \frac{(30)^2}{15} = 60$$

Sum of squares of samples

$$\Sigma A_1^2 = (-5)^2 + 0^2 + 6^2 + 8^2 + 1^2 = 126$$

Similarly, $\Sigma A_2^2 = 315,\ \Sigma A_3^2 = 69$

Total sum of squares $= 126 + 315 + 69 - (\text{C.F.}) = 510 - 60 = 450$

$$\text{Machine sum of squares (between samples)} = \frac{(10)^2}{5} + \frac{(35)^2}{5} + \frac{(-15)^2}{5} - \text{C.F.}$$

$$= 310 - 60 = 250$$

Error sum of squares (within samples)

= Total sum of squares – Machine sum of squares

= 450 – 250 = 200

Degrees of freedom for total sum of squares = $n - 1 = 15 - 1 = 14$

Degrees of freedom for Machines = $k - 1 = 3 - 1 = 2$

Degrees of freedom for Error $= n - k = 15 - 3 = 12$

ANOVA TABLE

Source of variation	*Sum of squares*	*Degrees of freedom*	*Mean sum of squares*	*Variance ratio of F*
Machine (between samples)	250	2	$\frac{250}{2} = 125$	$F_{2, 12} = \frac{125}{16.67}$
Error (within samples)	200	12	$\frac{200}{12} = 16.67$	= 7.498
Total	450	14	—	—

The tabular value of F at 5% level of significance with $v_1 = 2$, $v_2 = 12$ is 3.89 (given).

Conclusion: Since $F_{calculated} > F_{tabulated}$, the null hypothesis is rejected and the difference is significant and we conclude that there is a significant difference in the mean speed of machines.

TEST YOUR KNOWLEDGE

1. To test the significance of the variations of the retail prices in the commodity in three principal cities : Mumbai, Bangalore and Chennai, the four shops were chosen at random in each city and prices observed in rupees were as follows:

Mumbai	16	8	12	14
Bangalore	14	10	10	6
Chennai	4	10	8	8

Do the data indicate that the prices in the three cities are significantly different?

2. Below are given the yield kg for four varieties of tablets. Prepare ANOVA table and test that varieties differ significantly.

A	*B*	*C*	*D*
20	25	24	23
19	23	20	20
21	21	22	20

[**Hint.** Shift the origin at 20]

3. The following table gives the yields on 15 sample plots under three varieties of seeds:

Variety I:	20	21	23	16	20
Variety II:	18	20	17	15	25
Variety III:	25	28	22	28	32

Show that the seed varieties show variations more than could be covered by sampling variations. Given the tabulated value of F for $v_1 = 2$ and $v_2 = 12$ at 5% level of significance is 3.88.

4. The following figures relate to the production in kg of three varieties of wheat A, B and C used on 15 plots.

Wheat Variety	*Yields (kg)*					
RR 21–A	14	17	16	16		
K.68–B	15	11	13	15	13	14
Sonalika–C	18	16	18	19	15	

Test whether there is any significant difference in the production of three varieties.

[**Hint.** Shift the origin at 16]

5. The following figures relate to the production in kg of three varieties I, II, III of wheat shown in 12 plots:

Variety I:	14	16	18		
Variety II:	14	13	15	22	
Variety III:	18	16	19	19	20

Is there any significant difference in the production of three varieties? Given the tabulated value of F for $v_1 = 2$ and $v_2 = 9$ at 5% level of significance is 4.26.

6. A trucking company wishes to test the average life of each of the four brands of tyres. The company uses all brands on randomly selected trucks. The records showing the lives (thousands of miles) of tyres are as given in the table given below:

Brand I:	20	23	18	17	
Brand II:	19	15	17	20	16
Brand III:	21	19	20	17	16
Brand IV:	15	17	16	18	

Test the hypothesis that the average life for each brand of tyres is the same. Assume $\alpha = 0.01$.

7. What is analysis of variance? Mention its importance and uses.

8. Write short notes on:

   (*i*) ANOVA [*G.B.T.U. 2011*]

   (*ii*) One-way classification.

9. Under what conditions do we require to use one way Analysis of Variance ? Give examples and explain the method involved by taking a hypothetical example.

Answers

1. No **2.** Varieties donot differ significantly.
4. The difference is significant. **5.** No
6. No significant difference between the average lives.

5.28 STATISTICAL QUALITY CONTROL [*G.B.T.U. (MCA) 2010*]

A quality control system performs inspection, testing and analysis to ensure that the quality of the products produced is as per the laid down quality standards. It is called

"Statistical Quality Control" when statistical techniques are employed to control, improve and maintain quality or to solve quality problems. Building an information system to satisfy the concept of prevention and control and improving upon product quality requires statistical thinking.

Statistical quality control (S.Q.C.) is systematic as compared to guess-work of haphazard process inspection and the mathematical statistical approach neutralizes personal bias and uncovers poor judgement. S.Q.C. consists of three general activities:

(1) Systematic collection and graphic recording of accurate data

(2) Analyzing the data

(3) Practical engineering or management action if the information obtained indicates significant deviations from the specified limits.

Modern techniques of statistical quality control and acceptance sampling have an important part to play in the improvement of quality, enhancement of productivity, creation of consumer confidence and development of industrial economy of the country.

Following statistical tools are generally used for the above purposes :

(*i*) **Frequency distribution:** Frequency distribution is a tabulation of the number of times a given quality characteristic occurs within the samples. Graphic representation of frequency distribution will show :

(*a*) Average quality

(*b*) Spread of quality

(*c*) Comparison with specific requirements

(*d*) Process capability.

(*ii*) **Control chart:** Control chart is a graphical representation of quality characteristics, which indicates whether the process is under control or not.

(*iii*) **Acceptance sampling:** Acceptance sampling is the process of evaluating a portion of the product/material in a lot for the purpose of accepting or rejecting the lot on the basis of conforming to a quality specification.

It reduces the time and cost of inspection and exerts more effective pressure on quality improvement than it is possible by 100% inspection.

It is used when assurance is desired for the quality of materials/products either produced or received.

(*iv*) **Analysis of data:** It includes analysis of tolerances, correlation, analysis of variance, analysis for engineering design, problem solving technique to eliminate cause to troubles. Statistical methods can be used in arriving at proper specification limits of product, in designing the product, in purchase of raw-material, semi-finished and finished products, manufacturing processes, inspection, packaging, sales and also after sales service.

5.29 ADVANTAGES OF STATISTICAL QUALITY CONTROL

1. Efficiency: The use of statistical quality control ensures rapid and efficient inspection at a minimum cost. It eliminates the need of 100% inspection of finished products because the acceptance sampling in S.Q.C. exerts more effective pressure for quality improvement than it.

2. Reduction of scrap: It uncovers the cause of excessive variability in manufactured products forecasting trouble before rejections occur and reducing the amount of spoiled work.

3. Easy detection of faults: In statistical quality control after plotting the control, charts ($\overline{X}$, R, P, C, U) etc. when the points fall above the upper control limits or below the lower control limit, it gives an indication of deterioration in quality. Necessary corrective action may be then taken immediately.

4. Adherence to specifications: So long as a statistical control continues, specifications can be accurately predicted for future by which it is possible to assess whether the production processes are capable of producing the products with the given set of specifications.

5. Increases output and reduces wasted machine and man hours.

6. Efficient utilization of personnel, machines and materials results in higher productivity.

7. Creates quality awareness in employees. However, it should be noted that statistical quality control is not a panacea for assuring product quality.

8. Provides a common language that may be used by all three groups designers, production personnel and inspectors in arriving at a rational solution of mutual problems.

9. Points out when and where 100% inspection, sorting or screening is required.

10. Elimination of bottlenecks in the process of manufacturing.

It simply furnishes 'perspective facts' upon which intelligent management and engineering action can be based. Without such action, the method is ineffective.

Even the application of standard procedures is also very dangerous without adequate study of the process.

5.30 REASONS FOR VARIATIONS IN THE QUALITY OF A PRODUCT

Two extremely similar things are rarely obtained in nature. This fact holds good for production processes as well. No production process is good enough to produce all items of products exactly alike. The variations are due to two main reasons:

(*i*) **Chance or random causes:** Variations due to chance causes are inevitable in any process or product. They are difficult to trace and to control also even under best conditions of production.

These variations may be due to some inherent characteristic of the process or machine which functions at random.

If the variations are due to chance factors alone, the observations will follow a "normal curve". The knowledge of the behaviour of chance variation is the foundation on which control chart analysis rests. The conditions which produce these variations are accordingly said to be **"under control"**. On the other hand, if the variations in the data do not conform to a pattern that might reasonably be produced by chance causes, then in this case, conditions producing the variations are said to be **"out of control"** as it may be concluded that one or more assignable causes are at work.

(*ii*) **Assignable causes:** The variations due to assignable causes possess greater magnitude as compared to those due to chance causes and can be easily traced or detected. The power of the shewhart control chart lies in its ability to separate out these assignable causes of quality variations *e.g.* in length thickness, weight or diameter of a component.

The variations due to assignable causes may be because of following factors :

(*i*) Differences among machines

(*ii*) Differences among workers

(*iii*) Differences among materials

(*iv*) Differences in each of these factors over time

(*v*) Differences in their relationship to one another.

These variations may also be caused due to change in working conditions, mistake on the part of operator etc.

5.31 TECHNIQUES OF STATISTICAL QUALITY CONTROL

To control the quality characteristics of the product, there are two main techniques :

1. Process Control: It is a process of monitoring and measuring variability in the performance of a process or a machine through the interpretation of statistical techniques and it is employed to manage in-process quality. This technique ensures the production of requisite standard product and makes use of control charts.

2. Product control: This technique is concerned with inspection of already produced goods to ascertain whether they are fit to be despatched or not. To achieve the objectives, it makes use of sampling inspection plans.

5.32 CONTROL CHART

A control chart is a graphical representation of the collected information. It detects the variation in processing and warns if there is any departure from the specified tolerance limits. In other words, control charts is a device which specifies the state of statistical control or is a device for attaining quality control or is a device to judge whether the statistical control has been attained.

The control limits on the chart are so placed as to disclose the presence or absence of the assignable causes of quality variation which makes the diagnosis possible and brings substantial improvements in product quality and reduction of spoilage and rework.

Moreover, by identifying chance variations, the control chart tells when to leave the process alone and thus prevents unnecessarily frequent adjustments that tend to increase the variability of the process rather than to decrease it.

There are many types of control charts designed for different control situations. Most commonly used control charts are

(*i*) **Control charts for variables:** They are useful to measure quality characteristics and to control fully automatic process. It includes $\overline{X}$ and R-charts and charts for $\overline{X}$ and σ.

(*ii*) **Control charts for attributes:** It includes P-chart for fraction defective. A fraction defective control chart discloses erratic fluctuations in the quality of inspection which may result in improvement in inspection practice and inspection standards.

It also includes C-chart for number of defects per unit.

5.33 OBJECTIVES OF CONTROL CHARTS [*G.B.T.U. (C.O.) 2010)*]

Control charts are based on statistical techniques.

1. $\overline{X}$ and R or $\overline{X}$ and σ charts are used in combination for control process. $\overline{X}$-chart shows the variation in the averages of samples. It is the most commonly used variables chart. R-chart shows the uniformity or consistency of the process *i.e.,* it shows the variations in the ranges of samples. It is a chart for measure of spread. σ-chart shows the variation of process.

2. To determine whether a given process can meet the existing specifications without a fundamental change in the production line or to tell whether the process is in control and if so, at what dispersion.

3. To secure information to be used in establishing or changing production procedures.

4. To secure information when it is necessary to widen the tolerances.

5. To provide a basis for current decisions or acceptance or rejection of manufactured or purchased product.

6. To secure information to be used in establishing or changing inspection procedure or acceptance procedure or both.

5.34 CONSTRUCTION OF CONTROL CHARTS FOR VARIABLES

First of all, a random sample of size n is taken during a manufacturing process over a period of time and quality measurements $x_1, x_2, \ldots\ldots, x_n$ are noted

Sample mean $$\bar{x} = \frac{x_1 + x_2 + \ldots\ldots + x_n}{n} = \frac{1}{n}\sum_{i=1}^{n} x_i$$

Sample range $\text{R} = x_{\text{max.}} - x_{\text{min.}}$

If the process is found stable, k consecutive samples are selected and for each sample, $\bar{x}$ and R are calculated. Then we find $\bar{\bar{x}}$ and $\bar{\text{R}}$ as

$$\bar{\bar{x}} = \frac{\bar{x}_1 + \bar{x}_2 + \ldots\ldots + \bar{x}_k}{k} = \frac{1}{k}\sum_{i=1}^{k} \bar{x}_i$$

and $$\bar{\text{R}} = \frac{\text{R}_1 + \text{R}_2 + \ldots\ldots + \text{R}_k}{k} = \frac{1}{k}\sum_{i=1}^{k} \text{R}_i$$

For $\bar{\text{X}}$-chart

$$\text{Central line} = \left.\begin{array}{l}\bar{\bar{x}}, \text{ when tolerance limits are not given} \\ \mu, \text{ when tolerance limits are given}\end{array}\right\}$$

where $$\mu = \frac{1}{2}\,[\text{LCL} + \text{UCL}]$$

LCL is lower control limit and UCL is upper control limit

Now, LCL (for $\bar{\text{X}}$-chart) $= \bar{\bar{x}} - \text{A}_2\bar{\text{R}}$ and UCL (for $\bar{\text{X}}$-chart) $= \bar{\bar{x}} + \text{A}_2\bar{\text{R}}$ are set.

A_2 depends on sample size n and can be found from the following table :

Sample size (n)	2	3	4	5	6	7	8	9	10	11	12	13	14	15	16	17	18	19	20
A_2	1.88	1.02	0.73	0.58	0.48	0.42	0.37	0.34	0.31	0.29	0.27	0.25	0.24	0.22	0.21	0.20	0.19	0.19	0.18

For R-chart Central line (CL) $= \bar{\text{R}}$

Now, LCL (for R-chart) $= \text{D}_3\bar{\text{R}}$ UCL (for R-chart) $= \text{D}_4\bar{\text{R}}$ are set.

where D_3 and D_4 depend on sample size and are found from the following table :

Sample size (n)	D_3	D_4	d_2
2	0	3.27	1.13
3	0	2.57	1.69
4	0	2.28	2.06
5	0	2.11	2.33
6	0	2.00	2.53
7	0.08	1.92	2.70
8	0.14	1.86	2.85
9	0.18	1.82	2.97
10	0.22	1.78	3.08
11	0.26	1.74	3.17
12	0.28	1.72	3.26
13	0.31	1.69	3.34
14	0.33	1.67	3.41
15	0.35	1.65	3.47
16	0.36	1.64	3.53
17	0.38	1.62	3.59
18	0.39	1.61	3.64
19	0.40	1.60	3.69
20	0.41	1.59	3.74

To compute upper and lower process tolerance limits for the values of x, we have

$$\text{LTL} = \bar{\bar{x}} - \frac{3\bar{R}}{d_2} \qquad \text{UTL} = \bar{\bar{x}} + \frac{3\bar{R}}{d_2}$$

where d_2 is found from the above table.

Moreover, the process capability is given by $6\sigma = 6\dfrac{\bar{R}}{d_2}$ where σ is standard deviation.

While plotting the $\bar{X}$-chart the central line on the $\bar{X}$ chart should be drawn as a solid horizontal line at $\bar{X}$. The upper and lower control limits for $\bar{X}$ chart should be drawn as dotted horizontal lines at the computed values.

Similarly, for R-chart, the central line should be drawn as a solid horizontal line at $\bar{R}$. The upper control limit should be drawn as dotted horizontal line at the computed value of UCL_R. If the subgroup size is 7 or more, the lower control limit should be drawn as dotted horizontal line at LCL_R. However, if the subgroup size is ≤ 6, the lower control limit for R is zero.

Plot the averages of subgroups in $\bar{X}$-chart, in the order collected and ranges in R-chart which should be below the $\bar{X}$-chart so that the subgroups correspond to one-another in both the charts. Points outside the control limits are indicated with cross (×) on $\bar{X}$-chart and the points outside the limits on R chart by a circle (⊙).

5.35 CONTROL CHARTS FOR ATTRIBUTES

Following control charts will be discussed here

(*i*) P chart (*ii*) *np* chart (*iii*) C chart

As an alternative to $\overline{X}$ and R chart and as a substitute when characteristic is measured only by attribute, a control chart based on fraction defective $\overline{p}$ is used, called P-chart.

$$\overline{p} = \frac{\text{No. of defective articles found in any inspection}}{\text{Total no. of articles actually inspected}}.$$

(*i*) **Control limits (3σ limits) on p-chart:** We know that for binomial distribution, the mean value of total number of defectives in a sample n is np and standard deviation is $\sqrt{npq}$ or $\sqrt{np(1-p)}$.

$\therefore$ Mean value of fraction defective is $\overline{p}$ and S.D.

$$\sigma_p = \frac{1}{n}\sqrt{n\overline{p}(1-\overline{p})} = \sqrt{\frac{\overline{p}(1-\overline{p})}{n}}$$

$$\therefore \quad \text{CL} = \overline{p}$$

The upper and lower limits for P-chart are, $\text{UCL}_\text{P} = \overline{p} + 3\sigma_p = \overline{p} + 3\sqrt{\frac{\overline{p}(1-\overline{p})}{n}}$

and $\text{LCL}_\text{P} = \overline{p} - 3\sigma_p = \overline{p} - 3\sqrt{\frac{\overline{p}(1-\overline{p})}{n}}$.

Due to the lower inspection and maintenance costs of P-charts, they usually have a greater area of economical applications.

(*ii*) **Control limits for *np* chart:** Whenever subgroup size is variable, P-chart is used but if it is constant, the chart for actual number of defectives called *np* chart is used.

$$\text{CL} = n\overline{p} \quad \text{where} \quad \overline{p} = \frac{\Sigma np}{\Sigma n}$$

$$\text{UCL}_{np} = n\overline{p} + 3\sigma_{np} = n\overline{p} + 3\sqrt{n\overline{p}\,(1-\overline{p})} \qquad (\text{where } \sigma_{np} = n\sigma_p)$$

and

$$\text{LCL}_{np} = n\overline{p} - 3\sqrt{n\overline{p}\,(1-\overline{p})}.$$

Note. In case of $\overline{X}$ and R chart, it may not be necessary to draw lines connecting the points which represent the successive subgroups. But incase of P-chart, a line connecting the points is usually helpful in interpretation of the chart. Such a line assists in the interpretation of trends.

(*iii*) **Control limits for C chart**

(*a*) **Difference between a defect and defective**

An item is called defective if it fails to conform to the specifications in any of the characteristics. Each characteristic that does not meet the specifications is a defect. An item is defective if it contains atleast one defect. The *np* chart applies to the number of defectives in subgroups of constant size while C chart applies to the number of defects in a subgroup of constant size.

(*b*) **Basis for control limits on C chart**

Control limits on C chart are based on **Poisson distribution:** Hence two conditions must be satisfied. The first condition specifies that the area of opportunity for occurrence of defects should be fairly constant from period to period. Second condition specifies that opportunities for defects are large while the chances of a defect occurring in any one spot are small.

(*c*) **Calculation of control limits on C chart**

Standard deviation $\sigma_c = \sqrt{\overline{C}}$

Thus 3σ limits on a C chart are $UCL_c = \overline{C} + 3\sqrt{\overline{C}}$ and $LCL_c = \overline{C} - 3\sqrt{\overline{C}}$

and central line $CL = \overline{C}$

where $$\overline{C} = \frac{\text{Number of defects in all samples}}{\text{Total number of samples}}.$$

ILLUSTRATIVE EXAMPLES

Example 1. *The following are the mean lengths and ranges of lengths of a finished product from 10 samples each of size 5. The specification limits for length are 200 ± 5 cm. Construct $\overline{X}$ and R-chart and examine whether the process is under control and state your recommendations.*

Sample No.	*1*	*2*	*3*	*4*	*5*	*6*	*7*	*8*	*9*	*10*
Mean ($\overline{X}$)	*201*	*198*	*202*	*200*	*203*	*204*	*199*	*196*	*199*	*201*
Range (R)	*5*	*0*	*7*	*3*	*3*	*7*	*2*	*8*	*5*	*6*

Assume for $n = 5$, $A_2 = 0.58$, $D_4 = 2.11$ and $D_3 = 0$.

Sol. (*i*) **Control limits for $\overline{X}$ chart:**

Central limit CL = 200 | ∵ Tolerance / specification limits are given ∴ $\mu = 200$

$$UCL_{\bar{x}} = \bar{\bar{x}} + A_2\overline{R} = \mu + A_2\overline{R}$$

$$LCL_{\bar{x}} = \bar{\bar{x}} - A_2\overline{R} = \mu - A_2\overline{R}$$

where $$\overline{R} = \frac{R_1 + R_2 + \ldots\ldots + R_{10}}{10} = \frac{46}{10} = 4.6$$

Then, $$UCL_{\overline{X}} = 200 + (0.58 \times 4.6) = 202.668$$

$$LCL_{\overline{X}} = 200 - (0.58 \times 4.6) = 197.332.$$

(*ii*) **Control limits for R chart:**

Control limit $CL = \overline{R} = 4.6$

$$UCL_R = D_4\overline{R} = 2.11 \times 4.6 = 9.706$$

$$LCL_R = D_3\overline{R} = 0 \times 4.6 = 0$$

The $\overline{X}$ and R-charts are drawn below:

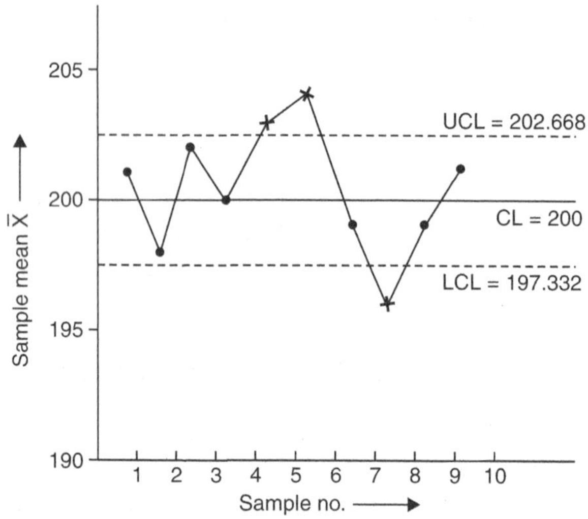

$\overline{X}$-Chart

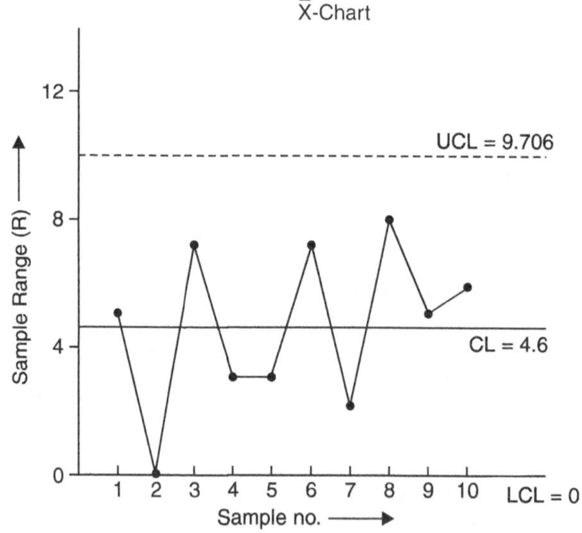

R-Chart

It is noted that all points lie within the control limits on R chart. Hence the process variability is under control. But in X-chart, points corresponding to sample number 5, 6 and 8 lie outside the control limits. Therefore the **process is not in statistical control.** The process should be halted and it is recommended to check for any assignable causes. Fluctuation will remain until these causes, if found, are removed.

Example 2. *A drilling machine bores holes with a mean diameter of 0.5230 cm and a standard deviation of 0.0032 cm. Calculate the 2-sigma and 3-sigma upper and lower control limits for means of sample of 4.* [G.B.T.U. (C.O) 2010]

Sol. Mean diameter $\overline{\overline{x}} = 0.5230$ cm

$$\text{S.D. } \sigma = 0.0032 \text{ cm}$$

$$n = 4$$

(*i*) 2-sigma limits are as follows:

$$CL = \bar{\bar{x}} = 0.5230 \text{ cm}$$

$$UCL = \bar{\bar{x}} + 2\frac{\sigma}{\sqrt{n}} = 0.5230 + 2\times\frac{0.0032}{\sqrt{4}} = 0.5262 \text{ cm}$$

$$LCL = \bar{\bar{x}} - 2\frac{\sigma}{\sqrt{n}} = 0.5230 - 2\times\frac{0.0032}{\sqrt{4}} = 0.5198 \text{ cm.}$$

(*ii*) 3-sigma limits are as follows:

$$CL = \bar{\bar{x}} = 0.5230 \text{ cm}$$

$$UCL = \bar{\bar{x}} + 3\frac{\sigma}{\sqrt{n}} = 0.5230 + 3\times\frac{0.0032}{\sqrt{4}} = 0.5278 \text{ cm}$$

$$LCL = \bar{\bar{x}} - 3\frac{\sigma}{\sqrt{n}} = 0.5230 - 3\times\frac{0.0032}{\sqrt{4}} = 0.5182 \text{ cm.}$$

Example 3. *In a blade manufacturing factory, 1000 blades are examined daily. Draw the np chart for the following table and examine whether the process is under control?*

Date	: 1	2	3	4	5	6	7	8	9	10	11	12	13	14	15
No. of defective blades	: 9	10	12	8	7	15	10	12	10	8	7	13	14	15	16

[G.B.T.U. (C.O) 2010]

Sol. Here, $n = 1000$

Σnp = total number of defectives = 166

Σn = total number inspected = 1000 × 15

$$\therefore \quad \bar{p} = \frac{\Sigma np}{\Sigma n} = \frac{166}{1000\times 15} = 0.011$$

$$\therefore \quad n\bar{p} = 1000 \times 0.011 = 11$$

Control limits are $CL = n\bar{p} = 11$

$$UCL_{np} = n\bar{p} + 3\sqrt{n\bar{p}(1-\bar{p})} = 11 + 3\sqrt{11(1-0.011)} = 20.894$$

$$LCL_{np} = n\bar{p} - 3\sqrt{n\bar{p}(1-\bar{p})} = 11 - 3\sqrt{11(1-0.011)} = 1.106$$

The *np* chart is drawn in the figure. Since all the points lie within the control limits, the process is under control.

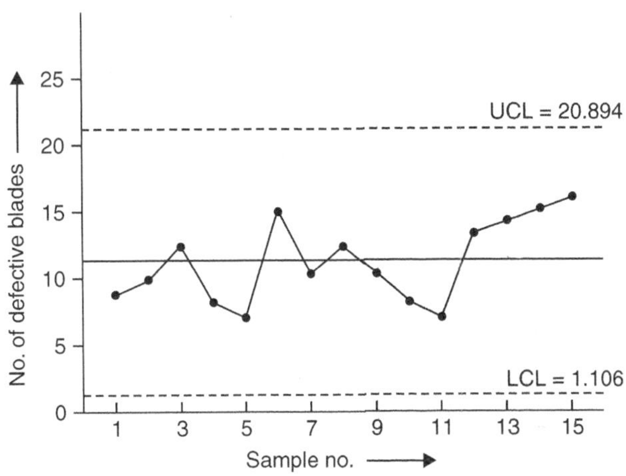

(*np*-chart)

Example 4. *In a manufacturing process, the number of defectives found in the inspection of 20 lots of 100 samples is given below:*

Lot no.	*No. of defectives*	*Lot no.*	*No. of defectives*
1	5	11	7
2	4	12	6
3	3	13	3
4	5	14	5
5	4	15	4
6	6	16	2
7	9	17	8
8	15	18	7
9	11	19	6
10	6	20	4

(i) Determine the control limits of p-chart and state whether the process is in control.

(ii) Determine the new value of mean fraction defective if some points are out of control. Compute the corresponding control limits and state whether the process is still in control or not.

(iii) Determine the sample size when a quality limit not worse than 9% is desirable and a 10% bad product will not be permitted more than three times in thousand.

Sol. (*i*)
$$\bar{p} = \frac{\text{Total number of defectives}}{\text{Total number of items inspected}} = \frac{120}{20 \times 100} = 0.06$$

$$\text{UCL}_p = \bar{p} + 3\sqrt{\frac{\bar{p}(1-\bar{p})}{n}} = 0.06 + 3\sqrt{\frac{0.06\,(1-0.06)}{100}} = 0.13095$$

$$\text{LCL}_p = \bar{p} - 3\sqrt{\frac{\bar{p}(1-\bar{p})}{n}} = 0.06 - 3\sqrt{\frac{0.06\,(1-0.06)}{100}} = -\,0.01095$$

Since the fraction defective cannot be (–) ve

$\therefore$ $\text{LCL}_p = 0$

After observing the values of defectives in the given example, it is clear that only 8th lot having fraction defective $\frac{15}{100} = 0.15$ will go above UCL_p.

(*ii*) After eliminating the 8th lot,

$$\text{Revised value of } \bar{p} = \frac{120-15}{100 \times 19} = 0.056$$

Revised control limits will be

$$\text{UCL}_p = 0.056 + 3\sqrt{\frac{0.056\,(1-0.056)}{100}} = 0.125$$

$$\text{LCL}_p = 0.056 - 3\sqrt{\frac{0.056\,(1-0.056)}{100}} = -\,0.013 \text{ } i.e., \text{ zero.}$$

It is clear that all the points are within control limits.

$\therefore$ Revised quality level $\bar{p} = 0.056$

(*iii*) Since a probability that a defective worse than 9% defective quality will not be permitted is more than 3 times in thousand (0.3%) is corresponding 36 limits

$$\therefore \qquad \bar{p} + 3p = 0.09$$

$$0.056 + 3\sqrt{\frac{0.056\,(1-0.056)}{n}} = 0.09 \Rightarrow \sqrt{\frac{0.056 \times 0.944}{n}} = \frac{0.034}{3}$$

Squaring, $$\frac{0.056 \times 0.944}{n} = \left(\frac{0.034}{3}\right)^2 = (0.01133)^2$$

$$n = \frac{0.056 \times 0.944}{0.01133 \times 0.01133} = 333.$$

Example 5. *Determine the control limits for $\bar{X}$ and R charts if $\Sigma\bar{X} = 357.50$, $\Sigma R = 9.90$, number of subgroups = 20. It is given that $A_2 = 0.18$, $D_3 = 0.41$, $D_4 = 1.59$ and $d_2 = 3.736$. Also, find the process capability.*

Sol. $$\bar{X} = \frac{\Sigma\bar{X}}{N} = \frac{357.50}{20} = 17.875, \quad \bar{R} = \frac{\Sigma R}{N} = \frac{9.90}{20} = 0.495$$

$$UCL_{\bar{X}} = \bar{\bar{X}} + A_2\bar{R} = 17.875 + (0.18 \times 0.495) = 17.9641$$

$$LCL_{\bar{X}} = \bar{\bar{X}} - A_2\bar{R} = 17.875 - (0.18 \times 0.495) = 17.7859$$

$$UCL_R = D_4\bar{R} = 1.59 \times 0.495 = 0.78705$$

$$LCL_R = D_3\bar{R} = 0.41 \times 0.495 = 0.20295$$

$$\sigma = \frac{\bar{R}}{d_2} = \frac{0.495}{3.735} = 0.13253$$

$\therefore$ Process capability = $6\sigma = 6 \times 0.13253 = 0.79518$.

Example 6. *If the average fraction defective of a large sample of a product is 0.1537, Calculate the control limits given that sub-group size is 2000.*

Sol. Average fraction defective

$$\bar{p} = 0.1537$$

Sub-group size is 2000

$$\therefore \qquad n = 2000$$

Central line $$CL = n\bar{p} = 2000 \times 0.1537 = 307.4$$

$$UCL_{np} = n\bar{p} + 3\sigma_{np} = n\bar{p} + 3\sqrt{n\bar{p}(1-\bar{p})}$$

$$= 307.4 + 3\sqrt{307.4\,(1-0.1537)}$$

$$= 355.787742$$

and $$LCL_{np} = n\bar{p} - 3\sqrt{n\bar{p}\,(1-\bar{p})} = 307.4 - 48.38774204$$

$$= 259.012258$$

TEST YOUR KNOWLEDGE

1. A company manufactures screws to a nominal diameter 0.500 ± 0.030 cm. Five samples were taken randomly from the manufactured lots and 3 measurements were taken on each sample at different lengths. Following are the readings:

Sample no.	*Measurement per sample x(in cm)*		
	1	*2*	*3*
1	0.488	0.489	0.505
2	0.494	0.495	0.499
3	0.498	0.515	0.487
4	0.492	0.509	0.514
5	0.490	0.508	0.499

Calculate the control limits of $\overline{X}$ and R charts. Draw $\overline{X}$ and R charts and examine whether the process is in statistical control? [Take $A_2 = 1.02, D_4 = 2.57, D_3 = 0$ for $n = 3$]

2. Discuss how control charts can be used in quality control of industrial products. The average percentage of defectives in 27 samples of size 1500 each was found to be 13.7%. Construct P-chart for this situation. Explain how the control chart can be used to control quality.

[**Hint.** $\overline{p} = 0.137$] *(G.B.T.U. 2010)*

3. The following data shows the value of sample mean $\overline{X}$ and range R for 10 samples of size 5 each. Calculate the values for central line and control limits for $\overline{X}$-chart and R chart and determine whether the process is under control.

Sample no. :	1	2	3	4	5	6	7	8	9	10
Mean $\overline{X}$:	11.2	11.8	10.8	11.6	11	9.6	10.4	9.6	10.6	10
Range R :	7	4	8	5	7	4	8	4	7	9

Assume for $n = 5$, $A_2 = 0.577$, $D_3 = 0$ and $D_4 = 2.115$.

4. It was found that when a manufacturing process is under control, the average number of defectives per sample batch of 10 is 1.2. What limits would you set in a quality control chart based on the examination of defectives in sample batches of 10?

[**Hint.** $\overline{p} = 0.12, n\overline{p} = 1.2$]

5. In a manufacturing process, the number of defective items found in the inspection of 15 lots of 400 items each are given below:

Lot No. :	1	2	3	4	5	6	7	8	9	10	11	12	13	14	15
No. of defectives :	2	5	0	14	3	0	1	0	18	8	6	0	3	0	6

6. (*i*) What are statistical quality control techniques? Discuss the objectives and advantages of statistical quality control.

   (*ii*) Give statistical quality control methods. Explain one of them with example.

7. Write short notes on the following :

   (*i*) Use of statistical techniques/methods in quality control [*G.B.T.U. (MCA) 2010*]

   (*ii*) P-chart, *np*-chart and *C*-chart *(G.B.T.U. 2010)*

8. Distinguish between the *np*-chart and *p*-chart. Following is the data of defectives of 10 samples of size 100 each. Construct *np*-chart and give your comments.

Sample no. :	1	2	3	4	5	6	7	8	9	10
No. of difectives :	6	9	12	5	12	8	8	16	13	7

(G.B.T.U. 2010)

9. In a factory producing spark plug, the number of defectives found in inspection of 20 lots of 100 each is given below. *(G.B.T.U. 2011)*

Lot No.	1	2	3	4	5	6	7	8	9	10	11	12	13	14	15	16	17	18	19	20
No. of defectives	5	10	12	8	6	4	6	3	3	5	4	7	8	3	3	4	5	8	6	10

Construct p-chart and state whether the process is in statistical control.

10. The data below given is the number of defective bearing in samples of size 150. Construct np-chart for these data. If any point(s) lie outside the control limits, assume that assignable cause can be found and determine the revised control limits :

Sample No.	1	2	3	4	5	6	7	8	9	10	11	12	13	14	15	16	17	18	19	20
No. of defectives	12	7	5	4	1	5	9	0	15	6	7	4	1	3	6	8	10	5	2	7

Determine metrial control limits for np chart and state whether the proven is in control.

11. The following table shows the number of missing rivets observed at the time of inspection of 12 aircrafts. Find the control limits for the number of defects chart and comment on the state of control.

Air craft number :	1	2	3	4	5	6	7	8	9	10	11	12
No. of missing rivets :	7	15	13	18	10	14	13	10	20	11	22	15

12. The number of customer complaints received daily by an organization is given below :

Day :	1	2	3	4	5	6	7	8	9	10	11	12	13	14	15
Complaints :	2	3	0	1	9	2	0	0	4	2	0	7	0	2	4

Does it mean that the number of complaints is under statistical control? Establish a control scheme for the future.

Answers

1. $CL_{\overline{X}} = 0.4988$, $UCL_{\overline{X}} = 0.5172$, $LCL_{\overline{X}} = 0.4804$, $CL_R = 0.018$, $UCL_R = 0.0463$, $LCL_R = 0$. The process is in control.

2. $CL_P = 0.137$, $UCL_P = 0.164$, $LCL_P = 0.110$

3. $CL_{\overline{X}} = 10.66$, $UCL_{\overline{X}} = 14.295$, $LCL_{\overline{X}} = 7.025$, $CL_R = 0.3$, $UCL_R = 13.32$, $LCL_R = 0$; The process is under control

4. $CL_{np} = 1.2$, $UCL_{np} = 2.175$, $LCL_{np} = 0.225$

5. $p = 0.011$, CL = 4.4, $UCLnp = 10.6581$, $LCLnp = 0$; No

8. Process is in statistical control.

9. Process is in statistical control. $UCL_p = 0.131$, $LCL_p = 0$

10. $CL_{np} = 5.36842$, $UCL_{np} = 12.19385$, $LCL_{np} = 0$

11. $UCL_C = 25.23$, $LCL_C = 2.77$. The process is in control.

12. $CL_C = 2.4$, $UCL_C = 7.05$, $LCL_C = 0$, the process is not under control.

ASSIGNMENT-V

1. Write the alternative hypothesis if the null hypothesis is $H_0 : \mu = 50$.
2. What is the number of degree of freedom for a contingency table of order 5×4?.
3. If the critical value of Z is 1.96, find the significance level of two tailed test.
4. What is the test statistic for testing $H_0 : \mu = \mu_0$ when sample size n is small?
5. When is the test statistic $F = \dfrac{S_1^2}{S_2^2}$ used?
6. While testing the significance of difference of two sample means incase of small samples, how is degree of freedom calculated?
7. When a true null hypothesis is rejected, what is the error made, known with?
8. In ANOVA, when calculated value of F is greater than the tabulated value, then what happens to null hypothesis?
9. For testing the significance of difference between two means, when population variance is not known, how do we read the statistic z?
10. In one way classified data involving 4 classes each having 12 observations, find the degrees of freedom associated with error sum of squares.
11. What is chi-square test?
12. What are the applications of χ^2–test?
13. What are the conditions for applying χ^2–test?
14. What do you mean by degrees of freedom?
15. Write a short note on:
    (*i*) Null hypothesis (*ii*) Alternative hypothesis (*iii*) Level of significance
16. Describe the method of sampling with examples.
17. Explain the usefulness of control charts for variables.
18. Write a short note on:
    (*i*) Confidence limits (*ii*) Test statistic
19. What do you mean by statistical quality control (SQC)?
20. What is F-test? Write its application.
21. Write a short not on 'ANOVA'.
22. Explain one way ANOVA classification.

Answers

1. $H_1 : \mu \neq 50$ **2.** 12 **3.** 0.05 or 5% level

4. $t = \dfrac{\bar{x} - \mu_0}{s/\sqrt{n}}$ **5.** $S_1^2 > S_2^2$ **6.** $d.f. = n_1 + n_2 - 2$

7. type I error **8.** rejected **9.** $\dfrac{\bar{x} - \bar{y}}{\sqrt{\dfrac{s_1^2}{n_1} + \dfrac{s_2^2}{n_2}}}$

10. 8.

EXAMINATION PAPERS

A.K.T.U., LUCKNOW

RAS–301

B. Tech. [SEMESTER-III]

THEORY EXAMINATION, 2018–2019
MATHEMATICS -III

Time: 3 Hours *Total Marks: 70*

Note: *Assume any Missing Data.*

SECTION–A

1. Attempt ***all*** parts of the following: **(7 × 2 = 14)**

(*a*) The function $f(x) = e^x(\cos y + i \sin y)$ is holomorphic or not.

(*b*) Find the residue of $\dfrac{z^2}{(z-1)(z-2)^2}$ at pole $z = 2$.

(*c*) Formula of Measure of Kurtosis $\beta_2 = \ldots$.

(*d*) The first three central moments of a distribution are 0, 15, –31. Find the moment coefficient of skewness.

(*e*) Obtain the function whose first difference is $9x^2 + 11x + 5$.

(*f*) Find the normal equation of a curve $y = ax + bx^2$

(*g*) Let $f(z) = u(r, \theta) + iv(r, \theta)$ be an analytic function. If $u = -r^3 \sin 3\theta$, then find v.

SECTION–B

2. Attempt any **Three** parts of the following: **(3 × 7 = 21)**

(*a*) From the following table of values of x and y, obtain $\dfrac{dy}{dx}$ for $x = 1.2, 2.2, 1.6$.

x:	1.0	1.2	1.4	1.6	1.8	2.0	2.2
y:	2.7183	3.3201	4.0552	4.9530	6.0496	7.3891	9.0250

(*b*) Using Runga-Kutta method of fourth order, find $y(0.8)$ correct to 4 decimal places if $\dfrac{dy}{dx} = y - x^2$, $y(0.6) = 1.7379$, taking $h = 0.1$.

(*c*) Using complex integration method, evaluate $\displaystyle\int_0^{2\pi} \frac{\cos 2\theta}{5 + 4\cos\theta}\, d\theta$.

(*d*) The equations of two regression lines, obtained in a correlation analysis of 60 observations are: $5x - 6y = 24$, $768x - 1000y = 3608$. What is the correlation coefficient? Show that the ratio of coefficient of variability of x to that of y is $\dfrac{5}{24}$. What is the ratio of variances of x and y?

(*e*) The pressure of the gas corresponding to various volumes V is measured, given by the following data:

V (cm^3)	50	60	70	90	100
P($kg\ cm^{-2}$)	64.7	51.3	40.5	25.9	78

Fit the data to the equation $PV^{\gamma} = c$.

SECTION–C

3. Attempt any **Two** parts of the following: **(2 × 3.5 = 07)**

(*a*) Find the unique polynomial P(*x*) of degree 2 such that: P(1) = 1, P(3) = 27, *P*(4) = 64, use Lagrange method of interpolation.

(*b*) Using Simpson's 3/8th rule on integration, evaluate $\int_0^6 \frac{1}{1+x}\,dx$.

(*c*) Expand $\frac{1}{z^2 - 3z + 2}$ in the region $1 < |z| < 2$.

4. Attempt any **Two** parts of the following: **(2 × 3.5 = 07)**

(*a*) If the probability of hitting a target is 10% and 10 shots are fired independently. What is the probability that the target will be hit at least once?

(*b*) A die is thrown 276 times and the results of these throws are given below:

No. appeared on the die	1	2	3	4	5	6
Frequency	40	32	29	59	57	59

Test whether the die is biased or not.

[Tabulated value of c^2 at 5% level of significance for 5 degree of freedom is 11.09]

(*c*) By Residue method, find the inverse Z-transform of $\frac{z}{z^2 + 7z + 10}$.

5. Attempt any **Two** parts of the following: **(2 × 3.5 = 07)**

(*a*) The following data regarding the heights (*y*) and weights (*x*) of 100 college students are given:

$$\sum x = 15000, \sum x^2 = 2272500, \sum y = 6800, \sum y^2 = 463025, \sum xy = 1022250$$

Find the equation of regression line of height on weight.

(*b*) Solve $x^3 - 5x + 3 = 0$ by using Regula-Falsi method correct up to four decimal places.

(*c*) From the table, estimate the number of students who obtained marks between 40 and 45.

Marks:	30–40	40–50	50–60	60–70	70–80
No. of students:	31	42	51	35	31

6. Attempt any **Two** parts of the following: **(2 × 3.5 = 07)**

(*a*) Find the residue of $f(z) = \dfrac{z^3}{(z-1)^4\,(z-2)\,(z-3)}$ at its pole and hence evaluate

$\int_c f(z)dz$, where C is the circle $|z| = 5/2$.

(*b*) Determine the largest Eigen value and corresponding Eigen vector of the matrix

$A = \begin{bmatrix} 2 & -1 & 0 \\ -1 & 2 & -1 \\ 0 & -1 & 2 \end{bmatrix}$ till three approximations.

(*c*) Verify Cauchy theorem by integrating e^{iz} along the boundary of the triangle with the vertices at the points $1 + i$, $-1 + i$ and $-1 - i$.

7. Attempt any **Two** parts of the following: **(2 × 3.5 = 07)**

(*a*) Use Picard's method to obtain y for $x = 0.2$. Given: $\dfrac{dy}{dx} = x - y$ with initial condition $y = 1$ when $x = 0$ correct up to four decimal places.

(*b*) In a normal distribution, 31% of the items are under 45 and 8% are over 64. Find the mean and standard deviation of the distribution. It is given that if

$$f(t) = \frac{1}{\sqrt{2\pi}} \int_0^t e^{-\frac{1}{2}x^2}\, dx \text{ then } f(0.5) = 0.19,\ f(1.4) = 0.42.$$

(*c*) Prove that: $h\nabla = -\log(1 - \nabla) = \sinh^{-1}(\mu\delta)$.

A.K.T.U., LUCKNOW

RAS–301

B. Tech. [SEMESTER-III]

THEORY EXAMINATION, 2017–2018

MATHEMATICS -III

Time: 3 Hours | *Total Marks: 70*

Notes: 1. *Attempt All Sections. If require any missing data; then choose suitably.*
2. *Any special paper specific instruction.*

SECTION–A

1. Attempt *all* questions in brief. **(2 × 7 = 14)**

(*a*) Define analytic function with an example.

(*b*) Define the Binomial distribution with mean and variance.

(*c*) Write the normal equations for the curve $y = \frac{a}{x} + bx$.

(*d*) Give comparison between Regula-Falsi method and Newton-Raphson method.

(*e*) Write the relation between n^{th} divided difference and n^{th} forward difference.

(*f*) What do you mean by initial value problem?

(*g*) Find $Z^{-1}\left(\frac{5}{5z-1}\right)$.

SECTION–B

2. Attempt any *three* of the following: **(7 × 3 = 21)**

(*a*) Give an example of a function in which Cauchy-Riemann equations are satisfied yet the function is not analytic at the origin. Justify your answer.

(*b*) Find the measurement of Skewness and Kutosis based on moments for the following distribution and draw your conclusion.

Marks:	5–15	15–25	25–35	35–45	45–55
No. of students:	1	3	5	7	4

(*c*) Decompose $A = \begin{bmatrix} 5 & -2 & 1 \\ 7 & 1 & -5 \\ 3 & 7 & 4 \end{bmatrix}$ in the form LU, where L is lower triangular matrix and U is upper triangular matrix and hence solve the system of equations:

$$5x - 2y + z = 4$$
$$7x + y - 5z = 8$$
$$3x + 7y + 4z = 10.$$

(*d*) Express the function $f(x) = \begin{cases} 1 \text{ when } |x| \le 1 \\ 0 \text{ when } |x| > 1 \end{cases}$, as a Fourier Integral. Hence evaluate $\int_0^{\infty} \frac{\sin \lambda \cos \lambda x}{\lambda} d\lambda$.

(*e*) Given the initial value problem $\frac{dy}{dx} = x^3 - y^3, y(0) = 1$.

Find the numerical solution of differential equation at $x = 0.6$ with $h = 0.2$ by using Runge-Kutta method of Fourth order.

SECTION–C

3. Attempt any *one* of the following: **(7 × 1 = 7)**

(*a*) Evaluate the integration: $\int_0^{\pi} \sin^4 \theta d\theta$.

(*b*) State and prove the Cauchy intergal formula. Also, evaluate $\int_C \frac{1}{(z^2+4)^2} dz = \frac{\pi}{16}$ where C is the circle $|z - i| = 2$.

4. Attempt any *one* part of the following: **(7 × 1 = 7)**

(*a*) Find Fourier cosine transform of $\frac{1}{1+x^2}$ and hence find Fourier sine transform of $\frac{x}{1+x^2}$.

(*b*) Find the inverse *Z*-transform of $F(z)$ where $F(z)$ is given by:

(*i*) $\frac{z}{(z+2)(z+3)}$ (*ii*) $\frac{7z - 11z^2}{(z-1)(z-2)(z+3)}$.

5. Attempt any *one* part of the following: **(7 × 1 = 7)**

(*a*) In a partially distributed laboratory record of an analysis of a correlataion data the following result are legible:

Variance of $x = 9$

Regression equation, $8x - 10y + 66 = 0$, $40x - 18y = 214$.

What were (*a*) the mean of x and y(*b*) the standard deviation of y and the correlation coefficient of x and y?

(*b*) Find the mean and variance of normal distribution.

6. Attempt any *one* part of the following: **(7 × 1 = 7)**

(*a*) Find the real root of the equation $x^3 - 2x + 5 = 0$ by method of False position correct to three decimal places.

(*b*) State and prove the Lagrange Interpolation formula. Find the interpotating polynomial by Lagrange interpolation formula for the given data:

x	5	6	9	11
y	12	13	14	16

7. Attempt any *one* part of the following: **(7 × 1 = 7)**

(*a*) Apply Simpson's 3/8th rule to obtain approximate value of

(*i*) $\int_0^{\pi/2} e^{\sin x} dx$

(*ii*) $\int_0^{0.3} (2x - x^2)^{1/2} \, dx$ using Simpson's rule with 6 interval.

(*b*) Find x for which y is maximum and find the maximum value of y.

x	1.2	1.3	1.4	1.5	1.6
y	0.9320	0.9636	0.9855	0.9975	0.9996

A.K.T.U., LUCKNOW

RAS–404

B. Tech. [SEMESTER-IV]

THEORY EXAMINATION, 2017–2018

MATHEMATICS -III

Time: 3 Hours *Total Marks: 70*

Note: *Attempt all Sections. If require any missing data; then choose suitably.*

SECTION–A

1. Attempt *all* questions in brief. **(7 × 2 = 14)**

(*a*) Write Simpson's 1/3 rule.

(*b*) Define average operator and central operator and find relationship between them.

(*c*) Define binomial distribution and write their mean and standard distributions.

(*d*) Find the missing term in the following table:

x	0	4	8	12	16	20
y	5	10	–	15	21	30

(*e*) State Cauchy integral theorem.

(*f*) Find the Z-transform of the sequence $\{a_n\}$

(*g*) Find the inverse Z-transform of $\left\{\dfrac{3}{3z-1}\right\}$.

SECTION–B

2. Attempt any *three* of the following: **(7 × 3 = 21)**

(*a*) Using Euler's modified method solve $dy/dx = \log(x + y)$ with the initial condition that $y = 1$ when $x = 0$, find y for $x = 0.1$ and $x = 0.2$.

(*b*) Use Runga-Kutta method of fourth order, solve $\dfrac{dy}{dx} = \dfrac{y^2 - x^2}{y^2 + x^2}$ with $y(0) = 1$, at $x = 0.2$ taken $h = 0.1$.

(*c*) Using method of Least squares, find the curve $y = a + bx + cx^2$ that best fit the Following data:

x	1	2	3	4	5
y	5	10	15	20	25

(*d*) If $f(z) = \begin{cases} \dfrac{x^3 y (y - ix)}{x^6 + y^2}, & z \neq 0 \\ 0, & z = 0 \end{cases}$

Then discuss $\dfrac{df}{dz}$ at $z = 0$.

(*e*) Find the Fourier sine transform of $\frac{e^{-ax}}{x}$, $a > 0$. Hence find Fourier sine transform of $\frac{1}{x}$.

SECTION–C

3. Attempt any *one* part of the following: **(7 × 1 = 7)**

(*a*) By using Newton-Raphson method, find the root $x^3 - 6x + 4 = 0$ correct to three places of decimal.

(*b*) Using Lagrange's interpolation formula, find the values of y corresponding to $x = 20$ from the following table:

x	6	15	21	26
y	21	25	31	36

4. Attempt any *one* part of the following: **(7 × 1 = 7)**

(*a*) Solve by Crout's method, the following system of equation:

X + Y + Z = 3, 2X – Y + 3Z = 16, 3X + Y – Z = – 3.

(*b*) Using Picard's method, find a solution of $\frac{dy}{dx} = 1 + xy$ upto third approximation, when $x_0 = 0, y_0 = 0$.

5. Attempt any *one* part of the following: **(7 × 1 = 7)**

(*a*) In a certain factory turning out razor blades, there is a small chance of 0.002 for any blade to be defective. The blades are supplied in packets of 10. Use appropriate and suitable distribution to calculate the approximate number of packets containing no defective, one defective and two defective blades respectively in a consignment of 50000 packets.

(*b*) A sample of 100 dry battery cells tested to find the length of life produced the following results $\bar{x} = 10$ hours, $\sigma = 3$ hours. Assuming the data be normally distributed, what percentage of battery cells are expected to have life

(*i*) more than 15 hours

(*ii*) between 10 and 14 hours.

6. Attempt any *one* of the following: **(7 × 1 = 7)**

(*a*) Evaluate $\int \frac{z^2 - 2z}{(z+1)^2 (z^2+4)} dz$, where C is the circle $|z| = 10$.

(*b*) Evaluate $\int \frac{e^z}{(z-1)(z-4)} dz$, where C is the circle $|z| = 2$ by using Cauchy's integral formula.

7. Attempt any *one* of the following: **(7 × 1 = 7)**

(*a*) Find the Fourier transform of e^{-ax^2}, where $a > 0$.

(*b*) Find the Z-transform of $\cosh\left(\frac{k\pi}{2} + \alpha\right)$.

A.K.T.U., LUCKNOW

NAS–401

B. Tech. [SEMESTER-IV]

THEORY EXAMINATION, 2017–2018
MATHEMATICS -III

(PAPER ID: 199418)

Time: 3 Hours *Total Marks: 100*

Note: *Attempt all Sections. If require any missing data; then choose suitably.*

SECTION–A

1. Attempt *all* questions in brief. **(2 × 10 = 20)**

(*a*) Determine analytic function $f(z)$ in terms of z whose real part is $x^3 - 3xy^2$.

(*b*) Write Cauchy's Reimann conditions in polar coordinates system.

(*c*) If the Fourier transform of e^{-x^2} is $\sqrt{\pi}\, e^{-p^{2/4}}$, then find Fourier transform of $e^{-5(x-2)^2}$.

(*d*) Find Z transform of a^k.

(*e*) Write normal equations to fit the curve $y = ax^2 + b$ by method of least square.

(*f*) Find mean and variance of Poisson distribution.

(*g*) Write the Newton Raphson iteration formula.

(*h*) Prove that third divided difference of $\left(\frac{1}{a}\right)$ is $-\frac{1}{abcd}$.

(*i*) Discuss diagonal dominant property for system of linear equations.

(*j*) Use Picard's method to obtain $y(0.2)$ up to two iterations. Given: $\frac{dy}{dx} = x - y$ with the condition $y(0) = 1$.

SECTION–B

2. Attempt any *three* of the following: **(10 × 3 = 30)**

(*a*) By contour integration, find: $\int_0^\infty \frac{x \sin x}{x^2 + a^2}\, dx$; $a > 0$.

(*b*) Using Z-transform solve the difference equation: $y_{k+2} + 4y_{k+1} + 3y_k = 3^k$, given $y_0 = 0$, $y_1 = 1$.

(*c*) In a normal distribution 31% of the items are under 45 and 8% are over 64. Find mean and standard deviation of the distribution. It is given that $f(t) = \frac{1}{\sqrt{2\pi}} \int_0^t e^{-\frac{1}{2}x^2}\, dx$, then $f(0.5) = 0.19$ and $f(1.4) = 0.42$.

(*d*) Find the real root of equation $3x + \sin x - e^x. = 0$ by the method of False-position correct to three decimal places.

(*e*) Find the value of $y(1.1)$ using Runge-Kutta method of fourth-order, given that $\frac{dy}{dx} = y^2 + xy, y(1) = 1$, take $h = 0.05$.

SECTION–C

3. Attempt any *one* part of the following: (10 × 1 = 10)

(*a*) State and prove Cauchy's Residue Theorem. Hence or otherwise evaluate

$$\oint_c \frac{z^2 - 2z}{(z+1)^2(z^2+4)} dz, \text{ where C is } |z| = 10.$$

(*b*) Expand $\frac{7z-2}{z^3 - z^2 - 2z}$ in the regions (*i*) $1 < |z+1| < 3$ (*ii*) $0 < |z+1| < 1$ (*iii*) $|z+1| > 3$.

4. Attempt any *one* part of the following: (10 × 1 = 10)

(*a*) Find Fourier cosine transform of $\frac{1}{1+x^2}$ and hence find Fourier sine transform of $\frac{x}{1+x^2}$.

(*b*) The temperature u in the semi-infinite rod $0 \le x < \infty$ is determined by the differential equation $\frac{\partial u}{\partial t} = k\frac{\partial^2 u}{\partial x^2}$ subject to the conditions;

(*i*) $u = 0$ when $t = 0, x \geq 0$ (*ii*) $\frac{\partial u}{\partial x} = -\mu$

(a constant) when $x = 0$ and $t > 0$. Making use of cosine transform, show that

$$u(x, t) = \frac{2\mu}{\pi}\int_0^{\infty} \frac{\cos px}{p^2}(1 - e^{-kp^2 t})dp$$

5. Attempt any *one* part of the following: (10 × 1 = 10)

(*a*) The first four moments of a distribution about $x = 4$ are 1, 4, 10 and 45. Calculate the moments about the mean and comment upon the skewness and kurtosis of the distribution.

(*b*) To test the effectiveness of inoculation against cholera, the following table was obtained:

	Attacked	Not attacked	Total
Inoculated	30	160	190
Not inoculated	140	460	600
Total	170	620	790

(The figures represent the number of persons)

Use χ^2-test to defend or refute the statement that the inoculation prevents attack from cholera. ($\chi_{0.05}2$ for 1 *d.f.* = 3.841).

6. Attempt any *one* part of the following: (10 × 1 = 10)

(*a*) By means of Newton's divided difference formula, find the value of $f(15)$ from the following table:

x	4	5	7	10	11	13
$f(x)$	48	100	294	900	1210	2028

(*b*) Estimate from the table, the number of students who obtained marks between 40 and 45.

Marks:	30–40	40–50	50–60	60–70	70–80
No. of students:	31	42	51	35	31

7. Attempt any *one* part of the following: (10 × 1 = 10)

(*a*) Solve by Crout's method, the following system of equations:

$x + y + z = 3,\ 2x - y + 3z = 16,\ 3x + y - z = -3$

(*b*) Find the approximate value of $\int_0^{\frac{\pi}{2}} \sin x \, dx$ by Simpson's rule.

A.K.T.U., LUCKNOW

RAS–401

B. Tech. [SEMESTER-IV]

THEORY EXAMINATION, 2017–2018

MATHEMATICS -III

Time: 3 Hours *Total Marks: 70*

Note: *Attempt all Sections. If require any missing data, then choose suitably.*

SECTION–A

1. Attempt *all* questions in brief. **(2 × 7 = 14)**

(*a*) Discuss Singularity and its types.

(*b*) Write Cauchy-Riemann equation in polar co-ordinates.

(*c*) The life of army shoes is normally distributed with mean 8 months and standard deviation 2 months. If 5000 pairs are insured, how many pairs would be expected to need replacement after 12 months? Given that $P(z \geq 2) = 0.0228$.

(*d*) Determine moment generating function of Binomal distribution.

(*e*) Prove that: $E^{\frac{1}{2}} = \mu + \frac{1}{2}\delta$.

(*f*) Write Newton-Cote's quadrature formula.

(*g*) Find Z-transform of $f(k) = u(-k)$.

SECTION–B

2. Attempt any *three* of the following: **(7 × 3 = 21)**

(*a*) Determine an analytic function $f(z)$ in terms of z if $u + v = 2\frac{\sin 2x}{e^{2y}} + e^{-2} - 2\cos 2x$.

(*b*) Find the mean and variance of Poisson distribution.

(*c*) Find $\int_0^6 \frac{e^x}{1+x}\,dx$ using (*i*) Trapezoidal rule, (*ii*) Simpson's 1/3rd rule and (*iii*) Simpson's 3/8rd rule.

(*d*) A rod is rotating in a plane. The following table gives the angle θ (in radians) through which the rod turned for various values of time t (in seconds).

t:	0	0.2	0.4	0.6	0.8	1.0	1.2
θ:	0	0.12	0.49	1.12	2.02	3.20	4.67

Calculate the angular velocity and angular acceleration at $t = 0.2$ and $t = 1.2$ second.

(*e*) Find Fourier cosine transform of $\frac{1}{1+x^2}$, hence find Fourier sine transform of $\frac{1}{1+x^2}$.

SECTION–C

3. Attempt any *one* part of the following: **(7 × 1 = 7)**

(*a*) Verify Cauchy theorem by integrating e^{iz} along the boundary of the triangle with the vertices at the points $1 + i$, $-1 + i$ and $-1 - i$.

(*b*) Evaluate $\int_0^{\infty} \frac{\sin mx}{x} dx$, $m > 0$.

4. Attempt any *one* part of the following: **(7 × 1 = 7)**

(*a*) The following table represents the height of a batch of 100 students. Calculate skewness and kurtosis:

Height (in cm)	59	61	63	65	67	69	71	73	75
No. of students	0	2	6	20	40	20	8	2	2

(*b*) Use the method of least squares to fit the curve $y = \frac{c_0}{x} + c_1\sqrt{x}$ to the following table of values:

x	0.1	0.2	0.4	0.5	1	2
y	21	11	7	6	5	6

5. Attempt any *one* part of the following: **(7 × 1 = 7)**

(*a*) Find the root of the equation $xe^x = \cos x$ using Regula-Falsi method correct to four decimal places.

(*b*) Find Newton's divided difference polynomial for the following data:

x:	–3	–1	0	3	5
$f(x)$:	–30	–22	–12	330	3458

6. Attempt any *one* part of the following: **(7 × 1 = 7)**

(*a*) Solve the initial value problem $u' = -2tu^2$, $u(0) = 1$ with $h = 0.2$ on the interval [0, 0.4]. Use Runge-Kutta fourth order method and compare your result with exact solution.

(*b*) Solve the following system of linear equations by Matrix decomposition method taking $t = 1$ for $1 \le t \le 3$.

$3x - y + 2z = 12;\ x + 2y + 3z = 11;\ 2x - 2y - z = 2.$

7. Attempt any *one* part of the following: **(7 × 1 = 7)**

(*a*) Using Z-transform, solve the following difference equation:

$$y_{k+2} + 4y_{k+1} + 3y_k = 3^k \text{ given that } y_k = 0 \text{ and } y_1 = 1.$$

(*b*) The temperature u in the semi-infinite rod $0 \le x < \infty$ is determined by the differential equation $\frac{\partial u}{\partial t} = k\frac{\partial^2 u}{\partial x^2}$ subject to conditions.

(*i*) $u = 0$ when $t = 0, x \ge 0$ (*ii*) $\frac{\partial u}{\partial x} = -\mu$ (a constant) when $x = 0$ and $t > 0$, (*iii*) $u(x, t)$ is bounded. Determine the temperature $u(x, t)$.

APPENDIX

Table 1: NORMAL TABLE
AREAS UNDER THE STANDARD NORMAL

$$\text{CURVE} = \frac{1}{\sqrt{2\pi}} \int_0^z e^{-\frac{z^2}{2}}\, dz$$

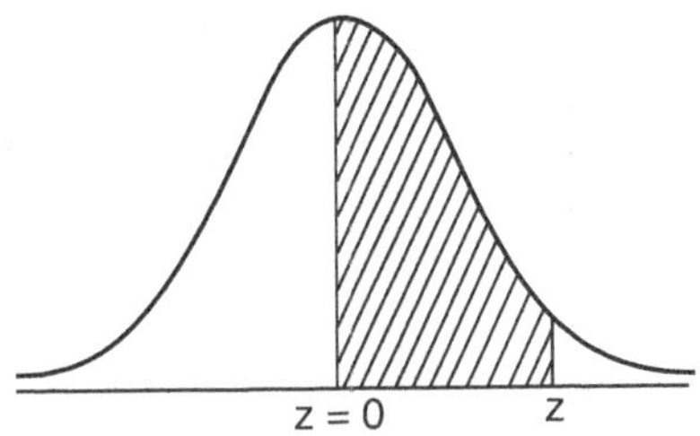

z	.00	.01	.02	.03	.04	.05	.06	.07	.08	.09
0.0	.0000	.0040	.0080	.0120	.0160	.0199	.0239	.0279	.0319	.0359
0.1	.0398	.0438	.0478	.0517	.0557	.0596	.0636	.0675	.0714	.0754
0.2	.0793	.0832	.0871	.0910	.0948	.0987	.1026	.1064	.1103	.1141
0.3	.1179	.1217	.1255	.1293	.1331	.1368	.1406	.1443	.1480	.1517
0.4	.1554	.1591	.1628	.1664	.1700	.1736	.1772	.1808	.1844	.1879
0.5	.1915	.1950	.1985	.2019	.2054	.2088	.2123	.2157	.2190	.2224
0.6	.2257	.2291	.2324	.2357	.2389	.2422	.2454	.2485	.2517	.2549
0.7	.2580	.2611	.2642	.2673	.2704	.2734	.2764	.2794	.2823	.2852
0.8	.2881	.2910	.2939	.2967	.2995	.3023	.3051	.3078	.3106	.3133
0.9	.3159	.3186	.3212	.3238	.3264	.3289	.3315	.3340	.3365	.3389
1.0	.3413	.3438	.3461	.3485	.3508	.3531	.3554	.3577	.3599	.3621
1.1	.3643	.3665	.3686	.3708	.3729	.3749	.3770	.3790	.3810	.3830
1.2	.3849	.3869	.3888	.3907	.3925	.3944	.3962	.3980	.3997	.4015
1.3	.4032	.4049	.4066	.4082	.4099	.4115	.4131	.4147	.4162	.4177
1.4	.4192	.4207	.4222	.4236	.4251	.4255	.4279	.4292	.4306	.4319
1.5	.4332	.4345	.4357	.4370	.4382	.4394	.4406	.4418	.4429	.4441
1.6	.4452	.4463	.4474	.4484	.4495	.4505	.4515	.4525	.4535	.4545
1.7	.4554	.4564	.4573	.4582	.4591	.4599	.4608	.4616	.4625	.4633
1.8	.4641	.4649	.4656	.4664	.4671	.4678	.4686	.4693	.4699	.4706
1.9	.4713	.4719	.4726	.4732	.4738	.4744	.4750	.4756	.4761	.4767
2.0	.4772	.4778	.4783	.4788	.4793	.4798	.4803	.4808	.4812	.4817
2.1	.4821	.4826	.4830	.4834	.4838	.4842	.4846	.4850	.4854	.4857
2.2	.4861	.4864	.4868	.4871	.4875	.4878	.4881	.4884	.4887	.4890
2.3	.4893	.4896	.4898	.4901	.4904	.4906	.4909	.4911	.4913	.4916
2.4	.4918	.4920	.4922	.4925	.4927	.4929	.4931	.4932	.4934	.4936
2.5	.4938	.4940	.4941	.4943	.4945	.4946	.4948	.4949	.4951	.4952
2.6	.4953	.4955	.4956	.4957	.4959	.4930	.4961	.4962	.4963	.4964
2.7	.4965	.4966	.4967	.4968	.4969	.4970	.4971	.4972	.4973	.4974
2.8	.4974	.4975	.4976	.4977	.4977	.4978	.4979	.4979	.4980	.4981
2.9	.4981	.4982	.4982	.4983	.4984	.4984	.4985	.4985	.4986	.4986
3.0	.4987	.4987	.4987	.4988	.4988	.4989	.4999	.4989	.4990	.4990
3.1	.4990	.4991	.4991	.4991	.4992	.4992	.4992	.4992	.4993	.4993

Table 2 : SIGNIFICANT VALUES t_v (α) OF t-DISTRIBUTION (TWO TAIL AREAS) [| t | > t_v(α)] = α

d.f.	*Probability (Level of Significance)*					
(v)	*0.50*	*0.10*	*0.05*	*0.02*	*0.01*	*0.001*
1	1.00	6.31	12.71	31.82	63.66	636.62
2	0.82	0.92	4.30	6.97	6.93	31.60
3	0.77	2.32	3.18	4.54	5.84	12.94
4	0.74	2.13	2.78	3.75	4.60	8.61
5	0.73	2.02	2.57	3.37	4.03	6.86
6	0.72	1.94	2.45	3.14	3.71	5.96
7	0.71	1.90	2.37	3.00	3.50	5.41
8	0.71	1.80	2.31	2.90	3.36	5.04
9	0.70	1.83	2.26	2.82	3.25	4.78
10	0.70	1.81	2.23	2.76	3.17	4.59
11	0.70	1.80	2.20	2.72	3.11	4.44
12	0.70	1.78	2.18	2.68	3.06	4.32
13	0.69	1.77	2.16	2.05	3.01	4.22
14	0.69	1.76	2.15	2.62	2.98	4.14
15	0.69	1.75	2.13	2.60	2.95	4.07
16	0.69	1.75	2.12	2.58	2.92	4.02
17	0.69	1.74	2.11	2.57	2.90	3.97
18	0.69	1.73	2.10	2.55	2.88	3.92
19	0.69	1.73	2.09	2.54	2.86	3.88
20	0.69	1.73	2.09	2.53	2.85	3.85
21	0.69	1.72	2.08	2.52	2.83	3.83
22	0.69	1.72	2.07	2.51	2.42	3.79
23	0.69	1.71	2.07	2.50	2.81	3.77
24	0.69	1.71	2.06	2.49	2.80	3.75
25	0.68	1.71	2.06	2.49	2.79	3.73
26	0.68	1.71	2.06	2.48	2.78	3.71
27	0.68	1.70	2.05	2.47	2.77	3.69
28	0.68	1.70	2.05	2.47	2.76	3.67
29	0.68	1.70	2.05	2.46	2.76	3.66
30	0.68	1.70	2.04	2.46	2.75	3.65
∞	0.67	1.65	1.96	2.33	2.58	3.29

Table 3 : F-Distribution

Values of F for F-Distributions with 0.05 of the Area in the Right Tail

	Degrees of freedom for numerator																		
	1	2	3	4	5	6	7	8	9	10	12	15	20	24	30	40	60	120	∞
1	161	200	216	225	230	234	237	239	241	242	244	246	248	249	250	251	252	253	254
2	18.5	19.0	19.2	19.2	19.3	19.3	19.4	19.4	19.4	19.4	19.4	19.4	19.4	19.5	19.5	19.5	19.5	19.5	19.5
3	10.1	9.55	9.28	9.12	9.01	9.94	8.89	8.85	8.81	8.79	8.74	8.70	8.66	8.64	8.62	8.59	8.57	8.55	8.53
4	7.71	6.94	6.59	6.39	6.26	6.16	6.09	6.04	6.00	5.96	5.91	5.86	5.80	5.77	5.75	5.72	5.69	5.66	5.63
5	6.61	5.79	5.41	5.19	5.05	4.95	4.88	4.82	4.77	4.74	4.68	4.62	4.56	4.53	4.50	4.46	4.43	4.40	4.37
6	5.99	5.14	4.76	4.53	4.39	4.28	4.21	4.15	4.10	4.06	4.00	3.94	3.87	3.84	3.81	3.77	3.74	3.70	3.67
7	5.59	4.74	4.35	4.12	3.97	3.87	3.79	3.73	3.68	3.64	3.57	3.51	3.44	3.41	3.38	3.34	3.30	3.27	3.23
8	5.32	4.46	4.07	3.84	3.69	3.58	3.50	3.44	3.39	3.35	3.28	3.22	3.15	3.12	3.08	3.04	3.01	2.97	2.93
9	5.12	4.26	3.86	3.63	3.48	3.37	3.29	3.23	3.18	3.14	3.07	3.01	2.94	2.90	2.86	2.83	2.79	2.75	2.71
10	4.96	4.10	3.71	3.48	3.33	3.22	3.14	3.07	3.02	2.98	2.91	2.85	2.77	2.74	2.70	2.66	2.62	2.58	2.54
11	4.84	3.98	3.59	3.36	3.20	3.09	3.01	2.95	2.90	2.85	2.79	2.72	2.65	2.61	2.57	2.53	2.49	2.45	2.40
12	4.75	3.89	3.49	3.26	3.11	3.00	2.91	2.85	2.80	2.75	2.69	2.62	2.54	2.51	2.47	2.43	2.38	2.34	2.30
13	4.67	3.81	3.41	3.18	3.03	2.92	2.83	2.77	2.71	2.67	2.60	2.53	2.46	2.42	2.38	2.34	2.30	2.25	2.21
14	4.60	3.74	3.34	3.11	3.96	2.85	2.76	2.70	2.65	2.60	2.53	2.46	2.39	2.35	2.31	2.27	2.22	2.18	2.13
15	4.54	3.68	3.29	3.06	3.90	2.79	2.71	2.64	2.59	2.54	2.48	2.40	2.33	2.29	2.25	2.20	2.16	2.11	2.07
16	4.49	3.63	3.24	3.01	2.85	2.74	2.66	2.59	2.54	2.49	2.42	2.35	2.28	2.24	2.19	2.15	2.11	2.06	2.01
17	4.45	3.59	3.20	2.96	2.81	2.70	2.61	2.55	2.49	2.45	2.38	2.31	2.23	2.19	2.15	2.10	2.06	2.01	1.96
18	4.41	3.55	3.16	2.93	2.77	2.66	2.58	2.51	2.46	2.41	2.34	2.27	2.19	2.15	2.11	2.06	2.02	1.97	1.92
19	4.38	3.52	3.13	2.90	2.74	2.63	2.54	2.48	2.42	2.38	2.31	2.23	2.16	2.11	2.07	2.03	1.98	1.93	1.88
20	4.35	3.49	3.10	2.87	2.17	2.60	2.51	2.45	2.39	2.35	2.28	2.20	2.12	2.08	2.04	1.99	1.95	1.90	1.84
21	4.32	3.47	3.07	2.84	2.68	2.57	2.49	2.42	2.37	2.32	2.25	2.18	2.10	2.05	2.01	1.96	1.92	1.87	1.81
22	4.30	3.44	3.05	2.82	2.66	2.55	2.46	2.40	2.34	2.30	2.23	2.15	2.07	2.03	1.98	1.94	1.89	1.84	1.78
23	4.28	3.42	3.03	2.80	2.64	2.53	2.44	2.37	2.32	2.27	2.20	2.13	2.05	2.01	1.96	1.91	1.86	1.81	1.76
24	4.26	3.40	3.01	2.78	2.62	2.51	2.42	2.36	2.30	2.25	2.18	2.11	2.03	1.98	1.94	1.98	1.84	1.79	1.73
25	4.24	3.39	2.99	2.76	2.60	2.94	2.40	2.34	2.28	2.24	2.16	2.29	2.01	1.96	1.92	1.87	1.82	1.77	1.71
30	4.17	3.32	2.92	2.69	2.53	2.42	2.33	2.27	2.21	2.16	2.09	2.01	1.93	1.89	1.84	1.79	1.74	1.64	1.62
40	4.08	3.23	2.84	2.61	2.45	2.34	2.25	2.18	2.12	2.08	2.00	1.92	1.84	1.79	1.74	1.69	1.64	1.58	1.51
60	4.00	3.15	2.76	2.53	2.37	2.25	2.17	2.10	2.04	1.99	1.92	1.84	1.75	1.70	1.65	1.59	1.53	1.47	1.39
120	3.92	3.07	2.68	2.45	2.29	2.18	2.09	2.02	1.96	1.91	1.83	1.75	1.66	1.61	1.55	1.50	1.43	1.35	1.25
∞	3.84	3.00	2.60	2.37	2.21	2.10	2.01	1.94	1.88	1.83	1.75	1.67	1.57	1.52	1.46	1.39	1.32	1.22	1.00

Table 4 : CHI-SQUARE

Significant Values χ^2 (α) of Chi-Square Distribution Right Tail Areas for Given Probability α,

$P = P_r (\chi^2 > \chi^2 (\alpha)) = \alpha$

And υ is Degrees of Freedom (d.f.)

Degree of freedom (ν)	*Probability (Level of Significance)*						
	0.99	*0.95*	*0.50*	*0.10*	*0.05*	*0.02*	*0.01*
1	.000157	.00393	.455	2.706	3.841	5.214	6.635
2	.0201	.103	1.386	4.605	5.991	7.824	9.210
3	.115	.352	2.366	6.251	7.815	9.837	11.341
4	.297	.711	3.357	7.779	9.488	11.668	13.277
5	.554	1.145	4.351	9.236	11.070	13.388	15.086
6	.872	2.635	5.348	10.645	12.592	15.033	16.812
7	1.239	2.167	6.346	12.017	14.067	16.622	18.475
8	1.646	2.733	7.344	13.362	15.507	18.168	20.090
9	2.088	3.325	8.343	14.684	16.919	19.679	21.669
10	2.558	3.940	9.340	15.987	18.307	21.161	23.209
11	3.053	4.575	10.341	17.275	19.675	22.618	24.725
12	3.571	5.226	11.340	18.549	21.026	24.054	26.217
13	4.107	5.892	12.340	19.812	22.362	25.472	27.688
14	4.660	6.571	13.339	21.064	23.685	26.873	29.141
15	4.229	7.261	14.339	22.307	24.996	28.259	30.578
16	5.812	7.962	15.338	23.542	26.296	29.633	32.000
17	6.408	8.672	15.338	24.769	27.587	30.995	33.409
18	7.015	9.390	17.338	25.989	28.869	32.346	34.805
19	7.633	10.117	18.338	27.204	30.144	33.687	36.191
20	8.260	10.851	19.337	28.412	31.410	35.020	37.566
21	8.897	11.591	20.337	29.615	32.671	36.343	38.932
22	9.542	12.338	21.337	30.813	33.924	37.659	40.289
23	10.196	13.091	22.337	32.007	35.172	38.968	41.638
24	10.856	13.848	23.337	32.196	36.415	40.270	42.980
25	11.524	14.611	24.337	34.382	37.65	41.566	44.314
26	12.198	15.379	25.336	35.363	38.885	41.856	45.642
27	12.879	16.151	26.336	36.741	40.113	41.140	46.963
28	13.565	16.928	27.336	37.916	41.337	45.419	48.278
29	14.256	17.708	28.336	39.087	42.557	46.693	49.588
30	14.933	18.493	29.336	40.256	43.773	47.962	50.892

Note. For degrees of freedom (ν) greater than 30, the quantity $\sqrt{2\chi^2} - \sqrt{2\nu - 1}$ may be used as a normal variate with unit variance.